Lecture Notes in Bioengineering

Lecture Notes in Bioengineering (LNBE) publishes the latest developments in bioengineering. It covers a wide range of topics, including (but not limited to):

- Bio-inspired Technology & Biomimetics
- Biosensors
- Bionanomaterials
- Biomedical Instrumentation
- Biological Signal Processing
- Medical Robotics and Assistive Technology
- Computational Medicine, Computational Pharmacology and Computational Biology
- Personalized Medicine
- Data Analysis in Bioengineering
- Neuroengineering
- Bioengineering Ethics

Original research reported in proceedings and edited books are at the core of LNBE. Monographs presenting cutting-edge findings, new perspectives on classical fields or reviewing the state-of-the art in a certain subfield of bioengineering may exceptionally be considered for publication. Alternatively, they may be redirected to more specific book series. The series' target audience includes advanced level students, researchers, and industry professionals working at the forefront of their fields.

Indexed by SCOPUS, INSPEC, zbMATH, SCImago.

Garry Kuan · Yu-Kai Chang · Tony Morris ·
Teo Eng Wah · Rabiu Muazu Musa ·
Anwar P. P. Abdul Majeed
Editors

Advancing Sports and Exercise via Innovation

Proceedings of the 9th Asian South Pacific Association of Sport Psychology International Congress (ASPASP) 2022, Kuching, Malaysia

Springer

Editors
Garry Kuan
Exercise and Sports Science Programme
School of Health Sciences
Universiti Sains Malaysia
Kubang Kerian, Kelantan, Malaysia

Yu-Kai Chang
Department of Physical Education
and Sport Sciences
National Taiwan Normal University
Taipei City, Taiwan

Tony Morris
Victoria University
Melbourne, VLC, Australia

Teo Eng Wah
University of Malaya
Kuala Lumpur, Malaysia

Rabiu Muazu Musa
Centre for Fundamental and Continuing
Education
Universiti Malaysia Terengganu
Kuala Nerus, Terengganu, Malaysia

Anwar P. P. Abdul Majeed
School of Robotics
Xi'an Jiaotong—Liverpool University
Suzhou, China

ISSN 2195-271X ISSN 2195-2728 (electronic)
Lecture Notes in Bioengineering
ISBN 978-981-19-8158-6 ISBN 978-981-19-8159-3 (eBook)
https://doi.org/10.1007/978-981-19-8159-3

This Springer imprint is published by the registered company Springer Nature Singapore Pte Ltd.
The registered company address is: 152 Beach Road, #21-01/04 Gateway East, Singapore 189721, Singapore

The Scientific Committees

Prof. Dr. Denny Ng Kok Sum, Heriot-Watt University Malaysia
Prof. Dr. Lam Su Shiung, Universiti Malaysia Terengganu, Malaysia
Prof. Dr. Yu-Kai Chang, National Taiwan Normal University, Taiwan
Prof. Dr. Frank Lu, Chinese Culture University, Taiwan
Prof. Dr. Wee Eng Hoe, Tunku Abdul Rahman University College, Malaysia
Assoc. Prof. Dr. Chih-Min Wu, Cheng Shiu University, Taiwan
Assoc. Prof. Dr. Ler Hui Yin, Tunku Abdul Rahman University College, Malaysia
Assist. Prof. Dr. Feng Tzu Chen, China Medical University, Taiwan
Dr. Kueh Yee Cheng, Universiti Sains Malaysia, Malaysia
Assoc. Prof. Dr. Anwar P. P. Abdul Majeed, Xi'an Jiaotong-Liverpool University, China
Dr. Rabiu Muazu Musa, Universiti Malaysia Terengganu, Malaysia
Dr. Chin Ngien Siong, Institute of Teacher Education Batu Lintang Campus, Malaysia
Dr. Teng Kie Yin, Institute of Teacher Education Tun Abdul Razak Campus, Malaysia
Mr. Gordon Nicolaus Jemat Anchang, UiTM Sarawak, Malaysia
Miss. Kho Chung Wei, Institute of Teachers Education Batu Lintang Campus, Malaysia

Preface

The International Congress on Sport and Exercise Psychology is the 9th Edition of the successful congresses organised by the Asian-South Pacific Association of Sports Psychology (ASPASP) on 9–11 August 2022. The ASPASP congress has been organised once every four years since the establishment of ASPASP in Singapore in 1989. ASPASP has held eight congresses since 1991. Presently, the ASPASP congresses are highly regarded and recognised by the sport and exercise communities in terms of its quality, standards and impact across the globe.

This congress aims to share current research findings and good practices and provide opportunities for collaboration and networking with colleagues, fellow counterparts and students within the Asian-South Pacific regions. It also aims to amplify our expected role in meeting the psychological and mental health needs of athletes and non-athletes and practitioners to share their best practices and challenges within the sports and exercise psychology disciplines.

The main theme of this conference was Fostering Partnership and Collaboration through Applied Technological Innovation: Advancing Sport and Exercise Psychology in Asia. The overarching sub-themes were (1) Applied Sport and Social Psychology; (2) Health and Exercise Psychology; (3) Counselling and Clinical Psychology in sports; (4) Motor Control and Learning and (5) Biomechanics, Data Mining and Machine Learning in sports.

The conference brought a new outlook on cutting-edge issues shared through keynote speeches by Prof. Dr. Robert Schinke (EdD, ISSP-R, CSPA; Laurentian University, Canada), Prof. Dr. Claudia Voelcker-Rehage (University of Munster, Germany), Prof. Dr. Tsung Min Hung (National Taiwan Normal University, Taiwan), Assoc. Prof. Dr. Ying Hwa Kee (Nanyang Technological University, Singapore), Asst. Prof. Dr. Wimonmas Prachakul (Kasetsart University, Thailand), Prof. Emeritus Dr. Tony Morris (University, Melbourne, Australia), Prof. Dr. Low Wah Yun@Sarinah Low Binti Abdullah (Universiti Malaya, Malaysia), Assoc. Prof. Dr. Garry Kuan (Universiti Sains Malaysia, Malaysia) and Assoc. Prof. Dr. Lim Boon Hooi (Universiti Malaya, Malaysia).

More than 135 submissions were received for the 9th ASPASP International Congress and were reviewed in a double-blinded manner, and 50 papers were agreed upon by the reviewers to be accepted and published in this volume of Lecture Notes in Bioengineering. The editors would like to express their gratitude to all the authors who submitted their papers.

Finally, the editors hope that readers find this volume informative, and we would like to thank Springer Nature for undertaking the publication of this volume. We would also like to express our gratitude to the Organising Committee and the Scientific Committee for their hard work in ensuring the success of the conference.

Kubang Kerian, Malaysia
Taipei City, Taiwan
Melbourne, Australia
Kuala Lumpur, Malaysia
Kuala Nerus, Malaysia
Suzhou, China

Assoc. Prof. Dr. Garry Kuan
Prof. Dr. Yu-Kai Chang
Prof. Emeritus Dr. Tony Morris
Dr. Teo Eng Wah
Dr. Rabiu Muazu Musa
Assoc. Prof. Dr. Anwar P. P. Abdul Majeed

Contents

About the Editors

Assoc. Prof. Dr. Garry Kuan is a lecturer at the Exercise and Sports Science Programme, School of Health Sciences, Universiti Sains Malaysia, Malaysia and a research fellow at the Department of Life Sciences, Brunel University London, UK. Currently, Dr. Garry is the secretary-general of the Asian-South Pacific Association of Sport Psychology (ASPASP), the secretary-general of the Malaysian Sport Psychology Association (MASPA), an executive board member of the Asian Council of Sports Science (ACESS), and the scientific committee of the World Exercise Medicine (WCEM). Since 2019, Dr. Garry was appointed as the chair of the International Network of the Young Scientist Network-Academy of Science Malaysia (YSN-ASM) and the Sport Psychology Panelist for the National Coaching Academy of Malaysia. Garry has published over 200 scholarly manuscripts, with over 90 high-impact ISI journals and served as an editor and a reviewer for numerous international journals. His research had won numerous awards internationally. In 2021, he is the recipient of the Ten Outstanding Young Malaysians in Academic Leadership and accomplishment, International Society of Sport Psychology (ISSP) Developing Scholar Award, the Outstanding Academician of the Year by Asia Awards and National Outstanding Community Service Award (3rd prize). In addition, he also received the 2020 IFPEFSSA International Eminent Educator Award, the A-CIPA Young Researcher Award at the 27th International Congress of Applied Psychology, the International Scholars Award at the ICSEMIS pre-Olympic conference and the Atsushi Fujita Research Scholarship at the 6th ASPASP conference. During his social time, he plays the first violin professionally and teaches communities to play various musical instruments. e-mail: garry@usm.my

Prof. Dr. Yu-Kai Chang is a distinguished professor in Department of Physical Education at National Taiwan Normal University, Taiwan, and the director of Physical Activity and Cognitive Neuroscience Laboratory. Dr. Chang is also the vice president of Society for Sport and Exercise Psychology of Taiwan (SSEPT) as well as a managing council member in both International Society of Sport Psychology (ISSP) and Asian-South Pacific Association of Sport Psychology (ASPASP). Dr. Chang's research focuses on Sport and Exercise Psychology, specifically in the exercise and cognition through experimental psychology and cognitive neuroscience approaches. To date, Dr. Chang has published more than 150 research papers in reputed journals including *Sport Medicine, Health Psychology, Medicine and Science in Sports and Exercise, NeuroImage, Journal of Sport and Exercise Psychology (JSEP), Psychology for Sport and Exercise, Psychophysiology*, etc. Dr. Chang is the recipient of 2018 Distinguished Research Award (Ministry of Science and Technique, Taiwan), 2014 Early Career Distinguished Scholar Award of North American Society for the Psychology of Sport and Physical Activity (NASPSPA); 2013 Developing Scholar Award of ISSP; 2012 Chinese Young Scholar Award of *International Journal of Sport and Exercise Psychology*; 2012 Da-You Wu Memorial Research Award (National Science Council, Taiwan), and 2008 Outstanding Dissertation Award of UNCG. Dr. Chang is an editor in *Physical Education Journal (TSSCI)*, an associate editor in *Journal of Sport and Health Science (SSCI/SCI)*, a section editor in *International Journal of Sport and Exercise Psychology* (Scopus) and an editorial board member in JSEP (SCI/SSCI).

Prof. Dr. Tony Morris is an emeritus professor of sport, exercise and healthy psychology at Victoria University, Melbourne, Australia. He was born in Leeds, England, and has lived in Melbourne, Australia, for over 25 years. Trained in psychology, with a Ph.D. in education, he previously worked in motor learning and sport psychology for eight years in the UK. Tony has published seven books and monographs, four as editor and three authors, and has authored more than 150 chapters, journal articles and refereed conference proceedings manuscripts. He has presented over 100 invited lectures around the world, including over ten keynote presentations at International conferences. e-mail: Tony.Morris@vu.edu.au

Dr. Teo Eng Wah graduated from University of Malaya (UM) with a Bachelor of Sport Science (Sport Management) degree. Later, he completed his Master of Education (Physical Education) from UM. Under the SLAI scholarship, he obtained his Master's and Ph.D. in Kinesiology (Testing and Measurement) at the University of Illinois, Urbana-Champaign, USA. Currently, Dr. Teo is a senior lecturer in Centre for Sport and Exercise Science, University of Malaya. His areas of interest are in sports testing and measurement, psychometrics, questionnaire development and validation, sport psychology, health and fitness and wildlife conservation. Currently, he is involved in sport studies in the area of motivation, self-esteem, ego resilience and sport commitment in Malaysian settings. He has authored five books and over 20 articles in various international journals. e-mail: vteo2@um.edu.my

Dr. Rabiu Muazu Musa holds a Ph.D. degree in Sports Science from Universiti Sultan Zainal Abidin (UniSZA), Malaysia. He obtained his M.Sc. in Sports Science from UniSZA and his B.Sc. in Physical and Health Education at Bayero University Kano, Nigeria. His research activity focused on the development of multivariate and machine learning models for athletic performance. His research interests include sports performance analysis, health promotion, sport and exercise science, talent identification, test and measurement as well as machine learning in sports.

He obtained several research grants both as a principal and a co-investigator. He is an active reviewer to many reputable journals and currently serving as an academic editor for *PLOS ONE* as well as an executive guest editor for the *Open Sports Science Journal* and *SN Applied Sciences*. He authored many publications in myriad journals, conference proceedings, books as well as chapters in books. He is currently a senior lecturer at the Centre for Fundamental and Continuing Education, Universiti Malaysia Terengganu. e-mail: rabiu.muazu@umt.edu.my

Dr. Anwar P. P. Abdul Majeed graduated with a first-class honours B.Eng. in Mechanical Engineering from Universiti Teknologi MARA (UiTM), Malaysia. He obtained an M.Sc. in Nuclear Engineering from Imperial College London, UK. He then received his Ph.D. in Rehabilitation Robotics from the Universiti Malaysia Pahang (UMP). He is currently serving as an associate professor at the School of Robotics, XJTLU. Prior to joining XJTLU, he was a senior lecturer (Assistant Professor) and was the head of Programme (Bachelor of Manufacturing Engineering Technology (Industrial Automation)) at the Faculty of Manufacturing and Mechatronics Engineering Technology, UMP. He is also currently serving as an adjunct lecturer at UCSI University, Malaysia. Dr. Anwar is also a visiting research fellow at EUREKA Robotics Centre, Cardiff Metropolitan University, UK. Dr. Anwar is a chartered engineer, registered with the Institution of Mechanical Engineers (IMechE), UK, a member of the Institution of Engineering and Technology (IET), UK, as well as a senior member of the Institute of Electrical and Electronics Engineers (IEEE). He is an active research member at the Innovative Manufacturing, Mechatronics and Sports Laboratory (iMAMS), UMP. His research interest includes rehabilitation robotics, computational mechanics, applied mechanics, sports engineering, renewable and nuclear energy, sports performance analysis as well as machine learning. He has authored over 60 papers in different journals, conference proceedings as well as books. He serves as a reviewer in a number of prolific journals such as *IEEE Access*, *Frontiers in Bioengineering and Biotechnology*, *SN Applied Sciences*, *PeerJ Computer Science*, *Applied Computing and Informatics* amongst others. He has also served as a guest editor for SN Applied Sciences, MDPI, Frontiers as well as an editor for several Springer book series. He is currently serving as an academic editor for *PLOS ONE*, a review editor for *Frontiers in Robotics and AI* and a section editor for *Mekatronika* (UMP Press). Dr. Anwar is also a member of the Young Scientists Network of the Academy of Sciences Malaysia (YSN-ASM). With regard to learned/civil society activities, he is an active member of the IET Malaysia Local Network as well as acting as a Liaison officer for the Imperial College Alumni Association Malaysia. e-mail: Anwar.Majeed@xjtlu.edu.cn

Commitment in Sport from the Perspective of Malaysian Adults

Arthur Ling, Teo Eng Wah, Ngien Siong Chin, Kho Lee Sze, Michelle Wong, and Holyvia Lau

Abstract Involvement in sport on regular basis is essential to ensure healthy lifestyle. Therefore, it is important to understand the factors that influence commitment in sport. Up to date, Sport Commitment Questionnaire-2 (SCQ-2) (Scanlan et al. in Psychol Sport Exerc 22:233–246, 2016) often utilised to comprehend the motivation of sport users to maintain their participation in sports. The objective of the study is to examine the differences in sports commitment and reasons that motivate coaches to participate in sports. A quantitative study with random sampling method using questionnaire was administered to Malaysian adults who play various sports. They have completed the questionnaire that consists of demographic information and SCQ-2 (Scanlan et al. in Psychol Sport Exerc 22:233–246, 2016). The SCQ-2 scale consists of 58 items with twelve subscales and structured on a five-point Likert scale and descriptive items. The independent t-test and one-way ANOVA were conducted to examine the differences in sports commitment among gender and age groups. There was a total of 389 respondents with a mean age of 36.41 years old (SD = 10.46) which completed the questionnaire. The participants consist of 254 (65.3%) male and 135 (34.7%) were female. The independent sample t-test showed that there was significant difference in enthusiastic commitment ($p = 0.01$) across gender, but there is no significant difference in constraints commitment ($p = 0.29$) across gender. An analysis of variance (ANOVA) on enthusiastic commitment and constraints commitment shows significant variation among age groups. A post hoc Tukey will be shown in the discussion part to show the difference. Lastly, the results showed that sport enjoyment is the main reason nurturing the motivation to participate in sport among working adults in Malaysia. Whereas, other opportunities was not the main concern for Malaysian adult to take part in sport. Adult Malaysians reviewed that sport enjoyment encourage them to participate in sport. It shows that

A. Ling (✉) · T. Eng Wah
Centre of Sport and Exercise Sciences, University Malaya, Kuala Lumpur, Malaysia
e-mail: ling_wei87@hotmail.com

N. S. Chin · M. Wong · H. Lau
Institute of Teacher Education Batu Lintang Campus, Kuching, Sarawak, Malaysia

K. L. Sze
Institute of Teacher Education Tun Abdul Razak Campus, Kuching, Sarawak, Malaysia

G. Kuan et al. (eds.), *Advancing Sports and Exercise via Innovation*, Lecture Notes in Bioengineering, https://doi.org/10.1007/978-981-19-8159-3_1

Malaysian adults enjoy the moment of relaxation that sports could bring to them as well as element of enjoyment despite of hectic working schedule. This promotes healthy lifestyle and preparing productive workers.

Keywords Sport commitment model · Malaysian sport · Sport enjoyment

1 Introduction

Studies have proven that active involvement in sport promotes healthy lifestyle [1–3]. Unfortunately, sport dropout rate increases to 70% over the years [4, 5]. The main cause of dropouts is that they tended to spend more time engaging in activities outside of their sport [6]. This had led to health deterioration among those who discontinued sports [7]. For example, those who withdraw from sport participation reported higher body fat and high depression, lower health-related quality of life and having greater psychological difficulties than their counterparts who still actively involve in sports [5, 8, 9]. Therefore, the need to study factors related to sport commitment is essential in this area.

With the interest to discover the factors influencing sport commitment, sport commitment model (SCM) was created by Scanlan and colleagues [10]. The initial sport commitment model consists of five factors which are sport enjoyment, involvement alternatives, personal involvement, social constraints and valuable opportunities. To further understand the elements that drove sport commitment, two additional components were added to SCM which were social support and desire to excel. Besides, enthusiastic commitment and constrained commitment were included in the model to further distinguish between when describing the types of commitments that driven sport commitment [11, 12].

Sport enjoyment refers to the fun or pleasure feelings towards the sport [10]. Overall, studies have conferred that sport enjoyment is the key element to sport commitment [10, 11, 13–16]. On the other hand, involvement alternatives (other activities that would divert the attention from participating in sport) were the least influential factor affecting sport commitment [16, 17]. Although most of the studies agreed that sport enjoyment ranked the first factor influencing sport commitment, studies [16, 18] also suggested that personal investment also plays important role in predicting sport commitment. The fourth factor in SCM, social constraints are the social expectations, standards or norms that compel someone to continue participating in a sport [10]. However, studies revealed that there was no relationship existing between social restrictions and sport commitment [19, 20]. Valuable opportunities mean possibilities or valuable things that one can obtain by continuing to engage in sports activities for example friendship, future career and travel time [10, 12]. Social support was included in the SCM because Scanlan et al. [17] revealed that encouragement or support given to the players has increased the sport involvement among rugby players. Lastly, desire to excel refers to one's ability to develop, strive for perfection, and achieve goals. Study has reported that rugby players have put in a lot of effort in training in order to win games [12].

The main objective of current study focuses on the sport commitment of Malaysian adults. This study utilised SCM as the framework to investigate the factors underlying sport commitment. Sport Commitment Questionnaire-2 (SCQ-2) was administered to Malaysian adults who involve in various sports.

2 Methods

2.1 *Sampling Procedures and Instrument*

The Sport Commitment Questionnaire-2 was administered to Malaysian adults who play sports. Adults recruited for this study were ranging from 21 years old to 65 years old. Data collection began following receipt of ethics board approval from the University of Malaya (UM. TCN2/RCH&E/UMREC-14). Prior to study, we write to the relevant sports bodies and organisations across Malaysia to request permission to perform this study on their members.

The participants were briefed on the confidentiality prior to answering the questionnaire. This study's participation was based on voluntary principles. Therefore, participants were free to withdraw from the study at any moment if they wish to. The questionnaire consists of two parts where Part A is used to gather demographic information and Part B is the SCQ-2 which consists of 58 items where all the items were structured based in five-point Likert scale that range from 1 (strongly disagree) to 5 (strongly disagree). Enthusiastic commitment and desire to excel (mastery achievement) were measured by six items, constrained commitment, sport enjoyment, other priorities, personal investment-loss, social support-informational and desire to excel-social achievement were measured by five items. Valuable opportunities, personal investment-quantity, social constraints and social support-emotional were measured by four items. The questionnaire was tested and showed good validity and reliability in the sport setting [11].

2.2 *Data Analysis*

The responses were all key in SPSS version 25.0 for analysis purposes. Descriptive analysis, T-test and one-way ANOVA were performed to address all the objectives of this study.

3 Result

3.1 Demographic Information

See Table 1.

Approximately, 500 questionnaires were distributed across Malaysia. 463 questionnaires were returned, and 74 questionnaires were eliminated during the data cleaning process. The total of respondents was 389 with a mean age of 36.41(SD = 10.46) years old. There were 254 (65.3%), male and 135 (34.7%) were female.

See Table 2.

The results were analysed using descriptive analysis to find the main factor that predict sport commitment. Table 2 shows that sport enjoyment was the most important factor for adult Malaysian to stay active in sport.

See Table 3.

Table 1 Demographic information of participants

Variable	*N*	Percentage
Gender	389	
Male	254	65.3
Female	135	34.7
Age group		
20–29	119	30.6
30–39	121	31.1
40–49	102	26.2
> 50	47	12.1

Table 2 Descriptive analysis of subscales

Subscales	Mean	SD
Enthusiastic commitment	4.19	0.61
Constrained commitment	2.08	0.90
Sport enjoyment	4.59	0.47
Valuable opportunities	3.37	0.75
Other priorities	2.50	1.02
Personal investment-loss	3.17	0.66
Personal investment-quantity	3.59	0.70
Social constraints	2.92	0.78
Social support-emotional	3.05	0.82
Social support-informational	3.17	0.80
Desire to excel-mastery achievement	3.78	0.71
Desire to excel-social achievement	3.67	0.69

Table 3 Comparison of sport commitment subscales between genders

Subscales	Male		Female		*u* diff	*p*
	M	SD	*M*	SD		
Enthusiastic commitment	4.24	0.61	4.08	0.59	0.16	0.01*
Constrained commitment	2.04	0.88	2.15	0.93	0.11	0.29
Sport enjoyment	4.61	0.46	4.56	0.48	0.05	0.27
Valuable opportunities	3.37	0.75	3.35	0.75	0.02	0.81
Other priorities	2.43	1.06	2.63	0.94	0.2	0.06
Personal investment-loss	3.17	0.67	3.15	0.66	0.02	0.76
Personal investment-quantity	3.64	0.70	3.51	0.69	0.13	0.08
Social constraints	2.93	0.83	2.92	0.69	0.01	0.95
Social support-emotional	3.04	0.86	3.06	0.78	0.02	0.82
Social support-informational	3.12	0.83	3.26	0.74	0.14	0.11
Desire to excel-mastery achievement	3.82	0.72	3.70	0.70	0.10	0.15
Desire to excel-social achievement	3.72	0.70	3.59	0.67	0.13	0.08

* $p < 0.05$

The result shows that male has higher enthusiastic commitment compared to female t(387) = 2.59, $p = 0.01$. There were no significant differences for the other subscales in the sport commitment model.

See Table 4.

A one-way ANOVA was used to compare the effect of age groups on factors that predict sport commitment. The result revealed that there were significant differences in constrained commitment (F(3, 385) = 10.70), other priorities (F(3, 385) = 0.00) and personal investment-loss (F(3, 385) = 0.04). Tukey HSD test for multiple comparisons found that the mean value of constrained commitment was significantly different between age groups of 20–29 and 20–29, 20–29 and 40–49, 20–29 and above 50 years old. Besides, Tukey HSD test for multiple comparisons also shows significant differences in other priorities between age group of 20–29 and 30–39, 20–29 and 40–49. Lastly, significant differences also found in personal investment-loss between age groups of 20–29 and 30–39.

4 Discussion

The descriptive analysis showed that sport enjoyment emerged as the major factor that predicted sport commitment among Malaysian adults. Generally, the result is in line with studies that proposed sport enjoyment is the main factor for sport commitment [11, 14–16, 21]. This has indicated that Malaysian adults involvement in sport was due to enjoyment element that sport could bring to them.

Table 4 Comparison of sport commitment subscales between age groups

Type of commitment	Mean				Tukey's HSD comparisons			
	Mean	SD	*F*	*P*	20–29	30–39	40–49	> 50
Enthusiastic commitment								
20–29	4.1	0.7	1.28	0.28		0.4	0.31	0.66
30–39	4.22	0.57			0.4		1	1
40–49	4.24	0.56			0.31	1		1
> 50	4.22	0.55			0.66	1	1	
Constrained commitment								
20–29	2.45	1.01	10.7	0.00*		0.00*	0.00*	0.03*
30–39	1.87	0.79			0.00*		0.99	0.68
40–49	1.91	0.74			0.00*	0.99		0.84
> 50	2.04	0.93			0.03*	0.68	0.84	
Sport enjoyment								
20–29	4.5	0.46	2.39	0.07		0.09	0.37	0.17
30–39	4.64	0.45			0.09		0.93	0.99
40–49	4.61	0.51			0.37	0.93		0.87
> 50	4.67	0.4			0.17	0.99	0.87	
Valuable opportunities								
20–29	3.48	0.89	1.91	0.13		0.1	0.45	0.97
30–39	3.26	0.66			0.1		0.9	0.57
40–49	3.33	0.63			0.45	0.9		0.89
> 50	3.43	0.82			0.97	0.57	0.89	
Other priorities								
20–29	2.87	1.19	8.1	0.00*		0.00*	0.00*	0.06
30–39	2.3	0.88			0.00*		0.99	0.84
40–49	2.34	0.85			0.00*	0.99		0.95
> 50	2.44	1.04			0.06	0.84	0.95	
Personal investment-loss								
20–29	3.28	0.74	2.75	0.04*		0.05*	0.2	1
30–39	3.06	0.6			0.05*		1	0.34
40–49	3.11	0.55			0.2	1		0.57
> 50	3.26	0.8			1	0.34	0.57	
Personal investment-quantity								
20–29	3.59	0.8	0.16	0.92		1	1	0.95
30–39	3.58	0.66			1		1	0.91
40–49	3.58	0.64			1	1		0.93
> 50	3.66	0.64			0.96	0.93	0.93	

(continued)

Table 4 (continued)

Type of commitment	Mean				Tukey's HSD comparisons			
	Mean	SD	*F*	*P*	20–29	30–39	40–49	> 50
Social constraints								
20–29	3	0.83	0.54	0.66		0.7	0.7	0.93
30–39	2.89	0.64			0.7		1	0.99
40–49	2.88	0.78			0.69	1		0.98
> 50	2.93	0.97			0.96	0.99	0.98	
Social support-emotional								
20–29	3.19	0.88	1.71	0.17		0.22	0.26	0.47
30–39	2.99	0.72			0.22		1	1
40–49	2.99	0.78			0.25	1		1
> 50	2.98	1.03			0.47	1	1	
Social support-informational								
20–29	3.29	0.86	1.28	0.28		0.25	0.53	0.69
30–39	3.09	0.7			0.25		0.97	0.99
40–49	3.14	0.76			0.53	0.97		1
> 50	3.14	0.97			0.69	0.99	1	
Desire to excel-mastery achievement								
20–29	3.8	0.77	0.12	0.95		0.99	0.96	0.96
30–39	3.79	0.66			0.99		0.98	0.99
40–49	3.75	0.68			0.95	0.98		1
> 50	3.75	0.8			0.96	0.99	1	
Desire to excel-social achievement								
20–29	3.71	0.77	0.18	0.91		0.96	0.92	0.95
30–39	3.66	0.65			0.96		1	1
40–49	3.65	0.62			0.92	1		1
> 50	3.64	0.73			0.95	1	1	

* $p < 0.05$

Independent samples t-test was conducted to look into the differences of sport commitment between genders. Some studies suggested that there was no significant difference in commitment across gender [22, 23]. This might be due to the nature of the sports that provide equal opportunities to both genders, the exposition of sports and its popularity among both genders. However, the findings of this study have reported different point of view where females perceived lower enthusiastic commitment compared to males (refer to Table 3). This might be due to female having other commitment or priorities [24]. These could be the commitment to the family and jobs where female spend more time and commitment rather than in sport. According to Coakley and White [25], females have often withdrawn from sport because they perceived sports that were irrelevant to their life [25]. Another explanation could be the nature of certain sports which might not provide equal opportunities to females which lowered the commitment among females [22].

This study also uses the analysis of variance (i.e. ANOVA) to explore the differences of sport commitment between age groups. The findings reported significant difference in constrained commitment, other priorities and personal investment-loss across age groups (refer Table 4). Result shows that adults in age group 20–29 have the highest constrained commitment. The result supports Weiss [26] study where it reports levels of commitment for young adults were low compared to older age groups. However, the findings were not inclusive as some studies [14, 22] reported that older age groups are more committed in sports. This indicates that commitment towards sport might be varied due to different sport activities.

Study also reported that age group 20–29 years old adults has the highest mean score for other priorities among all the age groups. This is because they have other activities and other commitment [22, 27] which might be driven them away from participating in sport. In this age group, most of them just started their career after high school or university or still pursuing their studies. Thus, they might want to focus on their career or studies rather than in sport. Besides, personal investment has played an important role in predicting sport commitment among adults aged 20–29. In this age group, some of them might be involved in the competitive level which required them to spend more resources such as time and money. Weiss [26] revealed competitive athletes had higher perceptions of investments in relation to sport commitment.

4.1 Implication for the Promotion of Sport in Malaysian Adult

Across the globe, studies had been done to understand the motivation of people participating in sports and promote healthy lifestyle. For example, studies reviewed enjoyment, fun and social interaction as important factors that underpin sport commitment [6, 10, 11, 14, 15, 22, 27]. With the result of the study showing that sport enjoyment is the main predictor to sport commitment, the government, sport authorities or school could design sport programme in which there are more fun elements to engage people to participate in sport among Malaysian adults. Promoting sports to females is also essential in Malaysia as females need to be encouraged to participate in sports. This can be done through exposition of benefits of sports in school and also by the Ministries such as Ministry of Youth and Sports and Ministry Women, Family and Community Development. Teacher in school should play important role to encourage female students to enjoy in PE class or in extra-curricular activities. Another approach is to provide females friendly environment such as female coaching sessions, female class and female coaches to encourage them to engage in sport.

4.2 Limitation of the Study

There were several limitations on this study. First, some participants were struggled to correctly respond to the items in SCQ-2 as Bahasa Melayu, Chinese language and other different native languages that are more widely spoken among Malaysians instead of English. Therefore, responses may not accurately reflect on participants' thoughts. In order to overcome this situation, extra clarification and explanation were given to those who needed them. Lastly, it is unknown how many people play sports in Malaysia because neither the National Sports Body nor the Ministry kept accurate records. In order to determine the required sample size, G-power calculation was used. The minimum recommended sample size by G-power software was 151. Overall, the sample size of this study was 389 which met the minimum requirement.

5 Conclusion

In general, sport enjoyment aroused as the most important factor that predicts sport commitment across gender and age groups. Besides, Malaysian male adult has higher enthusiastic commitment compared to female. In addition, adults aged 20–29 years old are having the risk of being drop out from sport because they have the highest constrained commitment and other priorities compared to other age groups. Therefore, the findings of this study help sport administrators to understand the differences in sport commitment across gender and age groups. Hence, future sport programmes designed need to focus on elements that could draw Malaysian adults' interest to participate in sport continuously. This could promote healthy lifestyle and reduce health-related problems in the community.

References

1. Ahn S, Fedewa AL (2011) A meta-analysis of the relationship between children's physical activity and mental health. J Pediatr Psychol 36(4):385–397. https://doi.org/10.1093/jpepsy/jsq107.10
2. Biddle SJ, Asare M (2011) Physical activity and mental health in children and adolescents: a review of reviews. Br J Sports Med 45(11):886–895. https://doi.org/10.1136/bjsports-2011-090185.11
3. Eime RM, Young JA, Harvey JT, Charity MJ, Payne WR (2013) A systematic review of the psychological and social benefits of participation in sport for children and adolescents: informing development of a conceptual model of health through sport. Int J Behav Nutr Phys Act 10(1):98–119. https://doi.org/10.1186/1479-5868-10-98
4. Woods R (2011) Social issues in sport, 2nd edn. Human Kinetics, Champaign, IL
5. Vella SA, Cliff DP, Magee CA, Okely AD (2015) Associations between sports participation and psychological difficulties during childhood: a two year follow up. J Sci Med Sport 18:304–309

6. Malinero O, Salguero A, Tuero C, Alvarez E, Marques S (2006) Dropout reasons in young spanish athletes: relationships to gender, type of sport and level of competition. J Sport Behav 29:255–269
7. Crane J, Temple V (2015) A systematic review of dropout from organized sport among children and youth. Eur Phy Educ Rev 21:114–131
8. Howie EK, McVeigh JA, Smith AJ, Straker LM (2016) Organized sport trajectories from childhood to adolescence and health associations. Med Sci Sports Exerc 48:1331–1339
9. Vella SA, Cliff DP, Magee CA, Okely AD (2014) Sports participation and parent-reported health-related quality of life in children: longitudinal associations. J Pediatr 164:1469–1474
10. Scanlan TK, Carpenter PJ, Lobel M, Simons JP (1993) Sources of enjoyment for youth sport athletes. Pediatr Exerc Sci 5(3):275–285
11. Scanlan TK, Chow GM, Sousa C, Scanlan LA, Knifsend CA (2016) The development of the sport commitment questionnaire-2 (English version). Psychol Sport Exerc 22:233–246. https://doi.org/10.1016/j.psychsport.2015.08.002
12. Scanlan TK, Russell DG, Scanlan LA, Klunchoo TJ, Chow GM (2013) Project on elite athlete commitment (PEAK): IV. Identification of new candidate commitment sources in the sport commitment model. J Sport Exerc Psychol 35(5):525–535
13. Weiss WM, Weiss MR (2003) Attraction-and entrapment-based commitment among competitive female gymnasts. J Sport Exerc Psychol 25(2):229–247
14. Young BW, Medic N (2011) Examining social influences on the sport commitment of masters swimmers. Psychol Sport Exerc 12(2):168–175. https://doi.org/10.1016/j.psychsport.2010.09.004
15. Iñigo MM, Podlog L, Hall MS (2015) Why do athletes remain committed to sport after severe injury? An examination of the sport commitment model. Sport Psychol 29(2):143–155
16. Frayeh AL, Lewis BA (2017) Sport commitment among adult recreational soccer players: test of an expanded model. Int J Exerc Sci 10(1):4–24
17. Scanlan TK, Russell DG, Beals KP, Scanlan LA (2003) Project on elite athlete commitment (PEAK II): a direct test and expansion of the commitment model with elite amateur sportsmen. J Sport Exerc Psychol 25:337–401
18. Casper JM, Gray DP, Babkes SM (2007) A sport commitment model perspective on adult tennis players' participation frequency and purchase intention. Sport Manag Rev 10:253–278. https://doi.org/10.1016/S1441-3523(07)70014-1
19. Guillet E, Sarrazin P, Carpenter PJ, Trouilloud D, Curry F (2002) Predicting persistence or withdrawal in female handballers with social exchange theory. Int J Psychol 37(2):92–104
20. Ryan RM, Deci EL (2000) Self-determination theory and the facilitation of intrinsic motivation, social development, and well-being. Am Psychol 55:68–78. https://doi.org/10.1037/0003-066X.55.1.68
21. Carpenter PJ, Scanlan TK (1998) Changes over time in the determinants of sport commitment. Pediatr Exerc Sci 10:356–365
22. Casper JM, Stellino MB (2008) Demographic predictors of recreational tennis participants' sport commitment. J Park Recreat Admi 26(3):95–115
23. Jess S (2009) Examining sport commitment and intentions to participate in intramural sports: application of the sport commitment model and the theory of planned behaviour in a campus recreational sport setting. Master dissertation. Retrieved from ProQuest Dissertations & Theses Global (Accession No. 305169065)
24. Wigglesworth JC, Young BW, Medic N, Grove JR (2012) Examining gender differences in the determinants of masters swimmers' sport commitment. Int J Sport Exerc Psychol 10(3):236–250
25. Coakley J, White A (1992) Making decisions: gender and sport participation among British adolescents. Sociol Sport J 9:20–35
26. Weiss WM (2015) Competitive-level differences on sport commitment among high school- and collegiate-level athletes. Int J Sport Exerc Psychol 13(3):286
27. Weiss WM, Neibert PJ (2014) Changes over time in the predictors of athletic training program commitment. Athl Train Educ J 9(2):64–71

Effectiveness of Brain Breaks® for Students: A Systematic Review and Meta-Analysis

Mengyuan Zhao, Garry Kuan, Yee Cheng Kueh, Ke Zhou, and Shiyue Wang

Abstract Background: Physical activity (PA) is a key factor of a healthy lifestyle. However, many students do not acquire sufficient PA due to numerous internal and external factors that influence students' PA levels. Brain Breaks Physical Activity Solution (Brain Breaks®) is an online PA video project based on classroom environment, which is multimedia video technology. It may provide an opportunity to increase school-based PA and other possible benefits. The aim of this systematic review and meta-analysis was to evaluate the impact of Brain Breaks® for students. Methods: A systematic search of electronic databases (PubMed, Web of Science, OVID, ScienceDirect, EBSCOhost, and Scopus) was performed between February 10, 2022, and March 30, 2022. Studies that investigated the classroom-based Brain Breaks® among students were included. Meta-analyses were conducted in Review Manager 5.3, with effect sizes calculated separately for each subscale outcome assessed. Results: Fifteen articles met the inclusion criteria for the review, and seven provided sufficient data and appropriate design for inclusion in the meta-analyses. Studies mainly investigated different physical activity-related outcomes. Results of the meta-analyses showed Brain Breaks® had a positive effect on students' attitudes toward physical activity. Including: Benefits (MD = 0.29, 95% CI: 0.12, 0.46, $p =$ 0.0006), Importance (MD = 0.28, 95% CI: 0.17, 0.39, $p < 0.00001$), Learning (MD = 0.75, 95% CI: 0.40, 1.10, $p < 0.0001$), Self-efficacy (MD = 0.60, 95% CI: 0.28, 0.93, $p = 0.0003$), Fun (MD = 0.29, 95% CI: 0.18, 0.40, $p < 0.00001$), Fitness (MD = 0.26, 95% CI: 0.16, 0.36, $p < 0.00001$), and Personal Best (MD = 0.27, 95% CI: 0.16, 0.37, $p < 0.00001$). Conclusions: Brain Breaks® may have a positive impact on attitudes toward physical activity. However, it is not possible to draw definitive conclusions due to the level of heterogeneity in Attitudes toward Physical Activity

M. Zhao (✉) · G. Kuan · Y. C. Kueh · S. Wang
Universiti Sains Malaysia, 16150 Kubang Kerian, Kelantan, Malaysia
e-mail: zhaomy@student.usm.my

M. Zhao · S. Wang
Jiangsu Medical Vocational College, Yancheng 224006, China

K. Zhou
Henan University, Kaifeng 475001, China

G. Kuan et al. (eds.), *Advancing Sports and Exercise via Innovation*, Lecture Notes in Bioengineering, https://doi.org/10.1007/978-981-19-8159-3_2

Scale (APAS) subscales outcomes assessed. Future studies should consider more high methodological intervention designs to achieve meaningful evidence.

Keywords Students · Brain Breaks® · Physical activity · Classroom-based intervention

1 Introduction

1.1 Background

Lack of physical activity (PA) is a global public health problem. It is estimated that 31.1% of the world's adult population was below the American College of Sports Medicine (ACSM) recommendations (150 min of moderate exercise (PA) per week or 75 min of vigorous exercise (PA) per week) in 2017 [1]. A longitudinal study showed that people who engaged in 150 min of moderate-intensity aerobic PA, 75 min of high-intensity PA per week, or two or more sessions of muscle strengthening training per week had a lower all-cause and cause-specific mortality compared with those who did not achieve the recommended level of activity [2, 3]. It was found in a two-year follow-up study that students' daily PA time decreased and sedentary time increased, suggesting that the amount of time spent studying increases and the amount of time applied to PA decreases with age [3]. Interventions to improve physical health need to be implemented early, especially in the child population, for whom healthy development is particularly important during the growth phase [4]. Health-related attitudes and behavior patterns are formed in early childhood, and they can often be maintained until the adolescent stage, so early health interventions may have more potential impact [5, 6]. A sedentary lifestyle is one of the reasons for the decline in PA levels, and the development of this lifestyle is linked to a number of factors, such as lack of available leisure time due to heavy school workloads, lack of desire to engage in physical exercise, and prolonged internet time, all of which may affect children's PA levels [7]. Various factors can limit regular extra-curricular sports activities, there are many external reasons affecting students' PA levels [8], such as the weather conditions, where extra-curricular activities may be canceled because of bad weather such as storms or heavy snow. At this time, students are unable to engage in daily physical activity and physical activity levels are reduced [9]. It is worth considering increasing PA levels by increasing students' interest in physical activity and increasing opportunities to be active.

The recent development of the internet has also brought innovation to the sports world. Interactive internet-based video games can increase children's interest in PA, increase PA levels, and promote healthy lifestyles. The recent development of the internet has also brought innovation to the sports world. Children's interest in physical activity, as well as their PA levels and healthy lifestyles, can be improved through interactive internet-based video games. The proper application of internet technology is an effective way to motivate students to exercise [10]. HOPSports has developed a classroom-based physical activity program, Brain Breaks Physical Activity Solutions (Brain Breaks®) that makes the most of recess or class time to enhance PA opportunity [11]. It has a wide selection of physical activity-related videos that students can follow. Researchers in several countries and regions have already put Brain Breaks® into classroom use [12–14]. A self-report instrument used by many researchers in the intervention was the Attitudes toward Physical Activity Scale (APAS), and the results of the subscales were used to illustrate the effect of the Brain Breaks® intervention [15]. Zhou et al. [16] showed statistical differences in the Benefits (F1), Importance (F2), Fun (F5), and Personal Best (F7) subscales, but not in the remaining groups. Glapa et al. [17] showed statistically significant differences in the Self-efficacy on learning with video exercises subscale and no statistical differences between the experimental and control groups for the other subscales; in the Balasekaran et al. [18] experiment, the results showed statistically significant differences on all seven scales of the APAS. The effect of the Brain Breaks® intervention is unclear as there are different studies with different results.

To date, no meta-analysis has evaluated the effectiveness of Brain Breaks® in the student population. This activity may be beneficial to the student population, so there is a need to systematically evaluate the appropriateness and effectiveness of Brain Breaks® promotion in this target population. The findings have important implications for informing and supporting future use in student groups. Moreover, it may give teachers another way of thinking about teaching in the classroom.

Therefore, the aim of this systematic review was to examine the effectiveness of Brain Breaks® intervention among students. A search of the published literature on the use of Brain Breaks® in students was conducted, and meta-analysis was used to calculate the combined effect size and to analyze the possible effects of this intervention.

2 Methods

2.1 Registration and Protocol

To provide a comprehensive review and improve the quality of meta-analysis articles, this study followed the PRISMA statement [19]. This review was registered in the International Prospective Register of Systematic Reviews (PROSPERO) (record # CRD42022315938).

2.2 Search Strategy

The literature search was conducted in accordance with the search requirements in the PRISMA statement. Six databases of PubMed, Web of Science, OVID, ScienceDirect, EBSCOhost, and Scopus were searched for relevant literature on the impact of Brain Breaks® interventions on students while tracking references cited by relevant systematic review. As the first Brain Breaks®-related article was published in 2008, we set the time search to 2008–2022 and the last search date to March 30, 2022. A supplementary search was conducted in Google Scholar. Using keywords plus free words (synonyms): "Brain Breaks", student*, classroom-based physical activity.

2.3 Criteria

Inclusion criteria:
The papers for this systematic review were chosen using a set of predetermined inclusion criteria. Each study needed to meet the following criteria:

1. Experimental study design;
2. The experimental intervention was Brain Breaks® only, with no other physical activity interventions, which of course did not include normal physical education classes or activities;
3. Participants are healthy students (including students at any stage, e.g., primary, secondary, or university students), and the subjects do not have special student populations (e.g., obese, physically dysfunctional, etc.);
4. Completeness of study data (clear means and standard deviations of results);
5. Published articles are in English and Chinese and can be fully accessed.

Exclusion criteria:

1. Overview articles, commentaries, surveys, etc. non-intervention study design;
2. Incomplete data for the outcome indicator, such as only mean value without standard deviation;
3. The difference between the base of the experimental and control groups was too large; the experimental design was not reasonable, etc.

2.4 Study Selection

This study generated 215 publications from a search of six databases, PubMed, Web of Science, OVID, ScienceDirect, EBSCOhost, and Scopus, with two additional publications from Google Scholar (Fig. 1). After removing duplicates ($n = 61$), 154 papers were initially retained. By reading the title, abstract, and keywords of the article, 123 articles inconsistent with the theme were deleted. This process was completed by one author. A total of 31 articles were identified as potentially relevant based on the inclusion criteria. Two authors independently read the full text of the remaining articles to conduct further screening. When there were different views on the results, the two authors discussed, giving their respective reasons. Sixteen papers were excluded for different reasons, while reference lists were checked in all included papers, but no additional studies met the inclusion criteria. Finally, 15 papers were included in the qualitative analysis.

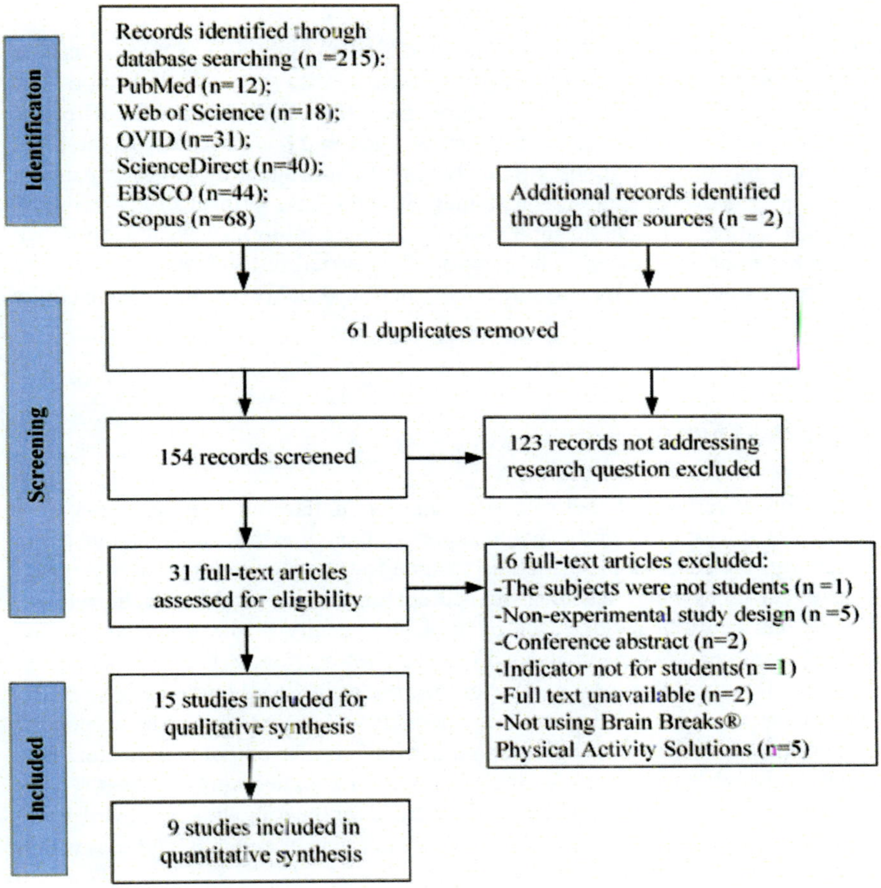

Fig. 1 PRISMA flow diagram showing flow of studies through the review process

2.5 Data Extraction

According to the needs of the study, data were extracted from the included literature by a searcher, including the first author of the literature, year of publication, sample size, gender, age, study design, duration, and outcome measures.

2.6 Methodological Quality

The Effective Public Health Practice Project (EPHPP) tool (2008) [20] was used to assess the methodological quality of the included studies across six indicators: (A) selection bias; (B) study design; (C) confounders; (D) blinding; (E) data collection methods; and (F) withdrawals and dropouts. Each entry is rated on a three-point scale, choosing "strong", "moderate", or "weak". Following the tool's accompanying instructions, the overall score was assigned based on these ratings: strong (no weak component ratings), moderate (one weak component rating), or weak (more than one weak component rating). The process was carried out independently by the two authors, where there was disagreement, there was deliberation until reaching a consensus.

2.7 Meta-Analyses

Meta-analyses were conducted when the same or similar outcome indicators were presented in at least three of the included papers. Outcome indicators include physical activity-related outcomes (such as attitude toward physical activity, PA level, and physical fitness) and academic-related outcomes (i.e., academic achievement, memory, classroom behavior). Due to the heterogeneity of the study design, the included research experiments needed to have control groups (i.e., RCT or quasi-experimental with control group). One of the articles included in the review section of this study was not included in the meta-analysis because there was no control group in the experimental design, although it had similar outcome indicators (PA level) as in the other two. In this systematic review, 15 studies were included, 9 were included in the meta-analysis, and 6 were excluded for the following reasons: 1 study had no control group ($n = 1$); there were no greater than or equal to 3 articles which used identical or similar instruments ($n = 5$).

2.8 Statistical Analysis

The results of this study were calculated using the Review Manager 5.3 software. A fixed effects model was used to test homogeneity across multiple studies, and a random effects model was used if heterogeneity existed. After examination, it was found that there was heterogeneity among the studies, and the random effects model was adopted. If the results were measured using the same measurement tool, effect sizes were computed as the mean difference (MD); otherwise, the standardized mean difference (SMD) was used.

Heterogeneity test: The heterogeneity of the included studies was measured by I^2, with higher I^2 statistics indicating greater heterogeneity, with I^2 values of 25, 50, and 75% representing low, moderate, and high heterogeneity, respectively. If $I^2 > 50\%$, there was more significant heterogeneity [21].

Combined statistic test: If $p \leq 0.05$, the combined statistic of multiple studies was statistically significant; if $p > 0.05$, it is not statistically significant [22].

Consolidated subgroup data: The synthetic mean and synthetic standard deviation were calculated using the following formula.

$$\overline{X}_T = \frac{\sum n_i \overline{X}_i}{\sum n_i} \tag{1}$$

$$S_T = \sqrt{\frac{\sum n_i S_i^2 + \sum n_i d_i^2}{\sum n_i}} \tag{2}$$

$$d_i = \overline{X}_T - \overline{X}_i \tag{3}$$

$\overline{X}_T$ is the total mean, $\overline{X}_i$ is the mean of each group, n_i is the number of participants in each group, S_T is the total standard deviation, S_i^2 is the variance of each group, and S_i is the standard deviation of each group.

3 Results

3.1 Included Studies

Fifteen articles were ultimately included in this study, with an overall sample size of 6524 participants. Most of the papers were published between 2018 and 2021 ($n = 14$), and only one was published in 2014. Research sites across multiple countries and regions, including one in China, Macedonia, Poland, Singapore, Lithuania, Croatia, Kaunas, and Bratislava, as well as one in eight different countries, two in South Africa, and four in Malaysia. Most of the studies were conducted over a period of 3–4 months, with the minimum duration of one intervention being 6 weeks. Interventions are focused on 3 to 5 min or 5 to 10 min on every school day. The studies focused on primary school students, with the oldest participants having a mean age of 11.18 and 11.71 years. The measures were focused on physical activity attitude assessment, and APAS was used in seven papers (Table 1).

3.2 Methodological Quality

Of the 15 papers included, four achieved a strong rating score [13, 16, 17, 23], six achieved a moderate rating [14, 18, 24–27], and five received a weak rating score [12, 28–31]. The moderate and weak ratings were mainly due to the lack of reporting blinding of participants and researchers and the dropout rate of the studies (Table 1).

3.3 Publication Bias

The amount of literature ultimately included in meta-analyses was too small ($n = 9$) to allow for a test of bias. Therefore, it is unclear whether there was publication bias between individual studies [32].

3.4 Meta-Analyses

Overall Effect Test

An overall effect test on the entire sample found that Brain Breaks had a positive effect on the change in students' attitudes towards physical activity. An overall homogeneity test of the included literature showed $I^2 = 83\%, p < 0.00001$, indicating heterogeneity among the nine studies, so a random effects model was used, reflecting the possibility of potential moderating variables. According to Cohen [33], SMD < 0.2 is a small effect size, $0.2 \leq$ SMD < 0.5 is a fair effect size, $0.5 \leq$ SMD < 0.8 is a medium effect size, and SMD ≥ 0.8 is a large effect size. The overall effect size showed that the SMD was 0.47, reaching a fair effect size. The results of the two-tailed test ($p < 0.00001$) indicated that the combined statistic was statistically significant, with a 95% confidence interval of [0.29, 0.64]. The above data suggest that Brain Breaks had a positive impact on the change in students' attitudes towards physical activity (Fig. 2).

Moderating Effects

In addition to examining the effect of Brain Breaks on attitudes towards physical activity, this study also examined potential moderating variables that influence the relationship between the two. In this meta-analysis, the four variables of time, duration, sample size, and grade were tested for moderating effects (Table 2).

Table 1 Study characteristics

Paper	Sample size	Gender (M/F)	Age (mean ± SD)	Study design	Duration	Outcome measures	Study quality
Zhou et al. 2021 China [16]	$n =$ 704	EG:	EG:	3–5 min each school day, 30 min per week	3 months	APAS	Strong
		190/163	9.41 ± 0.91				
		CG:	CG:				
		180/171	9.44 ± 0.92				
Popeska et al. 2018 Macedonia [24]	$n =$ 283	EG:	EG:	3–5 min each school day, five days per week	3 months	APAS	Moderate
		86/66	8.24 ± 0.82				
		CG:	CG:				
		69/62	9.18 ± 1.13				
Glapa et al. 2018 Poland [17]	$n =$ 326	EG:	EG:	3–5 min each school day, 2 times per day	4 months	APAS	Strong
		132/132	9.6 ± 1.08				
		CG:	CG:				
		38/24	10.1 ± 0.92				
Bonnema et al. 2020 South Africa [13]	$n =$ 114	EG:	EG:	5 to 10 min every day during school time	3 months	APAS	Strong
		44/31	11.4 ± 0.54				
		CG:	CG:				
		27-Dec	11.71 ± 0.49				
Balasekaran et al. 2021 Singapore [18]	$n =$ 113	EG:	EG:	3–5 min daily, five days per week	10 weeks	APAS	Moderate
		22/26	9.71 ± 0.99				
		CG:	CG:				
		25/40	9.66 ± 0.94				
Emeljanovas et al. 2018 Lithuania [14]	$n =$ 181	EG:	8.54 ± 1.10	5–9 min each school day	3 months	APAS	Moderate
		50/43					
		CG:					
		48/40					
Mok et al. 2020 Eight countries [12]	$n =$ 3036	EG:	8–11 years	3–5 min, two times per day, 5 days each week	4 months	APAS	Weak
		958/956					
		CG:					
		538/584					
Hajar et al. 2019 Malaysia [25]	$n =$ 335	EG: CG	10.51 ± 0.50	5 min each school day, five times per week	4 months	PALMS-Y-M	Moderate
		183:152					
		M:F					
		159:176					
Hajar et al. 2019 Malaysia [26]	$n =$ 335	EG: CG	10.51 ± 0.50	5 min per day, five times per week	12 weeks	Short-term memory (Digit span test)	Moderate
		183:152					
		M:F					
		159:176					

(continued)

Table 1 (continued)

Paper	Sample size	Gender (M/F)	Age (mean ± SD)	Study design	Duration	Outcome measures	Study quality
Podnar et al. 2018 Croatia [23]	$n = 98$	–	6–10-year-old	5 min per day	12 weeks	On-task behavior and physical activity	Strong
Rizal et al. 2019 Malaysia [28]	$n =$ 322	EG:CG: 177/145 M:F 159/163	10.53 ± 0.50	30 min per week	12 weeks	TTM, (process of change, decisional, balance, and self-efficacy)	Weak
Rizal et al. 2020 Malaysia [27]	$n =$ 287	EG:CG: 128/159 M:F 148/139	10.51 ± 0.50	Accumulated 30 min per week, five times per week	12 weeks	Processes of Change questionnaire	Moderate
van Stryp et al. 2021 South Africa [29]	$n = 48$	EG: 28/20	6.6 ± 0.4	10 min per day, twice a week, a total of 24 times	6 weeks	PA patterns (sedentary behavior, PA level)	Weak
Tumynaitė et al. 2014 Kaunas [30]	$n =$ 113	EG: 31/31 CG: 29/22	8.24 ± 1.10	5–9 min every school day	3 months	Physical fitness tests	Weak
Tománek et al. 2019 Bratislava [31]	$n =$ 229	EG: 65/58 CG: 50/56	11.18 ± 0.77	3–5 min every school day	3 months	Attitudes of the secondary school pupils toward physical activity	Weak

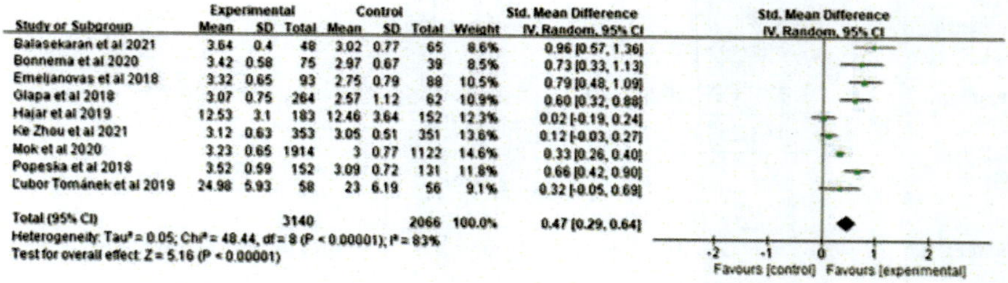

Fig. 2 Forrest plot of overall effect test

Table 2 Moderating effect test of meta-analysis of Brain Breaks® on attitudes toward PA

Moderators	Homogeneity test			Category	ES, 95% CI	Two-tailed test		*N*
	Chi^2	*p*	I^2 (%)			*Z*	*p*	
Time	0.82	0.36	0	3–5 min	0.39[0.09,0.70]	2.51	0.01	5
				> 5 min	0.58[0.32,0.84]	4.32	< 0.0001	4
Duration	1.86	0.17	46.4	≤ 3 months	0.58[0.28,0.88]	3.75	0.0002	6
				4 months	0.31[0.05,0.56]	2.36	0.02	3
Sample size	13.01	0.0003	92.3	≤ 200	0.82[0.61,1.03]	7.81	< 0.00001	3
				> 200	0.33[0.15,0.50]	3.73	0.0002	6
Grade	0.08	0.78	0	≤ grade 5	0.46[0.26,0.66]	4.48	< 0.00001	7
				grade 6	0.52[0.12,0.92]	2.57	0.14	2

Note N refers to the number of articles included

In terms of activity time, Brain Breaks had a significant effect on physical activity attitudes in both subgroups, although the effect of using Brain Breaks for longer than 5 min per day had a medium effect (ES = 0.58), the difference between the two groups was not statistical significance ($p = 0.36$).

In terms of intervention duration, meta-analysis results showed that Brain Breaks had a significant effect on physical activity attitudes in both groups. The intervention duration of 3 months or less produced a larger effect size (ES = 0.58) than 4 months (ES = 0.31). However, it is important to note that the effect sizes were not significantly different between the two subgroups ($p = 0.17$).

In terms of sample size, Brain Breaks produced significant effects in both subgroups. There was a significant difference between the effect sizes of the two groups ($Chi^2 = 13.01, p = 0.0003$). The group with a sample size of less than 200 produced a larger effect size (ES = 0.82) than the group with a sample size of 200 or more (ES = 0.33).

In terms of grade, the results showed that Brain Breaks produced a significant effect ($p < 0.01$) in the group of grade 5 or below and not in the grade 6 group. However, the effect values for the two groups were not significantly different ($Chi^2 = 0.08, p = 0.78$).

APAS Subscale Results Analysis

The indicators in the seven included articles were APAS subscales, which were continuous outcome variables. Therefore, the mean difference (MD) was chosen as the effect size. Meta-analysis results showed that heterogeneity tests F1 ($I^2 = 94\%$, $p < 0.00001$), F2 ($I^2 = 86\%, p < 0.00001$), F3 ($I^2 = 98\%, p < 0.00001$), F4 ($I^2 = 98\%, p < 0.00001$), F5 ($I^2 = 85\%, p < 0.00001$), F6 ($I^2 = 83\%, p < 0.00001$), F7 ($I^2 = 82\%, p < 0.0001$), all of which were highly heterogeneous, so a random effects model was used. The results of the combined effect size analysis showed that F1 (MD = 0.29, 95% CI: 0.12, 0.46, $p = 0006$), F2 (MD = 0.28, 95% CI: 0.17, 0.39, $p < 0.00001$), F3 (MD = 0.75, 95% CI: 0.40, 1.10, $p < 0.0001$), F4 (MD = 0.60, 95% CI: 0.28, 0.93, $p = 0.0003$), F5 (MD = 0.29, 95% CI: 0.18, 0.40, $p < 0.00001$),

Table 3 Brain Breaks® on APAS subscale

Subscale	Homogeneity			MD [95%CI]	Two-tailed test	
	Chi²	*p*	I^2 (%)		*Z*	*p*
Benefits (F1)	95.86	< 0.00001	94	0.29[0.12,0.46]	3.44	0.0006
Importance (F2)	42.27	< 0.00001	86	0.28[0.17,0.39]	4.96	< 0.00001
Learning (F3)	343.38	< 0.00001	98	0.75[0.40,1.10]	4.18	< 0.0001
Self-efficacy (F4)	227.93	< 0.00001	98	0.60[0.28,0.93]	3.62	0.0003
Fun (F5)	41.34	< 0.00001	85	0.29[0.18,0.40]	5.39	< 0.00001
Fitness (F6)	35.39	< 0.00001	83	0.26[0.16,0.36]	5.14	< 0.00001
Personal best (F7)	32.82	< 0.0001	82	0.27[0.16,0.37]	5.00	< 0.00001

F6 (MD = 0.26, 95% CI: 0.16, 0.36, $p < 0.00001$), and F7 (MD = 0.27, 95% CI: 0.16, 0.37, $p < 0.00001$) (Table 3), suggesting that Brain Breaks physical activity improved all aspects of physical activity attitudes.

4 Discussion

Brain Breaks is a physical activity program based in the classroom environment, which is already used in many countries and regions in classroom teaching [7]. The main objective of this study was to examine the effect of Brain Breaks on physical activity and academic-related indicators among students in general. However, there was only one article on academic-related indicators (on-task behavior) [23], short-term memory [26], and physical fitness tests [30], two on processes of change in physical activity [27, 28], so the number of above similar outcomes that could be combined was too small to merge. In addition, van Stryp et al. study [29] was an experimental intervention, but without a control group, therefore, it also did not meet the inclusion criteria, so seven of the final included papers used the APAS. Tománek et al. [31] used the scale of attitudes of the secondary school pupils toward physical activity, Hajar et al. [25] used Physical Activity and Leisure Motivation Scale-Youth-Malay (PALMS-Y-M), respectively. The results of the overall effect test for these nine texts indicated a positive effect on attitudinal change toward physical activity in students. Attitudinal changes may bring about behavioral changes and therefore may increase physical activity levels, leading to a reduction in sedentary behavior and an improvement in physical condition.

As there was a high heterogeneity, a sensitivity test was conducted, and the results were found to be stable. This study provided an analysis of possible moderating effects. The relationship between variable *X* and variable *Y* is influenced by variable *M*, which is referred to as the moderating variable [34]. By analyzing the moderating variables, it is possible to discover what potential variables influence the relationship between the two and thus provide guiding recommendations for classroom-based physical activity interventions. The results of the moderating effect test in this study suggested that the sample size may be a potential moderating variable. In previous meta-analyses of the effects of physical activity interventions, moderating variables may be considered as physical activity duration, intervention period, age, sample size, etc. [35, 36].

In this study, the results of the temporal subgroup analysis showed no statistical difference between the two research conditions. Of the literature included in the meta-analysis, five papers used a once-a-day intervention of 3–5 min of exercise following a video, and four papers used a 5–10-min daily intervention of exercise following a video. For physical activity time, an additional maximum of 10 min of physical activity per day is still minimal compared to the recommended PA level [37], so different physical activity times between the two groups may not have an effect on the results.

The least intervention period in this study was 10 weeks ($n = 1$), 3 months for 5 studies and 4 months for 3 interventions. Subgroup analysis of the intervention duration was divided into two subgroups according to a 3-month cut-off. Although the results of the subgroup analysis showed no statistical significance, the inclusion of more literature needs to be explored, considering that no more than 3 months group only covered three papers, which may result in an uneven distribution of literature.

In the moderating effect test for sample size, there was high heterogeneity in the effect of Brain Breaks on physical activity attitudes in both groups ($I^2 = 92.3\%$), indicating that there was an effect of study sample size on the relationship between both Brain Breaks intervention and physical activity attitudes. The 200 and below sample size group produced a larger effect, which was consistent with the ease of deriving better intervention effects in experiments with fewer subjects, but only three papers were included in this group, so a conclusion drawn from a small number of studies may be somewhat biased [35].

Among the grade moderating variables, four of the papers included students in grades 3–5, one in grade 1–4 and grade 4–5, two in grade 6, with the two groups divided by grade 5. Balasekaran et al. [18] did not mention grade level but stated that the age range of the participants was 8 to 11 years, with a mean age of 9.69, which was closer to the age of the population in the below grade 5 group, it was therefore included in this subgroup. Although the results indicated no statistical difference between the two groups, they were not sufficient to suggest that age was not a potential moderating variable in the effect of Brain Breaks on attitudes toward physical activity. One reason is that the year 6 group contained only two papers, and the data were not sufficient. The other is that in addition to the grade groups included in the meta-analysis, there are additional grade groups for which the effect of Brain Breaks intervention is not yet known. So, whether grade is a potential moderating variable for the effect of Brain Breaks on attitudes toward physical activity needs to be clarified by more analysis of data from the literature at different age levels. Overall, there is a need to include more studies for the analysis of moderating variables due to the limitations of the data in the literature, which may produce false positive or false negative results.

Attitudes towards physical activity as whether individuals tend to favor or disapprove of physical activity behavior [38]. It has been shown that children's perceptions of their attitudes toward physical activity were related to whether regular physical activity could enhance their self-satisfaction and sense of achievement [17, 39]. APAS is a self-rating scale developed by Mok et al. [15]. The scale consists of seven components, all related to physical activity behaviors, including (F1) Benefits: Promoting holistic health; (F2) Importance of exercise habits; (F3) Learning: Self-efficacy in learning with video exercises; (F4) Self-efficacy: Self-efficacy in using video exercises; (F5) Fun: Exercise motivation and enjoyment; (F6) Fitness: Self-confidence in physical fitness, and (F7) Personal Best: Trying to do your personal best. Data collection and analysis of the subscales in this paper were using Mok et al. [12] study as the standard.

The results of all subscales in the seven included papers showed significant differences between the experimental and control groups ($p < 0.05$), indicating that Brain Breaks had a positive impact on various aspects of physical activity attitudes. The most significant effects were found for F3 (ES = 0.75) and F4 (ES = 0.60), but results for all subscales showed a high degree of heterogeneity ($I^2 > 75\%$). For each study, not all subscales showed statistical differences, for example, Zhou et al. [16] results illustrated that Brain Breaks had a positive effect on the component of four subscales of the scale, Benefits (F1), Importance (F2), Fun (F5), and Personal Best (F7). But there were no statistical differences for the other three subscales. Although these seven papers had a high degree of similarity in experimental design and data statistics, there were still factors that may cause differences in the results, such as subjects. The participants were of similar age, but in different countries and regions, which may cause cultural differences to affect the results.

It is worth noting that the measurement instruments used were all post-translational culturally adapted APAS. The original English version of the APAS had 57 entries, but after cultural adaptation in China, the scale used by Zhou et al. [16] was left with only 35 entries and 7 dimensions. The other five papers used the post-cultural adaptation APSA in each country, all of which had seven dimensions. However, Glapa et al. [17] used a scale with only six dimensions and no Self-efficacy in using video exercises (F4), so the meta-analysis of this subscale was included in six papers. The scale used by Emeljanovas et al. [14] had eight dimensions. Due to the factors Learning and Health in this scale were the same as the questions under the F3 entry in the APAS, the data from these two subscales were combined and both grouped into F3 in the meta-analysis. Future studies could include more similar literature for analysis, with further subgroup analyses to obtain more moderating variables and explore the impact of Brain Breaks on the results of each subscale.

5 Limitations

Due to the small number of literature included in this study, the analysis results may not fully reflect the objective situation; the only outcome indicator included in the meta-analysis was the Physical Activity Attitude Scale, with no other indicators related to physical activity or learning, so other possible benefits of Brain Breaks could not be analyzed; the oldest group included was only adolescents in grade 6 (mean age 11.71); the effectiveness of its application in other populations is not yet known; there is a wide variety of videos on the Brain Breaks website, and the intensity of exercise varies from exercise to exercise. Some experiments did not explicitly mention how intense the exercise was at the time of the intervention, which may have had an impact on the accuracy of the results of the meta-analysis.

6 Conclusions

Brain Breaks intervention may provide a practical, low-cost, and effective strategy to improve young students' attitudes toward physical activity, completing the transition from psychological recognition to behavior. Given the small amount of literature included, the findings of this systematic review should be interpreted with caution. There is scope for further exploration of the effects of Brain Breaks intervention, which needs to be supported by more experimental data. Brain Breaks could be implemented with a wider range of students than just teenagers. It is suggested that it could be applied to different groups such as high school and university students. Also, when implementing the intervention activities, specifying the intensity of the different video campaigns beforehand may bring more reliable and convincing evidence to the experimental results.

Acknowledgements The authors would like to acknowledge Ming-Kai Chin for his guidance and for providing the Brain Breaks platform, under the GCH and HOPSports® Inc. This research was supported by the Ministry of Higher Education Malaysia for Fundamental Research Grant Scheme (FRGS) with Project Code: FRGS/1/2020/SKK06/USM/03/13.

References

1. Tao K, Liu W, Xiong S, Ken L, Zeng N, Peng Q, Yan X, Wang J, Wu Y, Lei M, Li X, Gao Z (2019) Associations between self-determined motivation, accelerometer-determined physical activity, and quality of life in Chinese college students. Int J Environ Res Public Health 16(16):2941
2. Zhao M, Veeranki SP, Magnussen CG, Xi B (2020) Recommended physical activity and all cause and cause specifc mortality in US adults: prospective cohort study. BMJ 370:20–31
3. Kallio J, Hakonen H, Syväoja H, Kulmala J, Kankaanpää A, Ekelund U, Tammelin T (2020) Changes in physical activity and sedentary time during adolescence: gender differences during weekdays and weekend days. Scand J Med Sci Sports 30(7):1265–1275
4. Robinson LE, Stodden DF, Barnett LM, Lopes VP, Logan SW, Rodrigues LP, D'Hondt E (2015) Motor competence and its effect on positive developmental trajectories of health. Sports Med 45(9):1273–1284
5. Tammelin R, Yang X, Leskinen E, Kankaanpaa A, Hirvensalo M, Tammelin T, Raitakari OT (2014) Tracking of physical activity from early childhood through youth into adulthood. Med Sci Sports Exerc 46(5):955–962
6. Best JR (2010) Effects of physical activity on children's executive function: contributions of experimental research on aerobic exercise. Dev Rev 30(4):331–351

7. Kuan G, Rizal H, Hajar MS, Chin MK, Mok MMC (2019) Bright sports, physical activity investments that work: implementing brain breaks in Malaysian primary schools. Br J Sports Med 53(14):905–906
8. Dong J, Zhang S, Xia L, Yu Y, Hu S, Sun J, Zhou P, Chen P (2018) Physical activity, a critical exposure factor of environmental pollution in children and adolescents health risk assessment. Int J Environ Res Public Health 15(2):176
9. Wang L, Tang Y, Luo J (2017) School and community physical activity characteristics and moderate-to-vigorous physical activity among Chinese school-aged children: a multilevel path model analysis. J Sport Health Sci 6(4):416–422
10. González CS, Gómez N, Navarro V, Cairós M, Quirce C, Toledo P, Marrero-Gordillo N (2016) Learning healthy lifestyles through active videogames, motor games and the gamification of educational activities. Comput Human Behav 55:529–551
11. Chin MK, Edginton CR, Tang MS (2012) School physical education and health: a model of best practice, integrating local context with global trends. The Global J Health Phys Educ Pedagogy 1(4):251–282
12. Mok MMC, Chin MK, Korcz A, Popeska B, Edginton CR, Uzunoz FS, Podnar H, Coetzee D, Georgescu L, Emeljanovas A, Pasic M, Balasekaran G, Anderson E, Durstine JL (2020) Brain breaks® physical activity solutions in the classroom and on attitudes toward physical activity: a randomized controlled trial among primary students from eight countries. Int J Environ Res Public Health 17(5):1666
13. Bonnema J, Coetzee D, Lennox A (2020) Effect of a three-month HOPSports brain breaks® intervention programme on the attitudes of Grade 6 learners towards physical activities and fitness in South Africa. J Phys Educ Sport 20:196–205
14. Emeljanovas A, Mieželienė B, Mok MMC, Chin MK, Česnaitienė V, Fatkulina N, Trinkuniene L, Sanchez GFL, Suárez AD (2018) The effect of an interactive program during school breaks on attitudes toward physical activity in primary school children. Anales de Psicología 34(3):580–586
15. Mok MM, Chin MK, Chen S, Emeljanovas A, Mieziene B, Bronikowski M, Laudanska-Krzeminska I, Milanovic I, Pasic M, Balasekaran G, Phua KW, Makaza D (2015) Psychometric properties of the attitudes toward physical activity scale: a rasch analysis based on data from five locations. J Appl Meas 16(4):379–400
16. Zhou K, He S, Zhou Y, Popeska B, Kuan G, Chen L, Chin MK, Mok M, Edginton CR, Culpan I, Durstine JL (2021) Implementation of brain breaks® in the classroom and its effects on attitudes towards physical activity in a chinese school setting. Int J Environ Res Public Health 18(1):272
17. Glapa A, Grzesiak J, Laudanska-Krzeminska I, Chin MK, Edginton CR, Mok MMC, Bronikowski M (2018) The impact of brain breaks classroom-based physical activities on attitudes toward physical activity in polish school children in third to fifth grade. Int J Environ Res Public Health 15(2):368
18. Balasekaran G, Ibrahim AAB, Cheo NY, Wang PK, Kuan G, Popeska B, Chin MK, Mok MMC, Edginton CR, Culpan I, Durstine JL (2021) Using Brain-Breaks® as a technology tool to increase attitude towards physical activity among students in Singapore. Brain Sci 11(6):784
19. Moher D, Liberati A, Tetzlaff J, Altman DG, PRISMA Group TP (2009) Preferred reporting items for systematic reviews and meta-analyses: the PRISMA statement. Ann Intern Med 151(4):264–269
20. National collaborating centre for methods and tools. Quality assessment tool for quantitative studies (2008) Retrieved from http://www.nccmt.ca/resources/search/14. Accessed on 21 May 2022
21. Wang D, Zhai JX, Mou ZY, Zong HX, Zhao XD, Wang XY, Gu P (2009) Discussing on the research of heterogeneity in meta-analysis. Chin J Evid-Based Med 9(10):1115–1118
22. Wen J, Li YP (2007) The selection of a summary statistic for use in meta-analysis. Chin J Evid-Based Med 7(8):606–613
23. Podnar H, Novak D, Radman I (2018) Effects of a 5-minute classroom-based physical activity on on-task behaviour and physical activity levels. Kinesiol 50(2):251–259

24. Popeska B, Jovanova-Mitkovska S, Chin MK, Edginton CR, Mok MMC, Gontarev S (2018) Implementation of brain breaks® in the classroom and effects on attitudes toward physical activity in a Macedonian school setting. Int J Environ Res Public Health 15(6):1127
25. Hajar MS, Rizal H, Kueh YC, Muhamad AS, Kuan G (2019) The effects of brain breaks on motives of participation in physical activity among primary school children in Malaysia. Int J Environ Res Public Health 16(13):2331
26. Hajar MS, Rizal H, Muhamad AS, Kuan G (2019) The effects of brain-breaks on short-term memory among primary school children in Malaysia. Int Conf MoHE pp 493–502
27. Rizal H, Hajar MS, Muhamad AS, Kuan G (2020) The effects of brain breaks physical activity solutions on processes of change in physical activity among the Malaysian primary school children. In: Enhancing health and sports performance by design: proceedings of the 2019 movement, health & exercise (MoHE) and international sports science conference (ISSC). Springer Nature, p 481
28. Rizal H, Hajar MS, Muhamad AS, Kueh YC, Kuan G (2019) The effect of brain breaks on physical activity behaviour among primary school children: a transtheoretical perspective. Int J Environ Res Public Health 16(21):4283
29. van Stryp O, Africa E, Kidd M, Duncan MJ (2021) The effect of active brain-breaks during a typical school day on the in-school physical activity patterns of Grade 1 children in the Western Cape, South Africa. Edu 3–13:1–12
30. Tumynaitė L, Miežienė B, Mok MMC, Chin MK, Putriūtė V, Rupainienė V, Emeljanovas A (2014) Effects of intervention "Hopsport Brain Breaks" program on physical fitness and sedentary behaviour in primary school. Balt J Sport Health Sci 3(94):57–66
31. Tománek Ľ, Cihová I, Luptáková G, Antala B, Chin MK, Šagát P (2019) Effect of technology based programme "brain breaks" on the pupils' attitudes towards physical activity in secondary schools. Soc Sci Hum 27(S3):47–60
32. Sterne JA, Sutton AJ, Ioannidis JP, Terrin N, Jones DR, Lau J, Carpenter J, Rücker G, Harbord RM, Schmid CH, Tetzlaff J, Deeks JJ, Peters J, Macaskill P, Schwarzer G, Duval S, Altman DG, Moher D, Higgins JP (2011) Recommendations for examining and interpreting funnel plot asymmetry in meta-analyses of randomised controlled trials. BMJ 342:d4002
33. Cohen J (1988) Statistical power analysis for the behavioral sciences. Erlbaum, Hillsdale
34. Zhang L, Lin YC (2011) Moderation and mediation in experimental study. Manag Sci 24(01):108–116
35. Kang YJ, Yu GY (2019) Effects of exercise on low back pain: a meta-analysis. Sports Sci 39(02):63–72
36. Xu L (2014) Effects of physical activity on subjective well-being- evidence from the meta-analysis of experiment studies. Sports Sci 34(10):29–38
37. Hills AP, Dengel DR, Lubans DR (2015) Supporting public health priorities: recommendations for physical education and physical activity promotion in schools. Prog Cardiovasc Dis 57(4):368–374
38. Tran Y, Yamamoto T, Sato H, Miwa T, Morikawa T (2020) Attitude toward physical activity as a determinant of bus use intention: a case study in Asuke. Japan. IATSS Res 44(4):293–299
39. Kueh YC, Abdullah N, Kuan G, Morris T, Naing NN (2018) Testing measurement and factor structure invariance of the physical activity and leisure motivation scale for youth across gender. Front Psychol 9:1096

The Fundamental Motor Skills Performance of Children in Kapit, Sarawak

Jeswenny Fresshila John, Ngien Siong Chin, Yee Cheng Kueh, and Garry Kuan

Abstract Children's physical activity (PA) is determined by their motor skills, with increased fundamental motor skills (FMS) allowing them to participate in a wider range of PA, games, and sports. The purpose of this study was to examine the FMS in children aged 10 to 12 years old using the Canadian Agility and Movement Skill Assessment (CAMSA). The participants were in years 4, 5, and 6 ($N = 487$; male $= 241$; female $= 246$) from two primary schools in Kapit, Sarawak. CAMSA was used to measure children's FMS performance (jumping, sliding, catching, throwing, skipping, hopping, and kicking). Descriptive statistics were used to analyse the data. The results showed that 99.6% of the participants were unable to achieve the recommended level of the total CAMSA scores based on the Canadian norm. Given the positive correlation between children's FMS performance and PA, it is recommended that FMS intervention can be implemented to improve their FMS performance as well as sustain their engagement in PA.

Keywords Physical activity · Fundamental motor skills · Children · CAMSA

1 Introduction

Physical activity (PA) has always played a significant role in determining children's quality of life, physiologically or psychologically. In particular, the benefits of PA in physical, social, and cognitive development among children have been well documented for decades. For instance, engaging in PA will increase overall health [1],

J. F. John (✉) · G. Kuan
Exercise and Sports Science Programme, School of Health Sciences, Universiti Sains Malaysia, Kubang Kerian, Kelantan, Malaysia
e-mail: jeswenny@student.usm.my

N. S. Chin
Institute of Teacher Education Batu Lintang Campus, Kuching, Sarawak, Malaysia

Y. C. Kueh
Biostatistics and Research Methodology Unit, School of Medical Sciences, Universiti Sains Malaysia, Kubang Kerian, Kelantan, Malaysia

G. Kuan et al. (eds.), *Advancing Sports and Exercise via Innovation*, Lecture Notes in Bioengineering, https://doi.org/10.1007/978-981-19-8159-3_3

better concentration and attention [2], lessen the risk of non-communicable diseases (NCDs), reduce depression, anxiety, and stress [3] as well as develop a sense of trust and trustworthiness [4]. Given the positive relationship between PA and children's quality of life, PA is widely promoted worldwide, notably being promoted in the Sustainable Development Goals (SDGs) by the United Nations, whereby 7 out of 17 SDGs (e.g. good health and well-being (SDG 3); gender equity (SDG 5); industry, innovation and infrastructure (SDG 9); reduced inequalities (SDG 10); sustainable cities and communities (SDG 11); climate action (SDG 13); peace, justice and strong institutions (SDG 16)) are linked to PA promotion strategies [5].

Alternatively, failure to engage in PA will result in children's motor deficiency and negative consequences to their health. One of the negative consequences of inadequate PA engagement is the higher prevalence of overweight and obesity [6, 7]. Childhood overweight and obesity continue to pose a serious threat to public health worldwide [8, 9]. Specifically, the prevalence of overweight and obesity among Chinese primary school children was 15.2 per cent and 11.7 per cent, respectively, from 2014 to 2017 [10], while in Ghana, it was reported that 16.4 per cent of children aged 8 to 11 were overweight or obese [11]. Furthermore, one-third (30%) of Malaysian children between the ages of 5 and 17 were overweight or obese [12]. Several studies [13, 14] have found that PA is an effective measure to combat this issue. However, there is evidence that children and adolescents do not engage in PA [7, 12].

The prevalence of physically active Malaysian secondary school students was only 19.8 per cent [15], whereas 39 per cent were physically inactive [12]. Children's PA is determined by their motor skills [16], with increased fundamental motor skills (FMS) allowing them to participate in a wide range of PA, games, and sports [17]. FMS is the fundamental building block of PA which includes object control, locomotor, and stability skills [18]. There is evidence that children worldwide, including in Malaysia, have low proficiency in FMS. In Australia, less than 50 per cent of year 6 students were able to demonstrate mastery in running, jumping, kicking, and throwing skills [19]. Similarly, Singaporean children have been reported to exhibit "average" or "below average" locomotor skills as well as "poor" or "below average" object control skills [20]. In Malaysia, children aged 8 and 9 years have a low motor performance level compared to their chronological age [21].

In this regard, children's FMS performance must be monitored and assessed regularly to ensure children acquire adequate proficiency in FMS, particularly in comparison to their chronological age. One of the tools used in assessing children's FMS is the Canadian Agility and Movement Skill Assessment (CAMSA). CAMSA has been used widely in different regions and cultural backgrounds to determine children's FMS performance, including in Spain, China, England, and Australia [22–25].

Thus, this study aimed to examine the FMS performance of children aged 10 to 12 years from two primary schools in Kapit, Sarawak, using the CAMSA. The hypothesis in this study is that most of the children will be unable to achieve the recommended level of CAMSA. This study will enable physical education (PE) teachers to plan and implement FMS intervention as strategic measures for improving children's FMS performance.

2 Methods

2.1 Participants

The participants were 487 primary school students (years 4, 5, and 6) from two primary schools in Kapit, Sarawak, Malaysia. The participants comprised 241 males (49.5%) and 246 females (50.5%). As for age, 162 participants were 10 years old, 154 participants were 11 years old, and 171 participants were 12 years old. Participation in this study was completely voluntary, in which a consent form was given to and signed by respective parents while ensuring the participants' confidentiality. All participants were physically healthy and were free from any form of congenital physical defects.

2.2 Fundamental Movement Skill

The children's FMS performance was measured using the CAMSA. CAMSA is a tool developed by the Healthy Active Living and Obesity Research Group (HALO) [26] as a part of the Canadian Assessment of Physical Literacy Second Edition (CAPL-2) to measure FMS performance among children aged 8 to 12 years old. There are seven skills in CAMSA, namely 2-foot jump, side slide, catch, throw, skip, 1-foot hop, and kick, whereby children completed all the seven skills in succession (Fig. 1). Rather than assessing any deficits in FMS performance, CAMSA assesses how well children would perform in an active play situation. Other existing tools, namely Test of Gross Motor Development (TGMD), Movement Assessment Battery for Children (MABC), and Körperkoordinationstest Für Kinder (KTK), measured motor skills in isolation.

Children's CAMSA performance was determined by two components which are the criteria score (quality of the CAMSA skill performed) and time score (the time the children took to complete the seven CAMSA skills), which were measured simultaneously in a trial. In addition, the criteria score was based on process (e.g. "weight transfer and body rotation") and product (e.g. "ball hits the target") assessment. Both criteria and time scores have a maximum point of 14 points, respectively. In terms of criteria score, 1 point was given to correctly performed skill and 0 points were given to incorrectly performed skill. Meanwhile, for the time score, the time the participants took to complete the seven CAMSA skills was converted to a predefined point score ranging from 1 to 14, whereby a higher time score reflected a faster time for children to complete all the CAMSA skills. The criteria score and the time score were totalled to generate a total CAMSA score ranging from 1 to 28, which was then interpreted into "Beginning", "Progressing", "Achieving", and "Excelling".

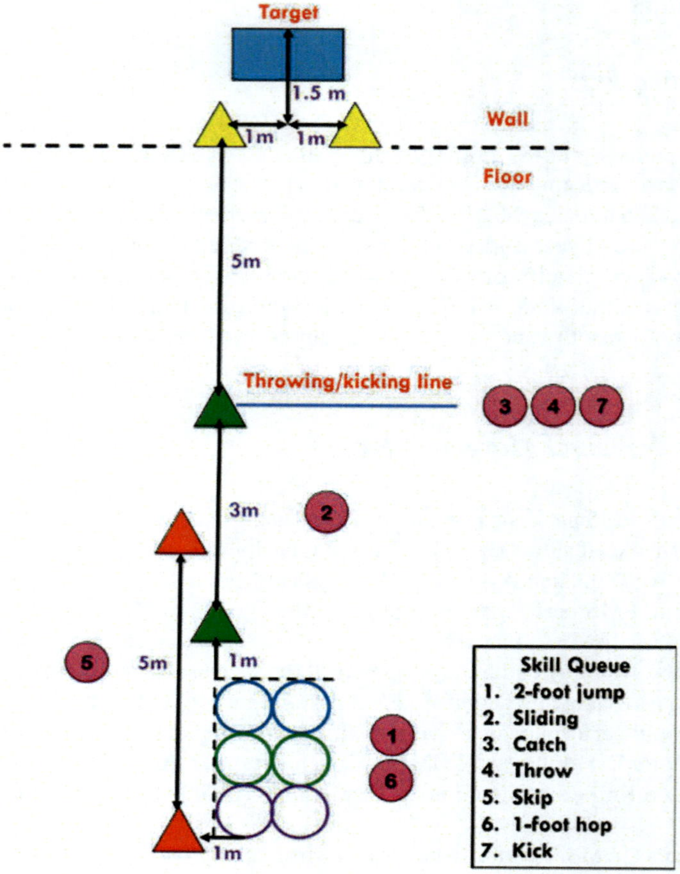

Fig. 1 CAMSA layout plan (CAPL-2 edition)

2.3 *Procedure*

The CAMSA was administered to the children by the researchers and PE teachers of the two primary schools in Kapit, Sarawak, as test administrators. All the test administrators involved in this study were provided training sessions, including setting up in the testing area, timing, verbal cue, positioning/throwing the ball, and running practice drills. The testing lasted for four weeks (two weeks per school) and was conducted in the school hall. Before the testing, the test administrators set up the testing area according to the CAMSA layout (Fig. 1) and prepared the required equipment: 6 hula hoops, 6 cones, 1 cardboard wall target, 1 football, and 1 softball.

Children observed two demonstrations by the test administrators before attempting the CAMSA. The test administrators completed the CAMSA in slow

Table 1 CAMSA skills' criteria

Skill	Criteria
2-foot jumping 1	Three 2-foot jumps in and out of the yellow/purple/blue hoops
2-foot jumping 2	No extra jumps and no touching of hoops
Sliding 1	Body and feet are aligned sideways when sliding in one direction
Sliding 2	Body and feet are aligned sideways when sliding in opposite direction
Sliding 3	Touch cone with low centre of gravity and athletic position
Catch	Catches ball (no dropping or trapping)
Throw 1	Uses overhand throw to hit target
Throw 2	Transfers weight and rotates body
Skip 1	Correct hop-step pattern
Skip 2	Uses arms appropriately (alternates arms and legs, arm swinging for balance)
1-foot hop 1	Land on one foot in each hoop
1-foot hop 2	Hops once in each hoop (no touching of hoops)
Kick 1	Smooth approach to kick ball and hit target
Kick 2	Elongated stride on last stride before impact

motion to demonstrate each skill criteria correctly while providing the verbal description of each skill and emphasising the cue words during the first demonstration. Meanwhile, the test administrators completed the CAMSA at full speed during the second demonstration, while performing each skill correctly. There were two test administrators involved in each trial. The first test administrator timed, gave the verbal cue, threw the ball during throwing skill, and positioned the ball for kicking skill, whereas the second test administrator assessed the skills performed by the participants based on the 14 criteria skills as shown in Table 1.

2.4 Data Analysis

The Statistical Package for the Social Sciences (SPSS) version 28.0 (IBM Corp., Armonk, NY, USA) was used to analyse the data. The frequency of children that obtained 1 mark for each CAMSA skill, children's CAMSA criteria and time score, children's total CAMSA scores, and interpretation of the total CAMSA scores based on the Canadian norms were determined using the descriptive analysis.

3 Results

Table 2 shows the demographic characteristics of the participants. The participants were 487 primary school students aged 10 to 12 from two primary schools in Kapit,

Sarawak. The participants comprised 241 males (49.5%) and 246 females (50.5%). Most of the participants were 12 years old (171 participants) followed by 10 years old and 11 years old with 162 and 154 participants, respectively.

Figure 2 shows that most of the participants obtained 1 point for 2-foot jumping 1 skill (389 participants), followed by 2-foot jumping 2 (379 participants), 1-foot hop 1 (372 participants), catch (367 participants), kick 1 (333 participants), kick 2 (302 participants), sliding 3 (286 participants), throw 2 (274 participants), skip 1 (237 participants), 1-foot hop 2 (201 participants), sliding 1 (198 participants), sliding 2 (192 participants), throw 1 (137 participants), and skip 2 (56 participants).

Figure 3 shows that 77 participants gained 8 points for criteria score, followed by 9 points (64 participants), 7 points (61 participants), 10 points (55 participants), 5 points (50 participants), 6 points (48 participants), 11 points (37 participants), 4 points (29 participants), 12 points (22 participants), 3 points (14 participants), 2 points (11 participants), 13 points (10 participants), 1 point (7 participants), and 14 points (2 participants).

Figure 4 shows that most of the participants (151 participants) gained a time score of 1, followed by 4 (72 participants), 3 (65 participants), 5 (63 participants), 2 (49 participants), 6 (33 participants), 7 (21 participants), 8 (16 participants), 9

Table 2 Demographic characteristics of the participants

Characteristics	Frequency	Percentage (%)
Gender		
Male	241	49.5
Female	246	50.5
Age		
10	162	33.3
11	154	31.6
12	171	35.1

Fig. 2 Number of participants that obtained 1 mark for each CAMSA skill criteria

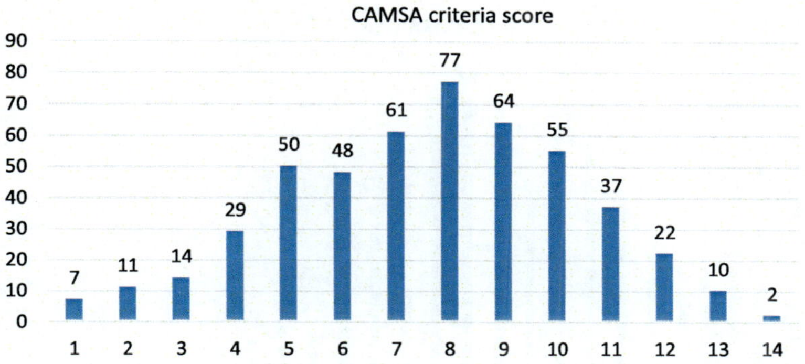

Fig. 3 Number of participants and CAMSA criteria score

(10 participants), 10 (6 participants), and 12 (1 participant). However, none of the participants obtained a time score of 11, 13, or 14.

Figure 5 shows that 47 participants respectively gained a score of 11 and 12, followed by 9 and 13 (41 participants respectively), 8 (40 participants), 7 and 10 (38 participants respectively), 14 (34 participants), 15 (30 participants), 6 (28 participants), 16 (21 participants), 5 (17 participants), 18 (12 participants), 17 (11 participants), 19 (10 participants), 20 (9 participants), 3 (8 participants), 4 (6 participants), 2 (4 participants), 22 (2 participants), and 21, 23, and 24 (1 participant respectively). However, none of the participants obtained a total CAMSA score of 1, 25, 26, 27, or 28.

Figure 6 shows that most of the participants (452 participants) were labelled as "Beginning", 33 participants as "Progressing", and 2 participants as "Achieving". However, none of the participants were ranked as "Excelling".

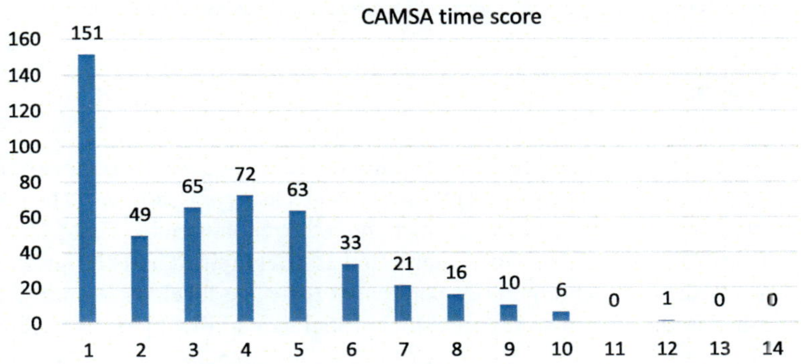

Fig. 4 Number of participants and CAMSA time score

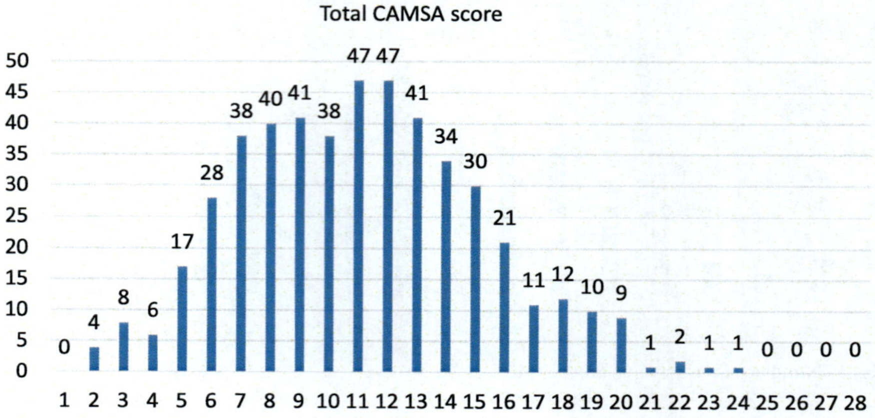

Fig. 5 Total CAMSA score

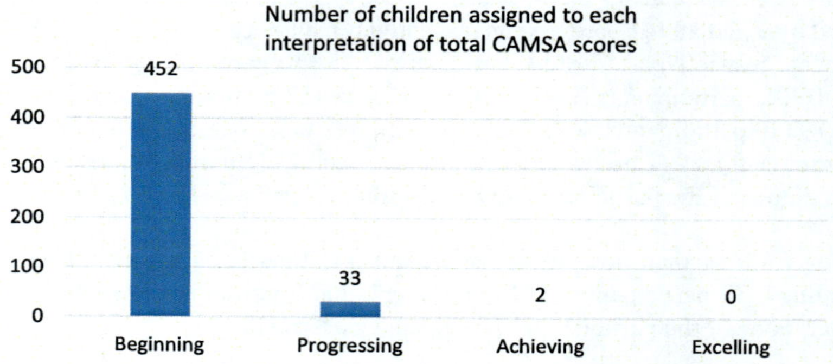

Fig. 6 Number of participants assigned to each interpretation of total CAMSA scores

4 Discussion

The study aimed to examine the fundamental motor skills (FMS) performance of children aged 10 to 12 years old from two primary schools in Kapit, Sarawak. The results showed that most of the children are falling behind the recommended CAMSA level, whereby their FMS performance was interpreted as "Beginning" and "Progressing", suggesting that the children could not perform at their age level. This finding is in line with other studies that also reported low proficiency in FMS [27, 28], which is very alarming given that proficiency in FMS is highly associated with PA engagement. Furthermore, no children were able to reach the "Excelling" rank for the total CAMSA scores, thus highlighting the lack of competency in FMS among children. However, some children gained relatively high scores for the skill score. In this regard, two children achieved a maximum of 14 points for skill scores but were unable to achieve the "Excelling" rank due to their poor time scores. The findings found that most of the children obtained 1 point for the time score, which implies that the children took more than 30 s to complete the CAMSA. The highest time score achieved was 12 points by only one child.

As previously mentioned, the CAMSA norm is a standardised norm for Canadian children in which the time score was developed based on data collected from more than 10,000 Canadian children. Therefore, the researchers proposed further research on the CAMSA norm, particularly its time score, to accommodate the Malaysian children population. This proposal is aligned with another study conducted among Chinese children using the CAMSA to measure FMS performance. Specifically, the development of the Chinese norm for CAMSA was suggested as the researchers found that Chinese children generally displayed low-level FMS performance when referred to the Canadian norm [23]. Cultural differences were also discussed in the study, further supporting the evidence on differences in motor skills' development among different regions [29]. Consequently, the development of Malaysia's own CAMSA norm will help researchers to collect more reliable data, thus reflecting the actual Malaysian's children FMS performance.

Moreover, it was noted that most of the children performed well in jumping, catching, and kicking skills. In terms of kicking skills, it is believed that children performed well due to their participation in football games with their peers during their playtime, which is greatly influenced by the immense popularity of football in Malaysia [30]. Nonetheless, it was undetermined which gender or age performed better for the kicking skill. Another significant finding from this study is that majority of the children displayed poor performance in skip skills, particularly for "Uses arms appropriately; alternates arms and legs, arm swinging for balance" criteria (skip 2) as only 56 out of 487 participants gained a point for the skill. As mechanical components are associated with potential movement development, these results point to a lack of coordination and the mechanism of force production [18]. In conclusion, these findings added further to limited studies available on FMS performance among Malaysian primary school children.

There were some limitations in this study. This study involved children from two primary schools in Kapit, Sarawak; thus, the results of this study were limited and were not generalisable to the whole 10- to 12-year-old children in Sarawak. In addition, as the COVID-19 pandemic was still ongoing during the data collection, the data collection was administered in a manner where everyone was ensured to comply and adhere to the Standard Operation Procedure (SOP).

5 Conclusion

Based on the findings of this study, the FMS performance of primary school students in Kapit, Sarawak, is at a poor level as the children were unable to achieve the recommended level of the total CAMSA scores. In that sense, further research on FMS performance among primary school children in other regions in Malaysia is needed to generalise the overall performance of Malaysian primary school children and add to the literature review on the topics. Assessing the FMS performance will help PE teachers plan and implement FMS interventions to improve children's FMS performance and sustain their engagement in PA.

Acknowledgements This research was supported by the Ministry of Higher Education Malaysia for Fundamental Research Grant Scheme (FRGS) with Project Code: FRGS/1/2020/SKK06/USM/03/13.

References

1. Crumbley CA, Ledoux TA, Johnston CA (2018) Physical activity during early childhood: the importance of parental modeling. Am J Lifestyle Med 14(1):32–35. https://doi.org/10.1177/1559827619880513
2. Buchele Harris H, Cortina KS, Templin T, Colabianchi N, Chen W (2018) Impact of coordinated-bilateral physical activities on attention and concentration in school-aged children. BioMed Res Int 2018:1–7. https://doi.org/10.1155/2018/2539748
3. Kohl III HW, Cook HD Committee on Physical Activity and Physical Education in the School Environment, Food and Nutrition Board & Institute of Medicine (eds) (2013) Educating the student body: taking physical activity and physical education to school. The National Academies Press, Washington DC
4. Di Bartolomeo G, Papa S (2019) The effects of physical activity on social interactions: the case of trust and trustworthiness. J Sports Econom 20(1):50–71. https://doi.org/10.1177/1527002517717299
5. Salvo D, Garcia L, Reis RS, Stankov I, Goel R, Schipperijn J, Hallal PC, Ding D, Pratt M (2021) Physical activity promotion and the united nations sustainable development goals: building synergies to maximize impact. J Phys Act Health 18(10):1163–1180. https://doi.org/10.1123/jpah.2021-0413
6. Chan YY, Lim KK, Lim KH, Teh CH, Kee CC, Cheong SM, Khoo YY, Baharudin A, Ling MY, Omar MA, Ahmad NA (2017) Physical activity and overweight/obesity among Malaysian adults: findings from the 2015 national health and morbidity survey (NHMS). BMC Public Health 17(1):733. https://doi.org/10.1186/s12889-017-4772-z
7. Guthold R, Stevens GA, Riley LM, Bull FC (2018) Worldwide trends in insufficient physical activity from 2001 to 2016: a pooled analysis of 358 population-based surveys with 1·9 million participants. Lancet Glob Health 6(10):e1077–e1086. https://doi.org/10.1016/S2214-109X(18)30357-7
8. Sahoo K, Sahoo B, Choudhury AK, Sofi NY, Kumar R, Bhadoria AS (2015) Childhood obesity: causes and consequences. J Family Med Prim Care 4(2):187–192. https://doi.org/10.4103/2249-4863.154628
9. Williams SE, Greene JL (2018) Childhood overweight & obesity: affecting factors, education and intervention. J Child Obes 3(2):9
10. Zhang X, Zhang F, Yang J, Yang W, Liu W, Gao L, Peng Z, Wang Y (2018) Prevalence of overweight and obesity among primary school-aged children in Jiangsu Province, China, 2014–2017. PLoS ONE 13(8):e0202681. https://doi.org/10.1371/journal.pone.0202681
11. Adom T, De Villiers A, Puoane T, Kengne AP (2019) Prevalence and correlates of overweight and obesity among school children in an urban district in Ghana. BMC Obes 6:14. https://doi.org/10.1186/s40608-019-0234-8
12. Institute for Public Health (IPH), National Institutes of Health, Ministry of Health Malaysia (2020) National health and morbidity survey (NHMS) 2019: Vol. I: non-communicable

diseases: risk factors and other health problems. Retrieved from https://iku.moh.gov.my/images/IKU/Document/REPORT/NHMS2019/Report_NHMS2019-NCD_v2.pdf. Assessed on 14 May 2014
13. Li XH, Lin S, Guo H, Huang Y, Wu L, Zhang Z, Ma J, Wang HJ (2014) Effectiveness of a school-based physical activity intervention on obesity in school children: a nonrandomized controlled trial. BMC Public Health 14:1282. https://doi.org/10.1186/1471-2458-14-1282
14. Wiklund P (2016) The role of physical activity and exercise in obesity and weight management: time for critical appraisal. J Sport Health Sci 5(2):151–154. https://doi.org/10.1016/j.jshs.2016.04.001
15. Institute for Public Health (IPH) (2017) National health and morbidity survey (NHMS) 2017: adolescent health survey 2017. Malaysia. Retrieved from https://iku.moh.gov.my/images/IKU/Document/REPORT/NHMS2017/AHSReportNHMS2017.pdf. Assessed on 14 May 2014
16. Stodden DF, Goodway JD, Langendorfer SJ, Roberton MA, Rudisill ME, Garcia C, Garcia LE (2008) A developmental perspective on the role of motor skill competence in physical activity: an emergent relationship. Quest 60(2):290–306. https://doi.org/10.1080/00336297.2008.10483582
17. Cohen KE, Morgan PJ, Plotnikoff RC, Callister R, Lubans DR (2014) Fundamental movement skills and physical activity among children living in low-income communities: a cross-sectional study. Int J Behav Nutr Phys Act 11(1):49. https://doi.org/10.1186/1479-5868-11-49
18. Gallahue DL, Ozmun JC, Goodway JD (2012) Understanding motor development: infants, children, adolescents, adults, 7th edn. McGraw-Hill Education, New York
19. Hardy LL, Barnett LM, Espinel P, Okely AD (2013) Thirteen-year trends in child and adolescent fundamental movement skills: 1997–2010. Med Sci Sports Exerc 45(1):1965–1970. https://doi.org/10.1249/MSS.0b013e318295a9fc
20. Mukherjee S, Ting JLC, Fong LH (2017) Fundamental motor skill proficiency of 6- to 9-year-old Singaporean children. Percept Mot Skills 124(3):584–600. https://doi.org/10.1177/0031512517703005
21. Jakiwa J, Suppiah PK (2020) Perbezaan tahap prestasi motor kanak-kanak berdasarkan etnik dan umur kronologi. Malays J Mov Health Exerc 9(1):159–172. https://doi.org/10.15282/mohe.v9i1.399
22. Menescardi C, Villarrasa-Sapiña I, Lander N, Estevan I (2021) Canadian agility and movement skill assessment (CAMSA) in Spanish context: evidences of reliability and validity. Meas Phys Educ Exerc Sci 26(3):245–255. https://doi.org/10.1080/1091367X.2021.2020794
23. Cao Y, Zhang C, Guo R, Zhang D, Wang S (2020) Performances of the Canadian agility and movement skill assessment (CAMSA), and validity of timing components in comparison with three commonly used agility tests in Chinese boys: an exploratory study. PeerJ 8:e8784. https://doi.org/10.7717/peerj.8784
24. Grainger F, Innerd A, Graham M, Wright M (2020) Integrated strength and fundamental movement skill training in children: a pilot study. Children (Basel) 7(10):161. https://doi.org/10.3390/children7100161
25. Lander N, Morgan PJ, Salmon J, Logan SW, Barnett LM (2017) The reliability and validity of an authentic motor skill assessment tool for early adolescent girls in an Australian school setting. J Sci Med Sport 20(6):590–594. https://doi.org/10.1016/j.jsams.2016.11.007
26. Canadian Assessment of Physical Literacy (2017) Manual for test administration second edition. Retrieved from https://www.activehealthykids.org/wp-content/uploads/2022/04/capl-2-manual-en.pdf. Assessed on 16 May 2022
27. Duncan MJ, Roscoe CMP, Noon M, Clark CCT, O'Brien W, Eyre ELJ (2019) Run, jump, throw and catch: how proficient are children attending English schools at the fundamental motor skills identified as key within the school curriculum? Eur Phys Educ Rev 26(4):814–826. https://doi.org/10.1177/1356336X19888953
28. Rodrigues D, Avigo EL, Barela JA (2015) Proficiency of fundamental motor skills in children from a public school in the city of São Paulo. Braz J Mot Behav 9(1):21–30. https://doi.org/10.20338/bjmb.v9i1.53

29. Luz C, Cordovil R, Rodrigues LP, Gao Z, Goodway JD, Sacko RS, Nesbitt DR, Ferkel RC, True LK, Stodden DF (2019) Motor competence and health-related fitness in children: a cross-cultural comparison between Portugal and the United States. J Sport Health Sci 8(2):130–136. https://doi.org/10.1016/j.jshs.2019.01.005
30. Radzi H, Buhari NRM (2021) Dialek dalam gaya bahasa lagu pasukan bola sepak di Malaysia. J Linguistik 25(1):133–147

An Open Trial Targeting Weight-Related Psychological Difficulties Among Young Adults with Overweight or Obesity During COVID-19 Lockdown

Patricia Pawa Pitil and Siti Raudzah Ghazali

Abstract Weight-related psychological difficulties (WRD) are associated with overweight and obesity including among young adults. We developed a program called ACT-EX (Acceptance and Commitment Therapy with Exercise), which incorporates six processes to increase weight-related psychological flexibility: acceptance, cognitive defusion, being present, self-as-context, values, and committed action for young adults who are overweight or obese (OW/OB). This open trial evaluated the feasibility, acceptability, and efficacy of the ACT-EX program. Fifty young adults who were OW/OB (78% female, 22% male; mean age = 21.32 years (SD = 1.2); Body Mass Index (BMI) = 30.01 kg/m^2 (SD = 4.63); 36% Malay; 28% Bumiputera Sarawak; 20% Chinese; and Bumiputera Sabah and Indians, 4% each) participated in a 6-week intervention and 6-week follow-up study. The WRD was measured by the Acceptance and Action Questionnaire for Weight-Related Difficulties Revised-18 (AAQW-R18). The ACT-EX program was well accepted. Significant decreases were seen in WRD in the post-intervention program and at the 6-week follow-up time ($p < 0.05$) and BMI ($p < 0.05$). Participants who participated in the intervention program showed a significant weight loss. This trial provides preliminary evidence for the feasibility, acceptability, and efficacy of the ACT-EX program among young adults who are OW/OB. In conclusion, the ACT-EX showed beneficial effects on reducing WRD and improving BMI. A controlled and randomized trial is warranted to further examine the effectiveness of ACT intervention in behavioral and weight management.

Keywords Acceptance and commitment therapy · Exercise · Overweight · Obese · Young adults

P. P. Pitil (✉) · S. R. Ghazali
Faculty of Medicine and Health Sciences, Universiti Malaysia Sarawak, 94300 Kota Samarahan, Sarawak, Malaysia
e-mail: patriciapawa@uitm.edu.my

P. P. Pitil
Faculty of Sports Science and Recreation, Universiti Teknologi MARA Sarawak Branch, 94300 Kota Samarahan, Sarawak, Malaysia

G. Kuan et al. (eds.), *Advancing Sports and Exercise via Innovation*, Lecture Notes in Bioengineering, https://doi.org/10.1007/978-981-19-8159-3_4

1 Introduction

World Health Organization (WHO) reported that the prevalence of obesity is increasing every year, and it is the ten most common health problem in the world. The body weight category is determined by the body mass index (BMI), which is computed by dividing a person's weight in kilograms by the square of their height in meters (kg/m^2). For the Asian cutoff point, overweight ranges between 23.0 kg/m^2 to 27.4 kg/m^2, and obesity is from 27.5 kg/m^2 to 34.9 kg/m^2 [1]. Despite various intervention programs in terms of health intervention of diet and exercise, the trend seems not to be decreasing. Being obese, the person is susceptible to physical risks such as orthopedic injuries, cardiovascular disease, metabolic disease, and psychological risks [2, 3].

University students, who are young adults, underwent significant adjustments when they transitioned from high school to a more self-disciplined, independent, and learning-focused environment [4]. During those years, they undergo phases where peer acceptance and the tendency to behave according to the environment and society are important in this early adulthood phase. Most of the youth population is represented by university students aged from 18 years old to their mid-twenties. University students must adjust to new academic and social interests, as well as physical and mental changes that occur with the transition to living on their own, in order to be prepared for a future career. These new experiences might sometimes have an impact on their physical and psychological well-being. It is reported that 38.5% of university students in Malaysia are overweight and obese [5]. Recently, the prevalence rate of overweight and obesity was 22.19 and 16.88%, respectively [6]. This accounts for at least 4 out of 10 university students being overweight or obese, which is worrying.

As most exercise professionals claim that exercise and a healthy diet would tackle the obesity epidemic problem, the trend keeps rising yearly. Despite current exercise trends such as intense training, group exercise, and modified eating behavior, the alarm could not be denied every year. Therefore, it might be the best time to focus on the psychological side of this population, which may involve evaluating their behavior in relation to body weight. The key to losing weight is these three components' combination: good nutrition, physical activity, and behavior modification. One promising behavioral intervention in weight management is the Acceptance and Commitment Therapy (ACT).

ACT is a cognitive behavioral approach to target avoidance of individuals when experiencing difficult thoughts and feelings [7]. It is a third-generation cognitive-behavioral approach that employs acceptance and mindfulness processes, commitment, and behavior change processes to develop psychological flexibility by defusing unpleasant thoughts and accepting difficult emotions while sustaining value-based action [8]. This therapy includes six psychological processes: acceptance, defusion, self-as-context contact with the present moment, values, and committed action [7]. ACT might effectively reduce weight self-stigma and impact health-related behaviors in overweight and obese individuals. It is a promising acceptance-based intervention to manage those with overweight or obese.

Although most studies in Malaysia have identified the causes of obesity, it has been limited to mostly on epidemiological analyses; consequently, there is a gap in the research. The intervention involving a psychological approach to weight loss is still in infancy in Malaysia, even though this type of treatment is widely accepted elsewhere [9, 10]. Weight loss is evidenced by the influence of a high level of physical activity [11]. However, the exercise intervention combined with behavioral treatment is yet to be explored. To address this gap, we developed a program called ACT-EX (Acceptance and Commitment Therapy with Exercise), which incorporates six processes to improve WRD; acceptance, cognitive defusion, being present, self-as-context, values, and committed action for young adults who are overweight or obese (OW/OB). This open trial assessed the feasibility and preliminary efficacy of the ACT-EX program on WRD and BMI outcomes among university students who are OW/OB. We hypothesized that participants would improve on WRD as the primary outcome as well as the BMI. We also hypothesized the program would be feasible as indicated by low or zero attrition and adherence to the intervention components.

2 Methods

2.1 Participants

During the COVID-19 lockdown, participants were recruited from two public universities in Kota Samarahan, Sarawak, Malaysia, using a purposive sampling technique. The advertisement was posted online between January and March 2021. This is a non-probability sampling method in which the target population is chosen based on the purpose of the study, with the hope that each participant will give the study unique and useful information [12]. The inclusion criteria of a participant included: (a) person with BMI category of overweight to obese (overweight = 23.0–27.4 kg/m^2, obese = $\geq$ 27.5 kg/m^2); (b) does not have any chronic disease (hypertension, diabetes) or orthopedic injury; (c) has good access to the internet for full videoconference sessions for intervention. The exclusion criteria of a participant included: (a) a person who is currently on medication and has chronic disease history; (b) orthopedic injury patient; and (c) currently participating in a supervised exercise program. Fifty participants with the inclusion and exclusion criteria were invited to participate in the study (see Table 1).

Table 1 Characteristics of the participants

Demographics	*Mean (SD)*
BMI	30.01 (4.63)
Age	21.32 (1.20)
Sex	*n (%)*
Male	11 (22)
Female	39 (78)
Race	*n (%)*
Malay	18 (36)
Bumiputera Sarawak	14 (28)
Bumiputera Sabah	4 (8)
Chinese	10 (20)
Indian	4 (8)
Study level	*n (%)*
Diploma	3 (6)
Bachelor degree	47 (94)

2.2 *Measures*

2.2.1 Anthropometry

The participants' self-reported body weight and height were gathered. For weight classification, a young individual's self-reported body weight and height can be used to figure out their BMI [13–15]. Thus, the gathered data were considered reliable for analysis in this study. The weight and height of the participants were recorded to the closest kilogram (kg) and centimeter, respectively. This section included instructions on how to self-weight with bathroom scales at home, which should be used for bodyweight records whenever possible. BMI was calculated by dividing weight in kilograms (kg) by height in meters (m^2), i.e., BMI = weight (kg)/height^2 (m^2). BMI cutoff values in the current study were 18.5 kg/m^2 (underweight), 18.5—22.9 kg/m^2 (normal weight), 23.0—27.4 kg/m^2 (overweight), and $\geq$ 27.5 kg/m^2 (obese) [1]. The participants who were in overweight or obese category were eligible to take part in the study.

2.2.2 Acceptance and Action Questionnaire for Weight-Related Difficulties Revised—18 (AAQW-R18)

Pitil and Ghazali [16] The AAQW-R18 is an 18-item self-report measure of WRD that examines weight-related thoughts, feelings, and physical sensations in relation to experiential avoidance. Items are rated on a 7-point scale (1 = "never true" or "not at all believable" and 7 = "always true" or "completely believable"). Higher scores reflect more weight-related experiential avoidance. There have been reports of good internal consistency, test–retest reliability, and validity [16]. Cronbach's alpha for the present study was 0.64, which is considered acceptable [17].

2.2.3 Demographic Characteristics

The demographic information included sex, age (in years), and study levels.

2.2.4 Design and Procedure

The study was an open trial of 50 participants who were allocated to ACT-EX program. To evaluate the feasibility and initial efficacy of other study, a single-group design was utilized. The intervention lasted for 12 weeks. Groups met were weekly for ACT intervention, and 3 times per week for exercise intervention over 6 weeks. There was no intervention contact between week 6 and the follow-up at week 12.

The research proposal was presented to the Medical Ethics Committees, Faculty of Medicine and Health Sciences, Universiti Malaysia Sarawak Research. The study protocol was executed following the ICH Good Clinical Practice Guidelines, Malaysia Good Clinical Practice Guidelines, and the Declaration of Helsinki (UNIMAS/NC-21.02/03–02 Jld.4 (02)). Participants who met inclusion criteria were provided the Physical Activity Readiness Questionnaire and consent form as they agreed to become participants. Before they sign the written consent form, they have informed about the study procedures and potential risks. The consent form described (i) the confidentiality and anonymity of their data, (ii) the purpose and procedure of the study, (iii) the decision to participate in the study, and (iv) the right to withdraw their data at any time. They complete the questionnaire, which includes self-reported weight and height, demographics, and the AAQW-18. Following the baseline assessment, participants attended a 6-week ACT-EX program and underwent post-intervention and 6-week follow-up assessments. At the 12-week follow-up, participants were given a token of appreciation for their time and effort.

2.2.5 ACT-EX Program Components

The ACT-EX program was a virtual by using the videoconferencing, group-based treatment. The participants attended the ACT sessions once a week for 1.5 h each session for 6 weeks. This ACT was designed to promote psychological flexibility that encourages openness, awareness, and engagement. The content of the sessions was adapted from an ACT intervention that has been designed for the context of weight loss used in previously published studies [18, 19]. The key intervention components were (a) values, to increase motivation; (b) cognitive defusion, to disrupt the linkages between food-and-exercise-related thoughts and behavior; (c) acceptance, to help participants tolerate negative emotions; (d) become more mindful; and (e) committed to weight loss-related values. Participants were allowed to take notes during the sessions and were given the theoretical component of the ACT model. At the end of the sessions, homework tasks were also included. In order to increase participation, the program's content addressed the participants' personally identified values (see Table 2 for the content of the ACT sessions).

The researcher supervised the exercise program, which consisted of three virtual videoconference sessions each week for 6 weeks. A minimum of 24 h between sessions was allocated to allow for recovery and to avoid injury. Each session consisted of 60 min of moderate-intensity exercise. The exercise guideline follows the guideline based on the American College of Sports Medicine. A minimum of 150 min of moderate intensity per week was suggested for an overweight or obese population for weight loss [20]. A total body workout was implemented for all sessions. The intensity of exercise was between 65 to 85% of maximum heart rate (HRmax) or Rate of Perceived Exertion (RPE) of 12—fairly light to 15—hard [21].

The percentage of HR maximum was utilized to calculate the target HR. Specifically, HRmax is equaled to 220 minus age. The study participants' age ranged from 18 to 25 years old; thus, their HRmax was between 195 and 202 bpm. The target HR was determined by the percentage of exercise intensity multiplied by 220 minus age. Table 3 shows the target HR by age and the percentage of the exercise intensity.

During each workout, the participants were instructed to assess their HR using a smartwatch or their RPE based on the Borg scale. The exercises consisted of body-weight exercises targeting the large muscle groups (e.g., push-ups, squats, lunges, crunches). A warm-up and cooling down were executed during the beginning and at the end of each session. The intervention was well attended, with 100% attendance. The feasibility of the virtual videoconference and group-based treatment contributed to this success.

2.2.6 Statistical Analysis

IBM SPSS Statistics version 26 was used to analyze the data. For demographics and outcome variables, descriptive statistics (frequency, percentages, means, and standard deviation) were presented. Repeated measure analysis of variance was used to investigate changes in outcome measures: WRD and BMI over time (baseline to post-intervention and baseline to week-12 follow-up). Cohen's *d* was calculated from baseline to post-intervention (week 6) and from baseline to follow-up (week 12) to compute effect sizes using the estimated values. A within-group effect size of 0.5, 0.8, and 1.1 was considered small, moderate, and large, respectively [22]. All analyses were set at a significant level of 0.05 ($p < 0.05$).

Table 2 Content of ACT sessions

Session/component	Aims	Key metaphors and exercises	Application to weight loss
1 Values	• To identify personal values • To evaluate if weight reduction would support such values	The choice point Values assessment rating	Enhance motivation to lose weight
2 Cognitive defusion	• To learn to detach themselves from stressful thoughts concerning bodyweight and to increase their commitment to health-related values	Push papers	• To be aware of when one's thoughts may jeopardize one's efforts to eat healthily and exercise • To aid in dissolving the connection between diet and food-related thoughts and actions
3 Acceptance/willingness	• To improve willingness to tolerate adverse internal states, the distress associated with physical exercise (e.g., weariness, physical discomfort, impulses to stop), and diet-related feelings (e.g., cravings)	Breathing exercise	To assist participants in coping with negative emotions (physical activity and diet-related feelings) rather than relying only on control or avoidance strategies
4 Mindfulness	• To build a sense of self that enables the free flow of thoughts and emotions • To enable the individual to maintain a nonjudgmental relationship with psychological and environmental phenomena as they occur	Observational self-exercise Be where you are exercise	To aid defusion and acceptance strategies

(continued)

Table 2 (continued)

Session/component	Aims	Key metaphors and exercises	Application to weight loss
5 Committed action	• To set goals aligned with the participant's values • To become aware of the necessity of adhering to values	Goals, barriers, and action exercise Living in accordance with values exercise	• To help the participant set and achieve weight loss goals • To facilitate long-term adherence to weight loss/maintenance strategies
6 Review	• Reviewing the participants' experiences over the previous weeks and assisting them in handling any concerns that may have occurred • To remind participants of key concept	Question and answer session Selection of exercises and metaphors from previous sessions	Same as before

Table 3 Target HR based on age and exercise intensity

Age	HRmax (220—age)	65% of HRmax	85% of HRmax
		Target HR (bpm) [% × (220—age)]	
18	202	131	172
19	201	131	172
20	200	130	170
21	199	129	169
22	198	129	168
23	197	128	167
24	196	127	167
25	195	127	166

Note HRmax: Heart rate maximum, Bpm: Beat per minute

3 Results

The participants reported significant decreases in WRD (Wilks' Lambda $= 0.155$, $F = 131.11$, $p = 0.001$, $np^2 = 0.86$) post-intervention and at the 6-week follow-up. The BMI also decreased over the 12th week of the study (Wilks' Lambda $= 0.371$, $F = 37.44$, $p = 0.001$, $np^2 = 0.63$), although it was not significant at the post-intervention ($p > 0.05$). A significant decrease was observed at baseline to follow-up ($p < 0.05$)

(see Table 4). Analysis also showed that the participants continued to show a decrease in WRD (see Fig. 1) and lose weight (see Fig. 2) after the program.

4 Discussion

The purpose of this study was to evaluate the effectiveness of ACT combined with structured exercise program (ACT-EX) in improving WRD and BMI among university students who were overweight or obese during COVID-19 lockdown. As far as we are aware, the present study is the first attempt to include structured exercise program in ACT-based intervention designed exclusively for weight loss in our country. The ACT-EX program provided therapy that addressed six major psychological processes: acceptance, defusion, self-as-context contact with the present moment, values, and committed action for psychological flexibility and weight control, plus the 3 sessions per week for 6 weeks of structured exercise program. Results showed that the ACT-EX program in this study could be a promising intervention for weight loss among OW/OB young adults.

There was zero attrition rate throughout the 12-week duration, from the baseline to post-intervention at the sixth week to the follow-up at week 12. This rate proves the feasibility of the videoconferencing method in delivering the intervention. Due to the limitation of movement caused by COVID-19, the online therapy and exercise intervention via video conferencing made the participants commit to the intervention. This result should be highlighted as successful as physical sessions, be it psychological therapy or exercise sessions.

This study suggests the effectiveness of ACT-EX in reducing WRD among the OW/OB participants. The intervention effect was large for both baseline to post-intervention and baseline to follow-up. The mean plot confirmed this finding (see Fig. 1). The evidence based for the ACT as an effective treatment for WRD was also found in the previous studies [23, 24]. However, this study is the first one that includes structured exercise in ACT intervention. We speculate that the exercise component might have contributed to the significant finding. Exercise promotes pleasant effects, such as improved self-esteem and body image [25], influencing the ACT's effectiveness in reducing experiential avoidance related to their weight. This indicates that successful intervention to reduce WRD for OW/OB university students was with ACT and the element of exercise.

The WRD was further improved even though the interventions have ended, indicating that the participants were more likely to practice what they had learned during the intervention. They may have continued the exercise by themselves with the ACT knowledge they obtained, thus making them more psychologically flexible. The exercise intervention may become a habit among the participants, where they underwent three sessions of moderate-intensity exercise for 6 weeks. Consistently exercising may facilitate more frequent exercise via habit formation [26], improving the participants' weight-related psychological flexibility.

Table 4 Analyses from baseline to post-intervention and baseline to week 12 follow-up on study outcome measures ($N = 50$)

	Observed mean (SD)			**Change from baseline to post-intervention**			**Change baseline to week-12 follow-up**		
	Baseline	Week 6 post-intervention	Week 12 follow-up	Mean difference (95% CI)	Effect size	*p*	Mean difference (95% CI)	Effect size	*p*
WRD	80.96 (10.69)	57.66 (7.25)	53.46 (9.11)	23.20 (18.93, 27.67)	2.18	0.001*	27.50 (23.31, 31.69)	2.57	0.001*
BMI	30.01 (4.63)	28.38 (5.36)	27.69 (4.71)	1.63 (−0.504, 4.074)	0.35	0.177	2.32 (1.66, 2.98)	0.50	0.001*

Note. WRD: weight-related difficulties; BMI: Body mass index.
*Significant at 0.05 ($p < 0.05$).

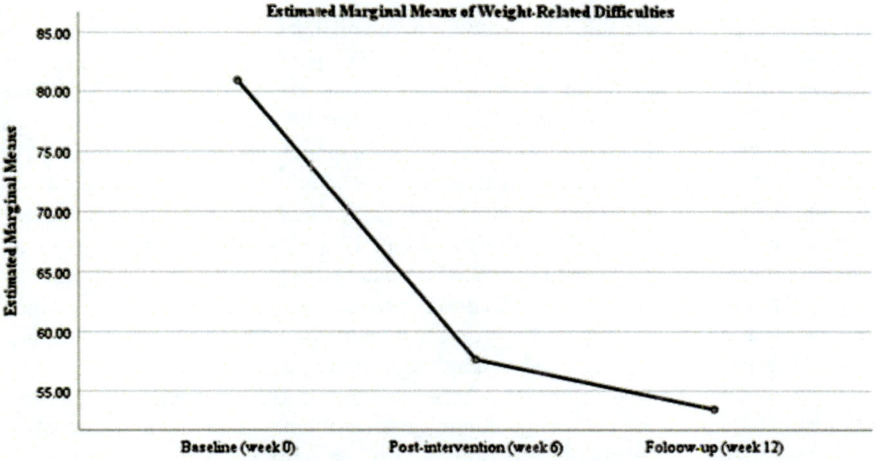

Fig. 1 The changes of WRD across baseline to post-intervention and follow-up

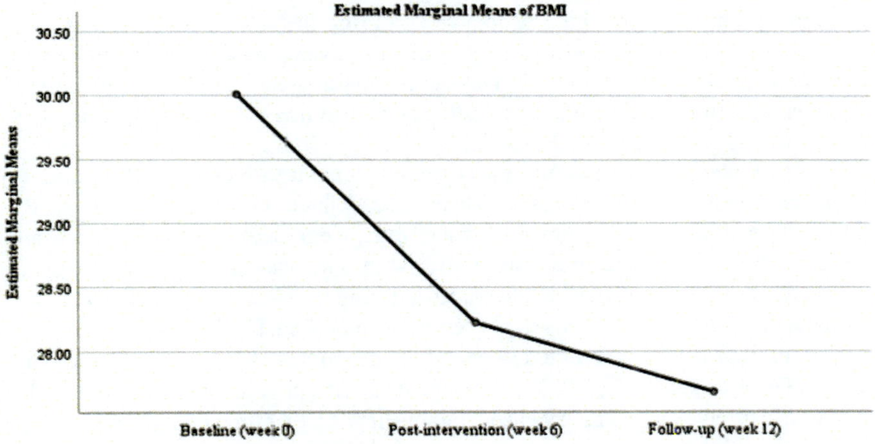

Fig. 2 The changes of BMI across baseline to post-intervention and follow-up

Despite the fact that this was an open trial, the results are encouraging, since the participants reported experiencing weight loss after the intervention (5.9%) and the follow-up (1.83%). The intervention effect was very small from baseline to post-intervention but increased from baseline to follow-up in small magnitude. The trend can be seen in the means' plot for the changes from baseline to follow-up (see Fig. 2). This finding suggests that the ACT-EX was effective in reducing BMI. The regular exercise sessions per week contributed to the weight loss. It is advisable to engage in moderate-intensity exercises for at least three days per week of accumulation of 150 min, which was achieved in this study. Following this guideline would bring health benefits [20]. As found in this study, it is known that exercise intervention could promote weight loss and improve BMI [27, 28]. The post-intervention to follow-up reduction of BMI shows that exercise did affect the sustainability of weight loss. The BMI was reduced by 7.7% from baseline to follow-up. When the program ended, the weight loss program aimed to ensure that the participants kept improving their BMI even in a small gradual reduction. It is always challenging to see participants keep attaining their weight loss after a program ends [29], which would likely be achieved in this study. Even though the follow-up lasted for 12 weeks, which was equivalent to three months, the gradual decrease in BMI is sustainable. This finding suggests that the participants have the potential to continue to lose weight over time. Therefore, the structured exercise integrated with ACT benefits of BMI improvement among the OW/OB participants and could benefit them in the long run.

In the context of weight loss, ACT promotes that healthy behaviors be aligned with firmly held personal values and commitment to them, which serve as an anchor when confronted with difficult situations and boost motivation to keep behaving healthily [30, 31]. The introduction of structured exercises in the ACT intervention improves weight management. The participants in this study were likely to continue their exercise routine after regularly engaging in three sessions per week for a 6-week intervention. The ACT was held once a week during the 6-week intervention. The exercise may become a routine and can be practiced even after the intervention ends. Consequently, weight loss was assisted with the maintenance of physical exercises. ACT intervention has appeared to promote short-term increases in physical activity [18] and was also seemed possible in this study.

The ACT intervention taught the acceptance element to increase willingness to experience aversive internal states, the distress associated with physical activity. By doing the breathing exercise, physical discomfort, weariness, and other unpleasant feelings that occur during exercise were alleviated. This was to help the participants increase their tolerance toward the discomfort during the exercise. By increasing the tolerance, the dropout toward physical exercise is lessening. Fatigue experience from exercise sessions has been reported as one of the dropout reasons in physical activity among university students [32]. The present study has proven zero dropouts in the 6-week structured exercise intervention, again showing that they had increased their psychological flexibility.

This is the first study to integrate ACT with a structured exercise program to treat overweight and obese young adults. This ACT-EX program demonstrated feasibility and efficacy in decreasing WRD and BMI. This study also demonstrated the acceptability of videoconferencing method of delivering the ACT intervention and exercise sessions. However, some limitations need to be noted. Most importantly, this study did not include a control group, and we cannot conclude the model compared to a control group of young adults in the overweight and obesity category. The study relied on self-report body weight and height measurement for the BMI. These variables are likely to be biased in a self-report format, and future studies may improvise using proper equipment (e.g., weighing scale and stadiometer) for more precise measurement. This sample was inclusive of racial diversity, which involved the majority of races in Malaysia, but predominantly female, which might limit the generalizability of our findings.

5 Conclusions and Future Study

ACT-EX is the first program that integrates a structured exercise program in ACT intervention for overweight and obese young adults, especially university students. This open trial contributes to the expanding body of research demonstrating that ACT-EX is highly acceptable and offers encouraging data regarding WRD and weight loss. More studies are needed to examine the variables associated with the overweight and obese population. More research is needed to evaluate whether the ACT-EX program can effectively treat WRD and weight management, how it can be integrated into an exercise in a weight loss program and whether doing so improves long-term outcomes for overweight and obese individuals. Future research should compare the ACT-EX to a control condition with a longer follow-up period, as this could result in a sustainable effect on WRD and weight loss.

Acknowledgements Authors would like to thank the Medical Ethics Committees, Faculty of Medicine and Health Sciences, and Universiti Malaysia Sarawak Research for the ethical approval. Not to forget the university students who participated in this study.

References

1. Zainudin S, Daud Z, Mohamad M, Boon ATT, Wan Mohamed WMI (2014) A summary of the Malaysian clinical practice guidelines on management of obesity 2004. J ASEAN Fed Endocr Soc 26(2):101–104
2. Lin HY, Huang CK, Tai CM, Lin HY, Kao YH, Tsai CC, Hsuan CF, Lee SL, Chi SC, Yen YC (2013) Psychiatric disorders of patients seeking obesity treatment. BMC Psychiatry 13(1):1–8
3. Abilés V, Rodríguez-Ruiz S, Abilés J, Mellado C, García A, Pérez de la Cruz A, Fernández-Santaella MC (2010) Psychological characteristics of morbidly obese candidates for bariatric surgery. Obes Surg 20(2):161–167
4. Stok FM, Renner B, Clarys P, Lien N, Lakerveld J, Deliens T (2018) Understanding eating behavior during the transition from adolescence to young adulthood: a literature review and perspective on future research directions. Nutrients 10(6):667–683
5. Wan Mohamed Radzi C, Salarzadeh Jenatabadi H, Alanzi A, Mokhtar MI, Mamat MZ, Abdullah NA (2019) Analysis of obesity among Malaysian University students: a combination study with the application of Bayesian structural equation modelling and Pearson correlation. Int J Environ Res Public Health 16(3):492
6. Pitil PP, Ghazali SR (2022) Overweight and obesity: a study among university students in Sarawak, Malaysia. Int J Health Promot Educ 1–13
7. Hayes SC, Strosahl KD, Wilson KG (1999) Acceptance and commitment therapy: an experiential approach to behavior change. The Guilford Press, New York
8. Hayes SC, Luoma JB, Bond FW, Masuda A, Lillis J (2006) Acceptance and commitment therapy: model, processes and outcomes. Behav Res Ther 44(1):1–25

9. Jin J (2018) Behavioral interventions for weight loss. JAMA 320(11):1210
10. US Preventive Services Task Force, Curry SJ, Krist AH, Owens DK, Barry MJ, Caughey AB, Davidson KW, Doubeni CA, Epling JW, Jr Grossman DC, Kemper AR, Kubik M, Landefeld CS, Mangione CM, Phipps MG, Silverstein M, Simon MA, Tseng CW, Wong JB (2018) Behavioral weight loss interventions to prevent obesity-related morbidity and mortality in adults: US preventive services task force recommendation statement. JAMA 320(11):1163–1171
11. Johnson NA, Sultana RN, Brown WJ, Bauman AE, Gill T (2021) Physical activity in the management of obesity in adults: a position statement from exercise and sport science Australia. J Sci Med Sport 24(12):1245–1254
12. Black K (2010) Business statistics: contemporary decision making. 6 edn. Wiley
13. Olfert MD, Barr ML, Charlier CM, Famodu OA, Zhou W, Mathews AE, Byrd-Bredbenner C, Colby SE (2018) Self-reported vs. measured height, weight, and BMI in young adults. Int J Environ Res Public Health 15(10):2216
14. Quick V, Byrd-Bredbenner C, Shoff S, White AA, Lohse B, Horacek T, Kattelmann K, Phillips B, Hoerr SL, Greene G (2015) Concordance of self-report and measured height and weight of college students. J Nutr Educ Behav 47(1):94–98
15. Pun KP, Thangiah N, Abdul Majid H (2021) Validity of self-reported weight, height and body mass index among Malaysian late adolescents. Amerta Nutr 5(4):377–386
16. Pitil PP, Ghazali SR (2021) Psychometric properties of acceptance and action questionnaire (AAQ) for weight-related difficulties questionnaire among university students. Glob Bus Manage Res 13(2):168–176
17. Pallant J (2020) SPSS survival manual: a step by step guide to data analysis using IBM SPSS. Routledge
18. Butryn ML, Forman E, Hoffman K, Shaw J, Juarascio A (2011) A pilot study of acceptance and commitment therapy for promotion of physical activity. J Phys Act Health 8(4):516–522
19. Tapper K, Shaw C, Ilsley J, Hill AJ, Bond FW, Moore L (2009) Exploratory randomised controlled trial of a mindfulness-based weight loss intervention for women. Appetite 52(2):396–404
20. World Health Organization (2010) Global recommendations on physical activity for health. World Health Organization
21. Borg GA (1982) Psychophysical bases of perceived exertion. Med Sci Sports Exerc 14(5):377–381
22. Cohen J (2013) Statistical power analysis for the behavioral sciences. Routledge
23. Potts SA (2018) Putting weight in context: acceptance and commitment therapy (ACT) guided self-help for weight self-stigma. Utah State University
24. Petersen JM, Durward C, Levin M (2021) Weight-related psychological inflexibility as a mediator between weight self-stigma and health-related outcomes. Bull Menninger Clin 85(3):316–330
25. Zamani Sani SH, Fathirezaie Z, Brand S, Pühse U, Holsboer-trachsler E, Gerber M, Talepasand S (2016) Physical activity and self-esteem: testing direct and indirect relationships associated with psychological and physical mechanisms. Neuropsychiatri Dis Treat 12:2617–2625
26. Hagger MS (2019) Habit and physical activity: theoretical advances, practical implications, and agenda for future research. Psychol Sport Exerc 42:118–129
27. Cox CE (2017) Role of physical activity for weight loss and weight maintenance. Diab Spectr Publ Am Diab Assoc 30(3):157–160
28. Bellicha A, van Baak MA, Battista F, Beaulieu K, Blundell JE, Busetto L, Carraça EV, Dicker D, Encantado J, Ermolao A, Farpour-Lambert N, Pramono A, Woodward E, Oppert JM (2021) Effect of exercise training on weight loss, body composition changes, and weight maintenance in adults with overweight or obesity: an overview of 12 systematic reviews and 149 studies. Obes Rev 22(Suppl 4):e13256
29. Hall KD, Kahan S (2018) Maintenance of lost weight and long-term management of obesity. Med Clin N Am 102(1):183–197
30. Forman EM, Butryn ML (2015) A new look at the science of weight control: how acceptance and commitment strategies can address the challenge of self-regulation. Appetite 84:171–180

31. Lillis J, Kendra KE (2014) Acceptance and commitment therapy for weight control: model, evidence, and future directions. J Contextual Behav Sci 3(1):1–7
32. Sáez I, Solabarrieta J, Rubio I (2021) Reasons for sports-based physical activity dropouts in university students. Int J Environ Res Public Health 18(11):5721

The Effect of Virtual Reality Imagery on Motivation and Football Kicking Skill Performance Among Youth Football Players in Sarawak

Candrawati Ibrahim, Garry Kuan, Ayu Suzailiana Muhamad, Yee Cheng Kueh, and Ngien Siong Chin

Abstract Imagery has been used by athletes and exercisers to achieve a variety of affective, cognitive, and behavioural outcomes. Mental imagery of planned activities increases the motivation of participation. Expectations of reward are associated with significantly more vivid imagery. Therefore, there is a need to implement virtual reality imagery, a new innovation in sport imagery that has the potential to attract players' attention while reducing the time required to achieve an effective level of imagery. 60 Sarawakian youth football players (Mean $\pm$ SD $= 21.08 \pm 1.49$) volunteered to participate in this study. They were randomly assigned to one of the three research conditions: Virtual Reality Imagery Group (VRI; $n = 20$), Imagery Script Group (IS; $n = 20$), and Control Group (CG; $n = 20$). Both of the VRI and IS received kicking skills training for 12 weeks. No intervention was given to the CG. Sports Motivation Scale-6 Questionnaire (SMS-6) and the Mor-Christian Football Kicking Test were employed at both the pre- and post-test. As a result, VRI demonstrated higher performance than the other groups, $F(2.57) = 5.53, p = 0.006$. Hence, it can be concluded that VRI could aids in the improvement of the kicking skill performance among the Sarawak youth football players.

Keywords Imagery · Virtual reality · Football · Motivation · Young athletes

C. Ibrahim (✉) · G. Kuan · A. S. Muhamad · Y. C. Kueh
Universiti Sains Malaysia, Kubang Kerian, Kelantan, Malaysia
e-mail: candrawati@student.usm.my

N. S. Chin
Institute of Teacher Education, Batu Lintang Campus, Kuching, Sarawak, Malaysia

C. Ibrahim
Institute of Teacher Education, Tun Abdul Razak Campus, Kuching, Sarawak, Malaysia

G. Kuan et al. (eds.), *Advancing Sports and Exercise via Innovation*, Lecture Notes in Bioengineering, https://doi.org/10.1007/978-981-19-8159-3_5

1 Introduction

Athletes, coaches, and supporters at high-level athletic events realise the importance of psychological variables in obtaining optimal physical performance. Psychological elements, particularly during competitions, are crucial in deciding an athlete's success or failure. Training influences athletes' performances in four ways: physical, technical, tactical, and psychological. While physical, technical, and tactical qualities all contribute to a champion's development, psychological development is equally essential [1].

When it comes to improving athletic performance, players' psyche is just as important as their physical abilities. Despite its importance in determining performance in competitive sports, sports psychology is not as commonly practised as the technical, tactical, or physical components. Because an athlete's mental state has frequently a substantial influence on their physical state, the psychological components of performance cannot be separated from the physical aspects. In Western nations, there is an understanding that athletic success is attributable to variables other than physiological strength or fitness and biomechanical (technical) characteristics. This demonstrates that the role and use of psychological components of sports in assisting player success is increasing [2].

Successful athletes have shown how important it is to be physically, intellectually, and emotionally strong in order to perform at their best. Athletes, coaches, and anyone participating in high-level sports events recognise the relevance of psychological components in achieving great performance in a tournament [3]. For these reasons, addressing the psychological aspects that impact young football players is crucial.

1.1 Imagery Improve Motivation

Imagery is a collection of mental training techniques used to assist athletes to increase motivation, enhance skill performance, and facilitate motor learning. It is described as using all of one's senses to replicate or construct a mental experience [4]. Imagery may be utilised to achieve both cognitive and motivational objectives. The cognitive function of imaging is the use of mental images to practise certain sports skills and formulate competing tactics.

The utilisation of images to experience goal completion, successful coping, and arousal control requirements is the motivational function of imaging. According to the research on mental practise [5], imagining is an effective cognitive approach for improving motor skill development and performance. Photographs may be used to create one's own universe by including one's own ideas and mental imagery. Imagery has been acknowledged as strategies in coaching to improve athletic performance, with the capacity to influence an athlete's psychological state [4]. As a result, imaging has been effectively employed as one of the mental training ways for most athletes to add mental repetition capacity to an action or skill, perhaps helping them enhance their level of confidence and concentration when executing motor abilities [6]. Furthermore, sportsmen employ imaging not just as a psychological strategy, but also as a mental preparation before important games [7].

Some feel that imagining is more effective when all senses are included, including kinaesthetic experiences that players may have throughout a game. During an athlete's real performance, images such as the scorching sensation of lactic acid buildup in the muscles, a rapid heartbeat, and the perfume of a grass field, for example, can be fairly distinct. Furthermore, assuming the same attitude as he would execute in performance, as well as utilising any utensils and clothing appropriately, might increase the physical part of the imagery. As a consequence, athletes must visualise the required abilities, and if they are unclear of the ideal methods, the trainer may aid in providing the appropriate procedures. Its goal is to discourage athletes from engaging in improper practises [8].

Brain activity during imagining, according to a research conducted by a group of neurologists, can give vital information for the individual to perform more vivid visualisation [8]. Furthermore, the majority of sports psychologists believed that visualisation may help coaches and athletes achieve peak performance [9, 10]. Furthermore, [11] shown that visualising not only increases athlete performance but also overcomes the mental barrier that is the root reason of poor performance.

More specifically, imagery can improve not just athlete performance but also a range of other areas such as technical skill development, learning techniques, strategy layout, and mental process improvement [12–14]. Visualisation has been proven in several studies to have an ergogenic effect on athletes' motor performance [11].

Furthermore, a research discovered that mental imagery benefits athletes in four different ways. For starters, it helps athletes stay healthy since it involves both mental and physical training. Second, it can help athletes move from the bottom to the top of the fitness rankings. Third, it can provide athletes more explicit instructions on how to apply their talent during training or competition. Lastly, it can enhance athletes' motivation and confidence in achieving their goals. Athletes who have fulfilled all of their physical and mental preparation are expected to win [15]. As a result, there is a need for innovation that can capture the attention of players while also lowering the time necessary to generate excellent images.

1.2 Virtual Reality (VR) Device and Sport Performance

The using of proper technological equipment, such as a helmet with a screen inside or gloves with sensors, can interact with a virtual reality (VR) simulation of a three-dimensional image or environment in a way that is ostensibly real or tangible [16]. Virtual reality in sports offers individualised training of technical-tactical as well as motor abilities against a specific opponent or condition independent of time and place [17]. The usage of virtual reality (VR) technology by players, coaches, and other sports professionals is accelerating. Numerous industries, including phobia therapy, civil engineering, architecture, medical training, military training, and entertainment, have found extensive usage for virtual reality technology [18]. It is, however, rarely utilised to enhance athletic performance, particularly in football, for example, the picture of individuals who experience agony due to the creation of lactic acid, the sensation when the heart rate increases, or the picturesque meadow that is particularly inspirational during the player's real performance. As a result, it is critical for the player to visualise themselves practising skills essential to the correct technique, so that if they are confused whether the method they are using is correct, they may consult with their coach. This is to guarantee that players do not focus on ineffective approaches.

During training and competition sessions, virtual reality allows coaches to examine their players' performance and conduct from various viewpoints. According to [19], virtual reality consists of three steps where the first is the recognition of player motions in a certain sport, the second is the adaptation to certain limits to modify elements of the simulation, and the third is the display of the virtual world. Virtual reality allows the user to feel as if they are in a real area by recreating a trustworthy artificial environment that can elicit a reaction comparable to when executing physical actions in a real setting. Virtual reality has been found to improve rugby decision-making [20], football abilities [21], and swimming [22].

Therefore, it is important to highlight and examine the role of motivation at the youth football players' level in order to try and identify what keeps the youth football players going and what increases the probability of achieving excellence in accordance with their true potential and how virtual reality imagery and motivational process develop youth football players in their football kicking skill. The findings of the study are also important to guide players and coaches to know the effect of using virtual reality imagery. Thus, this research is designed to discover and to understand how the virtual reality imagery works, to give information to both athletes and coaches to understand the potential effects of using virtual reality imagery on motivation and football kicking skill among youth football players in Sarawak.

1.3 Kicking Skill (Instep)

One of the most well-liked team sports in the world is football. The kicking technique is one of the fundamental skills that need to be mastered and competent in football. The key offensive action in this game is the football kick, and the team with the more kicks on target has a better chance of scoring and winning.

The instep kick or the "laces" kick is the single most significant kicking skill in football (Fig. 1). The instep is the bony structure on top of the foot. The arch of the foot is the curve between the heel and forefoot. The instep is indeed the area between the back of the toes and the front of the ankle. The arch extends from the rear of the toes all the way to the heel. For this main kick of football, players do not "toe" the ball, but rather use the upper part of the foot covered by the shoelaces. The instep drive requires the instep to force the football ball upward and forward. As a result, one of the goals of youth football training programmes is to develop instep

Fig. 1 Instep kick (point of contact) used at the youth football training programme

kick technique [23]. Another critical part of successful kicking is the location of the support foot behind and behind the ball. It has been recommended that the foot should fall 5 to 10 cm behind the ball and 5 to 28 cm beside it. It was recommended placing a foot support 30 cm to the side and 10 cm behind the ball [24].

2 Methodology

2.1 *Research Design*

A quantitative research was utilised in the form of quasi-experimental designs and surveys that include pre-test and post-test measurements that evaluate the effects of using virtual reality imagery on motivation (Sports Motivation Scale-6 Questionnaire (SMS-6) and football kicking skill (Mor-Christian Football Kicking Test) among youth football players in Sarawak.

The participants were divided into three research conditions: Virtual Reality Imagery Group (VRI; $n = 20$), Imagery Script Group (IS; $n = 20$), and Control Group (CG; $n = 20$). Participants in the VRI and IS groups involved in imagery training for a 12-week period (24 sessions). Pre-tests and post-tests for the assessment of football kicking skills will be conducted prior to the intervention and after participants completed their imagery sessions using virtual reality, imagery script, and kicking skills training. Participants will also complete the Sports Motivation Scale-6 (SMS-6) before and after the pre-test and post-test. In addition, the participants were asked to fulfil a form that contained their demographic information. This study obtained ethical approval from the Human Research Ethical Committee of the Universiti Sains Malaysia (Approval code: USM/JEPeM/21050348).

2.2 Study Area

The study was carried out at the Institut Pendidikan Guru Kampus Tun Abdul Razak (IPGKTAR) Campus in Samarahan, Sarawak. The questionnaire distribution and completion (pre-test and post-test) as well as the virtual reality imagery and imagery script (twice a week) sessions were conducted at the lecture class in gymnasium, IPGKTAR. Meanwhile, pre-test and post-test assessment of the football kicking test was performed on the football field at IPGKTAR.

2.3 Participants

Sixty participants (N=60) involved in this study consisted of youth football players in Sarawak. Participants were males between 18 and 23 years old and actively competing in local soccer leagues at the state level. They had at least 2 years of competitive experiences representing the state team and proficient in the Malay Language. They were divided into three research conditions: Virtual Reality Imagery Group (VRI; n = 20), Imagery Script Group (IS; n = 20), and Control Group (CG; n = 20). All participants were required to perform a demonstration of the kicking skills (instep kick). Participants' participation is voluntary and anonymous and their confidentiality is guaranteed. Only participants who have given consent by signing the consent form were allowed to participate in this study.

2.4 Research Tool

2.4.1 Demographic Form

A short demographic information form was used to collect the details of participants' age, height, weight, ethnic, education background, and achievements in football competition at different levels.

2.4.2 Sport Motivation Scale-6 (SMS-6)

The 28-item SMS-6 [25] is based on self-determination theory and was designed to assess contextual intrinsic motivation, extrinsic motivation, and amotivation. Players responded to the item "Why do you practice your sport?" with responses from a Likert-type scale that ranges from 1 (does not correspond at all) to 7 (corresponds exactly). The SMS consists of seven subscales with four items attached to each. The SMS-6 showed good validity and reliability in sports and physical education settings [26, 27]. The SMS-6 internal consistency values were 0.92 for intrinsic motivation, 0.84 for extrinsic motivation, and 0.82 for a motivation [26]. The SMS-6 was administrated to all participants at pre-test and post-test.

2.4.3 Football Kicking Skills Performance Assessment (Static)

Mor-Christian Football Kicking Test [28] is one of the football tests that evaluate kicking ability in football. This kicking test can verify the kicking ability exactly, especially for the football attackers, as it can test the object kicking accuracy. However, kicking power is not tested and scoring scheme has a contradiction. Kicking ability is mainly concerned with accuracy, power and speed, and the speed is a key factor to determine the quality and the chances of scoring, but this test just evaluates the subject's kicking accuracy only.

Mor-Christian Football Kicking Test (Fig. 2) was utilised to evaluate the kicking skill of the participants. The test was carried out at pre-test and post-test. For this assessment, obstacles of 4 round shapes with 1.21 m diameter are attached to the goalposts of the soccer goal to serve as targets, one in each corner of the goal. The player kicks at each of the 4 obstacles from 14.5 m distance 4 times. Accurate target is 10 points, missed target is 4 points. Total points are recorded after the tests. A line is drawn 16 yards (16.63 m) from the designated area and parallel to it. The participants kick a stationary ball with the preferred foot from any self-chosen point along the 16-yard (14.63 m) line. One practice trial is given for each target, then four consecutive balls are attempted at each of the four targets. The participants must start with the lower left target, proceed to the lower right target, then aim at the upper left target, and finish with the upper right target. When aiming towards a target, the participants receive a score of two points if the ball is kicked through the target, and a score of one point if the ball is shot through the adjoining target. Balls that rebound from the target are considered as successful trials. No points are given for balls that roll or bounce through the target area. The validity is 0.78 and reliability is 0.96.

Fig. 2 Mor-Christian Football kicking test

Fig. 3 Google daydream VR device

2.4.4 Imagery Tools

Despite the fact that each participant's intervention was unique, the script was theoretically grounded in bio-informational theory [29, 30]. Thus, the script contained stimulus and response propositions. Stimulus propositions described the situations being imagined [31]. The scripts included multiple senses (e.g. vision, audition, and kinaesthetic) and environmental (i.e. practice setting) and timing (i.e. correct timing) elements.

2.4.4.1 Virtual Reality Imagery

During an imagery session for the virtual reality imagery (VRI) group, participants watched a video using a virtual reality device which is Google Daydream (Fig. 3) and using a mobile phone that sets up on a virtual reality device; participants are standardised in a sitting position to watch video images using a virtual reality device (Fig. 4). The sessions were carried out twice a week for 12 weeks. The VR headset's goal is to have an immersive virtual reality experience and is adjusted to the tightening straps according to their comfort. A virtual reality Google Daydream headset is a head-mounted device made to wear over the eyes like goggles. This virtual reality device immerses the wearer throughout.

Mobile smartphones used during virtual reality sessions are linked to the Google Daydream device (Fig. 5). Video images were converted to mp4 format and were transferred to the mobile smartphones. Since all the mobile phones used in this study are smartphones, it is easier for the participants to see the video pictures more clearly. Fig. 6 shows the making of the virtual reality video kicking skill and Fig. 7 shows the completed video with voice recording in both Malay and English subtitles.

Fig. 4 Virtual reality imagery session

Fig. 5 Mobile smartphone attached to google daydream

2.4.4.2 Script Imagery (English and Malay Version)

Script imagery was employed to the SI group. The sessions were carried out twice a week for 12 weeks. The script was as follows: Look up before kicking the ball. Glance up at the field in front. Focus as much as possible on where the ball to go.

Fig. 6 The making of virtual reality video kicking skill (instep)

Fig. 7 Completed video with voice recording and Malay-English subtitles

Angkat kepala sebelum menendang bola. Lihatlah padang di hadapan. Fokus seberapa lama yang mungkin di mana mahu arahkan bola itu.

Keep your eye on the ball. Look down to the ball until you kick it.

Awasi bola. Lihat ke arah bola sehingga menendang bola tersebut.

Run towards the ball with even strides. Avoid stretching forward too far or taking a series of short steps, since this throws off your balance.

Berlari ke bola dengan langkah yang sama. Elakkan meregangkan kaki dan mengambil langkah pendek terlalu banyak, kerana ini akan menjejaskan keseimbangan.

Plant non-kicking foot next to the ball. Place it beside the ball, about shoulder width away from your other foot. A good, medium distance should feel comfortable to you. Keep your toes pointed straight ahead to the goal.

Letak kaki sokongan dekat dengan bola. Letakkan di sebelahnya dan jarakkan kaki untuk menendang bola selebar bahu. Arah jari kaki ke arah gol.

Bring your kicking leg back as far as possible. Bend your leg at the knee and keep your toes pointed downwards. In the perfect kicking position, your leg looks like it's in the shape of a "V". Stretch your legs often to increase your flexibility.

Ayunkan kaki ke belakang sejauh mungkin untuk menendang. Bengkokkan lutut dan biarkan jari kaki ke bawah. Kedudukan menendang yang sempurna adalah ketika kaki kelihatan seperti "V". Regangkan kaki untuk meningkatkan kelenturan.

Lock your ankle in place. Your foot should still be pointed downwards, straight and firm. Your ankle shouldn't wiggle at all as you kick the ball. Any movement weakens your shot.

Kunci pergelangan kaki. Kaki tetap diarah ke bawah, lurus dan lembut. Jangan goncang pergelangan kaki ketika menendang bola, kerana sebarang pergerakan akan melemahkan tendangan.

Stand straight as you kick.

Berdiri tegak semasa menendang.

Target the middle of the ball. Kicking it in the centre gives you control over your shot while also generating lots of power.

Sasarkan bahagian tengah bola. Menendang di bahagian tengah akan memberi kawalan ke atas tendangan dan juga akan menghasilkan banyak kuasa.

Instep contacting the ball. Hit the ball with the top part of your foot covered by the shoelaces to generate powerful but accurate shots.

Tendang bola menggunakan bahagian kekura kaki. Bawa kaki ke hadapan. Tendang bola dengan bahagian atas kaki atau bahagian tali kasut untuk menghasilkanten-dangan yang kuat tetapi tepat.

Follow through with your kicking leg. After striking the ball, don't stop moving. Bring your kicking leg through at a medium height. Step through and plant your leg on the ground.

Ikut lajak. Selepas menendang bola, jangan berhenti bergerak. Terus menghayun ke depan kaki menendang pada ketinggian sederhana dan mendarat.

2.5 Statistical analysis

Mixed-design ANOVA was performed to determine the significant mean difference in sport motivation and kicking performance scores between the groups (virtual reality imagery, imagery script, and control group) and within time effects (baseline and 12 weeks).

3 Results

Participants who use the Google Daydream devices for imagery outperform the imagery and control groups in terms of motivation. According to the findings, the influence of using virtual reality images on sports motivation has aided in enhancing football kicking skill performance for the aspect of groups (need to insert the results of pre and post test measurements for motivation and kicking test (can put in table). What was the p value?).

4 Discussions

The present study's aim was to investigate the effect of virtual reality imagery on motivation and football kicking skill performance among youth football players in Sarawak by assigning the participants to either virtual reality imagery (VRI), imagery script (IS), or the control groups. At both the pre- and post-test, the Sports Motivation Scale-6 Questionnaire (SMS-6) and the Mor-Christian Football Kicking Test were used to assess the participants' motivation and kicking performance. Furthermore, the outcome measurements were assessed at baseline and 12 weeks after training, and there was no significant mean difference in sport motivation or kicking performance between the study groups at baseline.

The study found a significant mean difference in sports motivation between the study groups after 12 weeks of training. The participants in the VRI group were more motivated than those in the IS group (121.85) and the control group (96.40). However, only the VRI group's motivation score goes up significantly from the baseline to the end of the 12 weeks. These findings illustrate VRI's effectiveness in increasing sports motivation. This was reflected in previous studies, which found that VRI boosts athletes' confidence and attention levels, enables them to overcome mental obstacles, and gives them the ability to mentally repeat an action or skill [6, 11]. Furthermore, the majority of sports psychologists stated VRI could aid athletes' and coaches' attaining peak performance [9, 10].

In terms of kicking performance score after 12 weeks, the VRI group had a considerably higher mean (68.0) than the IS and control groups (59.0 and 59.7, respectively). The results shows that both the VRI and IS groups' mean performance scores considerably increased from baseline to 12 weeks after treatment, whereas there was no significant change in the control group. This demonstrates how the VRI and SI have an impact on an athlete's kicking performance, which can improve their overall sporting performance. Although no comparable studies have been conducted in Malaysia to date, Taekwondo athletes' sports performances were found to be positively correlated with both internal and external imagery in a previous study [5]. Additionally, one of the crucial skills that must be developed to be proficient in football is the kicking technique [23].

The present study examined the effects of VRI and IS, which reflect how they influence athletes' sport motivation and kicking performance. More studies are needed into the design of customised VIR and IS aimed at the development of skills that reflect the learning and movement process instead of the outcome of the skill being learned. Finally, although the sample size of 60 may be considered adequate for this study, it should be acknowledged that future studies should consider larger sample sizes to have more power and detect any significant difference.

5 Conclusion

The present study confirms the beneficial effect of virtual reality imagery (VRI) and imagery script (IS) in promoting sports motivation and kicking performance skills among football players in Sarawak. A significant difference was found between the VRI, IS, and control groups after 12 weeks, with the VRI having a higher influence on the athletes' level of motivation and kicking performance skills. Additionally, this study is the first to examine the effectiveness of VRI and IS in Malaysia for kicking performance and sports motivation. These results show that more research needs to be done with different types of athletes and in different parts of Malaysia. The results indicated that VR imagery can help in motivating the players to improve their kicking skill technique. However, further study is needed to fully comprehend the link between VR imagery and motivation.

References

1. Moran AP (2004) Sport and exercise psychology: a critical introduction. Routledge, New York
2. Tajul Arifin Muhamad (2006) Hubungan motivasi, gaya penumpuan dan kebimbangan dengan pencapaian Pemain Golf Remaja Kebangsaan di Malaysia
3. Institut Sukan Negara Malaysia: Apa Itu 'Psikologi' Sukan (2015). Retrieved from http://isn.gov.my/
4. Arjeria GP, Kumar S (2015) Effect of mental imagery on penalty shooting in handball. Int J Phys Educ 8:24–27
5. Parnabas V, Parnabas J, Parnabas AM (2015) The influence of mental imagery techniques on sport performance among taekwondo athletes. Eur Acad Res II(11):14729–14734
6. Amasiatu AN (2013) Mental imagery rehearsal as a psychological technique to enhancing sports performances. Educ Res Int 1(2):69–77
7. Watt A (2003) Development and validation of the sport imagery ability measure. Faculty of Human Development, Victoria University of Technology
8. Afrouzeh M, Sohrabi M, Torbati HT, Gorgin F, Mallett CJ (2013) Effect of PETTLEP imagery training on learning of new skills in novice volleyball players. Life Sci J 10:231–238
9. Hall CR, Martin KA (1997) Measuring movement imagery abilities: a revision of the movement imagery. J Mental Imagery 21:143–154

10. Janssen JJ, Sheikh AA (1994) Enhancing athletic performance through imagery: an overview. In: Sheikh AA, Korn ER (eds) Imagery and sports physical performance. Baywood publishing, Amityville, NY, pp 1–22
11. Wakefield CJ, Smith D (2009) Impact of differing frequencies of PETTLEP imagery on netball shooting. J Imag Res Sport Physi Act 4(1):1–12
12. Ribeiro J, Madeira J, Dias C, Stewart LR, Corte-Real N, Fonseca A (2015) The use of imagery by Portuguese soccer goalkeepers. J Imag Res Sport Physi Act 10(1):9–17. https://doi.org/10.1515/jirspa-2014-0009
13. Munroe KJ, Giacobbi PR Jr, Hall C, Weinberg R (2000) The four Ws of imagery use: where, when, why, and what. Sport Psychol 14(2):119–137. https://doi.org/10.1123/tsp.14.2.119
14. Williams SE, Cumming J (2011) Measuring athlete imagery ability: the sport imagery ability questionnaire. J Sport Exerc Psychol 33(3):416–440
15. Rattanakoses R, Geok SK, Abdullah MC (2012) Effect of imagery practice program on imagery ability in Thailand adolescent cyclists. Int J Acad Res Bus Soc Sci 2(11):114–133
16. Oxford University Press (2018) Virtual reality: definition of virtual reality in English by oxford dictionaries. Retrieved from https://en.oxforddictionaries.com/definition/virtual_reality
17. Akbas A, Marszalek W, Kamieniarz A, Polechonski J, Slomka KJ, Juras G (2019) Application of virtual reality in competitive athletes – a review. J Hum Kinet 69:5–16
18. Stinson CA, Bowman DA, Gracanin D, Ollendick TH (2013) Virtual reality for sports training. Thesis Master, Virginia Polytechnic Institute and State University
19. Bideau B, Kulpa R, Vignai N, Brault S (2010) Using virtual reality to analyze sports performance. IEEE Comput Graph Appl 30:14–21. https://doi.org/10.1109/MCG.2009.134
20. Watson G, Brault S, Kulpa R, Bideau B, Butterfield J, Craig C (2011) Judging the 'passability' of dynamic gaps in a virtual rugby environment. Hum Mov Sci 30(5):942–956
21. Petit JP, Ripoll H (2008) Scene perception and decision making in sport simulation: a masked priming investigation. Int J Sport Psychol 39(1):1–19
22. Fels S, Kinoshita Y, Chen TPG, Takama Y, Yohanan S, Gadd A, Funahashi K (2005) Swimming across the pacific: a VR swimming interface. IEEE Comput Graph Appl 25(1):24–31
23. Weineck J (1997) Fuβballtraining. Teil 1: Konditionstraining des fussballspielers. Perimed: Spitta Verlag
24. McLean BD, Tumilty DM (1993) Left-right asymmetry in two types of soccer kick. Br J Sports Med 27:260–262
25. Mallett C, Kawabata M, Newcombe P, Otero-Forero A, Jackson S (2007) Sport motivation scale-6 (SMS-6): a revised six-factor sport motivation scale. Psychol Sport Exerc 8(5):600–614. https://doi.org/10.1016/j.psychsport.2006.12.005
26. Alexandris K, Tsorbatzoudis C, Grouios G (2002) Perceived constraints on recreational sport participation: investigating their relationship with intrinsic motivation, extrinsic motivation and a motivation. J Leis Res 34:233–252
27. Pelletier LG, Fortier MS, Vallerand RJ, Tuson KM, Brière NM, Blais MR (1995) Toward a new measure of intrinsic motivation, extrinsic motivation, and a motivation in sports: the sport motivation scale (SMS). J Sport Exerc Psychol 17(1):35–53
28. Mor D, Christian V (2018) The development of a skill test battery to measure general soccer ability. N C J Health Phys Educ 11:15. Oxford University Press
29. Lang PJ (1977) Imagery in therapy: an information processing analysis of fear. Behav Ther 8:862–886
30. Lang PJ (1979) A bio-informational theory of emotional imagery. Psychophysiology 17:495–512
31. Morris T, Spittle M, Anthony W (2005) Imagery in sport. Human kinetics

The Physical Fitness Benefits of Speed Agility Training Aids Set

Nurlyn Syifaa' Jacklyna Aban, Candrawati Ibrahim, Ismail Dikoh, Ung Siong Nguang, Garry Kuan, Ngien Siong Chin, and Amir Ahmad Dewan

Abstract Physical fitness innovation is important for improving speed and agility, which has the effect of boosting physical fitness among physical education students. In order to develop physical fitness among students and athletes, sports organisations and schools must constantly update and invent new original concepts. In other words, innovation has an impact on and aids in the improvement of physical fitness. To succeed in sports, they must excel in physical fitness, as well as speed and agility on the court or field. The aims of this study are to investigate the effect of Speed Agility Training Aids Set (SATAS) on physical fitness among physical education students in Sarawak. A total of 40 physical education students in Sarawak (19–22 years old) volunteered to participate in this study. The Southeast Missouri (SEMO) Agility Test (Kirby 1971) was used for the pre- and post-test. The intervention consisted of 16 sessions implemented for a duration of 8 weeks. A post-test was administered following the completion of the 16 SATAS sessions. The mean values of the pre-test were 10.47 (SD = 1.07) seconds and post-test were 10.02 (SD = 0.96) seconds. The findings demonstrated that SATAS is an effective innovation tool that deserves more attention in order to aid in the improvement of physical fitness performance.

Keywords Speed · Agility · Physical fitness · Physical education · Innovation

N. S. J. Aban (✉) · A. A. Dewan
SK (A) Ibnu Khaldun, Kota Samarahan, Sarawak, Malaysia
e-mail: nurlynsyifaaaban@gmail.com

C. Ibrahim · I. Dikoh · U. S. Nguang
Institute of Teacher Education Tun Abdul Razak Campus, Kuching, Sarawak, Malaysia

C. Ibrahim · G. Kuan
Exercise and Sports Science Programme, School of Health Sciences, Universiti Sains Malaysia, Kubang Kerian, Kelantan, Malaysia

N. S. Chin
Institute of Teacher Education Batu Lintang Campus, Kuching, Sarawak, Malaysia

G. Kuan et al. (eds.), *Advancing Sports and Exercise via Innovation*, Lecture Notes in Bioengineering, https://doi.org/10.1007/978-981-19-8159-3_6

1 Introduction

Physical fitness is broken down into different component categories that alter on a regular basis. Athletes can be trained in various areas of physical fitness in order to improve specific sports skills or performance. Building on specific components can lead to fitness, mental health, and lifestyle objectives. Physical fitness is divided into two types, namely health-related fitness and skill-related fitness. Cardiovascular fitness, muscular strength, muscular endurance, body composition, and flexibility are the health-related components, whereas agility, speed, power, balance, coordination, and response time are the six components of skill-related fitness. These skill-related components are movements that an individual must accomplish in order to exhibit a variety of motor skills and movement patterns successfully. Athletes or students who really are physically fit cannot only increase their speed and agility, but also reduce the effects of incidents and deaths caused by movement [1].

Physical fitness is defined as the absence of any visible disease in a person. Physical activity can also assist people who already have cardiovascular disease, according to [2]. It also promotes health and well-being, aids in the prevention of a variety of health problems, lowers the risk of diseases such as hypertension, diabetes, cancer, and others, and increases general quality of life [3–5]. Physical fitness has been linked to lower total and cardiovascular mortality [6] and even little improvements in fitness can lead to a reduction in mortality [7].

Physical activity reduces your risk of stress, tension, and depression while also making you feel better. Someone who is not physically fit cannot be deemed healthy. As a result, not only physical fitness but also overall fitness must be considered. Physical well-being is frequently defined as the absence of illness. In the current day, chronic diseases are major causes of death. The bulk of chronic diseases are caused by a lack of physical activity [8]. Physical well-being is defined as the absence of any obvious disease, such as despair or anxiety. According to [9], physical activity can help avoid depression.

1.1 Speed and Agility

Every coach must have proper training techniques to assist athletes in improving their speed and agility and motivating athletes [10]. Speed is an important fitness component for a variety of sports. The ability to move quickly is referred to as speed. The distance travelled in the shortest amount of time is referred to as speed. Individuals who focus on foot speed can run quickly, but individuals who focus on hand speed can throw or hit the ball quickly.

Speed is the most crucial part of fitness for some athletes, such as track and field sprinters, sprint swimmers, cyclists, and speed skaters. In many other sports, particularly team field sports, speed is an important component of total fitness. Speed is classified into two types: general and particular speed. General speed is defined as the ability to perform any movement (motor reaction) quickly, often across a short distance of 5–10 m, whereas specific speed is defined as the ability to perform certain actions at a specific time. Athletes who can outrun their opponents have an advantage. A speedier athlete, for example, may be able to reach a ball faster than a rival or even outrun a pursuer. As a result, athletes in almost all sports place a premium on speed.

Training for agility formation can be tailored to the needs of the type of game or activity, as each sport requires a different level of agility. The training focuses on agility elements while also improving physical fitness components based on motor behaviour such as speed, power, reaction time, coordination, and balance [11, 12]. Agility is the ability to change body position, stop and start movements quickly, and control and direct the movement of the whole body. Agility requires a good level of speed, flexibility, and coordination of movements to confuse or avoid the opponent. It is the foundation of motor behaviour that emphasises high performance in sports. It is an important component of physical fitness and is useful in a wide range of sports and physical activities. In team sports such as football, soccer, basketball, hockey, volleyball, and rugby, you must react quickly to the movement patterns of your teammates and the ball.

According to [13], it stated that reactive agility and change of direction speed are important determinants of success in football (soccer). Training factors have a significant impact on a person's dexterity. Because each sport necessitates a different level of agility, training for agility formation can be tailored to the specifics of the game or activity. Epley [14] also stated that any coach needs an effective training approach to help players improve their agility and motivation. The training emphasises agility while also improving physical fitness components based on motor behaviour.

1.2 Speed Agility Training Aids Set (SATAS)

The application of scientific knowledge study findings in the training of players may bring more ease. Technology advancements have also offered specialised groups with the opportunity to simply provide solutions to ensure that young athletes' performance continues to improve in order to make Malaysia famous all over the world. The advancement of technology, particularly in sport, is an unavoidable component of improving an athlete's achievement performance in the twenty-first century [15]. The introduction of innovative physical fitness training tools acknowledges the use of creative ways for sports skill training in order to increase performance and discovered that physical fitness and sports training can be completed more swiftly and efficiently.

The main issue with developing this innovation is the lack of up-to-date tools and methods for training athletes in Malaysia. The majority of them continue to use traditional methods to train young athletes at the school. Most physical education teachers or coaches only employ traditional training methods. Coaches of badminton players, for example, only show the directions to players with badminton rackets and without additional tools. The player will move in the direction indicated by the coach,

Fig. 1 Speed agility training aids set (SATAS)

but the player will not be heading in the proper area or in the direction desired by the coach due to the lack of a tracking device. Furthermore, there are no training aids that make it easier for coaches to monitor or record player performance.

The innovation of the Speed Agility Training Aids Set (Fig. 1) makes it possible for athletes to train on their own without the guidance of a teacher or coach in the training grounds. Those who also can enhance their own speed and agility using the laptops and sensors built into the boards. As just a result, the coach could use laptop to track the athletes' skills and talents. Teachers' and coaches' effectiveness and savings in time could be enhanced, and athletes can keep a record of one's record keeping on the laptop for reference in the future.

This innovation, which uses high-quality sensors to detect sound, can improve an athlete's ability to change direction and run quickly towards the right target in a set period of time. Athletes will be able to accelerate by touching the sound sensor on the board. When the sensor detects a sound, it sends a number counts to a computer for recording via the UNO output. Besides that, the use of SATAS can increase athletes' motivation to train harder. It can calculate the number of targets that an athlete can complete in a given period of time (timer). Given that SATAS is portable and simple to use, athletes can train without the assistance of a teacher or coach. Teachers and coaches can monitor the performance of their athletes by observing the results on the laptop.

The Fourth Industrial Revolution (4IR) is a term used to describe the blurring of the physical, digital, and biological worlds. It is the result of advancements in artificial intelligence (AI), robotics, the Internet of Things (IoT), 3D printing, genetic engineering, quantum computing, and other technologies. The skills of the

Fourth Industrial Revolution (4IR) were applied to the Speed Agility Training Aids Set (SATAS) (Fig. 1). The researcher expressed innovative thinking, original ideas, problem-solving skills, ideation, analysis, critical analysis, and interpersonal skills. These are some of the most critical 4IR-driven areas of expertise required to be successful in the evolving sport industry work environment.

SATAS implementation procedures state that (Fig. 2). In the front of touch board 1, the athlete takes a ready position (Fig. 3). The game begins with a buzzer (Fig. 4) and a timer (along with countdown timer) (Fig. 5). The athlete runs to the touch board 2 direction (Fig. 6). The presence of both the player is detected by 2 ultrasonic sensors. Sensors (Fig. 7) placed just on board that detect players' movement would then indicate an entry into the Arduino UNO R3 Advance Beginner Learning Sensor Starter Kit V5 (Fig. 8). An adding functional to the UNO board (Fig. 8) board is provided by a sensor can detect the player's touch on the board. The exit signal from the UNO board (Fig. 8) will be sent to display 1 to add a value of COUNT (Fig. 9) depending on how often the sensor detects the player's touch on the board. After the timer runs out (Fig. 5) and the buzzer sounds (Fig. 4), the final count value will be displayed (Fig. 5).

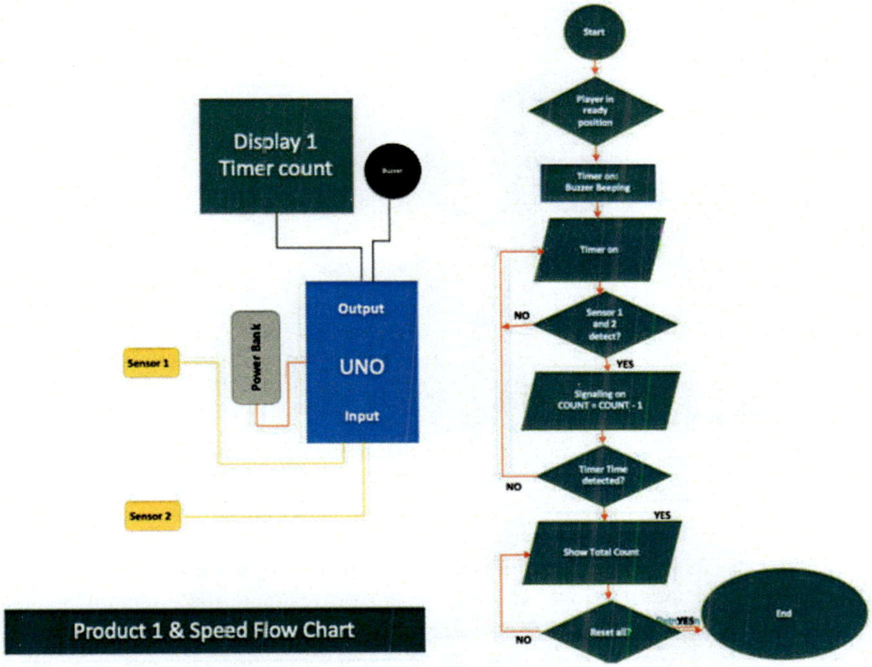

Fig. 2 SATAS flow chart

Fig. 3 Ready position (Touch Board 1)

Fig. 4 Buzzer

2 Methodology

2.1 Research Design

The study was carried out quantitatively using quasi-experimental designs that included pre-tests, interventions, and post-tests to investigate the effects of using SATAS innovation tools as an external perspective to obtain physical fitness (speed and agility) effectiveness among Sarawak Physical Education students aged 19–22 years old. This research is separated into three stages. Before taking the SEMO Agility Test, research participants fill out demographic information in the pre-test phase. Following the pre-test, research participants will move on to the intervention phase, which will last approximately for 8 weeks and consist of two sessions

Fig. 5 Laptop for timer and count value display

Fig. 6 Touch board 2

each week. After completing of the 8-week intervention phase, research participants will proceed to phase 3, the post-test phase, when they will refill their demographic information and take the SEMO Agility Test.

Fig. 7 Ultrasonic sensor

Fig. 8 Arduino UNO R3

Fig. 9 Numeric display

2.2 Research Location

The intervention was carried out on the premises of Institut Pendidikan Guru Kampus Tun Abdul Razak in Kota Samarahan, Sarawak. The facilities make it convenient for study participants to follow all stages because it is a training facility that allows the phases of the study to be done at the site chosen by the researcher.

2.3 Research Participants

Prior to conducting the study, permission had been acquired from the proper authorities. The Director of the Institute of Teacher Education Tun Abdul Razak Campus approved the research ethics. The approval was communicated to the Head of Department of Physical Education and Health at the Institute of Teacher Education Tun Abdul Razak Campus in order to ensure complete collaboration and coordination in the administration of the study.

Physical education students from Institut Pendidikan Guru Kampus Tun Abdul Razak took part in this study. The study included 40 male and female students. The total number of participants for the treatment and control groups was 15, according to [16] and [17]. However, after accounting for the 20% reduction rate, the overall sample size is 20.

Participants in the study were physical education students whose age ranged between the ages of 19 and 22 years old. Participants must also be in a good health and have no form of injuries. Participants were selected based on their ability to perform and have train regularly. They gave their firmed consent to participate in the study and have the option to withdraw at any time. There were no costs or incentives for the participants. As a show of gratitude, they received a certificate and a few trinkets.

2.4 Research Tool

2.4.1 SEMO Agility Test

The SEMO Agility Tests [18] was used to test individual dexterity to move the body forward, backward, and sideways. Movement in SEMO Agility Test is similar to movement in badminton. Figure 10 shows the SEMO agility test. This test was reported to have validity as high as 0.63 while reliability reached 0.88 for two trials. This proves that this test is accurate to measure the level of agility of athletes.

The SEMO Agility Test has a rectangular area measuring '3.6 m (horizontal) × 5.7 m (vertical)' feet and each runway will be marked with a skittle. The subject will be asked to stand at the starting point with his back towards the 'D' point. At

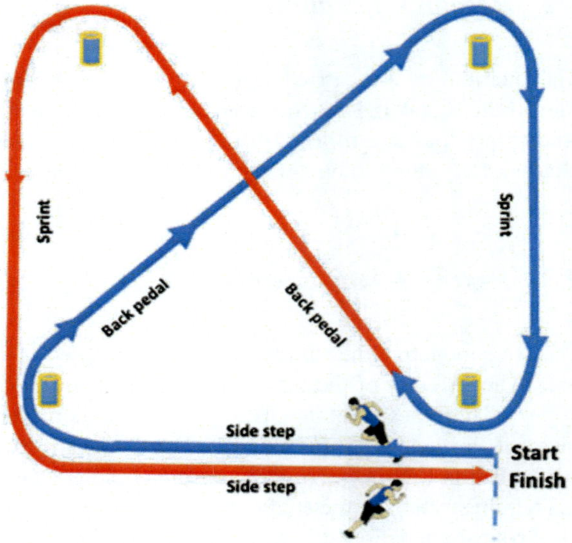

Fig. 10 SEMO Agility Test

the "Ready, Go" command signal, the subject starts a side run from A to B and passes outside the corner of skit B. The subject then runs backwards from B to D and passes outside the corner of skit D. After that, the subject is asked to sprint forward from D to A and then pass outside the corner of skit A. The subject then again runs backwards from A to C and then passes outside the corner of skit C. Next, the subject will sprint forward from C to B and will pass outside the corner of skit B. Finally, they will do a side run from B to the finish line A. Plus, while doing a side run, the legs should not cross at all. When running backwards, the subject's body should be upright in the opposite direction. Also, incorrect procedure is not acceptable, the subject needs to be tested until a perfect attempt. An attempt will be given before the test begins. The researcher will use a measuring map in preparing the SEMO Agility Test. After that, make sure the distance is correct according to the proper measurement. Each section has a different test run. To make it easier for subjects to change their running movements, the researcher can place skittles or cones of different colours or the researcher always gives instructions when wanting to change movements. The researcher should give the subject a trial period for the test in order to familiarise themselves with the test. Also make sure that the subject is given a break first after doing the ABCD experiment. Each subject was given two tests and the time of each trial was recorded accurately up to 0.1 s. A time value of less than two tests is the subject score.

Ranking	Man	Woman
Excellent	< 10.7	< 12.19
Good	10.47–11.49	12.20–12.99

(continued)

(continued)

Ranking	Man	Woman
Moderate	11.50–13.02	13.00–13.90
Weak	13.03–13.79	13.91–14.49
Very weak	> 13.8	> 14.50

3 Research Finding

The mean values of the pre-test for all participants were 10.47 (SD = 1.07) seconds and the post-test were 10.02 (SD = 0.96) seconds (Table 2). According to the findings, the Speed Agility Training Aids Set (SATAS) had aided in the improvement of speed and agility performance among Sarawak physical education students. According to the findings, the Speed Agility Training Aids Set (SATAS) had aided in the improvement of speed and agility performance among Sarawak physical education students (Figs. 11, 12 and 13) (Tables 1 and 2).

4 Discussion

Sports innovation is a new research area that integrates sport with many types of experiences and practises. Sport innovations include technological advancements, tools, and equipment, as well as innovative methods and coaching, consumer products and services, digital content, and venues. Social, institutional, and organisational

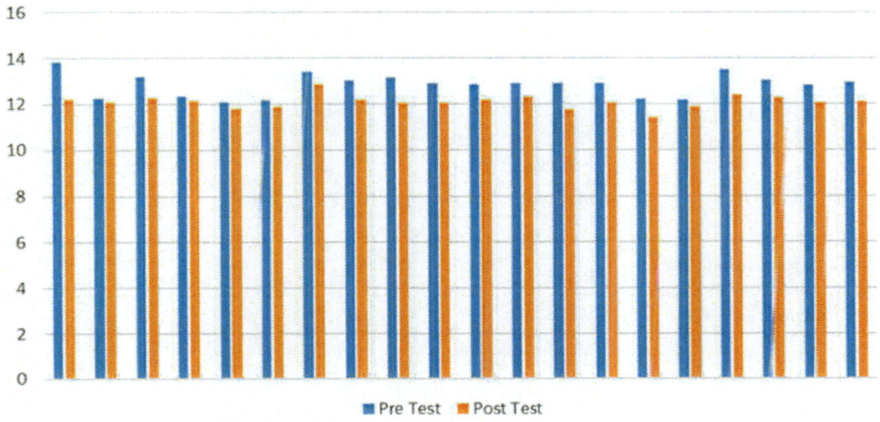

Fig. 11 Results pre- and post-test (Female)

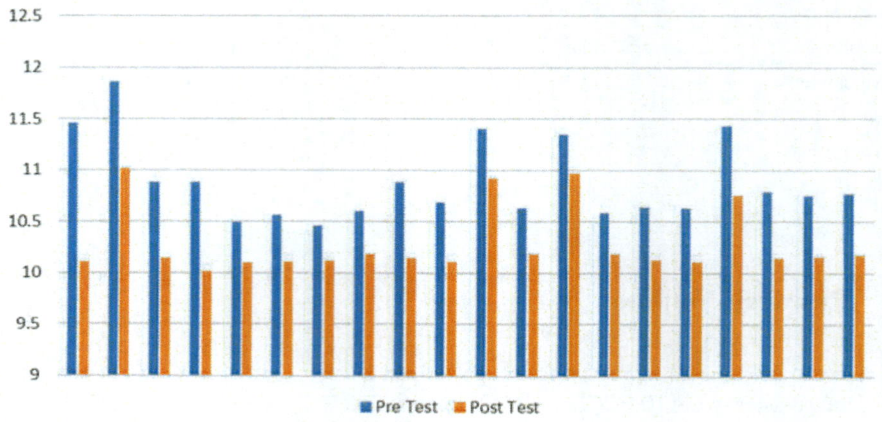

Fig. 12 Results pre- and post-test (Male)

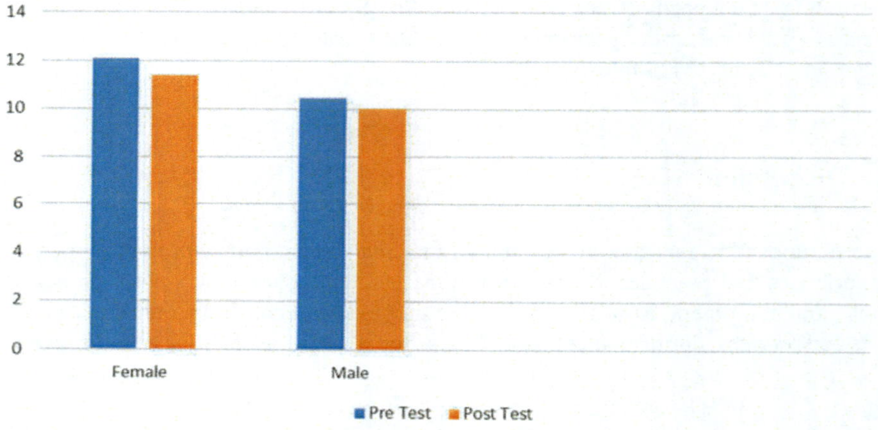

Fig. 13 Performance times of female and male participants

Table 1 SEMO test result of female and male participants

Gender	Pre-test (s)	Post-test (s)
Female (n = 20)	12.1 ± 0.49	11.39 ± 0.20
Male (n = 20)	10.47 ± 0.39	10.02 ± 0.33

Table 2 Mean values and standard deviation SEMO test of participants (N = 40)

Results	Pre-test (s)	Post-test (s)
Mean	10.47	10.02
Standard deviation	1.07	0.96

restrictions continue to limit sport innovation research. These viewpoints on innovative phenomena should be explored in the future sports innovation research. While developing ideas, sport equipment, and product lines are significant and contribute to the growth of modern sport, more research on sport innovation issues and difficulties, social innovations in sport, and effective administration and leadership of sport innovation will be necessary.

This Speed Agility Training Aids Set (SATAS) is intended to assess a player's ability to move in a sideways, forward, and backward motion. Students and athletes use SATAS to improve their physical fitness. Using this remarkable innovation training aids set can considerably improve physical fitness. Furthermore, the SATAS offers diversity, excitement, and enjoyment to the teaching and learning process, as well as exhilarating sports concept understanding. It is also applicable to all sports.

5 Conclusion

In the absence of a coach, SATAS is intended to make training more engaging and manageable. SATAS is also predicted to boost players' interest in, and incentive for, speed and agility training. In keeping with the present 4.0 industrial revolution, SATAS has revolutionised existing traditional speed and agility training techniques.

References

1. Xu B (2015) How to improve the athletes' physical fitness. In: Proceedings of the 2nd int conf civ mat environ sci. Atlantis Press, pp 282–284
2. Warburton DER (2006) Health benefits of physical activity: the evidence. Can Med Assoc J 174(6):801–809. https://doi.org/10.1503/cmaj.05135.1
3. Owen N, Sparling PB, Healy GN, Dunstan DW, Matthews CE (2010) Sedentary behavior: emerging evidence for a new health risk. Mayo Clin Proc 85(12):1138–1141. https://doi.org/10.4065/mcp.2010.0444
4. Lavie CJ, Ozemek C, Carbone S, Katzmarzyk PT, Blair SN (2019) Sedentary behavior, exercise, and cardiovascular health. Circ Res 124(5):799–815. https://doi.org/10.1161/CIRCRESAHA.118.312669
5. Jiménez-Pavón D, Carbonell-Baeza A, Lavie CJ (2020) Physical exercise as therapy to fight against the mental and physical consequences of COVID-19 quarantine: special focus in older people. Prog Cardiovasc Dis 63(3):386–388. https://doi.org/10.1016/j.pcad.2020.03.009
6. Blair SN, Kampert JB, Kohl HW, Barlow CE, Macera CA, Paffenbarger RS Jr, Gibbons LW (1996) Influences of cardiorespiratory fitness and other precursors on cardiovascular disease and all-cause mortality in men and women. JAMA 276(3):205–210
7. Erikssen G, Liestøl K, Bjørnholt J, Thaulow E, Sandvik L, Erikssen J (1998) Changes in physical fitness and changes in mortality. Lancet 352(9130):759–762
8. Booth F, Roberts CK, Laye M (2011) Lack of exercise is a major cause of chronic diseases. Compr Physiol 2(2):1143–1211
9. Schuch FB, Vancampfort D, Firth J, Rosenbaum S, Ward PB, Silva ES, Hallgren M, Ponce De Leon A, Dunn AL, Deslandes AC, Fleck MP, Carvalho AF, Stubbs B (2018) Physical

activity and incident depression: a meta-analysis of prospective cohort studies. Am J Psychiatry 175(7):631–648
10. Nazarudin MN, Waridi N, Saad I (2017) Kesan latihan ketangkasan reaktif ke atas peningkatan komponen kecergasan berasaskan lakuan motor dalam kalangan pemain ragbi (Novis). Int J Educ Psychol Couns 2(4):56–70
11. Kraemer WJ, Gomez AL (2001) Effect of resistance training on women's strength, power and occupational performances. J Med Sci Sports Exerc 33(6):1011–1025
12. Mackenzie B (2000) Lateral change of direction test. Retrieved from http://www.brianmac.co.uk/latchang.htm
13. Krolo A, Gilic B, Foretic N, Pojskic H, Hammami R, Spasic M, Uljevic O, Versic S, Sekulic D (2020) Agility testing in youth football (soccer) players; evaluating reliability, validity, and correlates of newly developed testing protocols. Int J Environ Res Public Health 17(1):294
14. Epley B (2004) Anaerobic, speed and agility training. The path to athletic power, the model conditioning programme for championship performance. Champaign, Human Kinetics, pp 265–296
15. Liang ZQ, Li JS (2018) Progresses of the badminton equipment relate to exercise: some training aspects. J Sports Med Ther 3:001–009. https://doi.org/10.29328/journal.jsmt.1001020
16. Cohen L, Manion L, Morrison K (2007) Research methods in education, 6th edn. Routllege Falmer, London, New York
17. Gall MD, Borg WR, Gall JP (1996) Educational research: an introduction. Longman Publishing
18. Kirby RF (1971) A simple test of agility. Coach Athlete 25(6):30–31

Attitude and Concerns Toward Visual Quality Improvement in Sport Engagement Among Student Athletes in Taiwan: A Qualitative Exploration

Yi-Ting Huang and Jen-De Chen

Abstract Visual quality (VQ), a critical function in sports, has become a focal point for athletes pursuing better performance. This study explored how athletes perceived the importance of VQ and their concerns upon VQ improvement behaviors. Optometry was performed with all participants by the first author prior to data collection to examine objective vision condition and needs. Adopting the qualitative approach using semi-structured interviews, the authors conducted in-depth interviews with four college student athletes needing VQ improvement. All interviews were converted into verbatim transcripts; content analysis was conducted based on open coding. Topics were identified and classified into categories then developed into domains based on relationship among categories. Results indicated that as follows: (a) Participants all agreed that VQ would affect their performance and wore contact lenses when engaging in sports. (b) While wearing contacts, participants experienced decrease in VQ indicators including vision, clarity, comfort, and stableness, which in term affects sports performance. (C) Very few athletes would take actions to improve VQ because of money and time cost, whereas trading for limited effects. Furthermore, lacking of professional service while selecting contact lenses contributed to compromised corneal curvature or comfort. Conclusion: Participants were in the contemplative stages and not yet in the "action stage" according to the Transtheoretical Model of change due to insufficient clinical knowledge. Furthermore, low motivation in improving VQ could be explained by the Expectancy–Value theory since all participants were not in the elite level and might have lower motivation to improve their VQ to enhance their sports performance.

Keywords Expectation · Value theory · Open coding · Transtheoretical model · Optometry · Sport vision · Content lenses

Y.-T. Huang
Graduate Institute of Sports and Health, National Changhua University of Education, 50042 Changhua, Taiwan

J.-D. Chen (✉)
Department of Sport, National Changhua University of Education, 50042 Changuha, Taiwan
e-mail: andiechen@gm.ncue.edu.tw

G. Kuan et al. (eds.), *Advancing Sports and Exercise via Innovation*, Lecture Notes in Bioengineering, https://doi.org/10.1007/978-981-19-8159-3_7

1 Introduction

Vision is a human sense essential for daily living tasks, and is particularly crucial for athletes, who require speed and accuracy to achieve favorable performance [1]. During competition, athletes must concentrate on the environment, the situation, the target, and the movements of their opponents and act responsively. Scientific methods have been widely used to improve sports performance; in addition to physical fitness, techniques, and sport psychology, these methods include those designed to improve visual quality [2–4].

Previous research has indicated that visual quality plays a key role in ball sports and martial arts [3, 5]. However, numerous athletes have applied incorrect prescriptions of inappropriate aids due to lack of relevant knowledge, which has a negative effect on their competition and training. Medically, visual quality involves various indices such as clarity, eye comfort, and eye moisture [6]. Visual quality is an abstract concept and has been evaluated using instruments (e.g., autorefractors, aberrometers, and slit lamps) and patients' subjective descriptions, which can lead to considerable assessment biases. Numerous famous athletes (e.g., NBA players Stephen Curry, Rudy, Gay, and Chris Paul; Olympic shooting gold medalist Wang Yifu; and Olympic diving gold medalist Guo Jingjing) lacked satisfactory visual acuity (VA) and must receive refractive correction; numerous athletes wear contact lens instead of glasses because glasses can negatively affect their sports performance [7]. However, because of convenience and cost considerations, numerous athletes have opted to use contact lens without first obtaining a professional prescription or engaging professional services [4].

Most athletes improved their sports performance by enhancing physical fitness and skills, but they are unaware that sports performance is also directly or indirectly affected by their vision, hearing, reactions, and even uniforms [3, 4]. Beckerman and Hitzeman [5] reported that 44.6% of athletes aged 5–19 years who participated in the Amateur Athletic Union AAU Junior Olympic Games have not received refractive correction, and they stressed the essential role of periodic eye examination for athletes because 25% of them have not received a comprehensive eye examination. Chang et al. [8] reported that numerous teenage athletes have not received refractive correction and proposed hypotheses on why these athletes have not done so; however, the hypotheses have not yet been verified.

At the time of writing, no studies have been conducted to examine the role of visual quality for athletes in the context of Taiwanese sports, and research on their VA is limited. The present study aimed to investigate the attitudes toward visual quality among Taiwanese athletes, the effect of poor visual quality on their sports performance, and the reasons for their inaction with respect to improving their vision. The results are expected to increase the awareness of athletes regarding visual quality in sports training and serve as a reference for sports teams that intend to improve the visual quality of their athletes. Hence, the purposes of the present study are to:

1. Understand the knowledge of athletes regarding their visual quality;

2. Explore the sports experiences of athletes with poor visual quality and the effect of such experiences on their sports performance;
3. Clarify how athletes respond to poor visual quality; and
4. Understand the experiences of athletes with respect to improving their visual quality in sports and why they have not been actively improving it.

2 Research Methods

2.1 Participants

Purposive sampling was performed to recruit four paraprofessional student athletes (with the aliases Olaf, Big, Elephant, and Pear) through social networks. All the participants had experienced or were experiencing impaired competitive and training performance because of poor visual quality at the time of the present study.

Three of the participants were volleyball players, and one was a soccer player. All of them had regularly engaged in specialty sports training and participated in official competitions for 4–15 years. They had engaged in 3–4 h of training for 3–5 days per week. These participants underwent optometric examinations performed by the first author, who was a licensed national professional optometrist. The examination results revealed that three of the participants exhibited a VA of < 1.0 with sports visual correction prescriptions. Among the three participants who used contact lenses, two had contact lenses with sliding levels that were outside the desired range, which could lead to blinking-induced vision blurriness or foreign object sensations. Table 1 lists the demographic information of the participants.

2.2 Data Collection

In the present qualitative study, in-depth, semi-structured interviews were conducted for data collection. The preliminary outline of the interviews was formulated in accordance with the research objectives of the present study and literature findings; after peer feedback was obtained and a pilot study was conducted, the interview guide was then revised and formalized. The participants could freely express their experiences during their interviews. The interviews were conducted in person with a focus on the participants' experiences and perspectives. The participants were encouraged to share their autonomous and unique personal experiences to ensure the reliability of the present research. The interview process was fully observed and recorded for data analysis.

The researchers were familiar with the essential elements of qualitative interviews. Researcher 1 is a professional optometrist who is knowledgeable about contact lens practices and skilled in assessing visual quality. However, this researcher refrained from bringing up personal experiences during the interviews to maintain truthfulness

Table 1 Demographic information and visual quality of participants

Alias	Olaf	Big	Elephant	Pear
Sex	Male	Male	Male	Male
Age	19	20	27	22
Specialty sports	Volleyball	Volleyball	Soccer	Volleyball
Dedicated time	4	5	14	5
Refractive status	OD:PL − 1.00 × 180 OS: − 0.75 DS	OD: − 4.50–0.75 × 015 OS: − 4.50–1.00 × 165	OD: − 3.25–1.50 × 010 OS: − 2.50–1.25 × 175	OD: − 5.75–0.25 × 060 OS: − 6.00–0.50 × 150
Static VA in sports	0.9 + 1	0.9 + 2	0.8 + 2	1.0 + 1
Vision problems	Visual acuity (VA) < 6/6; subpar tear quality	VA < 6/6; severely insufficient contact lens sliding	VA < 6/6	Excessive contact lens sliding
Sports visual correction prescription	Naked vision	OD − 4.25 DS OS − 4.25 DS	OD − 3.00 DS OS − 2.50 DS	OD − 6.00 DS OS − 6.00 DS

and avoid leading the participants' responses; thus, data comprehensiveness was ensured. Researcher 2 has experience as an instructor for qualitative research courses at a graduate institute and was tasked with supervising and controlling the data collection and analysis processes to maintain the quality of the present study and provide data analysis checks and recommendations. The interviews were conducted at locations with minimal disruptions, namely private spaces outside school for Olaf and Big and university research rooms for Elephant and Pear. Each interview lasted for an average of 1 h (Table 2).

Table 2 Time and locations of interviews

Participant	Big	Olaf	Elephant	Pear
Location	Private space	Private space	Gymnasium research room in a university in Changhua	Gymnasium research room in a university in Changhua
Date	2021/12/01	2021/12/01	2022/01/07	2022/01/07
Time	19:00–20:00	20:00–21:00	16:00–16:30	17:30–18:20

2.3 Research Procedure

The participants were informed of the objectives and procedures of the present study and asked to read the outline of the interviews before providing informed consent. Before the start of their interviews, the participants were briefly asked about their demographic information and received an eye examination at a university optometry laboratory to collect data for objective VA analyses. The interviews were conducted in accordance with the interview guide and recorded in full. The recordings were transcribed, and the participants verified that the transcripts accurately conveyed what they had said during the interviews. The data were then analyzed and generalized.

2.4 Data Analysis

The data were analyzed through data reduction, data display, and conclusion drawing and verification. They were systematically organized through open coding. Units were distinguished by extracting the meanings conveyed by the participants, which were then simplified as topics for preliminary coding. The relationships between the topics were identified and induced as themes for secondary coding. The themes were then organized and induced as the final results.

3 Results

The interview data consisted of five components, namely (1) the timing and frequency of refractive examinations, (2) the methods the participants had adopted for refractive correction in their daily lives and in sports activities, (3) the effect of poor visual quality on the participants' athletic performance, (4) refractive correction measures and their contact lens selection process during sports activities, and (5) the participants' knowledge and attitudes with respect to sports visual quality improvement.

3.1 Timing and Frequency of Refractive Examinations

None of the participants received regular refractive examinations on a semiannual basis. Three of them had visited optometric clinics for eyeglass or contact lens prescriptions within the 2 years prior to the start of the present study. Elephant had not undergone any refractive examination within the 7–8 years prior to the start of the present study but had been receiving the same visual correction prescription within the 8 years prior to the present study.

I had it for about 2 years ago, I looked at the arrows pointing up, down, left, and right and the colors red and green… I've done it for 2 years. Big (A004)

I studied hard in the my third year of high school and developed myopia, but the condition wasn't too bad. I have bought a pair of contact lenses at an optical shop but stopped wearing them because I wasn't used to it. It's been 2 years since that time. Olaf: (B004)

I've never done that! I've had (this pair of eyeglasses) since my first or second year of university (laughs). It's been 7 or 8 years since then. Elephant: (C004)

Eye examinations? The last time I had one was a half year ago when I was buying a pair of contact lens. I took the eye examination because it was convenient to do so then. (Did your eye prescription change?) Barely. Pear: (D004)

3.2 Refractive Correction Methods Adopted for Daily Life and Sports Activities

Most of the participants wore eyeglasses in their daily lives but wore contact lenses when they were training or competing in sports. Eyeglasses were more convenient and comfortable to wear in their daily lives, whereas contact lens were safer for sports activities, particularly high-intensity training and competitions.

I wear contact lenses most of the time for high-intensity sports activities. I used to wear eyeglasses, but one time my eyeglasses broke while I was playing (sports). Therefore, I have been wearing contact lenses since my second year of high school. Big: (A005)

(Do you wear eyeglasses or contact lenses in your daily life?) No. (Have you checked your eye prescriptions?) Yes, it was 0.8 in one eye and 1.0 in the other, which was confirmed through a physical examination. I used to wear contact lenses, but I stopped wearing them because they cause my eyes to dry out. Olaf: (B005)

I wear contact lenses. Monthly ones. (Do you also wear them in your daily life?) I wear glasses most of the time. Elephant: (C005)

I wear contact lenses… Both daily and monthly ones. I wear daily one now. I wear eyeglasses in class because I need to look at the computer, and I feel that eyeglasses provide better protection. (Were you wearing contact lenses when you started playing?) No, because my sports activities were not that intense yet. I started wearing contact lenses because my sports activities become more intense when I joined the varsity. Pear: (D005)

3.3 Effect of Poor Visual Quality on Sports Performance

Because all the participants except Olaf wore contact lenses for refractive correction, the comfort of contact lenses was crucial. During their interviews, the three participants indicated that the dryness and foreign body sensation caused by their contact lenses hindered their performance and reactions during sports activities.

I used to wear cheap contact lenses, but most of them were uncomfortable to wear. I don't know if it's because of their poor quality. Several of those lenses hurt my eyes and made me want to blink when I couldn't see clearly. (How did you deal with it?) I changed my lenses during a match but endured it when I was training. Big: (A006)

Nothing except excessive light (laughs). Olaf: (B006)

I sometimes experience discomfort, dryness, and pain when I wear contact lenses. (How do you deal with these problems?) I have tried to rinse my lenses with water, wipe my eyes, or blink, but I still couldn't see the ball. I asked someone who wore eyeglasses, and he said that my eyes were dry because I didn't sleep enough. Elephant: (C006)

When the ball is fast, my eyes get dry, and my contact lenses slide, cause blurriness, or felt like foreign objects in my eyes, and I cannot see when I blink all the time. When the ball is approaching, I need to blink repeatedly for a period of time. This happens when I wear contact lenses for approximately 8–9 h. I wear my contact lenses before I start my evening training. I used to wear them for approximately 8–9 h before training or a match but now I wear them just before training or a match. Pear: (D006)

3.4 *Refractive Correction Measures and Their Selection Process During Sports Activities*

Most of the participants (excluding Olaf) who wore contact lenses for sports activities and stopped wearing them subsequently did so because of the discomfort caused by their lenses. All the participants bought contact lenses on the basis of recommendations provided by optical shop employees. Olaf selected his contact lenses on the basis of his keratometry results, whereas the other three participants bought contact lenses after trying one or two pairs even though their lenses were not suitable for them because of moisture-, corneal curvature-, and base curve-related factors, which in turn were related to the low price of those lenses and their convenience in purchase.

I wear Bausch+Lomb nowadays. The clerk gave it to me when I started looking for contact lenses, but it was expensive, so I opted for cheaper ones. I only returned to wearing Bausch+Lomb because I have often ended up wearing uncomfortable lenses. Big: (A007)

I was thinking about whether I should wear eyeglasses or contact lenses. I chose contact lenses for sports, and I asked about contact lenses at an optical shop. I thought that I needed a pair immediately, so I bought and tried a few pairs. (Did you buy them for price or comfort?) Comfort. I bought more than one brand and excluded those that did not fit my eyes, but none of them were comfortable to wear; specifically, they caused dryness in my eyes after extended use. Therefore, I stopped wearing contact lenses. Olaf: (B007)

I looked for the ones that were cheap and comfortable to wear because the options of monthly lens options was limited. (So did you try them out?) No. I just told the shop that I wanted a monthly pair, and they introduced me to a specific brand, which I bought. Elephant: (C007)

I have tried two types of contact lenses. (Daily ones?) Yes. I chose the one more convenient for purchase. Pear: (D007)

3.5 *Participants' Knowledge and Attitudes with Respect to Sports Visual Quality Improvement*

The participants evaluated their VA and the clarity, comfort, and stability provided by their contact lenses; they also rated the importance of visual quality to them and their knowledge of the relationship between visual quality and sports performance. This section includes participant's perception and self-evaluation on visual quality, awareness of needs for visual quality improvement, and their attitudes toward improving visual quality as following:

Perception and Self-Evaluations on Visual Quality With the exception of Elephant, all the participants regarded their own visual quality as being above average. Nevertheless, they reported problems regarding the visual clarity and comfort of their contact lenses.

> (My vision is) not bad. I think my vision in the court is good. Everything is clear, and the ball isn't too blurry. The only problem is that my contact lenses feel uncomfortable when I put them on. Big: (A008)

> I think it's fine, but things become blurry when I cover my left eye and I can't read the text on a can if I do that. Olaf: (B008)

> My vision becomes blurry when I wear contact lenses, which are astigmatic and causes distant objects to appear blurrier than when I'm wearing eyeglasses. Recently, my eyesight has declined. I have been staring at computer screens and electronic products for extended periods of time, and my eyes feel uncomfortable and sometimes painful. Elephant: (C008)

> It's fine, except for when my contact lenses slide out of place, which causes things to become blurry and forces me to blink to keep them in place. (Does that feel uncomfortable?) Not very. Slightly, but not very. Pear: (D008)

Awareness of Needs for Visual Quality Improvement All the participants indicated their awareness that improving their visual quality leads to improved sports performance, and they expressed that VA can be improved through sufficient sleep, a reduction in contact lens wearing time, or the use of contact lenses that have high DK/t and are suitable for their eye shapes.

> It would be good if my contact lenses are comfortable. Otherwise, I would be afraid of putting on my contact lenses just before I start my sports activities. Comfortable lenses eliminate the need to blink my eyes. Big: (A009)

> Probably. It'd be okay if the vision can become clear, and the eyes better not be dry. I refused to wear contact lenses because they were so dry. Olaf: (B009)

> (Do you think that visual quality will affect your performance in sport?) Yes, absolutely! After a good sleep, my eyes become comfortable. Otherwise, it hurts sometimes when I wear contact lenses. Elephant: (C009)

> Yes, it makes a difference when I only put on my contact lenses just before my sports activities. It becomes less uncomfortable when I don't wear my contact lenses for excessively long periods of time. Pear: (D009)

Attitudes Toward Improving Visual Quality All the participants indicated that they regarded visual quality as a crucial factor and that improving visual quality improves their sports performance. However, they did not take active measures to improve their visual quality unless they experienced eye discomfort or eye-related-inconveniences. In addition to being habitual, they felt that the measures for improving their visual quality were limited in terms of their effectiveness.

I pay attention to my eyes only when they are affected or feel uncomfortable… Like I said, I only seek a diagnosis when it hurts to wear my contact lenses. Big: (A010)

I think visual quality is crucial, but I don't know how to improve it. The contact lenses that I have worn were uncomfortable. (Have you worn the silicon hydrogel ones?) I don't think so. (These lenses are more comfortable but more expensive. Are you willing to try them?) I'll think about it. I only play volleyball for a few hours and don't want to spend too much money. Olaf: (B010)

Yes, I pay attention to visual quality. The day before each match, I sleep early for comfortable eyes. (Then will you try toric contact lenses? They provide enhanced clarity.) Toric contact lenses? I don't know. I think I would. Elephant: (C010)

I want to improve my visual quality, but I haven't done it. I'm lazy. I thought my situation would improve, but it did not. (Not to a considerable degree?) No. Pear: (D010)

4 Discussion

The present study examined the attitudes of athletes toward visual quality in sports activities, the effect of visual quality on sports performance, and the athletes' attitudes to solve visual quality problems that negatively affect their sports performance. The results may provide a reference for teacher, coaches, and managers who are helping athletes to improve their sport performance. The findings of the present study are as follows:

4.1 *Visual Quality Affects Sports Performance*

The higher the level of competition that an athlete is involved in, the more crucial is the role of visual quality to the athlete's sports performance. Most of the participants agreed that poor visual quality negatively affects their sports performance. This finding is consistent with the position argued by Wilson [9].

4.2 *Wearing Contact Lenses Causes Problems Related to Visual Quality*

Visual quality comprises eye clarity, comfort, and moisture. Athletes who play specific sports do not wear eyeglasses because high speed of balls during high-intensity training or competitive matches increases the risk of sustaining injuries caused by eyeglasses. The present study focused on the visual quality of athletes who wear contact lenses. The interviews conducted in the present study revealed that although most athletes chose to wear contact lenses during sports activities, their lenses were a poor fit for them. Poorly fit contact lenses cause dry eyes and unsatisfactory sliding, leading to temporary blurred vision or foreign object sensation, which in turn causes athletes to blink frequently during sports activities. Furthermore, the participants' contact lenses differed slightly from their eyeglasses in terms of prescription parameters. Most of the participants' lenses were spherical and reduced VA and visual clarity, which directly increased their reaction time and indirectly reduced their sports performance. This finding is consistent with that reported by Appelbaum and Erickson [10].

4.3 Participants Did Not Take Active Measures to Improve Their Visual Quality Despite Their Awareness of Its Importance

Although the participants acknowledged the importance of visual quality to a moderate or high degree, few of them had taken active measures to improve it. Although the participants recognized that visual quality affects sports performance, few of them had actively sought professional help to improve their visual quality. This was because they did not expect to achieve a substantial improvement in their sports performance through an improvement in visual quality, which they regarded as a time- and cost-intensive endeavor. Furthermore, the participants exhibited biases in their perceptions of visual quality because of a lack of professional optometric knowledge. The dryness caused by contact lens can be solved by using silicon hydrogel lenses, which were revealed by tear test results to exhibit high oxygen permeability. The problems associated with contact lens sliding can be mitigated by performing a fitting on the basis of comprehensive corneal assessment results. The inadequacy of eye prescriptions can be detected through regular clinical checkups. Astigmatism can be mitigated by wearing toric contact lenses. However, most contact lenses that provide high visual quality are expensive, and finding the appropriate pair of contact lenses is a time-consuming process. The participants in the present study were not top Taiwanese athletes and exhibited a low level of motivation to substantially improve their sports performance; therefore, improving visual quality was not their first priority.

The participants regarded visual quality as having a moderate or high level of importance, but they rarely took active measures to improve it. Under the Transtheoretical Model developed by Prochaska and DiClemente [11], behavioral change occurs in the following stages: (1) precontemplation (i.e., no recognition of a problem or lack of intention to change), (2) contemplation (i.e., not knowing how to solve a problem despite having an awareness of its presence), (3) preparation (i.e., commitment to identifying methods, sources and support, and motivations for solving a problem), (4) action (i.e., start of behavioral change), and (5) maintenance (maintaining a change to achieved expected results). Most of the participants in the present study were in the stage of contemplation or preparation; they recognized that their visual quality needed improvement but did not know what they could do to improve it because of a lack of professional clinical knowledge or the high cost of making such a change (i.e., time and money).

The participants' inaction with respect to improving their visual quality can be explained by the Expectancy–Value theory of motivation [12]. Motivations for an action are based on an individual's expectations and evaluation of a task, which influence their behavioral choices, persistence, and performance. The willingness of individuals to commit to a task is determined by their expectations for success and the value of the task. When an individual expects to succeed in a task and perceive that the task is valuable, they are actively and persistently committed to completing it. The participants in the present study were not at the elite level and might hold low expectations for their sports outcomes. Therefore, they had low expectations for their visual quality in sports activities and did not perceive visual quality improvement as a valuable objective. Most of the participants chose to adapt to their current situations instead of actively improving their visual quality.

5 Conclusion

The results of the present study revealed that visual quality considerably affects the sports performance of athletes and the participants all acknowledged the essential role of visual quality in their sports performance, but most of them chose to ignore their visual quality problems and adapt to their current situations.

Improving visual quality is a time- and cost-intensive endeavor. Numerous athletes believe that improvements in visual quality only have a limited effect on improving their sports performance. Athletes who spent additional time and money on improving their visual quality have only achieved limited improvements in their sports performance, which is not attractive enough for athletes to improve their visual quality. Furthermore, most athletes do not receive comprehensive professional optometric examinations when they purchase their contact lenses; most of them chose to use lenses that were cheap or recommended by optical shop employees instead of those that fit the keratometric properties of their eyes and provide a high level of comfort. Today, the optometric industry is becoming an increasingly professional and delicate industry, and detailed and comprehensive assessments are mandatory for the fitting of contact lenses.

The results of the present study must be applied in consideration of the participants' characteristics. Notably, they were not elite athletes and specialized only in ball sports. Therefore, the results of the present study cannot be generalized to athletes who are involved in other types of sports and exhibit different skill levels relative to those of the participants. Moreover, because the sample was small, the results of the present study cannot be extrapolated to all athletes in Taiwan. Nevertheless, these results are relevant to sports practices. Sports trainers must regularly examine their athletes' VA, routinely track the suitability of their athletes' contact lenses, and provide their athletes with the correct VA education to improve their awareness of and attention to their visual quality. Committed actions must be taken to optimize the VA of athletes and improve their sports performance.

References

1. Bulson RC, Ciuffreda KJ, Hayes J, Ludlam DP (2015) Effect of retinal defocus on basketball free throw shooting performance. Clin Exp Optom 98:330–334
2. Chen QS (2014) In search of excellence in sports performance. J Sports Perform 1:1–5 (Chinese version)
3. Liu Y-H, See L-C, Huang C-C, Ju Y-Y, Chang S-T (2012) A comparison study of sports vision between elite Taekwondo contestants and university student athletes. Sports Coaching Sci 39–51. https://doi.org/10.6194/SCS.2012.25.04
4. Yi M (2006) The management of the contact lens market is also "invisible" Guangxi quality supervision herald 32
5. Beckerman S, Hitzeman SA (2003) Sports vision testing of selected athletic participants in the 1997 and 1998 AAU junior Olympic games. Optom 74:502–516
6. Visser E-S, Visser R, van Lier HJJ, Otten HM (2007) Modern scleral lenses part II: patient satisfaction. Eye Contact Lens 33:21–25
7. Zuhairi NA, Kuan YM, Manan FA, Knight VF, Omar R (2018) The importance of visual awareness among junior athletes in Klang Valley, Malaysia. Malays J Public Heal Med 1:124–129
8. Chang S-T, Liu Y-H, Lee J-S, See L-C (2015) Comparing sports vision among three groups of soft tennis adolescent athletes: normal vision, refractive errors with and without correction. Indian J Ophthalmol 63:716
9. Wilson TA, Falkel J (2004) Sportsvision: training for better performance. Human kinetics 1
10. Appelbaum LG, Erickson G (2018) Sports vision training: a review of the state-of-the-art in digital training techniques. Int Rev Sport Exerc Psychol 11:160–189
11. Prochaska JO, DiClemente CC (1982) Transtheoretical therapy: toward a more integrative model of change. Psychother Theory Res Pract 19:276
12. Wigfield A, Eccles JS (2000) Expectancy–value theory of achievement motivation. Contemp Educ Psychol 25:68–81

Mind-Mapping: Exploring Intraindividual Experiences Across Injury Rehabilitation

Jolly Roy, Jeykar T. Lazarus, and Tejasvi Karanwal

Abstract Sport psychology practitioners frequently look for new resources to provide quality services to athletes. The authors have highlighted the use of the mind-mapping technique in an athlete who underwent anterior cruciate ligament (ACL) reconstruction surgery. Mind-mapping was employed to elicit information on athletes' experiences across injury rehabilitation. The study provided the experiences of three different phases of injury rehabilitation, viz., onset, during rehabilitation, and return to sport. The content overlap values indicated that the experiences across the three phases were unique. A graphical representation of mind-mapping for all three phases is provided for readers' understanding.

Keywords Mind-mapping · Sports injury · Psychological experiences

1 Introduction

The mind-mapping approach (MMA) [1] is a technique that helps to tap the information creatively and organize it into meaningful units. In an injury context, MMA can be considered a supportive means for eliciting information and providing a road map to navigate across the rehabilitation phase. There are two approaches to mind-mapping: (1) the traditional approach, using paper and pencil, and (2) the modern approach using the mind-mapping software. Mind-mapping enables practitioners to delve deep into themselves and take responsibility for their own action tendencies. It creates more structured experiences and provides purposeful insights for committing to the rehabilitation process.

Very often a sport psychologist has to face challenging questions from injured athletes. This happens when the athletic identity is compromised and athletes face uncertainty about their recovery. Injured athletes also experience social isolation, loneliness, fear, anxiety, and a lack of inner drive, as they strive to seek the meaning

J. Roy (✉) · J. T. Lazarus · T. Karanwal
Faculty of Sports and Exercise Sciences, Sri Ramachandra Institute of Higher Education and Research, Chennai, India
e-mail: jollyroy@sriramachandra.edu.in

G. Kuan et al. (eds.), *Advancing Sports and Exercise via Innovation*, Lecture Notes in Bioengineering, https://doi.org/10.1007/978-981-19-8159-3_8

of their life without active engagement in sports. Safety in sports is a prerequisite for continued participation in sports [2]. We believed that understanding the pathway of how one navigated from the onset of the injury through the rehabilitation phase to return to sports will help in the future prevention and management of sports injuries. While providing professional support, we have observed that a sudden pause after injury forces an athlete to reflect and adapt to an unprecedented period in their active career. Most athletes initially deny their situation and move on to tolerate the injury condition before fully accepting their situation. From our experiences in the applied context, we have observed that those athletes who struggle to adapt to an 'injured situation' experience heightened anxiety and frustration leading to poor mental health. These challenges pose a deeply uncomfortable situation for the rehabilitation therapist, the sport psychologist, and the athlete. To overcome the monotony and barriers related to self-doubt during injury rehabilitation, strong professional bonding and innovative measure were required to tread the rehabilitation path. In trying to do this, an innovative method was warranted to create a comfortable environment for the athlete, while we navigate him through the rehabilitation phase. It was also important for us to ensure that the athlete does not develop any emotional trauma while treading through the memories of injury.

Among injuries, rupture of the ACL impacts an athlete's football career [3]. Despite the client having a general awareness of the different experiences during injury rehabilitation, there is little understanding of a structured process and support system the client experienced across the injury rehabilitation period. This study, therefore, sought to elicit information in a more structured way. We used the mind-mapping technique to draw information regarding injury rehabilitation experiences through the natural association of events. We then provided a graphical representation linking thoughts and events for easy understanding. Mind-mapping also helped to demonstrate the pathway toward recovery which helps a client for building confidence in a similar future context.

The sport psychologist initially used open-ended questions to elicit descriptive information about the injury onset and rehabilitation period. Since the information was a mix of experiences, mind-mapping was done to organize the information. The major focus of the interview question guiding the mind-mapping was: (a) Can you introspect a bit on the onset of the injury situation and describe your experiences? (b) Can you describe your personal experiences and the support you received during rehabilitation? (c) Could you elaborate on the influences of your perceived states toward the return to sport?

2 Methods

2.1 Participants

The football player involved in this study is referred to as TK (24 years old). Verbal consent was taken from the client to publish the work for educational purposes. TK was referred by a sports medicine doctor to instill self-confidence in the athlete toward a return to sport. TK is very passionate about football and has represented the country in the U-17 tournament. Typically, the athlete trained in the morning and evening.

Mind-mapping is a technique to capture thoughts and visually arrange them in a radiant format using the software. The unique elements used to create a mind-map include central title, branch, levels, and subtopics. Mind-mapping boosts efficiency that makes it easy to understand the structure and organize information.

2.2 Materials

'MindMaple Lite Version 1.3' is free software for creating mind-maps and organizing tasks. The software comes with six predefined templates but also allows the practitioner to add their own mind-map items such as text, icons, graphics, or objects which can be customized. These are easy to customize so that the practitioner can create the map most appropriate to the required needs. 'MindMaple' software is compatible with windows and Mac programs. The practitioner can present the ideas or content generated in different forms such as text, graphs, images, and charts. A computer with a basic i3 processor was used for this assessment.

2.3 Methods

The subject TK (24 years old) footballer was familiarized with the mind-mapping procedure and was asked to work on the central topic of three phases of his experiences during injury rehabilitation (onset, during rehabilitation, and return to sport). The central topic provided the thought process to spread out freely to stimulate memory and associations. The branches depicted by curved lines connect closely related topics which represent that thinking is natural. The levels represent the hierarchy of branches formed around the central topic. The subject was assisted to navigate through different experiences and create different mind-maps for three different phases.

3 Results

In the onset phase, six hierarchies of branches were identified with sub-topic: pre-injury (4), play (3), injury (3), physiological (5), psychological (6), and pre-surgery (3) (Fig. 1).

During the rehabilitation phase, four hierarchies of branches were identified with sub-topic—coping strategies (6), effort (3), support (2), and post-surgery (5) (Fig. 2).

In the return to sport phase, three hierarchies of branches were identified with subtopics—start to play (3), re-injury (4), and adaptation (3) (Fig. 3).

A few highlights of the feedback from the athlete are provided herewith.

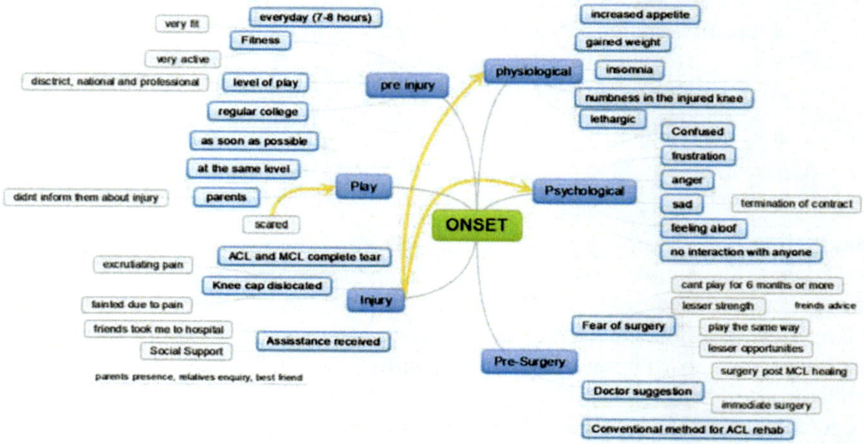

Fig. 1 Mind-map during the onset of injury

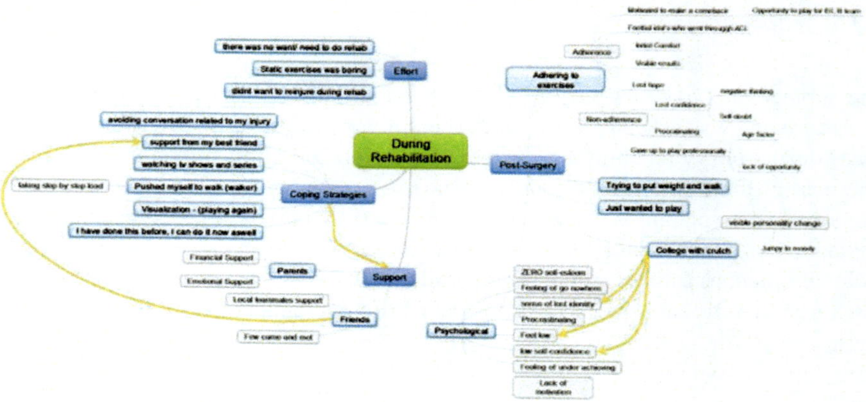

Fig. 2 Mind-map during the rehabilitation phase of injury

Fig. 3 Mind-map during return to sport

Fig. 4 Content overlap of experiences between different injury rehabilitation phases

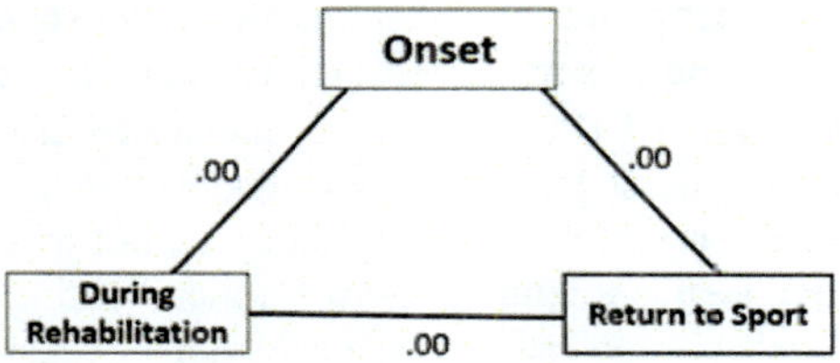

> I could elicit information naturally and my thoughts were flowing freely. The thought process was innovative and enjoyable.
>
> I enjoyed the process and graphical representation. I could navigate through the pathway of recovery with ease and build confidence.
>
> I could structure my thoughts and organize them systematically. Mind-mapping ushered me in solving the problem creatively during injury rehabilitation.

4 Discussion

This paper looks at how to use mind-mapping as a resource to elicit information from the client. It takes information in an innovative way and displays this information in a meaningful way. Mind-mapping is not often used or considered frequently in sport psychology context. Very often the clients look for 'something new' in all service delivery. Therefore, we decided to adopt this technique in a practical and usable way. This makes the process of taking the client through the pathway of recovery in a more convincing and enjoyable way. Three different mind-mappings (e.g., onset, during rehabilitation, and return to sport) were done with this client with three different central ideas placed in the center of the figure. Extending from each central point are the different branches which lead to different information. Taking information from several sources helped the sport psychologist to consolidate the different experiences the client went through in each stage of the rehabilitation.

Mind-mapping also engaged the client in creative thinking, which is evident from the feedback. Mind-mapping allowed the sport psychologist to understand how the experiences across injury rehabilitation interrelate.

Injury rehabilitation is a dynamic process where the experiences change constantly. These mind-maps can also be used for the future as a reminder of how the client navigated through the injury and recovery stage and helps the client to share these experiences with other athletes who get injured and impacts their identity. In this study, the athlete's mind-maps and the qualitative feedback indicated that he could clearly identify the issues and experiences he went through during the injury rehabilitation phases. The athlete's qualitative feedback expressing that he enjoyed the process shows how mind-mapping can be used to get the athlete involved in the process of eliciting information and retaining the interest in a more positive way to compensate for any painful experiences he had during injury rehabilitation pathway. From each hierarchical point, the athlete could explore the recovery experiences in-depth and this would provide the athlete and the sport psychologist with the length and breadth of understanding the athlete had about the pathway of recovery.

A crucial part of injury rehabilitation after ACL surgery is the lack of self-confidence in the athletes. Thus, dealing with the self-doubt of an athlete is vital. One such step is to emphasize the understanding of his natural coping strategies and the different support he experienced during rehabilitation. This mind-mapping technique offered this athlete a clear pathway of how his different experiences strengthened his inner core to deal with self-doubt and gain confidence.

The most appropriate means of interpretation of the participant's view of recovery was possible using mind-mapping. The content overlap values (Krahé 1986) were calculated to examine the content similarity between three different situational experiences. The content overlaps value of zero (.00) between 'onset and during rehabilitation', between 'onset and return to sport', and between 'during rehabilitation and return to sport', indicated that the athlete's experiences across the rehabilitation are unique. The individual overlap scores varied from 0 (all items were different) to 1.0 (all items were similar) [4].

The athlete's feedback revealed that the mind-mapping helped to elicit the information through natural thought. The athlete reported that the representation of thoughts was very simple and easy to organize. The mind-mapping demonstrated the pathway for confidence-building by navigating the athlete across his rehabilitation period. Mind-mapping can be utilized for creative problem-solving during injury rehabilitation.

5 Conclusion

This study provides an opportunity for sport psychologists to use mind-mapping in their practice. The study illuminated the different experiences that an athlete is exposed to across injury rehabilitation phases. It offers insights into the dynamic nature of individual experiences. Similar studies can be done to understand how wider contextual factors influence an individual not just in a rehabilitation context but in other areas also. This study highlights the importance of attending to contextual factors and structuring the experiences of the athlete. We believe that this study offers new insights into unexplored areas of sport psychology practice.

References

1. Leaf CM, Uys IC, Louw B (1993) Mind-mapping approach (MMA): a culture and language "free" technique. S Afr J Commun Disord 40(1):35–42
2. Verhagen EALM, van Stralen MM, van Mechelen W (2010) Behaviour, the key factor for sports injury prevention. Sports Med 40(11):899–906. https://doi.org/10.2165/11536890-000000000-00000
3. Volpi P, Bisciotti GN, Chamari K, Cena E, Carimati G, Bragazzi NL (2016) Risk factors of anterior cruciate ligament injury in football players: a systematic review of the literature. Muscles Ligaments Tendons J 6(4):480–485. https://doi.org/10.11138/mltj/2016.6.4.480
4. Krahé B (1986) Similar perceptions, similar reactions: an idiographic approach to cross-situational coherence. J Res Pers 20(3):349–361

An Experimental Study on Effect of Flipped Teaching on Learning Effectiveness and Learning Attitude of Secondary School Students in Macao

Chan Kit Ao and Qun Zhai

Abstract With the great impact of modern Internet information technology on traditional teaching, flipped teaching has attracted more and more attention. This study explored the effect of flipped teaching on the learning effectiveness and learning attitude of secondary school students in Macao. A teaching experimental research design was employed; the sample included 112 male secondary school students; they were randomly divided into an experimental group (n = 55) and a control group (n = 57). The experimental group received the flipped teaching while the control group received the traditional teaching, respectively. Both groups' training time was 18 weeks. The "*Physical Education Learning Effectiveness Scale" and the "The Physical Education Attitude Scale for Middle School Students"* were used to examine the effects of both teaching methods on secondary school physical education. The results showed that at the junior stage, there was no significant difference in the effects of two different teaching modes on the learning effectiveness and learning attitude of physical education. However, in the senior high school stage, though there was no significant difference in two different teaching methods on the learning effectiveness of physical education, the effect of flipped teaching on the learning attitude of physical education was significantly higher than that of traditional teaching mode.

Keywords Flipped teaching · PE · Learning effectiveness · Learning attitude

1 Introduction

As the society enters the information age, the use of information technology tools to assist teaching in the teaching activities of various subjects has become very popular, and the use of online teaching platforms is also relatively mature. Some studies have

C. K. Ao · Q. Zhai (✉)
Faculty of Health Sciences and Sports, Macao Polytechnic University, Macao SAR, China
e-mail: zqun@ipm.edu.mo

C. K. Ao
Yuet Wah College, Macao SAR, China

G. Kuan et al. (eds.), *Advancing Sports and Exercise via Innovation*, Lecture Notes in Bioengineering, https://doi.org/10.1007/978-981-19-8159-3_10

pointed out that the use of information technology tools in physical education can establish correct movement demonstrations and enable students to obtain significant results in cognition [1]. When students are assisted by information technology tools, they can indeed learn the technical actions that they do not know well through the teaching videos repeatedly, so that the learning effect can be better improved [2].

In recent years, flipped classroom has been popularized and applied in the teaching of various subjects, forming a trend of teaching reform, and many people in the education circle have responded, hoping to bring changes to modern education through this new teaching model [3]. It is proposed that flipped teaching can help students conduct in-depth learning, and can help improve students' ability to analyze, evaluate and create, and is more effective than traditional teaching. In the learning under the flipped teaching mode, students must preview the learning materials set by the teacher before class to acquire knowledge. In such a classroom, problems can be solved through discussions and exchanges between classmates and teachers, which can enhance students' interest in learning and make knowledge absorption and internalization more effective. Fengyan [4] pointed out in the research that the flipped teaching mode provides a suitable platform for the combination of information technology and education and teaching, so that the traditional classroom teaching form can be informatized and digitized. Through this platform, teachers can design targeted teaching to improve teaching effect. It was not until 2014 that the first research on the application of the flipped teaching model to physical education appeared. The research shows that the students are novel and attractive to the learning process under the flipped teaching; the learning atmosphere has become stronger, and the students are in a good mood. There are also more opportunities for communication and practice than before, which has a significant effect on improving teaching performance. A large number of research results show that the advantages of flipped teaching in physical education are mainly manifested in the following aspects: First, the teaching materials are set up on the online teaching platform, and students can watch teaching videos such as demonstration actions and basic theories before class. The lecture time is saved in the classroom, and the teacher can have more time to practice and correct the action. Secondly, when students encounter problems in class and cannot solve them immediately, they can watch the teaching videos repeatedly to understand the deficiencies and deal with them in a timely manner. Third, exchanges and discussions between classmates and teachers have been increased, making the relationship between teachers and students more harmonious, and the classroom atmosphere more active [5–7].

In addition to the significant effect on learning effectiveness, there are many places worth exploring in promoting learning attitudes. Caligaris et al. [8] research pointed out that students in the flipped teaching mode have a higher acceptance level and show a good learning attitude, which is different from the previous learning methods. Yang [9] pointed out that flipped teaching uses the network to build a teaching platform for effective communication between teachers and students, and allows students to learn before class through online teaching information. This new attempt has aroused students' freshness and stimulated their interest in physical education interest in learning. Pan [10] research shows that when the flipped teaching mode is integrated into the physical education classroom, the learning attitude is significantly different from the traditional physical education teaching mode. This feature can be used to improve the learning atmosphere of the classroom.

However, the application of information technology tools in the physical education classes of primary and secondary schools in Macao has not received much attention. Through various data searches, we found that there is no relevant research on the application of the flipped teaching model to middle school physical education in Macao. Therefore, this research is a teaching experiment attempt by applying the flipped teaching mode to the physical education teaching of middle school students in Macao. By studying the effect of flipped classroom on the physical education learning effect and learning attitude of Macao middle school students, it can provide reference for the implementation and promotion of flipped classroom in Macao middle school students' physical education teaching.

2 Methods

2.1 Participants

The participants of the study were four classes of form1 and form4 students, respectively, from Yuet Wah College. Fifty five students from form1 (year-1 middle school), 57 students from form4 (year-1 high school), and 112 students in total were dedicated in the study. All of them were male. It is divided into experimental group (using flipped classroom) and control group (using traditional teaching method). The experimental group (EG) consisted of 55 students. The control group (CG) consisted of 57 students. The students ranged in age from 12 to 16. They were all male.

2.2 Teaching Experiment Content Design

This research compares the results of PE learning effect and attitude of form1 and form4 students after flipped classroom. The CG adopts the traditional teaching method; the EG adopts flipped classroom. The EG and the CG had the same teaching objectives and tasks. The teaching of the EG is divided into three parts: pre-class, in-class, and after-class.

The basic teaching content before class includes setting up online teaching platform, independent learning materials before class, teachers in a specific time to answer students' questions, and urge students to complete the online learning content before class as required. The basic teaching content of the class includes routine physical education activities, technical demonstration and explanation of various sports events, teachers lead students to carry out technical movement exercises, students practice in groups and teachers' guidance, students discuss in groups, and teaching summarizes. The teaching method of the EG should be different from CG, which should focus on discussing and answering students' problems and puzzles in pre-class autonomous learning. The basic teaching content after class includes the teacher's reflection and summary of the class, improving instructional design, summarize the learning process and results of students. At the end of each class, the teacher proposes to the students the reflection in the teaching process, to consolidate and improve the teaching effect, and predicts the autonomous learning content of the next class.

The CG was taught by traditional teaching method. It mainly includes teacher's explanation, action demonstration, action correction, students' group practice, and teacher's guidance. Other teaching activities will also be arranged according to the actual teaching situation. At the end of each class, the teacher gives the students a preview of what they will learn and review in the next class.

The experiment period is from September 5, 2020 to December 31, 2020. A total of 18 weeks and 36 class hours, including 1 theoretical class (2 class hours) and 17 practical classes (34 class hours).

2.3 Measurements

"The Physical Education Learning Effectiveness Scale" and "The Physical Education Learning Attitude Scale for middle and high school students" were adopted as tools to research for the study.

"The Physical Education Learning Effectiveness Scale" of the study is reference from "The PE Learning Effectiveness Scale" compiled by Jin [11], which referring to the scale developed by different scholars [12–14]. The scale content includes three dimensions: affective learning, cognitive learning, and skill learning. The scale adopts five-point Likert-type scoring, and its reliability and validity are satisfactory. Cronbach's α values of the three dimensions in form1 scale were 0.841 (cognitive learning), 0.871 (skill learning), and 0.793 (affective learning), respectively, and 0.925 in total scale; Cronbach's α values of the three dimensions in form4 scale were 0.821 (cognitive learning), 0.837 (skill learning), and 0.724 (affective learning), respectively, and 0.919 in total scale, which indicates that the scale has good reliability.

"The Physical Education Learning Attitude Scale" of the study is referenced from [15]. The scale content includes three dimensions: learning style, interest and participation, and classroom performance. The scale adopts five-point Likert-type scoring, and its reliability and validity are satisfactory. Cronbach's α values of the three dimensions in form1 scale were 0.821 (learning style), 0.846 (interest and participation), and 0.764 (classroom performance), respectively, and 0.877 in total scale; Cronbach's α values of the three dimensions in form4 scale were 0.834 (learning style), 0.868 (interest and participation), and 0.658 (classroom performance), respectively, and 0.883 in total scale.

All indicate that the scale has good reliability. The validity test results of the scale showed that there was a significant positive correlation between the 20 items and the total amount table (correlation coefficient was 0.3–0.7). There was a significant positive correlation between the three dimensional factors and the total amount table (correlation coefficient is above 0.7). There was a significant positive correlation among the three dimensional factors (correlation coefficient is 0.3–0.7). It explains the rationality of the structure between the three dimensions. The results showed that the scale had a good structural validity.

Before the teaching experiment, the PE learning effect and PE learning attitude of the EG and the CG were tested uniformly. The statistical results of the pre-experiment investigation between the EG and the CG showed that there was no significant difference between the two groups in PE learning effectiveness and PE learning attitude ($p \geq 0.05$). The EG and CG before the experiment belong to the same level of experimental design requirements. The second questionnaire was issued after the experiment.

All questionnaires of students were distributed on class and collected simultaneously. 56 questionnaires were distributed to form1 students for the first time, of which 56 were recovered, 55 were effective, with an effective recovery rate of 98.2%; 56 questionnaires were distributed to form1 students in the second time, of which 56 were recovered and 53 were effective, with effective recovery rate of 94.6%. 57 questionnaires were distributed to senior form4 students for the first time, of which 57 were recovered and 57 were effective, with effective recovery rate of 100%; 57 questionnaires were distributed to form4 students in the second time, of which 57 were recovered and 57 were effective, with effective recovery rate of 100%.

3 Results

3.1 Middle School Survey Results (Form1)

Post-Experimental PE Learning Effectiveness and Learning Attitude This teaching experiment started from September 2020 to the end of December 2020, with a total of 18 weeks and 36 h. In this study, the flipped classroom method was applied to the teaching experiment in physical education class and compared with the traditional physical teaching method, so as to verify the differences in learning effectiveness and learning attitude.

The results of the post-experimental measurements of learning effectiveness and attitude of EG and CG in middle school (form1) are displayed in Table 1. The results showed that neither the flipped classroom nor the traditional teaching method had a significant promoting effect on middle school students' PE learning effectiveness and learning attitude. At the same time, compared with CG, EG had no significant difference in learning effectiveness (affective, cognitive, and skill) and learning attitude (learning style, interest and participation, classroom performance).

Learning Effectiveness and Learning Attitude Pre- and Post- Experiment Based on the post-experimental data analysis of the above two groups, we conducted a

Table 1 Middle school learning effectiveness and attitude in the EG and CG ($\overline{x} \pm s$)

Content	Dimension	Group		*t*	*p*
Learning effectiveness	Cognitive learning	EG	25.61 ± 4.41	1.371	0.176
		CG	24.07 ± 3.75		
	Skill learning	EG	24.46 ± 3.78	1.140	0.259
		CG	23.14 ± 4.54		
	Affective learning	EG	22.50 ± 3.95	− 0.171	0.865
		CG	22.66 ± 3.12		
	Total effectiveness	EG	72.57 ± 10.44	0.958	0.342
		CG	69.88 ± 9.97		
Learning attitude	Learning style	EG	15.00 ± 3.13	0.563	0.576
		CG	14.44 ± 3.97		
	Interest and participation	EG	13.92 ± 4.01	0.067	0.947
		CG	13.85 ± 3.73		
	Classroom performance	EG	19.26 ± 3.57	0.784	0.437
		CG	18.55 ± 3.04		
	Total attitude	EG	48.19 ± 9.84	0.527	0.600
		CG	46.85 ± 8.65		

comparative analysis of both groups measured before and after the experiment. The results of the learning effectiveness and attitude of the EG before and after the experiment are given in Table 2. It showed that the overall scores of learning effectiveness and attitude of the experimental group after the experiment were slightly higher than those before the experiment, but it showed no significant results, which may require further investigation and research. Table 3 shows the measurement results of the learning effectiveness and attitude of the CG before and after the experiment. The statistical results suggested that the 18-week teaching experiment intervention had no significant effect on the learning effectiveness and learning attitude of the control group.

3.2 High School Survey Results (Form4)

Post-Experimental PE Learning Effectiveness and Learning Attitude The results of the post-experimental measurements of PE learning effect and attitude of EG and CG in high school (form 4) are given in Table 4. The results showed that the form4 experimental group in learning effectiveness and learning attitude was better than that of the control group. In terms of learning effectiveness, sub-dimension cognitive learning was also close to the significant level ($p = 0.068$).

Table 2 Middle school EG's learning effectiveness and attitude pre- and post-experiment ($\overline{x} \pm s$)

Content	Dimension	Pre- or post-experiment		t	p
Learning effectiveness	Cognitive learning	Pre-	25.29 ± 3.33	− 0.297	0.767
		Post-	25.61 ± 4.41		
	Skill learning	Pre-	23.81 ± 3.35	− 0.659	0.513
		Post-	24.46 ± 3.78		
	Affective learning	Pre-	22.48 ± 2.76	− 0.020	0.984
		Post-	22.50 ± 3.95		
	Total effectiveness	Pre-	71.59 ± 7.78	− 0.390	0.698
		Post-	72.57 ± 10.44		
Learning attitude	Learning style	Pre-	14.96 ± 2.94	− 0.044	0.965
		Post-	15.00 ± 3.13		
	Interest and participation	Pre-	12.85 ± 3.88	− 0.986	0.329
		Post-	13.92 ± 4.01		
	Classroom performance	Pre-	18.25 ± 2.86	− 1.138	0.261
		Post-	19.26 ± 3.57		
	Total attitude	Pre-	46.07 ± 8.43	− 0.843	0.403
		Post-	48.19 ± 9.84		

Table 3 Middle school CG's learning effectiveness and attitude pre- and post-experiment ($\overline{x} \pm s$)

Content	Dimension	Pre- or post-experiment		t	p
Learning effectiveness	Cognitive learning	Pre-	25.14 ± 4.48	− 0.957	0.767
		Post-	24.07 ± 3.75		
	Skill learning	Pre-	24.25 ± 4.15	− 0.938	0.513
		Post-	23.14 ± 4.54		
	Affective learning	Pre-	22.60 ± 3.32	− 0.680	0.984
		Post-	22.66 ± 3.32		
	Total effectiveness	Pre-	72.00 ± 10.39	− 0.768	0.698
		Post-	69.88 ± 9.91		
Learning attitude	Learning style	Pre-	15.17 ± 3.01	− 0.773	0.443
		Post-	14.44 ± 3.97		
	Interest and participation	Pre-	13.07 ± 4.33	− 0.714	0.479
		Post-	13.85 ± 3.73		
	Classroom performance	Pre-	18.89 ± 2.43	− 0.454	0.651
		Post-	18.55 ± 3.04		
	Total attitude	Pre-	47.14 ± 8.21	− 0.128	0.899
		Post-	46.85 ± 8.65		

Table 4 High school learning effectiveness and attitude in the EG and CG ($\overline{x} \pm s$)

Content	Dimension	Pre- or post-experiment		t	p
Learning effectiveness	Cognitive learning	EG	26.07 ± 4.20	1.859	0.068
		CG	23.89 ± 4.60		
	Skill learning	EG	24.67 ± 4.09	0.850	0.399
		CG	23.75 ± 4.07		
	Affective learning	EG	23.32 ± 2.78	0.905	0.370
		CG	22.51 ± 3.82		
	Total effectiveness	EG	74.07 ± 10.11	1.371	0.176
		CG	70.17 ± 11.30		
Learning attitude	Learning style	EG	23.35 ± 4.20	2.858	0.006
		CG	19.89 ± 4.89		
	Interest and participation	EG	19.85 ± 3.36	0.383	0.704
		CG	19.41 ± 5.16		
	Classroom performance	EG	16.71 ± 1.86	1.511	0.136
		CG	15.86 ± 2.35		
	Total attitude	EG	59.92 ± 6.60	2.12	0.038
		CG	55.17 ± 9.92		

Thus, the flipped classroom had a positive effect on students' learning effectiveness, especially in cognitive learning.

Learning Effectiveness and Learning Attitude Pre- and Post-Experiment In order to analyze the influence of the two teaching methods on the physical education learning effectiveness and attitude of high school students, we also conducted a comparative analysis of both groups measured before and after the experiment.

The results of the pre- and post-experimental measures of learning effectiveness and attitude of high school (form4) EG are given in Table 5. The form4 experimental group showed improvements in both effectiveness and attitude averages. After the T-test, the results showed that the learning effect of all dimensions showed insignificant results, while the learning attitude was the opposite. Learning style ($p = 0.054$) is close to significant level; classroom performance ($p = 0.10$) showed significant results ($p < 0.05$); interest and participation ($p = 0.006$) showed very significant results ($p < 0.01$). Finally, the attitude score ($p = 0.002$) also showed a very significant level ($p < 0.01$); it shows that flipped classroom has positive effects on both learning effect and learning attitude, but the effect of learning attitude is more obvious.

Table 6 displays the results of learning effectiveness and attitude of the CG before and after the experiment. Different from the experimental group, the scores of the high school control group fluctuated in some dimensions, among which the average value of cognitive learning (23.93 > 23.89) of learning effectiveness decreased, while the average value of skill learning (23.03 > 23.75) and affective learning (21.17 > 22.51) increased. In terms of learning attitude, the average value of learning style

Table 5 High school EG's learning effectiveness and attitude pre- and post-experiment ($\overline{x} \pm s$)

Content	Dimension	Pre- or post-experiment		t	p
Learning effectiveness	Cognitive learning	Pre-	24.50 ± 2.89	− 1.627	0.109
		Post-	26.07 ± 4.20		
	Skill learning	Pre-	23.39 ± 3.60	− 1.248	0.218
		Post-	24.67 ± 4.09		
	Affective learning	Pre-	22.17 ± 2.17	− 1.709	0.093
		Post-	23.32 ± 2.78		
	Total effectiveness	Pre-	70.07 ± 7.55	− 1.677	0.099
		Post-	74.07 ± 10.11		
Learning attitude	Learning style	Pre-	20.92 ± 4.99	− 1.967	0.054
		Post-	23.35 ± 4.20		
	Interest and participation	Pre-	17.10 ± 3.87	− 2.837	0.006
		Post-	19.85 ± 3.36		
	Classroom performance	Pre-	15.28 ± 2.10	− 2.689	0.010
		Post-	16.71 ± 1.86		
	Total attitude	Pre-	53.32 ± 8.16	− 3.329	0.002
		Post-	59.92 ± 6.60		

(20.89 > 19.89) decreased, and both interest and participation (17.55) and classroom performance (19.41) improved. Although there was a change in the mean value, the results of learning effectiveness and attitude were not significant after T-test ($p > 0.05$).

4 Discussion

In recent years, the flipped teaching method has promoted the upsurge of teaching reform in the digital media era, which has brought different effects on students' learning effects and attitudes.

The statistical results of this study (Tables 5, 6) showed that the form4 EG had better learning results and attitudes than the form4 CG. In terms of learning effect, sub-dimension cognitive learning was also close to the significant level ($p = 0.068$). Thus, the flipped classroom had a positive impact on cognitive learning. This is consistent with existing research, Huang [16] proposed the flipped classroom can save a lot of time for teachers to explain, demonstrate, and increase the practice time of students, and watching teaching video can also help students to improve their weakness. To master motor skills more accurately, excite students' interest in learning and improve learning enthusiasm. At the same time, when students encounter problems

Table 6 High school CG's learning effectiveness and attitude pre- and post-experiment ($\overline{x} \pm s$)

Content	Dimension	Pre- or post-experiment		t	p
Learning effectiveness	Cognitive learning	Pre-	23.93 ± 3.49	0.032	0.975
		Post-	23.89 ± 4.60		
	Skill learning	Pre-	23.03 ± 4.12	– 0.672	0.504
		Post-	23.75 ± 4.07		
	Affective learning	Pre-	21.17 ± 3.45	– 1.406	0.165
		Post-	22.51 ± 3.82		
	Total effectiveness	Pre-	68.13 ± 9.41	– 0.745	0.460
		Post-	70.17 ± 11.30		
Learning attitude	Learning style	Pre-	20.89 ± 4.95	– 0.773	0.443
		Post-	19.89 ± 4.89		
	Interest and participation	Pre-	17.55 ± 4.56	– 1.455	0.151
		Post-	19.41 ± 5.16		
	Classroom performance	Pre-	14.79 ± 2.94	– 1.526	0.133
		Post-	15.86 ± 2.35		
	Total attitude	Pre-	53.24 ± 9.97	– 0.739	0.463
		Post-	55.17 ± 9.92		

in pre-class study and in-class practice, they should make use of the network communication platform set up by teachers or communicate with teachers and classmates in class and it also can make the classrrom atmosphere more active and interesting; students are more willing to communicate and discuss in that environment, thus to excite students' own potential, increase their proficiency in skills, and create a good learning atmosphere in class [17].

In the analysis of learning attitude, the results suggested form4 EG reached a significant level ($p < 0.05$), indicating that the flipped classroom has positive effect on learning attitude. Studies have found that the formation of learning attitude is social to a certain extent and is influenced by learning atmosphere and management atmosphere. In the flipped classroom, teachers and students are especially emphasized to discuss and communicate in class, to solve and improve problems together. It makes a certain role in improving the learning attitude, and also affects the learning effectiveness [18]. Compared with the traditional teaching method, the flipped classroom helps students to master the sports technical movements better. It can have a more positive impact on students' learning attitude [19].

In addition, teachers' teaching attitudes are also influential factors, set up positive learning example; let students have a good atmosphere for learning activities. Chem and Macredie [20] showed that teachers actively promote students' teaching activities in PE class with flipped classroom, which has a great impact on students' mentality and also makes a role in promoting their learning attitude.

Integrating the analysis results of different experimental groups in middle school (form1) and high school (form4), it could be seen that form4 students had better self-learning ability and better understanding and adaptation ability than form1 students in the process of learning. The results showed learning effectiveness and learning attitude were more significant in flipped classroom mode.

Although the advantages of flipped classroom mode in physical education teaching activities was acknowledged, the method to fully functioning advantages of flipped classroom in physical education needs to be further discussed. The key is the cultivation of students' independent learning ability in physical education teaching by flipped classroom model. The teaching result will be enhanced with appropriately improvement and targeted planning of teaching content. The research shows that flipped classroom also has its own shortcomings. Whether the implementation of flipped classroom can be effective must be based on physical education teaching objectives, teaching content, and students' basic situation to design the teaching process. Only in this way can it be combined with traditional physical education classes to bring the maximum effect [19].

5 Conclusion

The experimental results showed that compared with the traditional teaching method, the flipped classroom has no significant effectiveness on the learning effectiveness and attitude of the PE class in the form1 students of Yuet Wah College. The flipped classroom can promote the learning effectiveness of form4 students of Yuet Wah College. The effectiveness of cognitive learning effectiveness on learning effectiveness was significant, but the effectiveness of skill learning effectiveness and affective learning effectiveness was not significant; Compared with the traditional teaching method, the flipped classroom has an obvious promoting effectiveness on the learning attitude of form4 students of Yuet Wah College.

References

1. Huang QY (1994) The function and application of interactive multimedia - physical education. Instr Technol Media 16:21–26
2. Zhang YK (2001) Application of computer multimedia CAI in physical education teaching. Pingtung Univ Phys 5:119–135
3. Rui JC (2015) The past, present and future of the flipped classroom approach. Natl Acad Educ Res 1:21–23
4. Fengyan C (2014) Flipped classroom: the deep integration of information technology and education. Educ Rev 127–129
5. Xu L, Mu J (2017) Feasibility analysis of the implementation of flipped classroom in sports. China Educ Technol Equip 17:109–110

6. Li P, Wang L (2015) Opportunities and challenges: the enlightenment of flipped classroom for physical education teaching in colleges and universities in my country. J Nanjing Inst Phys Educ 14:122–128
7. Kong X (2019) How to introduce the flipped classroom model into physical education. Contemp Sport Sci Technol 9:141–191
8. Caligaris M, Rodríguez G, Laugero L (2016) A first experience of flipped classroom in numerical analysis. Procedia Social Behav Sci 217:838–845
9. Yang L (2019) Flipped classroom-innovative attempt of high school physical education teaching. Contemp Sport Sci Technol 9:74–75
10. Pan Y (2017) The application of flipped classroom into physical education teaching. China Sport Q 31:87–99
11. Jin ZL (2014) A study on learning satisfaction and learning effectiveness of physical education course between male and female junior high school students - the role of physical self-concept as a moderator variable. Natl Taiwan Univ Phys Educ Sport pp 139–143
12. Li QL (2012) A study on PE learning satisfaction and learning effectiveness of elementary school students in Taipei city. National Taipei University of Education
13. Hsu S-L (2009) Research on learning satisfaction and effectiveness of physical education in Kaohsiung senior high school students. National Taiwan Normal University
14. Huang MJ (2003) The relationship between senior high school & vocational high school students' learning satisfaction and effectiveness of physical education in Hsinchu. Taipei Physical Education College
15. Wang X-Z (2005) Study on the reform of PE learning evaluation in primary and secondary schools in China. East China Normal University
16. Huang LL (2017) Experimental research on flipped classroom in junior high school volleyball teaching. Shandong Normal University
17. Jian C (2019) Experimental research on the application of flipped classroom teaching mode in middle school basketball teaching under the background of campus basketball. Shanghai Normal University
18. Xiao Hong D, Ping R (2015) The research on learning attitudes theoretical courses or technical courses of P.E major students. J Hengyang Norm Univ 36:148–152
19. Wang GL (2016) A study on introducing flipped classroom into public PE teaching in colleges. Beijing Sports University
20. Chen SY, Macredie RD (2004) Cognitive modeling of student learning in web-based instructional programs. Int J Hum Comput Interact 17:375–402

An Experimental Study on the Effect of Blended Learning on the Learning Effectiveness and Learning Motivation of Secondary School Students in Macao

Ka Ho Cheong and Qun Zhai

Abstract The blended teaching and learning method that combine traditional face-to-face teaching with modern computer network information technology have attracted more and more attention and favor. This study explored the effects of blended teaching and learning methods on the learning effectiveness and learning motivation of middle school students in physical education. The results would provide the scientific basis and reference for the application and promotion of this new teaching and learning method in physical education and training. A teaching experimental research design was employed, the sample included 86 male middle school students, and they were divided into an experimental group ($n = 32$) and a control group ($n = 54$) by using the cluster sampling method. The experimental group received the blending teaching method, while the control group received the traditional teaching method for 8 weeks of training. The "Physical Education Learning Effectiveness Scale" and the "Middle School Students' Physical Education Participation Motivation Questionnaire" were used to examine the effects of both teaching methods on middle school physical education learning. The results showed that compared to the traditional method, the blended teaching method had significant effects on the learning effectiveness and learning motivation of the middle school students' physical training courses.

Keywords Blended learning · Learning effectiveness · Learning motivation · PE

1 Introduction

The COVID-19 epidemic has a huge impact on people's lives, studies, and work. In order to maintain having classes with the closure of schools and ensure the progress

K. H. Cheong · Q. Zhai (✉)
Faculty of Health Sciences and Sports, Macao Polytechnic University, Macao SAR, China
e-mail: zqun@ipm.edu.mo

K. H. Cheong
Macau Salesian High School, Macao SAR, China

G. Kuan et al. (eds.), *Advancing Sports and Exercise via Innovation*, Lecture Notes in Bioengineering, https://doi.org/10.1007/978-981-19-8159-3_11

and quality of education, online teaching and learning activities have become a necessity for education. In the post-epidemic era, the blended learning model, which combines the advantages of traditional teaching and modern online teaching methods, is gaining more and more attention and popularity. Blended learning, or B-learning, is a new approach to learning that has emerged from the beginning of this century, it generally refers to the combination of traditional face-to-face teaching and computer-assisted online learning [1]. According to this definition, a blended learning approach can be understood as "face-to-face teaching + online learning [2]". Some scholars consider "blended learning" to be the use of the right technology at the right time to allow students to develop their skills and thus achieve optimal teaching and learning outcomes [3]. Curtis J. Bonk summarizes the advantages of blended learning but also points out that there are many problems with blended learning, such as teachers' perceptions of blended learning, the increased burden of learning, learners' information technology literacy, and the degree to which learning content is integrated and understood, etc. [1]. This shows that blended learning is a new way of learning that places greater demands on teachers and students to adapt to the times. The teachers' ability of blended teaching, the ability of students to accept and adapt to blended learning, the level of development of information technology in teaching networks, and the feasibility and effectiveness of blended learning design will all directly affect the effectiveness and promotion of blended learning.

Physical Education (PE) is a course in which students use physical exercises as the main way to enhance their physical fitness, health, and physical literacy through a rational PE and a scientific process of physical exercise. The characteristics of PE lie in the fact that, in addition to equipping students with certain scientific knowledge, it enables students to master the knowledge, techniques, and skills of PE through repeated physical exercises, in which physical activities are closely integrated with mental activities for motor cognition. In the process of teaching PE, the teacher's straightforward demonstration and timely guidance to correct mistakes and feedback play a crucial role in the learning effect. Compared with teaching in other disciplines, the characteristics of physical education limit the applied research and practical promotion of blended teaching to a certain extent [3]. According to the results of our online search of domestic journals, research on the theory and practice of blended learning in PE only began in 2010, and they mostly focused on research and exploration of university PE courses [4–7]), with few research reports on PE learning for primary and secondary school students or adolescents. From these studies, we can see that blended learning, which brings together the perspective advantages of traditional teaching and modern e-learning, has obvious advantages and great potential in physical education [4]. Of course, this emerging teaching model and learning method in the practice of PE also reflect various problems that need to be further explored and improved [8].

In view of the above analysis of the research and practice of blended learning in physical education, this study applies the blended teaching model to a basketball class in a secondary school in Macao. Through a comparative analysis with the traditional PE teaching model, the positive effects of blended teaching and learning methods in terms of learning effectiveness and motivation are explored. This study provides a scientific basis and reference for the application and promotion of blended teaching and learning methods in PE.

2 Methods

2.1 Participants

In this study, 86 people from a secondary school (a boy school) in Macao were selected as the research participants. A cluster sampling method was employed to recruit participants. All of the participants were males. All participants were divided into experimental group (EG) and control group (CG) based on the actual situation and conditions of teaching; the basketball class of the school was selected as the EG ($n = 32$ students); while the volleyball class, which is closest to the basketball class in terms of the teaching content, was selected as the CG ($n = 54$ students).

2.2 Teaching Experiment Content Design

Video is an important medium for blended teaching, relying on exciting content to attract students' attention and enhance the educational effect, using topic discussions and tasks to enhance students' application and grasp of the relevant. Considering the popularity of online media and smartphones in today's society, the students in the EG first downloaded the WeChat application on their smartphones. Mobile phones allow for quick access to relevant messages and facilitate timely communication between students and teachers. The teacher added the 32 students in the EG to a WeChat group which was set up beforehand and sent the videos to the group before the class started, and the students were free to download and watch the videos in their spare time. After completing each task, the teacher would schedule a time to communicate and exchange ideas with students about the relevant skills. The basic structure of a physical training class in the blended learning model can be divided into a preparatory section, a basic section, and a closing section, similar to the structure of a traditional physical education class. To enable students' knowledge and skills to be truly internalized, in the basic section of the class, blended learning significantly compresses the time spent by the PE teacher explaining and demonstrating, while increasing the time spent by students on physical training. The main task of the teachers in the EG was to answer questions about the problems students encountered during the online learning sessions before class, to provide group and individual guidance according to the problems they raised with clear objectives, and to lead students to investigate the causes of and the solutions to their mistakes. The main task of the teachers in the CG was to follow the usual PE teaching procedures, i.e., to guide students to understand the basic concepts of the content through explanation and demonstrations, and to consolidate the knowledge through question and answer sessions and doing exercises. The EG and CG were identical and consistent in the content and organization of the preparatory and closing sections of the PE classes. This teaching experiment began in September 2020 and ended in December 2020, and it lasted for three months, with a total of 86 h. Students in the PE were asked to study the online materials teachers provided before each class for 10–15 min. Students from the CG were not asked to study online on their own before class.

2.3 Measurements

"The Scale of PE Learning Effectiveness" uses the PE learning effectiveness scale developed by Jinzhong [9], with 20 questions concerning three dimensions: affective learning, cognitive learning, and skill learning [10] Taking into account the cultural similarities and the differences in terminologies, we combined the understanding of idioms and concepts in the context of sports in Macao schools, and appropriately modified the wording of the scale, and named it the "PE Learning Effectiveness Scale". The results of the reliability test on the scale showed that the Cronbach's α values for the three-dimensional factors of the scale were 0.716 (cognitive learning), 0.677 (skill learning), and 0.738 (affective learning), and the Cronbach's α coefficient for the total scale was 0.822, indicating that this scale had good reliability. The results of the validity test on the scale showed moderately significant positive correlations between all 20 questions and the total scale ($r = 0.3$–0.7), highly significant positive correlations between the three dimensions and the total scale ($r = 0.8$ or more), and significant positive correlations between the three dimensions ($r = 0.6$–0.8), indicating good content and structural validity of the scale.

The "Middle School Students' PE Participation Motivation Questionnaire" was developed with reference to the "Behavioral Regulations in PE Questionnaire (BRPEQ)", which was revised and translated into Chinese from the BRPEQ developed by Nathalie et al. [11]. Earlier studies have shown the good internal reliability as well as the good factor and construct validity of this scale [12]. In the present study, the motivation dimension and relevant questions were removed as needed, the 16 questions of the two dimensions of autonomous motivation and controlled motivation were retained, and some wording in the questionnaire was modified according to the common choice of words used in the region. Hence, the "Middle School Students' PE Participation Motivation Questionnaire" was developed. The results of our reliability test on the questionnaire showed that the Cronbach's α values for the two dimensions were 0.884 (autonomous motivation) and 0.792 (controlled motivation), and the Cronbach's α coefficient for the total scale was 0.885, indicating that the questionnaire had good reliability. The results of the validity test of the questionnaire showed that there was a moderately significant positive correlation between all 16 questions and the total scale ($r = 0.3$–0.7), a highly significant positive correlation between the two dimensions and the total scale ($r = 0.8$ or more), and a significant positive correlation between the two dimensions (r = 0.7–0.8), indicating that the content and structural validity of the scale was good.

The first questionnaire was administered to students in the basketball class (PE) and the volleyball class (CG) prior to the start of the teaching experiment, and the second questionnaire was administered to both groups after the experimental intervention. All questionnaires were distributed and collected on site. Thirty-two questionnaires were distributed to students in the PE before and after the experiment, and 32 were returned, with a valid return rate of 100%. For the CG, 58 questionnaires were distributed to students before and after the experiment, and 54 were returned, with a valid return rate of 93%.

According to the requirements of the experimental design, a uniform test concerning PE learning effectiveness and motivation was conducted on both EG and CG prior the experiment, in order to determine the homogeneity of the two groups prior the experiment. The results of the pre-experimental test between EG and CG showed that there was no significant difference between the two groups in terms of PE learning effectiveness or motivation ($p \geq 0.0$ 5), which met the experimental design requirement that the two groups were in a state of homogeneity before the experiment.

3 Results

3.1 Post-experimental PE Learning Effectiveness and PE Learning Motivation

This teaching experiment was conducted in the first semester of the academic year 2020–2021, starting in September 2020 and ending on 30th December 2020. Throughout the experiment, the teaching procedures and teaching contents of the EG were designed and executed exactly according to the blended teaching model; while those of the CG were designed and executed exactly according to the traditional and conventional PE teaching model. At the end of the three-month experimental period, the PE learning effectiveness and motivation of the participants in both groups were measured again.

The results of the post-experimental measures of learning effectiveness and motivation in PE for the EG and CG are shown in Table 1. There were significant differences between the two groups on the learning effectiveness variables, with the EG having significantly higher levels of learning effectiveness (affective, cognitive, and skill) than the CG, indicating a significant effect of the experimental teaching intervention on the learning effectiveness of the EG. There were also significant differences between the two groups on the learning motivation variables, with the EG having significantly higher levels of motivation, including autonomous motivation and controlled motivation, than the CG, indicating that the experimental intervention also had a significant effect on the learning motivation of the EG.

Table 1 PE learning effectiveness and motivation in the EG and CG ($\overline{x}\pm s$)

Content	Dimension	Group		*t*	*p*
Learning effectiveness	Affective learning	EG	26.50 ± 2.31	5.61	0.00
		CG	23.44 ± 2.52		
	Cognitive learning	EG	28.69 ± 2.53	3.80	0.00
		CG	25.87 ± 3.72		
	Skill learning	EG	28.69 ± 2.99	3.83	0.00
		CG	25.74 ± 3.70		
	Total effectiveness	EG	83.88 ± 5.82	5.00	0.00
		CG	75.06 ± 8.89		
Learning motivation	Autonomous motivation	EG	36.16 ± 3.41	5.03	0.00
		CG	32.22 ± 3.56		
	Controlled motivation	EG	27.59 ± 2.78	2.37	0.02
		CG	26.11 ± 2.81		
	Total motivation	EG	63.75 ± 5.10	4.87	0.00
		CG	58.33 ± 4.92		

3.2 *Learning Effectiveness and Learning Motivation Pre- and Post- experiment*

In order to further observe and analyze the effects of different teaching models on students' PE learning effectiveness and motivation, we conducted a comparative analysis of the data of both groups measured before and after the experiment.

The results of the pre- and post-experimental measures of learning effectiveness and motivation of the EG are shown in Table 2. The post-experimental learning effectiveness scores of the EG were significantly higher than those of the pre-experimental scores, demonstrating that the experimental intervention had a significant effect on the learning effectiveness of the EG, and that the overall learning effectiveness of the EG was significantly improved after blended learning. The results of the learning motivation of the EG before and after the experiment showed that the post-experimental scores of motivations were significantly higher than those of the pre-experimental scores, which proved that the experimental intervention had a significant effect on the learning motivation of the EG as well.

The results of learning effectiveness and motivation of the CG before and after the experiment are shown in Table 3. The CG's PE learning effectiveness scores after the experiment were lower than those before the experiment, showing a decreasing trend in students' PE learning effectiveness. The results of the study showed that the traditional PE teaching model did not show a positive effect on the students' PE learning effectiveness, but rather a negative one. The reasons for the decline in the learning effectiveness of the CG need to be further investigated and studied. There was no significant change in the motivation of the CG before and after the

Table 2 EG's learning effectiveness and motivation pre- and post-experiment $(\overline{x}\pm s)$

Content	Dimension	pre- or post-experiment		t	p
Learning effectiveness	Affective learning	pre-	25.22 ± 3.07	−1.88	0.06
		post-	26.50 ± 2.31		
	Cognitive learning	pre-	27.28 ± 3.17	−1.96	0.05
		post-	28.69 ± 2.53		
	Skill learning	pre-	26.50 ± 2.71	−3.06	0.00
		post-	28.69 ± 2.99		
	Total effectiveness	pre-	79.00 ± 7.41	−2.93	0.00
		post-	83.88 ± 5.82		
Learning motivation	Autonomous motivation	pre-	33.72 ± 4.47	−2.45	0.01
		post-	36.16 ± 3.41		
	Controlled motivation	pre-	26.50 ± 2.44	−1.67	0.09
		post-	27.59 ± 2.78		
	Total motivation	pre-	60.22 ± 5.96	−2.54	0.01
		post-	63.75 ± 5.10		

experiment, indicating that the traditional PE teaching and training model did not have a significant effect on the PE learning motivation.

Table 3 CG's learning effectiveness and motivation pre- and post-experiment $(\overline{x}\pm s)$

Content	Dimension	pre- or post-experiment		t	p
Learning effectiveness	Affective learning	pre-	24.33 ± 2.67	1.78	0.07
		post-	23.44 ± 2.51		
	Cognitive learning	pre-	27.70 ± 3.63	2.59	0.01
		post-	25.87 ± 3.72		
	Skill learning	pre-	27.28 ± 3.52	2.21	0.03
		post-	25.74 ± 3.70		
	Total effectiveness	pre-	79.31 ± 8.47	2.54	0.01
		post-	75.06 ± 8.89		
Learning motivation	Autonomous motivation	pre-	33.63 ± 4.14	1.89	0.06
		post-	32.22 ± 3.56		
	Controlled motivation	pre-	59.98 ± 6.09	0.42	0.67
		post-	58.33 ± 4.92		
	Total motivation	pre-	59.98 ± 6.09	1.54	0.12
		post-	58.33 ± 4.92		

4 Discussion

Computer-mediated online teaching has been in practice for many years, but learning activities are difficult to be carried out anywhere or anytime due to the limitations of equipment and fixed locations. In recent years, mobile devices have become increasingly popular and have become the main channel for people to receive information because that they are not limited by time or space for information exchange, thus greatly changing the way people work and live. The changes in information and communication technology have also had a great impact on the field of education, as many academic disciplines have started to seek new teaching methods and models, going beyond the traditional teaching framework, and trying to create different and profound learning experiences with the support of new technologies and new media. PE is a major component of schooling, and in the past, traditional PE teaching used direct face-to-face coaching, with teachers placing more emphasis on demonstrations or explanation of rules for students to imitate and practice repeatedly [13]. In this type of traditional PE teaching, students are in a relatively passive learning state, rarely discovering and understanding relevant knowledge by themselves, and the knowledge or skills acquired mostly come from the classroom [14]. With the development of mobile technology, people are more willing to actively obtain information of their own interest from mobile devices, and students are more willing to start exploring and gradually or deeply absorb knowledge extensively through various learning platforms according to their own interests. In the context of PE, mobile technology brings about independent learning and exploration that allows students to gain motor skills, procedural knowledge, technical understanding, conceptual development, and experience, as well as in-class practice, practice fluency, or a certain level of motivation [15]. The results of this study corroborate with the significant advantages of the blended teaching model in PE and training, and in conjunction with the relevant studies that have been conducted, we can conclude that blended teaching in the context of PE has significant advantages in the following areas.

In terms of teaching procedures, compared to the traditional "teach first, learn later" approach, the "learn first, teach later" approach to blended learning not only improves the teacher's teaching efficiency in class but also improves the effectiveness of teaching. Most importantly, it helped teachers return to their roles which are the "most important" to their students—the guide to learning methods, and the helper of knowledge internalization [16]. The experimental intervention in this study was mainly manifested in the fact that the learning content of the EG was no longer obtained mainly from class, as in the traditional PE teaching model, but that the learning content and practice methods and guidelines were obtained before class through the online teaching network. Students were able to learn and communicate according to their own environment and conditions and could raise their questions in class, so that teachers could provide group or individual guidance with clear objectives. It is also clear from the results that the blended learning model has clear advantages in terms of improving learning effectiveness and motivation.

In terms of teaching methods, the EG could immediately ask the teacher questions through the online platform, and they could also practice repeatedly and correct anything that they got wrong during the process of blended learning. Students could also find relevant learning resources and communicate with each other through the online platform, which not only enriched their learning pathways and content, but also increased their interest in learning, and improved their problem-solving skills. Some studies point out that in the flipped (blended) teaching model, students who encounter problems they do not understand before class can actively communicate with their teachers and classmates using the online platform organized by the teachers. This strengthens the bond between teachers and students, making the atmosphere in class more harmonious, and students prefer to learn and communicate in such an environment, thus stimulating students' potential, increasing their familiarity with the content, and creating a better learning environment [7].

In terms of learning motivation, the positive boosting effect of the blended learning model on students' motivation compared to the traditional PE teaching model was evident in this study. In comparison with the CG, which used the traditional PE teaching model, the blended learning model had a significant boosting effect on the students' motivation to learn PE, especially for the autonomous learning motivation in the PE. During the experiment, the most profound experience of the teachers in the PE was the less restricted and more frequent communication and interaction between teachers and students. Once the students adapted to the new teaching model, the interaction between students and teachers, and among students, became more active and frequent, and the interaction between them was no longer restricted by time or environment. With the foundation of learning outside the classroom, teachers also have more time to provide group or individual guidance, and for interaction on problems, students encountered during their own learning, which in turn increases students' motivation and effectiveness in learning.

References

1. Zhan Z, Li X (2009) Blended learning: definition, strategies, current status and future trends—a conversation with Professor Curtis J. Bonk from Indiana university. USA China Educ Technol 257:1–5
2. Kekang H (2005) On new trend of theory for educational technology from blending learning. J Natl Acad Educ Adm 9:37–48
3. Liu Z (2016) Research of PE. Teaching combining network teaching with traditional classroom based on national excellent course resources. J Nanjing Sport Inst (Natural Sci 15:102–106
4. Zhao P, Li S, Sun Y, Wang J, Dong Y (2019) Instructional design and experimental Study of B-learning based on WeChat in physical education—a case study of yoga teaching in universities and colleges. Sport Res Educ, 53–58
5. Han T (2011) The construction of a "blended learning model" for physical education courses in higher education. Sport World Sch, 44–45
6. Wang T, Feng X (2019) Research progress on blended learning in college physical education teaching in the age of "internet +." J Anhui Sport Sci 195:62–66
7. Wang L (2017) An experimental study of "flipped classroom" in middle school volleyball teaching. Jinan Shandong Norm Univ, 21–22

8. Liu L, Gu N, Wang K (2020) A research on the application of blended learning model in the public physical education in colleges under the view of constructivism—taking tennis optional course of Yancheng normal university as an example. Liaoning Sport Sci Technol 220:151–154
9. Jin ZL (2014) A study on learning satisfaction and learning effectiveness of physical education course between male and female junior high school students—the role of physical self-concept as a moderator variable. Natl Taiwan Univ Phys Educ Sport, 139–143
10. Li J-Z (2014) A study on learning satisfaction and learning effect of physical education course between male and female junior high school students—the role of physical self-concept as a moderator variable. Taipei Natl Taiwan Univ Phys Educ Sport, 139–143
11. Aelterman N, Vansteenkiste M, Van Keer H, Van den Berghe L, De Meyer J, Haerens L (2012) Students' objectively measured physical activity levels and engagement as a function of between-class and between-student differences in motivation toward physical education. J Sport Exerc Psychol 34:457–480
12. Li G (2015) A study on female university students' physical education elective course and motivation and body self-esteem. Suzhou Soochow Univ, 15–16
13. Hill K (2014) Using mobile devices for motor-learning laboratory exercises. J Phys Educ Recreat Danc 85:20–26
14. Franklin R, Smith J (2015) Practical assessment on the run–iPads as an effective mobile and paperless tool in physical education and teaching. Res Learn Technol, 23
15. Crawford S, Fitzpatrick P (2015) Use of mobile digital technology and iPod touches in physical education. Handb Mob Teach Learn 50:499
16. Zheng X (2012) From "integration" to "unification". China Inf Technol Educ, 16–20

An Experimental Study on the Effect of Blended Teaching Method on Learning Basketball Skills and Learning Motivation for College Students

Bolun Ding and Qun Zhai

Abstract The blended teaching and learning method that combine traditional face-to-face teaching with modern computer network information technology have attracted more and more attention and favor. This study explored the effects of blended teaching and learning methods on the motor learning and learning motivation of college students in basketball learning. The results would provide the scientific basis and reference for the application and promotion of this new teaching and learning method in physical education and training. A teaching experimental research design was employed, the sample included 90 male college students, and they were divided into an experimental group ($n = 45$) and a control group ($n = 45$) by using the cluster sampling method. The experimental group received the blending teaching method, while the control group received the traditional teaching method for 8 weeks of training. General basketball skills assessment and the Behavioral Regulation in Physical Education Questionnaire (BRPEQ) were used to examine the effects of both teaching methods on basketball skills learning. The research results show that compared with the traditional teaching mode, the blended teaching mode has significant advantages in the mastery of college students' basketball skills and the improvement of learning motivation.

Keywords Blended learning · Learning basketball skills · Learning motivation

1 Introduction

Along with the development and popularization of network and mobile communication technology in recent years, the time and space constraints of information flow are significantly reduced. Online learning is becoming more and more acceptable and recognized. While classroom learning allows for a face-to-face connection, online learning also has its own benefits such as more convenience, flexibility, affordability, and earning opportunities [1]. Mobile and ubiquitous learning environment is in an

B. Ding · Q. Zhai (✉)
Faculty of Health Sciences and Sports, Macao Polytechnic University, Macao SAR, China
e-mail: zqun@ipm.edu.mo

G. Kuan et al. (eds.), *Advancing Sports and Exercise via Innovation*, Lecture Notes in Bioengineering, https://doi.org/10.1007/978-981-19-8159-3_12

attempt to make convenient and can share learning resources with each other, and it is used to improve the interaction between teachers and students so that students can simply receive learning resources at home and school but also allow teachers and parents more easily grasp the student's learning situation [2].

Blended learning model combines face-to-face and online activities in its application, and such learning reduces direct learning in the classroom. The goal of using this learning model is to help students become more independent and active in learning. The advantage of this learning model is that teachers can provide learning materials anytime and anywhere. Offline and online learning complement each other, learning becomes effective and efficient, teacher and student interaction are well established, accessibility is improved, and learning is flexible.

Self-determination means the ability or process of making one's own choices and controlling one's own life, and it has been proven to be a useful theoretical framework for studying motivational dynamics in the context of physical education (PE) [3]. The self-determination theory specifically distinguishes between two broader, qualitatively different forms of motivation, namely autonomous and controlled [4].

Autonomous motivation involves the regulation of behavior with the experiences of volition, psychological freedom, and reflective self-endorsement and consists of three subtypes. First, intrinsic motivation refers to engagement in a behavior for its own sake. The second form of autonomous motivation is integrated regulation, which involves engagement in a behavior because the behavior has been brought in alignment with the individual's other values and ideals. Controlled motivation refers to the pressured engagement in an activity and contains two subtypes. Introjected regulation involves participating in PE to avoid feelings of guilt, shame, and anxiety or to gain pride and ego enhancement. External regulation occurs when students engage in PE to obtain appreciation or rewards or to avoid punishments and criticism. Autonomous motivation and controlled motivation are contrasted with amotivation, which exists when people lack intentionality or engage in behaviors for reasons unknown [3].

Several studies demonstrated that autonomous motivation is associated with higher levels of self-reported physical activity, both during and outside the PE class [5, 6]. In contrast, controlled forms of motivation and amotivation are usually linked to negative outcomes such as boredom, unhappiness [7], and less or no intentions to participate in physical activity [8]. With respect to physical activity levels, controlled motivation has been found to be unrelated to objective physical activity in the exercise domain [9].

Motor skills have been always the main content of physical education (sports education), in the past, physical education used direct face to face as a guidance method, most teachers tend to demonstrate or interpret the rules, and let students actually repeat imitations and practice [10]. In terms of teaching strategy, fewer let students spontaneously find and understand the relevant knowledge, students are in a relatively passive learning position, and all they get knowledge or skill almost obtained from class, rarely expanded out of the teaching material.

Under the environment of college physical education in China, and because of the heterogeneous students, a traditional teaching method makes teacher difficult to take care of all students and also difficult to achieve further teaching effects. In order to identify an appropriate and effective teaching and learning method in basketball training among college students, it is significant to investigate the effectiveness of the blended teaching method in learning basketball skills and the learning motivation of the students.

2 Methods

2.1 Participants

The study participants were sophomore students in one of Mainland China universities. All students who selected the basketball training were randomly divided into two groups, such as the experimental group ($n = 45$) and the control group ($n = 45$), 20–22 years old, all were healthy males.

2.2 Teaching Strategy

The experimental group used the basketball teaching videos released by teachers on the WeChat public account as pre-class learning materials. Because the experimental group obtained the learning materials provided by the teacher through the WeChat public account video before the class for personalized online learning, the time for the teacher to explain and demonstrate during the face-to-face teaching was greatly reduced. Teachers can spend more time communicating and discussing with students purposively, and students can also get more practice time. The main task of the teachers of the experimental group in classroom teaching is to answer the difficult questions that students encountered in online learning before class and provide collective and individual guidance purposively. The main task of the teachers in the control group in the classroom was to follow the conventional physical education procedures to complete the teaching tasks, that is, the teachers let the students form the basic concepts of the learning content through explanations and demonstrations and become proficient and consolidate the learning content through answering questions and practicing.

Before the experimental intervention, in order to test whether the grouping of the experimental subjects meets the requirements of the homogeneity of the experimental groupings, the basketball skill level and learning motivation level of the experimental group and the control group were measured and tested. The results showed that there was no significant difference in the level of basketball skills and learning motivation between the two groups ($p < 0.05$), which met the requirements of the homogeneity of the experimental groupings.

2.3 Measurements

Basketball Skill Performance. Basketball skill performance was assessed by one teacher who has more than 15 years of experience in basketball teaching. The assessment includes two indicators: shooting and three-step layup. The shooting test is to set a 3.5 m radius distance from the center of the basket, complete 10 shots in one minute, and score 1 point for each shot in the hoop. The dribble three-step layup test uses the correct action, throws into the hoop to record 1 ball, and completes as many as possible within one minute.

Measurement of Learning Motivation. The Chinese version of the Behavioral Regulation in Sports Questionaire (C-BRPEQ) [11] "Motivation Scale for Regulation of Sports Behavior" [12] was used to measure and evaluate the motivation level of the subjects in basketball learning. "Regulating Sports Behavior Motivation Scale" has 20 items, which measure students' autonomous motivation (intrinsic motivation and identified regulation), controlling motivation (introjected regulation and external regulation), and non-motivation. The previous studies have confirmed that this scale has good reliability and validity [13–15]. In this study, the Cronbach's α of the Chinese version of BRPEQ's independent motivation, controlled motivation, and non-motivation was 0.86, 0.85, and 0.81, respectively.

2.4 Data Collection

The same questionnaire was distributed online among participants pre- and post-experimental intervention, respectively, with 8 weeks. Ninety pre-experiment questionnaires were distributed and returned with a 100% valid retuning rate, while only 87 were returned among 90 distributions of post-experiment with a 97% valid returning rate.

2.5 Data Analysis

The SPSS version 26 was used to analyze data, and descriptive analysis included mean, standard deviation, the independent samples t-test, and paired samples t-test was applied to compare different teaching strategies on the learning achievements.

Table 1 Intra-group effect of basketball skills in the EG

Content	Stage	*N*	*M*	SD	*t*	*p*
Shoot	Pre-expt	45	2.38	1.33	9.40	0.000
	Post-expt	45	4.73	1.54		
Layup	Pre-expt	45	0.71	1.05	20.35	0.000
	Post-expt	45	3.38	0.96		

3 Results and Analysis

3.1 Intra-Group Effect of the Experimental Group (EG)

Intra-group Effect Results of Basketball Skills in EG. Table 1 indicates that the basketball skill level of the EG, the mean score of shoot was 2.42 (pre-experiment) and 4.96 (post-experimental), and the mean score of layup was 0.71 (pre-experiment) and 3.33 (post-experimental). The results of the paired sample t-test on the pre- and post-experiment data showed that the difference between the basketball skill levels of the pre- and post-experiment was statistically significant, indicating that the blended learning mode adopted by the experimental group had a significant positive effect on the students' technical level.

Intra-group Effect of Learning Motivation in EG. The statistical results of the level of learning motivation in the EG (Table 2) show that the levels of intrinsic motivation, introverted regulation, and extrinsic regulation motivation have increased to varying degrees, the level of amotivation has been significantly reduced, and the level of identity regulation has been significantly reduced after the experimental intervention. The analysis of the results showed that the blended learning mode adopted by the experimental group was helpful to improve the students' controlling motivation, especially the introspective adjustment motivation, and reduce the amotivation, and had no significant effect on the autonomous learning motivation.

3.2 Intra-Group Effect of the Control Group (CG)

Intra-group effects of basketball skills in the CG. The statistical results of the basketball skill level of the CP (Table 3) show that the mean score of shoot was 2.31 (pre-experiment) and 4.04 (post-experimental), and the mean score of layup was 0.80 (post-experimental) and 2.38 (post-experimental). The results of the paired sample t-test on the data before and after the experiment showed that the difference between the basketball skill levels of the control group before and after the experiment was statistically significant, indicating that the traditional teaching model had a positive impact on the skills of the students in the control group.

Table 2 Intra-group effect of learning motivation in the EG

Content	Stage	*N*	*M*	SD	*t*	*p*
Intrinsic motivation	Pre-expt	45	17.72	1.59	1.346	0.186
	Post-expt	43	17.90	1.60		
Identified regulation	Pre-expt	45	17.41	1.63	0.260	0.796
	Post-expt	43	17.37	1.15		
Introjected regulation	Pre-expt	45	16.13	1.59	4.988	0.000
	Post-expt	43	17.11	1.67		
External regulation	Pre-expt	45	14.58	1.82	1.715	0.094
	Post-expt	43	14.83	1.58		
Amotivation	Pre-expt	45	9.69	2.57	11.299	0.000
	Post-expt	43	6.46	1.59		

Table 3 Intra-group effect of basketball skills in the CG

Content	Stage	*N*	*M*	SD	*t*	*p*
Shoot	Pre-expt	45	2.31	1.25	8.376	0.000
	Post-expt	45	4.04	1.47		
Layup	Pre-expt	45	0.80	1.19	9.142	0.000
	Post-expt	45	2.38	0.77		

Results of Intra-group Effects of Learning Motivation in the CG. The statistical results of the level of learning motivation in the CG (Table 4) show that the students' internal motivation, identification adjustment, and amotivation levels have significantly decreased, and the change of control motivation is not obvious after the experimental intervention.

Table 4 Intra-group effect of learning motivation in the CG

Content	Stage	*N*	*M*	SD	*t*	*p*
Intrinsic motivation	Pre-expt	45	17.72	1.96	5.220	0.000
	Post-expt	44	16.27	2.29		
Identified regulation	Pre-expt	45	17.04	2.51	1.988	0.053
	Post-expt	44	16.18	2.07		
Introjected regulation	Pre-expt	45	15.59	2.35	1.219	0.230
	Post-expt	44	15.86	2.08		
External regulation	Pre-expt	45	13.70	2.75	1.675	0.101
	Post-expt	44	14.38	1.89		
Amotivation	Pre-expt	45	9.59	2.52	3.620	0.001
	Post-expt	44	8.04	2.55		

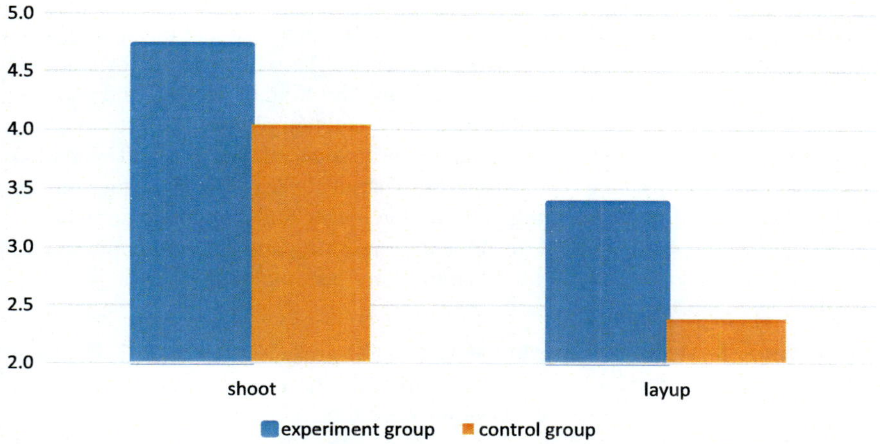

Fig. 1 Comparison of basketball skill of EP and CP after the experiment

3.3 Differences in Basketball Skills and Learning Motivation Between EG and CG

Differences in Basketball Skills Between EG and CG. Comparison of the basketball skill level of the two groups after the experimental intervention (Fig. 1) showed that the improvement of the basketball skill level of EP was significantly higher than that of CP. The score of EP in shooting and layup is significantly higher than that of CP ($p < 0.01$), indicating that compared with the traditional teaching model, the blended teaching model has more advantages in improving the level of basketball skills.

In order to compare and analyze the actual teaching effect of the two different teaching modes applied in physical education, we directly compared the learning results (technical test scores) of the two groups of students after the experimental intervention. The results show that although both the blended teaching model and the traditional teaching model have significantly improved the students' basketball skills, compared with the traditional teaching model, the blended teaching model has brought a greater improvement to the students in teaching basic basketball skills. The reason may be that because the students in the experimental group learned and mastered the physical education content in advance according to the online learning materials before class, they were more purposeful and pertinent when they returned to the classroom for offline learning and practice. This new teaching model stimulates students' interest in the course and encourages students to participate more in basketball learning [16, 17]. This model of introducing online teaching into the teaching process provides convenience for students' preview and review and further improves students' mastery of sports techniques [18]. This new model allows students to learn a wider range of basketball knowledge and plays an indispensable role in the learning and training of basketball skills [19]. In the traditional physical education classroom mode, although teachers put forward the requirements for students to preview before class and consolidate after class, this kind of preview and consolidation is not supervised and guided. Students only rely on physical education materials, the effect of learning and practice cannot be guaranteed, and physical education teachers cannot know the validity and progress of students' learning. Judging from the learning outcomes after the teaching intervention in this study, although both modes of teaching have a positive effect on students' mastering and improving basketball skills, the mixed teaching mode has brought more benefits to students in teaching basic basketball skills.

Differences in Learning Motivation Between EG and CG. The statistical results of the two groups' learning motivation levels after the experimental intervention (Fig. 2) showed that the scores of EG in intrinsic motivation, identified regulation, and introjected regulation were significantly higher than those of CG ($p < 0.01$), and the scores of amotivation were significantly lower than those of CG ($p < 0.01$). The results of the study show that, compared to the traditional teaching model, the blended teaching model has a significant positive effect on maintaining the motivation of each student.

Because this study used basketball elective courses as the experimental research content, most of the students who take the course have high learning motivation, but different teaching modes have different effects on the students' original learning motivation. The learning motivation of the students in the experimental group showed a trend of maintaining and increasing after the experimental intervention, while the learning motivation of the students in the control group showed a trend of maintaining and decreasing after the experimental intervention. After the experimental

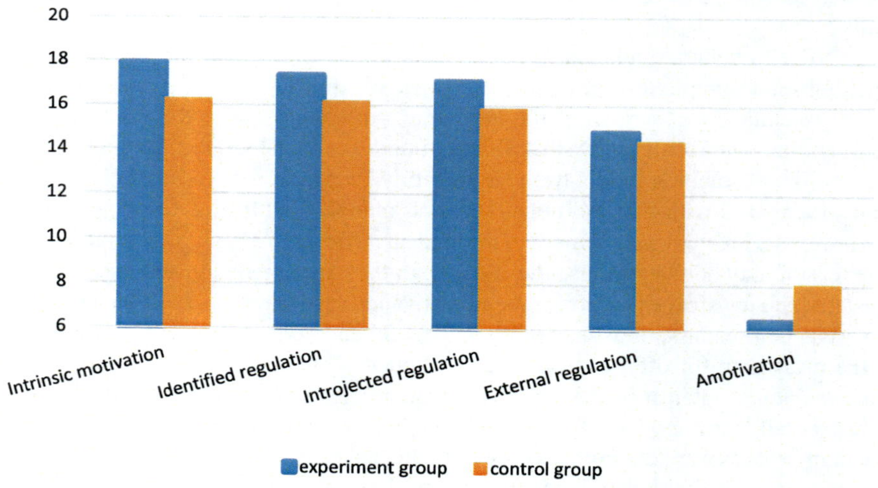

Fig. 2 Comparison of learning motivation of EG and CG after the experiment

intervention, there was a significant difference in learning motivation between the experimental group and the control group. The results of the study show that in the process of experiencing the diversified learning of the blended learning model, the students in the experimental group feel more convenience and freedom of learning than the pre-course cognition and have more experience with the teacher in the classroom, and not only having an efficient hook communication, but also feel the positive reform of the physical education classroom. In the new blended teaching model, students' enthusiasm for theoretical study, skill practice, and after-school review has been improved. In a harmonious, equal, and democratic teaching environment created by the new blended teaching model, students can provide more targeted guidance to teachers as soon as they encounter difficulties, or they can conduct repeated learning on the learning platform after class. In this teaching mode, teachers play a more guiding and correcting role in the classroom, which not only improves the efficiency of teaching but also highlights the main role of students in the learning process [20]. This means that the blended learning model rebuilds the relationship between teachers and students in the classroom, and teachers play a more guiding and correcting role in the classroom, which not only improves the efficiency of teaching, but also wins the respect of students. The traditional teaching mode pays too much attention to the overall teaching content and relatedness and the teachers in the classroom are not targeted enough to solve the problems. After the class, students are difficult to solve the learning problems they encounter in a timely and effective manner through other means due to the time influence of other courses. It reduces students' experience of the course and directly affects the maintenance of students' learning motivation.

4 Conclusion

Through the comparison of the teaching effects of different teaching modes, this research explores the actual learning effects of the blended teaching mode on students and the characteristics of internal learning dynamics and draws the following preliminary conclusions:

Both the blended learning model and the traditional teaching model have produced good teaching effects in the mastery and improvement of basketball skills in college physical education classrooms. Compared with the traditional teaching model, the blended teaching model has brought the basic skills of basketball to students' greater improvement. The blended learning model has obvious positive effects on the maintenance of students' autonomous learning motivation and the improvement of controlled learning motivation. The traditional teaching mode has a positive effect on maintaining students' controlled learning motivation, while it has a negative effect on students' autonomous learning motivation.

The results of this study support the research hypothesis that the blended learning model is more conducive to improving the learning effect and learning motivation of basketball sports skills than the traditional teaching model.

References

1. Chiang TH-C, Yang SJH, Yin C (2018) Effects of a mobile flipped classroom strategy and gender issues on students' learning achievement in three-on-three basketball competition teaching
2. Hwang G-J, Lai C-L, Wang S-Y (2015) Seamless flipped learning: a mobile technology-enhanced flipped classroom with effective learning strategies. J Comput Educ 2:449–473
3. Ryan RM, Deci EL (2000) Intrinsic and extrinsic motivations: classic definitions and new directions. Contemp Educ Psychol 25:54–67
4. Vansteenkiste M, Lens W, Deci EL (2006) Intrinsic versus extrinsic goal contents in self-determination theory: another look at the quality of academic motivation. Educ Psychol 41:19–31
5. Taylor IM, Ntoumanis N, Standage M, Spray CM (2010) Motivational predictors of physical education students' effort, exercise intentions, and leisure-time physical activity: a multilevel linear growth analysis. J Sport Exerc Psychol 32:99–120
6. Haerens L, Kirk D, Cardon G, De Bourdeaudhuij I, Vansteenkiste M (2010) Motivational profiles for secondary school physical education and its relationship to the adoption of a physically active lifestyle among university students. Eur Phys Educ Rev 16:117–139
7. Standage M, Duda JL, Pensgaard AM (2005) The effect of competitive outcome and task-involving, ego-involving, and cooperative structures on the psychological well-being of individuals engaged in a co-ordination task: a self-determination approach. Motiv Emot 29:41–68
8. Standage M, Duda JL, Ntoumanis N (2003) A model of contextual motivation in physical education: using constructs from self-determination and achievement goal theories to predict physical activity intentions. J Educ Psychol 95:97
9. Standage M, Sebire SJ, Loney T (2008) Does exercise motivation predict engagement in objectively assessed bouts of moderate-intensity exercise?: A self-determination theory perspective. J Sport Exerc Psychol 30:337–352
10. Hill K (2014) Using mobile devices for motor-learning laboratory exercises. J Phys Educ Recreat Danc 85:20–26
11. Aelterman N, Vansteenkiste M, Van Keer H, Van den Berghe L, De Meyer J, Haerens L (2012) Students' objectively measured physical activity levels and engagement as a function of between-class and between-student differences in motivation toward physical education. J Sport Exerc Psychol 34:457–480
12. Li G (2015) A study on female university students' physical education elective course and motivation and body self-esteem. Suzhou Soochow Univ, 15–16
13. Liu Z (2019) Research on the correlation between sports participation motivation and willpower among middle school students in Suzhou. Suzhou University
14. Qin D Indigenization of the Motivation Scale to Regulate Sports Behaviour. Suzhou University
15. Li G (2015) Investigation and research on the motivation and body self-esteem of female college students in sports option courses. Suzhou University
16. Jian C (2019) Experimental research on the application of flipped classroom teaching mode in middle school basketball teaching under the background of campus basketball. Shanghai Normal University
17. An F, Wang X, Chen J (2019) Design and application of mobile blended teaching mode. High Financ Educ Res 22:36–43
18. Zhao P (2011) "B-learning" learning theory and experimental research in sports. Northeast Normal University
19. Wang Y (2017) The application value of flipped classroom teaching mode in youth basketball teaching. Youth Sport, 91–92
20. Xiao Hong D, Ping R (2015) The research on learning attitudes theoretical courses or technical courses of P.E major students. J Hengyang Norm Univ 36:148–152

The Effects of PETTLEP Imagery Intervention on Learning of Badminton Skill for Beginners

Apanchanit Siripatt and Suebsai Boonveerabut

Abstract Badminton serves (BS) and return badminton serves (RBS) are very important for badminton sport, new player hard to control both skills. BS is a closed skill that takes place in a structured and static environment to serve the shuttlecock to serve area. While, RBS is an open skill that takes place in a dynamic and changing environment. Player trying to move around the court for return serve and multiple other variables that could affect them. This study explores the effects of PETTLEP imagery for learning badminton skill. Eighty university students who no experience in sport imagery and badminton competition were equally divided ($n = 20$) into four groups (G1; BS with PETTLEP training, G2; BS training only, G3; RBS with PETTLEP, and G4; RBS training). The data collecting from BS and RBS accuracy test, prior, and post-test on 4th and 8th week of training. Mean, SD, t-test, and one-way ANOVA with repeated measures, Bonferroni were used to analyze the data at significant .05 level. The results reveal that mean score of G1 and G2 after 8th wk. was increase significantly when compared with after the 4th wk. and before training ($p < 0.001$). There were no significant difference between G1 and G2 of BS group and between G3 and G4 of RBS group. However, the group of practiced badminton skill and PETTLEP imagery training (G1, G3) showed accuracy score better than the group of practiced badminton skill only (G2, G4). Therefore, PETTLEP imagery training showed a tendency to develop badminton skill for the beginner.

Keywords PETTLEP imagery · Badminton skill · Beginner

A. Siripatt (✉)
Faculty of Health Science Technology, Human Kinetics and Health School, HRH Princess Chulabhorn College of Medical Science, ChulabhornRoyalAcademy, Kamphaeng Phet 6 Road, Bangkok 10210, Thailand
e-mail: apanchanit.sir@cra.ac.th

S. Boonveerabut
Independent Scholar, Bangkok, Thailand

G. Kuan et al. (eds.), *Advancing Sports and Exercise via Innovation*, Lecture Notes in Bioengineering, https://doi.org/10.1007/978-981-19-8159-3_13

1 Introduction

Badminton is a popular sport, varied from leisure activity to pursuit excellent in national and international competitions. Many specific skills are important in badminton such as badminton service (BS) and returning badminton service skill (RBS). Sport skills can be divided into the categories "open skill" and "closed skill". In a badminton game, BS is a close skill. It is very important to score. Moreover, rally point is currently being used, which is different from the past games. A closed skill is a skill that takes place in a structured and static environment. Now picture that same badminton player serving the shuttlecock to serve area. They are distraction free and are given the opportunity to take an uninterrupted serve at the court. Good serve means accuracy of serving, which increases chances to score and an advantage in the competition. On the other hand, a serving may increase stress and anxiety due to pressure from making good serve in the game. Concentration and control as well as serving accuracy skill are needed.

The badminton skill is returning badminton service that increases a chance to score and get an advantage in the competition. Moreover, RBS is an open skill that takes place in a dynamic and changing environment. Picture a badminton player trying to move around the court for return serve, a crowd cheering, and multiple other variables that could affect the player. New badminton players have a variety of choices in returning serve. If there are many choices, the opponent decides to play hard. Badminton player must have a great reactions and accuracy of returning service to certain spots of the court. If they return serve well, it will be scored or enhance the chance to score points in the next stroke. Good return can force the server into a defensive shot and give the receiver a better chance of winning the point. Thus, the return serve is very important skill to score.

Sport psychology is the study about the process of creating and enhancing motivation for coaches and athletes to make a commitment for training and to develop the highest skill that can be used in competition to its fullest potential [1]. Many psychological skill techniques are used to enhance the ability of athletes. One of the popular ways is "sport imagery".

In sport psychology, imagery is a simulation of the moves that athletes experienced from the past and/or create new positive pictures or moves in their mind for the future. Imagery involves visual, auditory, olfactory, tactile, and kinesthetic senses as well as emotional sensibility [2]. It can be cognitive and motivational functions of imagery, both in general and specific contents. The imaging ability is based on the vividness and controllability of the pictures. It has been studied and compared between successful and less successful athletes and between elite and non-elite athletes. Imagery is the use of all the senses in order to create or recreate motion pictures in the mind. This allows a better understanding of the events. Imagery training was studied in many forms. Imagery starts with body relaxation then uses all senses to visualize or see things in an athlete's mind and feel how the body moves. Next, imagine someone such as an expert in that field. Try to imagine oneself as being the model.

In 2001, Holmes and Collins one's propose the PETTLEP imagery model [3]. It stands for the seven components of imagery, as follows: physical environment task timing learning emotion and perspective. This model has been studied abroad continuously for more than 15 years in a variety of sports. The PETTLEP model is a positive effect to practice individual sports, team sports, contact sports, and non-contact sports as it allows players to create their own movement based on sport skills clearly, to enhance physical skills, perceptual skill and psychological skills [4]. Therefore, we chose PETTLEP imaginary to test its effect on a group of badminton players at a university because PETTLEP enumerated categories, and the direction of the imagery is clear and easier for the player to follows. We hope that the PETTLEP was developed to enhance accuracy in BS and RBS for the player to concentrate and control their serve and return serve well.

2 Methodology

2.1 Participants

The participants were 80 male students (18–20 yrs.) from Srinakharinwirot University. They never competed in badminton competition and never practiced sports imagery. They volunteered and signed a research consent form. All participants selected and determined by simple random and divided into two groups (badminton serve and return serve).

2.2 Equipment

1. Badminton serve skills training program, returning badminton serve skills training program and scoring mats developed by the team of researcher and approved by three experts.
2. PETTLEP imagery model training program developed by the team of researcher and approved by three experts.

2.3 Procedure

2.3.1 First Step: Badminton Serve Skills

Forty students participated in badminton serve skill (BS) groups. They were asked to serve 20 times, before categorized into 2 matching groups. Group1 (G1; $n = 20$) practiced service skill with PETTLEP imagery. Group 2 (G2; $n = 20$) practiced service skill without imagery. Rating of badminton serving accuracy is 5 to 0. We

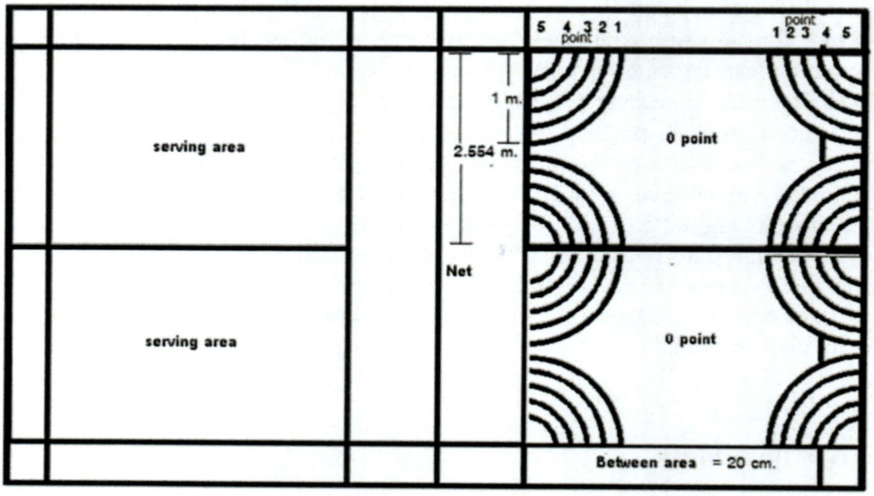

Fig. 1 Area of serving point in badminton court

used the guideline-scoring mat as shown in Fig. 1. For short ball, the nearer you are to the net, the higher the score will be. For serving a long ball, the farther you are from the net, the higher your score will be. The participants had to do 20 short balls and 20 long balls.

2.3.2 Second Step: Return Badminton Serves Skills

Forty students participated in return badminton serves skills (RBS) group. They were assigned to receive and return badminton services (RBS) 20 times by the research assistant, who is a badminton expert. Each RBS accuracy scores were collected when the shuttle cocks landed on the designated zones 1–5 points preferably either short or long services. These score zones were categorized by difficulty level in the court of short and long services. We used the guideline-scoring mat shown in Fig. 2, which shows different lines from Fig. 1. However, the concept is the same. RBS short ball started from the nearest to the net get 5–1 points, and RBS long ball started from to nearest based line to the inner area get 5–1 points consecutively.

They were divided into 2 groups: experimental and control. The experimental group (G3; $n = 20$) practices RBS and PETTLEP imagery model training. The control group (G4; $n = 20$) practices RBS only. It was an 8 week program, 3 times a week and 60 minutes per session. Between RBS testing, the research assistant was act to returning the shuttlecock for set under pressure. Both groups were tested their times: before training, at week 4 and at week 8. Mean, standard deviation, t-test, and one-way ANOVA with repeated measures were used to test the differences at level .05, and then, Bonferroni was conducted to identify the differences.

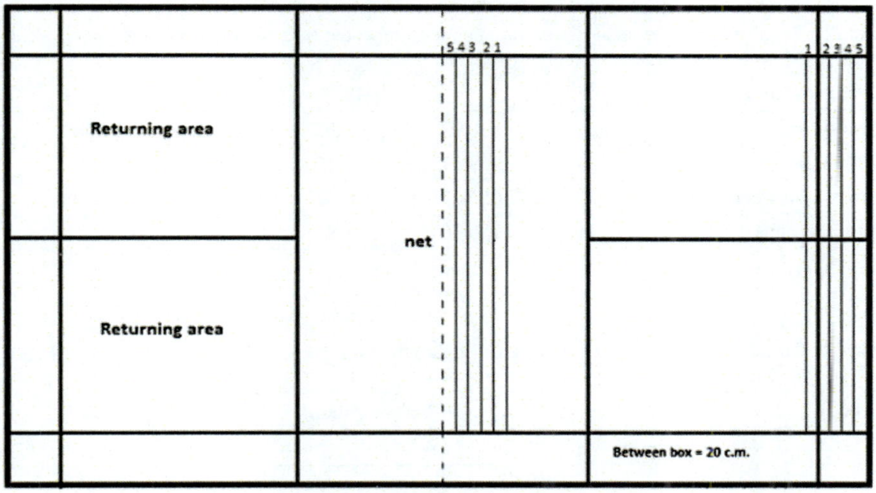

Fig. 2 Designated areas and scores of RBS, short, and long services

2.4 Results and Discussion

2.4.1 Badminton Service Skills

The result revealed that the mean score after the 8th week was significantly increased when compared with after the 4th week and before training for both groups (Group 1: (Before training, after the 4th, after the 8th week) $\bar{x} = 9.60 \pm 4.91, 19.55 \pm 7.80, 22.55 \pm 6.91$, Group2: $\bar{x} = 10.40 \pm 5.12, 18.15 \pm 6.54, 20.10 \pm 6.47$) (Fig. 3).

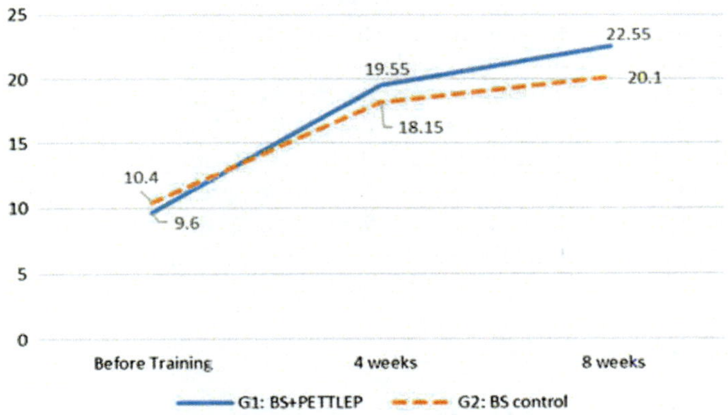

Fig. 3 Mean score of both groups in badminton serve skill (BS)

Table 1 Result of one-way analysis of variance with repeated measures in badminton serve skill score

Source of variance	df	SS	MS	F	p
Group 1Between groups	2	1838.03	919.02	129.36	0.000[c]
Within groups	38	269.97	7.10		
Group 2Between groups	2	1053.03	526.52	133.12	0.000[c]
Within groups	38	150.30	3.96		

[c] $p < 0.001$

Table 2 Bonferroni method was conducted compare the average pairwise within BS groups 1 and 2

Group	Time	$\bar{x}$	Before training	4 wk	8 wk
	Before training	9.60	–	9.95[a]	12.95[a]
Group 1	4 wk	19.55		–	3.00[a]
	8 wk	22.55			–
	before	10.40	–	7.75[a]	9.70[a]
Group 2	4 wk	18.15		–	1.95[a]
	8 wk	20.10			–

[a] $p < 0.05$

The result of one-way analysis of variance with repeated measures showed that differences within groups are significant at the 0.001 level (Table 1). The Bonferroni method was conducted to compare the average pairwise within groups 1 and 2 between before training and after the 4th week (G1:G2 = 9.95, 7.75), before training and after the 8th week (12.95, 9.70), and after the 4th week and 8th week (3.00, 1.95), the results were found the significant difference at the 0.05 level. The comparison between both groups showed that the mean score before training and after the 8th week was no significant difference (Table 2).

The comparison between both groups showed that there is no significant difference in the mean score before training and after the 8th week. In after, the 8th week result of BS accuracy is better than before training because both groups were BS accuracy trained continuously develop their skill. However, group 1 (BS with PETTLEP imagery model training) has better result than group 2 (BS only). This shows that PETTLEP imagery model training can be used to develop the skills of the athletes.

2.4.2 Returning Badminton Service

The result revealed that the mean score after the 8th week was significantly increased when compared with after the 4th week and before training for both groups (Group

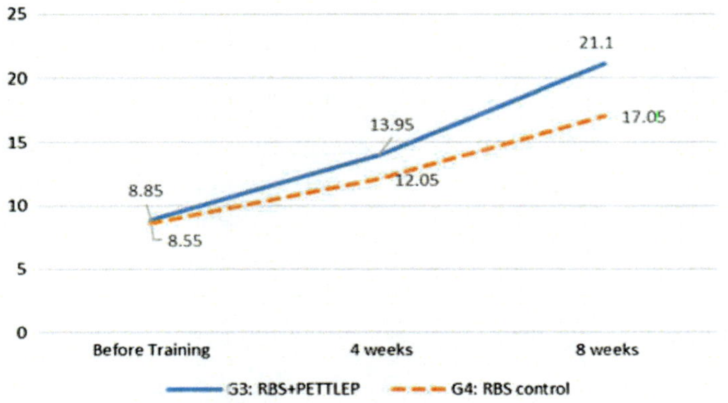

Fig. 4 Graph showed the mean score of RBS group

Table 3 Result of one-way analysis of variance with repeated measures in RBS groups 3 and 4

Source of variance	df	SS	MS	*F*	*p*
Group 3Between groups	2	1514.63	757.31	26.61	>0.001[c]
Within groups	38	1081.36	28.45		
Group 4Between groups	2	813.70	406.85	70.17	>0.001[c]
Within groups	38	220.30	5.79		

[c] $p < 0.001$

3: $\bar{x} \pm SD = 8.85 \pm 4.88, 13.95 \pm 4.97, 21.10 \pm 9.81$, Group 4: $\bar{x} \pm SD = 8.55 \pm 4.63, 12.05 \pm 4.37, 17.05 \pm 4.91$) (Fig. 4).

There was a significant difference within group, both experimental and control group at 0.001 level. Bonferroni method was conducted, comparison by Bonferoni post hoc test (before- after the 4th week, before-after the 8th week, and after the 4th week–8th week of training) revealed that the mean of accuracy returning badminton in group 3 was 5.10, 12.25, and 7.15, respectively, and in group 4 was 3.50, 8.95, and 5.45, respectively. The results were found the significant difference at the 0.05 level. There was no significant different between both groups between before and after the 8th week program participating (Tables 3 and 4).

3 Conclusion

In this study, we found that imagery helps enhance BS and RBS accuracy for badminton beginner. Experimental group who had PETTLEP imagery training with sport skill training shows high score more than group who training only sport skill. It is very easy to train specific skill because the distribution of the imaginary category

Table 4 Bonferroni method was conducted compare the average pairwise within RBS groups 3 and 4

Group	Time	$\bar{x}$	Before training	4 wk	8 wk
	Before training	8.85	–	5.10[a]	12.25[a]
Group 3	4 wk	13.95		–	7.15[a]
	8 wk	21.10			–
	before	8.55	–	3.50[a]	8.95[a]
Group 4	4 wk	12.05		–	5.45[a]
	8 wk	17.05			–

[a] $p < 0.05$

to see clearly help athletes understand easily. PETTLEP imagery model training combined with skills training allows athletes to train and learn new experiences in their mind. Moreover, they can concentrate and control their skill although they got more pressure. Therefore, it supports the training program that uses imagery to help new athletes develop better skills because it allows them feel like training twice in the meantime, review the skill has been trained and control their skill under pressure.

BS and RBS with PETTLEP imagery model training showed a tendency to perform better than BS and RBS only. Imagery is not only limited to still images. It involves a detailed motion image. Imagery is combination of perception and action similar to matching the complexity of environmental and motor skills. It is seen that the athlete needs ongoing imagery training to achieve proficiency [5]. PETTLEP imagery has an effect on the body because the mind constantly trains so that the body automatically reacts in an actual competition [6]. They suggest that PETTLEP imagery may be a viable alternative to physical practice in situations where further physical practice is not possible or advisable, such as through fatigue, boredom, or injury. Additionally, it should be remembered that a combination of physical practice and imagery is likely to prove more effective than either method on its own [6]. Therefore, we would recommend using physical practice complemented as much as possible with PETTLEP imagery for better effects.

References

1. Boonveerabut S (2009) Sport psychology. College of physical education Chonburi, Chon Buri
2. Weinberg R, Gould D (2015) Foundations of sport and exercise psychology, 6th edn. Human Kinetics, Champaign, IL
3. Holmes PS, Collins DJ (2001) The PETTLEP approach to motor imagery: A functional equivalence model for sport psychologists. J Appl Sport Psychol 13(1):60–83. https://doi.org/10.1080/10413200109339004
4. Siripatt A (2014) The PETTLEP model for sport. J Sport Sci Technol 15(2):1–12
5. Weinberg R (2008) Does imagery work? Effects on performance and mental skills. J Imag Res Sport Phys 3(1):1–21. https://doi.org/10.2202/1932-0191.1025

6. Simons J (2000) Doing imagery in the field. Human kinetics, Illinois
7. Wright CJ, Smith D (2009) The effect of PETTLEP imagery on strength performance. Int J Sport and Exerc Psychol 7(1):18–31. https://doi.org/10.1080/1612197X.2009.9671890

Validity, Reliability and Development of Soccer-Specific Battery Test

Ahmad Nadzmi, Ahmad Bisyri Husin Musawi Maliki, Jorrye Jakiwa, and Rabiu Muazu Musa

Abstract Soccer is also known as a complex sport as it involves physical, physiological and technical demand to be fit. Many professional soccer clubs actively investigate the factor to achieve high performance in soccer. Thus, the aim of this study was to establish the validity, reliability and development of soccer-specific battery tests. A total of 223 young male soccer players from several soccer positions between 13 and 17 years old participated in this study. However, 59 young male soccer players chosen randomly were performed in reliability test and accomplished 2 trials tests (run with ball, long passing, short passing, shooting to the right top corner (dead ball), shooting to the left top corner (dead ball), shooting from a pass (foot), heading mid-post and side-post). The data were tested using ANOVA to distinguish the conceivable systematic bias between 2 attempts of trial in each test, and also discriminant analysis (DA) was used to discriminate between elite and non-elite players from selected battery tests. The evaluated test revealed high intraclass correlation coefficients on run with ball (intraclass correlation coefficient > 0.90), meanwhile for other tests only at moderate levels (intraclass correlation coefficients, > 40) and small within-individual variation (coefficients variation, > 0.16%). DA successfully discriminates between elite and non-elite soccer players with seven significant parameters with elite tend to have greater aerobic functional capacity, more skilful and matured compared to non-elite ($p < 0.05$). Establishment of specific skill-related performance with moderate to high validity and reliability will ensure the distinction between elite and non-elite soccer players.

Keywords Intraclass correlation coefficient · Coefficient variation · Reliability test · Soccer

---

A. Nadzmi · A. B. H. M. Maliki (✉) · J. Jakiwa (✉)
Defence Fitness Academy, National Defence University of Malaysia, Kem Sungai Besi, Kuala Lumpur, Malaysia
e-mail: bisyri@upnm.edu.my

J. Jakiwa
e-mail: jorrye@upnm.edu.my

R. M. Musa
Center for Fundamental and Continuing Education, Universiti Malaysia Terengganu, Kuala Nerus, Terengganu, Malaysia

G. Kuan et al. (eds.), *Advancing Sports and Exercise via Innovation*, Lecture Notes in Bioengineering, https://doi.org/10.1007/978-981-19-8159-3_14

1 Introduction

Soccer also well-known as complex sport as it involves physical and physiological request to be fit in terms of aerobic. Accordingly, in the recent years, many professional soccer clubs increasingly investigate the factor of high performance, speed, agility and ball dribbling which have been shown in studies to be the greatest indicators of good soccer players and should be fostered from a young age [1]. Even, coaches are endlessly looking for the most efficient procedure for recognizing outstanding players to form an elite team [2]. As a result, the development of a soccer-specific battery has become essential for players to have the necessary levels of fitness (aerobic, speed, strength and agility) to continue competing [3–5].

Conventionally, the player selection process uses a specific test to assess player's criteria performance; otherwise, a mistaken selection can jeopardize a football team the championship and even miss millions of dollars if the player fails to attain the team's opportunities [5, 6]. Undeniably, this development battery test is associated to validity and reliability whereby it evaluates the effectiveness physical test during performance and classifies the advantages and disadvantages of each physical test, and it is verified that the test is practically applicable [7]. Hence, the main purpose of using validity and reliability is to identify the level of which protocol measures a hypothetical construct before we construct the hypothetical such as sport performance or a comparison of two different groups and to create the consistency of performance when carried out on more than occasion [8]. It allows for supervision as well as reference regardless of the long- or short-term and in future to those in need especially coaches and players to overcome numerous difficulties as well as strategic improvement in this era to attain goal [8].

Similarly, the strength and conditioning professional working with soccer team will also be able to manage a time-efficient, valid and reliable fitness test. Thus, the result can guide the strength and conditioning professionals and technical coaches in program designed, leading to more effective and efficient goal accomplishment [9]. Thus, the aim of this study is to classify the validity, reliability and the development of the specific battery test.

2 Methods

2.1 Player

The participants tested were 223 male adolescent athletes aged around 13 to 17 years old. However, on purpose of establishing the reliability of the soccer battery test, 59 players were randomly chosen. All the players were enlisted from all around Sport Schools in Malaysia that focused in soccer to progress more in their performance accomplishment since young age. Coaches and managers were informed ahead about the purpose of the research. Approvals were obtained with all the players signing consent forms.

2.2 Anthropometric, Growth and Maturation Test

In this test, researchers has measured the physical component whereby involved measurement specifically weight, height, sitting height and arm span. Height was assessed by using a wall mounted wooden stadiometer to the nearest 0.5 cm, while the weight was evaluated on a weight scale of kg [10, 11]. Meanwhile, arm span was quantified from fingertip to fingertip, whereas the back position is flat on the wall and arm stretched with palm facing the tester [12]. The equipment needed included of the measure tape at wall, measured in centimeter (cm) [13]. Skinfold was used to measure the triceps, biceps, subscapular and suprailiac to the nearest 0.01 mm, meanwhile the medial upper arm circumstances (MUAC) and calf circumstances (cc) were conducted by using measuring tape. All the measurements were performed based on the ISAK protocol, and the measurements were collected twice, and the mean score for the final score is recorded for additional analysis. Meanwhile, maturation was evaluated by applying tanner scale of maturity as suggested by the previous researchers and implementing the procedure and into current study [14].

2.3 Soccer Fitness Test

The test was completed according to the optional method for physical fitness tests [15]. The multistage 20 m shuttle run test was executed to attain the participant's maximal oxygen uptake while performing 5, 10 and 20 m. Each of the player kept running for how long they could stand in whatever length of time until could no longer keep pace with the velocity of the tape [16]. The result of VO2 max for every participant was stated by accomplished through checking the last level and ended shuttle number at the time when the participant voluntarily resigned from the test [17]. Although, drills of the participants might affect their scores, and it is still a valid test in evaluating maximum oxygen uptake and can be made in substantially many participants minimalizing expenses and time. Additionally, flexibility test is to measure the lower back and hamstrings through the sit and reach test. Meanwhile, agility was determined by the 505-agility test, and the purpose of the agility test is to identify the ability of speediness player to change the body position in the minimum requirement time.

2.4 Soccer Technical Skill Test

Running with the ball is one of technical skill test which is to measure the consistency of the player running while dribbling the ball from the starting point. The fewer the ball touches the foot the faster the ball goes, however, the player should have good control of the ball. Extensive study shows that running with the ball can increase the physiological stress rather to normal running [18]. Long pass test evaluated passing accurateness and shooting strength in long distance. The player delivered the ball from starting line to other players in long range scale according to the designated time whereby to identify their accurateness of the ball to reach to opposite player. Long passing is one of the most important and highly relevant technic and tactic skills needed in soccer whereas can determine the characteristics of the desired players [19]. Meanwhile, short-passing test inquires an approximation of the accuracy and the quickness the skill demonstrated in passing and only consider the ball that reaches the player within the specified time. In addition, short-passing capability is considered one of the most applicable skills for soccer player [20]. Although, the player will shoot from passing players and is demanded to score as many goals as possible where other player should to pass the ball to another player and that player needs to shoot to goal, while the ball is moving. However, in shooting to the right and to the left, test assesses a player's efficiency and speediness to score goals from different angles although basically this test was used to evaluate the accuracy and coordination of shooting ability [21]. The tester only considers goals that are designated time. Heading is one of the skills needed in soccer game that could be applied while performing in a game. This test determined the accuracy when the player head the ball.

2.5 Statistical Analysis

The means and standard deviations of all data are shown. Therefore, 95% confidence intervals are used to display the accuracy of population estimates (CIs). To identify any potential systematic bias between the two trial efforts in each test, all the data were examined using a one-way analysis of variance (ANOVA) test. Additionally, the statistical threshold was set at p 0.05. An intraclass correlation coefficient (ICC) was used as a temporary way to describe qualified variability [20]. Within-individual variability was fully captured by both coefficient variations and typical measurement error [21]. SPSS for Windows version 12.0 was used to conduct the statistical analysis (SPSS, Inc, Chicago, IL).

2.6 *Discriminant Analysis (DA)*

This test works to control the variables which divided into two or more clearly joined group or cluster. DA has been used in this study to determine whether the groups differ in variables and use those variables to predict group membership [22]. DA has been run for testing validation and separating another variable according to a significant cluster that contributes to different groups. Two groups were designated and attained from discriminant analysis whereby two sampling identified as elite and non-elite. The physical performance DA included in the raw data by using standard, backward stepwise and forward stepwise methods [23]. Also, in the forward stepwise method, variables are calculated in step by step starting with the very significant variable until no significant changes were obtained [18]. Meantime, in backward stepwise method, removed the variables step by step starting from less significant to no significant.

$$f(G_i) = k_i + \sum_{j=1}^{n} w_{ij} P_{ij}$$

where i is the number of groups (G), ki is the constant inherent to each group, n is the quantity of parameters utilized to categorize a set of data into a certain group, and wj is the mass coefficient assigned by DF analysis (DFA) to a given parameter (pj).

3 Result

Table 1 represents the result of all evaluated test through the descriptive statistics calculated across all 59 subjects for testing the validity and reliability, while 223 were tested in discriminant analysis of the athletes. Referring to the Table 1, it shows the minimum, maximum, mean, range and standard deviant of athletes in each test that has been carried out.

Table 1 Descriptive statistic of reliability soccer battery test

Statistic	Min	Max	Mean	SD	Range
Run w/ball (point)	1.93	7.58	9.23	1.39	5.65
Long passing (point)	0.00	12.00	0.88	1.21	12.00
Short passing (point)	1.00	15.00	4.85	1.92	14.00
Shooting TR (dead ball)	0.00	18.00	1.32	1.94	18.00
Shooting TL (dead ball)	0.00	13.00	1.93	2.27	13.00
Shooting From a pass (foot)	0.00	26.00	1.03	1.85	26.00
Heading (md_post) (point)	0.00	18.00	6.08	4.09	18.00
Heading (sd_post) (point)	0.00	18.00	3.66	3.64	18.00

Table 2 ANOVA from two consecutive trials

Variables	F value	Sig.
Runw/ball	8.05	0.006
LongPass	0.000	1.000
ShortPass	0.141	0.709
ShootingTR (DeadBall)	0.000	1.000
ShootingTL (DeadBall)	1.91	0.172
ShootingfromPass (Foot)	3.51	0.066
Heading (mdpost)	0.023	0.88
Heading (sdpost)	0.74	0.34

Table 3 Reliability statistics calculated from two consecutive trials (M ± SD)

Variables	Trial 1	Trial 2	ICC (CI)	TEM	CV/%
Runw/ball	4.65(0.70)	4.58(0.70)	0.98(0.97–0.99)	0.20	0.15
LongPass	0.44(0.75)	0.44(0.68)	0.57(0.32–0.76)	0.79	1.37
ShortPass	2.46(1.15)	2.39(1.22)	0.48(0.12–0.69)	1.39	0.40
ShootingTR	0.66(1.17)	0.66(1.17)	0.55(0.25–0.73)	1.30	1.47
ShootingTL	0.85(1.30)	1.08(1.33)	0.66(0.44–0.80)	1.32	1.18
ShootingfromPass (foot)	0.63(0.93)	0.89(1.19)	0.64(0.39–0.79)	1.11	1.21
Heading (mdpost)	3.12(2.19)	3.07(2.60)	0.60(0.33–0.76)	2.57	0.67
Heading (sdpost)	1.68(2.21)	1.98(2.34)	0.44(0.59–0.68)	2.73	1.00

Once applied on each test for the two trials, the ANOVAs discovered are 1.00–0.88 $(p > 0.05)$ which mean the result revealed the differences among two trials to be small and unstable meanwhile exception for run with ball test which is highly significance 0.006 $(p > 0.05)$ in Table 2.

Table 3 is focused on the reliability result whereby emphasize the performance averaged across the players for each recorded trial, together with the corresponding measures of reliability. Furthermore, the reliability as corresponding as ICC, for most of the tests, was just average. The only exception was the run with ball with the higher score (ICC, 0.98) in contrast with the result on behalf of coefficients variation (CV) which is low as expected of less than 1.49%.

In Table 4, the table showed the result of discriminant analysis on two different groups of players (eltie and non-elite) and revealed the sampling groups in misclassified mode has the total of 18 independent variables with 91.03%. Meanwhile, predefine mode has 18 independent variables and 93.72%. Other than that backward stepwise has 12 independent variables shown the highest amount with 96.41% while for forward stepwise (7 independent variables) with slightly differences amount between backward stepwise which is 95.96%. Box and whisker plot shown that there are 7 youth soccer performance in Table 4 of forward stepwise which mean 94 athletes were reported as elite players and 129 player categorized as non-elite players. Figure 1 represents box and whisker plot according to discriminant forward stepwise.

Table 4 Discriminant analysis on two different groups of players

Sampling groups	% Correct	Group assigned by DA	
		Elite	Non-elite
Misclassified mode			
Elite	88.66	86	11
Non-elite	92.86	9	117
Total	91.03	95	128
Predefined mode			
Elite	93.68	89	6
Non-elite	93.75	8	120
Total	93.72	97	126
Backward stepwise (12 significant variables)			
Elite	95.79	91	4
Non-elite	96.88	4	124
Total	96.41	95	128
Forward stepwise (7 significant variables)			
Elite	94.74	90	5
Non-elite	96.88	4	124
Total	95.96	94	129

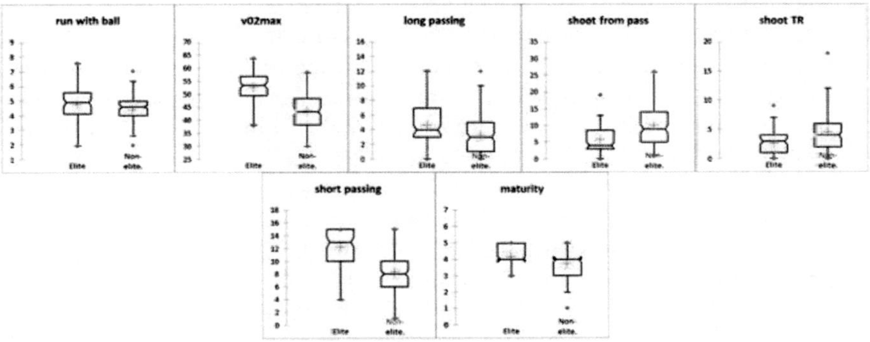

Fig. 1 Box and whisker plots based on forward stepwise

4 Discussion

The purpose of this study was to examine the validity and the reliability of the development soccer-specific battery test. Discussing about the results, only run with ball clearly shows that highly reliable significance test with (ICC, 0.98) and stated and compared to other tests [24]. Temporarily, heading md post in moderate level of reliable (ICC, 0.60) same situation with shooting from pass (foot) only score moderate level of reliable (ICC, 0.64) also shooting to the left (ICC, 0.66) and long pass too (ICC, 0.55). Meanwhile, heading sd post, shooting to the right and short past might not be sufficiently reliable with ICC > 0.55 and this situation mentioned in book whereby lack variability among the sampled size [25]. There are many factors influenced the reliability results such as the number of subjects, the number of completed trials, the participant's skills and what is the type of sample whether homogeneity or heterogeneity [22, 25].

The study involved 59 participants from all around Sport School in Malaysia. This number approximately associated to the number of players also carried out 2 trials in each of the tests. Normally, after getting used with the test procedure, therefore, it can be resolved that the test sample of players and the number of trials were relevant for the evaluation of the reliability of the designated soccer test; however unfortunately, the result was a disappointment [21]. It was not as expected as planned earlier. This reliable performance test with features such as one that has small changes in the mean, a small within-individual variation, and a high-test retest correlation did not produce result as expected as derived from the Table same as in Table 3 whereby the CV for all the test is shown as 0.16%–1.00% [26]. The tests assessed in this study established no significance differences among the intraclass correlation coefficient except for the run with ball test and individual variation (as expressed in CV) were low. Probability that the test is not suitable enough for the participants and perhaps the low number of participants in this test or possibly that the soccer player no enough experience and lack of confidence. This argued that agility tests could categorize elite soccer players from other papulation better than any other physical fitness test of power, flexibility or strength [1]. According to these studies, a current review of testing approaches applied in soccer determined that agility tests would be the best single indicator of overall soccer performance though none of agility test listed as one of the tests. Shooting skill also highlighted as one of the most vital roles in soccer especially when penalty kick [27]. The previous study on the picks of players through penalty kicks underlined mostly the horizontal direction whereby left, center or right of the kick or of the goalkeeper's jump [28]. Based on the result, all tests that involved with shooting are not strongly significance as stated.

Referring to discriminant analysis (DA) results, it divided into two groups such as elite and non-elite. Results displayed that elite groups dominate in run with ball and VO2max which mean elite group perform well in run with ball and VO2max without any problems. In the previous study also stated that elite group significantly correlates in VO2 max and endurance [29]. While, non-elite groups lead in short from pass and shooting TR with a high differential value compared to elite group.

However, elites and non-elites present the same results in short passing, long passing and maturity. This proved that non-elite groups can also perform well tests even though they do not have the skills compare to elite groups. After completion of this study, it is highly recommended to investigate more detailed as measured in this study but using experienced players such as professional players or elite because they have more experience in competition.

5 Conclusion

This study analyzed 8 soccer skills with 2 attempting of 59 male soccer athletes to examine the validity and reliability of soccer-specific skill. The analysis discovered that only run with ball is highly has significance difference (ICC > 0.98). even though the ICC value for other soccer skills that is shooting TL (to the left corner), shooting TR (to the right corner), heading (md post) and long passing are a bit lower than run with ball soccer skills (ICC > 0.58) but the value are still acceptable, showing that the other mentioned soccer skills earlier were still reliable to be applied. Surprisingly, for the heading (sd post) and short passing, the result is below expectation (ICC > 0.49). However, DA successfully discriminates 7 significant variables statistically between elite and non-elite which is elite group tend to be more expert in specific skill performance compare to non-elite. As conclusion, the result of this study does not attain high satisfaction as expected. Nevertheless, probability of getting strong significance can be enhanced with systematic procedure, more detail in exploration of the investigation and recommended that using experienced players, whereby they have more experience and more matured in facing competition.

References

1. Reilly T, Williams AM, Nevil A, Frank A (2000) A multidisciplinary approach to talent identification in soccer. J Sports Sci 18(9):695–702
2. Jones MA, Stratton G, Reilly T, Unnithan VB (2005) Biological risk indicators for recurrent non-specific low back pain in adolescents. British J Sports Med 39(3):137–140
3. Moreau D, Conway AR (2013) Cognitive enhancement: a comparative review of computerized and athletic training programs. Int Rev Sport Exerc Psychol 6(1):155–183
4. Malina RM, Ribeiro B, Aroso J, Cumming SP (2007) Characteristics of youth soccer players aged 13–15 years classified by skill level. British J Sports Med 41:5–290
5. Maliki AB, Abdullah MR, Juahir H, Abdullah F, Abdullah NA, Musa RM, Mat-Rasid SM, Adnan A, Kosni NA, Muhamad WS, Nasir NA (2018) A multilateral modelling of youth soccer performance index (YSPI). In: IOP conference series: materials science and engineering (vol 342, no 1, p 012057), IOP Publishing
6. Boon BH, Sierksma G (2003) Team formation: matching quality supply and quality demand. European J Oper Res 148(2):277–292
7. Carling C (2010) Analysis of physical activity profiles when running with the ball in a professional soccer team. J Sports Sci 28:319–326

8. Paul DJ, Nassis GP (2015) Physical fitness testing in youth soccer: issues and considerations regarding reliability validity sensitivity. Pediatr Exerc Sci 27:301–313
9. Hulse MA, Morris JG, Hawkins RD, Hodson A, Nevill AM, Nevill ME (2013) A field-test battery for elite, young soccer players. Int J Sports Med 34:302–311
10. Turner A, Walker S, Stembridge M, Coneyworth P, Reed G, Birdsey L, Barter P, Moody J (2011) A testing battery for the assessment of fitness in soccer players. Strength Conditioning J 3(5)
11. Maliki ABHM, Abdullah MR, Juahir H, Muhamad WSAW, Nasir NAM, Musa RM, Mat-Rasid SM, Adnan A, Kosni NA, Abdullah F, Abdullah NA (2018) The role of anthropometric, growth and maturity index (AGaMI) influencing youth soccer relative performance. In: IOP conference series: materials science and engineering (vol 342, no 1, p 012056), April 2018, IOP Publishing
12. Musa RM, Abdullah MR, Maliki ABHM, Kosni NA, Haque M (2016) The application of principal components analysis to recognize essential physical fitness components among youth development archers of Terengganu, Malaysia. Indian J Sci Technol 9(44):1–6
13. Maliki ABHM, Abdullah MR, Nadzmi A, Zainoddin MAR, Puspitasari IM, Jibril NFA, Nawi NA, Mat-Rasid SM, Musa RM, Suhaili Z, Kamarudin NA, Ali SKS (2021) Kids motor performances datasets. Data Brief 34:106582. https://doi.org/10.1016/j.dib.2020.106582
14. Faizatul N, Binti Jibril A, Abdullah MR, Bisyri A, Maliki HM, Puspitasari IM, Nawi NA, Nadzmi A, Amirur M, Zainoddin R, Yuningsih SH (2021) Comparison of body mass index class against motor performance towards 7-year-old boys in Malaysia. In: Proceedings of the first international conference on science, technology, engineering and industrial revolution (ICSTEIR 2020) 536:561–566. https://doi.org/10.2991/ASSEHR.K.210312.090
15. Hoff J, Wisløff U, Engen LC, Kemi OJ, Helgerud J (2002) Soccer specific aerobic endurance training. Br J Sports Med 36:218–221
16. Rampinini E, Impellizzeri FM, Castagna C, Couts AJ, Wisløff U (2009) Technical performance during soccer matches of the Italian Serie A league: effect of fatigue and competitive level. J Sci Med Sport 12(1):227–233
17. Nadzmi A, Abdullah MR, Maliki ABHM, Renaldi F, Sukono (2021) Comparison between gross motor skills performance and 7-years-old children with controlled effect of residential area. In: Proceedings of the first international conference on science, technology, engineering and industrial revolution (ICSTEIR 2020), 536. https://doi.org/10.2991/ASSEHR.K.210312.007
18. Taha Z, Haque M, Musa RM, Abdullah MR, Maliki ABHM, Alias N, Kosni NA (2017) Intelligent prediction of suitable physical characteristics toward archery performance using multivariate techniques. J Global Pharma Technol 9(7):44–52
19. Sajadi N, Rahnama N (2007) Analysis of goals in 2006 FIFA world cup. J Sport Sci Med 6(10):3
20. Abdullah MR, Maliki ABH, Musa RM, Kosni NA, Juahir H, Mohamed SB (2017) Identification and comparative analysis of essential performance indicators in two levels of soccer expertise. Int J Adv Sci Eng Inf Technol 7(1):305–314
21. Thomas JR, Nelson JK (2001) Research methods in physical activity. Human Kinetics, Champaign, IL
22. Hopkins WG (2000) Measures of reliability in sports medicine and science. Sports Med 30:1–5
23. Abdullah MR, Maliki ABHM, Musa RM, Kosni NA, Juahir H, Haque M (2016) Multi-hierarchical pattern recognition of athlete's relative performance as a criterion for predicting potential athletes. J Young Pharmacist. University Sultan Zainal Abidin. Terengganu, Malaysia
24. Markovic G, Dizdar D, Jaric S (2006) Evaluation of test of maximum kicking performance. J Sports Med Phys Fitness 46:215–220
25. Abd Hadi NH, Abdullah MR, Juahir H, Rahim AI, Elumalai G, Nadzmi A, Maliki ABHM (2021) Physical fitness key parameter for sport schools young athletes in Malaysia. Int J Acad Res Progressive Educ Dev 10(2). https://doi.org/10.6007/IJARPED/V10-I2/8799
26. Koo TK, Li MY (2016) A guideline of selecting and reporting intraclass correlation coefficients for reliability research. J Chiropr Med 15(2):155–163

27. Daly LE, Bourke GJ (2000) Interpretation and use of medical statistics. Blackwell Science Ltd., Oxford
28. Schimidt RA (1999) Motor and control learning: a behavioral emphasis. Human Kinetics, Champaign, IL
29. Svesson M, Drust B (2005) Testing soccer players. J Sports Sci 23:601–618

The Relationship Between the Female Athlete Triad and Dependence on Sports Coaches and Stress Responses

Chihiro Kemuriyama and Mitsuhiro Amazaki

Abstract Female athletes are susceptible to low energy levels, hypothalamic amenorrhea, osteoporosis, and a constellation of clinical conditions, which the American College of Sports Medicine defined as the female athlete triad (FAT). This study examined whether there is a difference in dependence on sport coaches, abnormal eating behavior, and athletes' stress responses related to subjective symptoms of FAT and explored the effects of dependence on sport coaches and abnormal eating behavior on athletes' stress response. The participants included 300 Japanese female athletes who completed a set of questionnaires on demographic variables, questions about FAT, the new version of the abnormal eating behavior scale, a dependence questionnaire, and the stress response scale for athletes. The results of a t-test revealed that helplessness, depression, all subscales of abnormal eating behavior, and all subscales of dependence on sport coaches were more prominent among the FAT group than non-FAT group. In addition, the results of multiple regression analysis highlighted that while inappropriate diet behavior was positively associated with helplessness, apprehension concerning to food intake was positively associated with physical fatigue, irritation and anger, distrust of people, and depression. Furthermore, binge eating was associated with helplessness and irritation and anger. Physical proximity was negatively associated with physical fatigue, and attention was positively associated with irritation and anger, and help was associated with irritation and anger and distrust of people. Thus, abnormal eating behaviors and dependence on sports coaches may exacerbate FAT symptoms and should be connected to consider ways to reduce the risk of FAT.

Keywords Female athlete triad · Dependence on sport coaches · Subjective symptoms of FAT · Abnormal eating behavior · Stress response

C. Kemuriyama (✉)
Gifu Shotoku Gakuen University, Gifu, Japan
e-mail: kemuriyama@gifu.shotoku.ac.jp

M. Amazaki
Aichi University, Toyohashi, Japan

G. Kuan et al. (eds.), *Advancing Sports and Exercise via Innovation*, Lecture Notes in Bioengineering, https://doi.org/10.1007/978-981-19-8159-3_15

1 Introduction

While female athletes have been active in international competitions such as the Olympics, the Female Athlete Triad (FAT) has been attracting attention [1, 2]. FAT refers to the interrelationships among energy availability, menstrual function, and bone mineral density, which may have clinical manifestations including eating disorders, functional hypothalamic amenorrhea, and osteoporosis [2]. Therefore, the development of effective prevention methods and remedial measures is urgently required.

In particular, a state of "low energy availability" in which energy intake from food and other sources is insufficient in relation to energy expenditure from exercise, can decrease hormone secretion, and increase the risk of causing amenorrhea and osteoporosis [3]. The beginning of FAT is considered to be the lack of available energy [4]. Nevertheless, female athletes consume a large amount of energy in their daily training and often lose a lot of weight and eat inappropriately, which increases the risk of FAT and exacerbates its symptoms [5]. In fact, female athletes, who became afraid to eat due to pressure to lose weight, have been reported to have experienced regret after eating or suffered from eating disorders due to unhealthy weight loss [6]. In addition, those with subjective symptoms of any or all of the FATs have a higher tendency toward abnormal eating behavior and a desire to lose weight than those without subjective symptoms. In addition, this may aggravate FATs, as they are more likely to suffer from unbalanced nutritional status and chronic energy deficiency caused by unhealthy eating behaviors [5].

The association between FAT onset status and psychological stress reactions has been confirmed. For example, those who have or are aware of FAT perceive higher physical and psychological stress reactions [7, 8]. In addition, in a survey of female rhythmic gymnasts, higher depression scores for psychological stress reactions were found in female athletes who were aware of FAT [5]. Furthermore, female athletes are characterized by greater dependence on their coaches than male athletes [9], and they typically rely more on their coaches for problem solving [10]. In addition, the presence of coaches and others is considered a factor that reinforces the state of eating disorders [11].

The purpose of this study was to examine whether there is a difference in dependence on sport coaches, abnormal eating behavior, and athletes' stress responses related to subjective symptoms of FAT and to explore the effects of dependence on sport coaches and abnormal eating behavior on athletes' stress response.

2 Methods

2.1 *Survey Participants*

The participants consisted of 300 female athletes aged 20–39 years (mean age 29.45 years, $SD = 5.44$) who competed at least three days per week and had participated in an official sports competition within the past five years. The mean height of the subjects was 158.05 cm ($SD = 8.66$), and the mean weight was 50.34 kg ($SD = 7.38$). The mean years of competition were 11.29 ($SD = 7.43$).

2.2 *Survey Period*

A Web-based survey was conducted through a research firm from December 2021 to January 2022.

2.3 *Survey Items*

2.3.1 Demographic Variables

Respondents were asked to provide their age, the type of sport they were currently competing in, the length of time they had been competing, their best results in the sport they were currently competing in, their height, and their weight.

2.3.2 Questions About Subjective Symptoms of FAT

Subjects were asked to select one of the three symptoms of low energy availability, exercise-associated amenorrhea, or osteoporosis that had been diagnosed by a physician in the past year as they were aware of the symptoms (e.g., exercise-associated amenorrhea, secondary amenorrhea in which menstruation has ceased for over three months and is considered to be caused by exercise).

2.3.3 Abnormal Eating Behaviors

The abnormal eating behavior scale-new version (AEBS-NV) [12] was used to measure the status of abnormal eating behavior in female athletes. This scale measures the tendency toward abnormal eating behavior and the psychological state characteristic of eating associated with it. This scale consists of 14 items in three factors: inappropriate diet behavior (6 items), apprehension concerning food intake

(5 items), and binge eating (3 items). The scale required responses on a 6-point scale wherein 1 = not at all true to 6 = completely true; higher total scores indicated more unusual eating behavior.

2.3.4 Dependence on Sports Coaches

To measure female athletes' dependence on their mentors, we used the dependence questionnaire developed by [13]. This scale measures the intensity of demand for interpersonal dependence and consists of 24 items in five subscales: physical proximity (3 items), attention (4 items), help (5 items), assurance (6 items), and psychic/spiritual (6 items). In this study, the subject of the demand that causes the dependent behavior was used as the sport coaches. The scale required responses on a 5-point scale wherein 1 = not true and 5 = true; higher total scores indicated more unusual dependence sports coaches.

2.3.5 Stress Response

The stress response scale for athletes (SRSA) [6] was used to measure the stress response of female athletes. This scale measures stress response in competition situations from psychological, physical, and behavioral aspects and consists of 15 items in five subscales: physical fatigue (3 items), helplessness (3 items), irritation and anger (3 items), distress of people (3 items), and depression (3 items). The scale required responses on a 5-point scale: from 1 = not at all to 5 = very much; higher total scores indicated greater stress response.

2.4 Ethical Considerations

This study was conducted with the approval of the Research Ethics Review Committee of Gifu Shotoku Gakuen University (Approval No. 2020–1).

2.5 Data Analysis

To examine whether female athletes' scores on abnormal eating behavior, dependence on sports coaches, and stress response differed depending on whether they were aware of FAT symptoms, t-tests with the AEBS-NV, the dependence questionnaire, and the SRSA subscale as dependent variables were considered. To examine the relationship among abnormal eating behavior, dependence on sports coaches, and stress response, multiple regression analysis was conducted with the presence of subjective symptoms on the FAT, the AEBS-NV, and the dependence questionnaire subscale as independent variables, and the SRSA subscale as the dependent variable.

The analysis was performed using IBM SPSS Statics 26.

3 Results

3.1 Examining Differences in Scores for Abnormal Eating Behavior, Dependence on Sport Coaches, and Stress Response with and Without FAT Subjective Symptoms

The results of a t-test revealed that the FAT group's score of inappropriate diet behavior (t [298] $= 4.705, p < 0.001$, Cohen's $d = 0.775$), apprehension concerning to food intake (t [298] $= 5.464$, $p < 0.001$, Cohen's $d = 0.893$), and binge eating (t [298] $= 2.675$, $p < 0.01$, Cohen's $d = 0.417$) of abnormal eating behavior was higher than non-FAT group. The FAT group's score of physical proximity (t [298] $=$ 3.502, $p < 0.01$, Cohen's $d = 0.562$), attention (t [298] $= 2.869$, $p < 0.01$, Cohen's $d = 0.404$), help (t [298] $= 2.949, p < 0.01$, Cohen's $d = 0.416$), assurance (t [298] $= 2.641, p < 0.01$, Cohen's $d = 0.372$), and psychic/spiritual (t [298] $= 3.517, p <$ 0.01, Cohen's $d = 0.496$) of dependence on sport coaches was higher than non-FAT group. Moreover, the FAT group's score of helplessness (t [298] $= 2.913, p < 0.01$, Cohen's $d = 0.411$) and depression (t [298] $= 2.421, p < 0.05$, Cohen's $d = 0.341$) of stress response was higher than non-FAT group (Table 1).

3.2 Examination of the Relationship Between Abnormal Eating Behavior, Dependence on the Instructor, and Stress Reactions

The results showed that the adjusted coefficient of determination for physical fatigue ($\triangle R^2 = 0.136$, $p < 0.001$), helplessness ($\triangle R^2 = 0.216$, $p < 0.001$), irritation and anger ($\triangle R^2 = 0.207$, $p < 0.001$), distrust of people ($\triangle R^2 = 0.224$, $p < 0.001$), and depression ($\triangle R^2 = 0.192, p < 0.001$) in the stress response was significant.

The standard partial regression coefficients for apprehension concerning to food intake ($\beta = 0.238, p < 0.05$) and physical proximity ($\beta = -0.234$, $p < 0.05$) were significant for physical fatigue. In addition, the results showed that the standard partial regression coefficient for inappropriate diet behavior ($\beta = 0.214, p < 0.05$) and binge eating ($\beta = 0.185, p < 0.05$) was significant for helplessness. Furthermore, the irritation and anger had a significant result for the standard partial regression coefficients for apprehension concerning food intake ($\beta = 0.222, p < 0.05$), binge eating ($\beta = 0.246, p < 0.01$), attention ($\beta = 0.227, p < 0.05$), and help ($\beta = -0.306$,

Table 1 Results of the t-test

	FAT group ($N = 64$)	Non-FAT group ($N = 236$)	t value	Effect size (d)
Abnormal eating				
Inappropriate diet behavior	20.77 (7.60)	15.96 (5.77)	4.705***	0.775
Apprehension concerning to food intake	18.14 (6.53)	13.33 (5.03)	5.464***	0.893
Binge eating	10.88 (4.26)	9.32 (3.57)	2.675**	0.417
Dependence on sport coaches				
Physical proximity	8.36 (3.40)	6.75 (2.70)	3.502**	0.562
Attention	12.91 (4.12)	11.34 (3.81)	2.869**	0.404
Help	16.44 (4.93)	14.48 (4.64)	2.949**	0.416
Assurance	20.34 (5.71)	18.29 (5.47)	2.641**	0.372
Psychic/spiritual	17.45 (6.37)	14.65 (5.45)	3.517**	0.496
Stress response				
Physical fatigue	10.31 (3.09)	10.02 (2.93)	0.698 n.s	0.098
Helplessness	9.34 (3.31)	8.04 (3.13)	2.913**	0.411
Irritation and anger	8.90 (3.93)	8.43 (3.44)	0.948 n.s	0.134
Distrust of people	8.95 (4.02)	7.88 (3.35)	2.170 n.s	0.306
Depression	9.77 (3.93)	8.55 (3.47)	2.421*	0.341

$^{***}p < 0.001$, $^{**}p < 0.01$, $^{*}p < 0.05$

$p < 0.05$). In addition, the standard partial regression coefficients for apprehension concerning food intake ($\beta = 0.392$, $p < 0.001$) and help ($\beta = -0.273$, $p < 0.05$) for interpersonal distrust and apprehension concerning food intake ($\beta = 0.353$, $p < 0.01$) for depression were significant (Table 2).

4 Conclusion

This study suggests that those with subjective symptoms of FAT are more likely to have abnormal eating behavior and dependence on their instructors and express a stronger stress response than those without subjective symptoms. These results support the suggestion that FAT symptoms are triggered by abnormal eating behavior [3] and that the involvement of sport coaches can aggravate FAT symptoms [11]. In particular, many sport coaches are experienced athletes—often successful athletes with a certain level of competitiveness. Therefore, customary and unhealthy methods may also be transferred from sport coaches to athletes as methods facilitating success [14]. In addition, there is a risk that excessive pressure to lose weight and unhealthy eating habits may be communicated by the sport coaches to the athlete, which may trigger a worsening of abnormal eating behavior. Furthermore, the fact that the FAT

Table 2 Results of multiple regression analysis

	Stress response (dependent variable)				
	Physical fatigue	Helplessness	Irritation and anger	Distrust of people	Depression
Subjective symptoms on the FAT	0.066 n.s	−0.029 n.s	0.074 n.s	0.032 n.s	−0.003 n.s
Abnormal eating					
Inappropriate diet behavior	0.056 n.s	0.214*	0.025 n.s	−0.037 n.s	−0.081 n.s
Apprehension concerning to food intake	0.238*	0.119 n.s	0.222*	0.392***	0.353**
Binge eating	0.154 n.s	0.185*	0.246**	0.114 n.s	0.126 n.s
Dependence on sport coaches					
Physical proximity	−0.234*	−0.105 n.s	−0.081 n.s	−0.054 n.s	− 0.133 n.s
Attention	0.225 n.s	0.170 n.s	0.227*	0.172 n.s	0.066 n.s
Help	−0.201 n.s	−0.209 n.s	−0.306*	−0.273*	− 0.074 n.s
Assurance	0.071 n.s	0.039 n.s	0.172 n.s	0.133 n.s	0.176 n.s
Psychic/spiritual	0.100 n.s	0.125 n.s	0.049 n.s	0.120 n.s	0.110 n.s
Adjusted coefficient of determination (ΔR^2)	0.136***	0.216***	0.207***	0.224***	0.192***

Note The values in the table are the standard partial regression coefficients (β) dependent variable: stress response, independent variable: abnormal eating, dependence on sport coaches
$^{***}p < 0.001$, $^{**}p < 0.01$, $^{*}p < 0.05$

group showed both abnormal eating behavior and higher scores for stress reactions such as helplessness and depression may be related to the fact that efforts to maintain or reduce body fat and weight are one of the psychological stressors for female athletes [15].

In this study, the results suggest that dependence on the sport coaches and tendency to abnormal eating behavior influenced stress reactions. Moreover, depending on the type of eating behavior abnormality and the type of dependence on the sport coaches, the associated stress responses were also found to be distinctive. In particular, oversensitivity to caloric intake and over eating can cause numerous stress reactions.

Therefore, it is important to support regular and well-balanced eating habits and to appropriate the athlete's cognition to ensure that they do not feel depressed after eating. However, when athletes dieted without regard to their health or engaged in binge eating in reaction, they were more likely to experience psychological stress responses such as helplessness and anger.

This study highlighted that physical fatigue was reduced when the patient attempted to feel safe with their sport coaches. In addition, the results showed that the more the athletes sought help from their leaders, the more they reduced their feelings of anger and distrust of others. However, the more the athletes wanted the sport coaches to focus on and understand their feelings, the more feelings of anger increased. The results suggest that athletes' stress responses decrease when they seek support that can be concretely implemented by the sport coaches. However, athletes may feel deficient with respect to invisible support such as acceptance and empathy, which may increase their feelings of anger. This requires that the desire to be psychologically close and empathetic be satisfied in the support of female athletes. When female athletes exhibit high demands to be understood and empathized with by their sport coaches, they are also at higher risk for inappropriate eating behaviors [5]. Therefore, the sport coaches' empathy and psychological support can both reduce stress reactions and prevent abnormal eating behavior.

Thus, FAT is assumed to be intimately related to abnormal eating behavior, dependence on the instructor, and stress reactions. In addition, abnormal eating behaviors and stress may exacerbate FAT symptoms. Therefore, future studies should examine ways to reduce the risk of abnormal eating behaviors and stress and prevent FAT.

Acknowledgements This work was supported by JSPS KAKENHI Grant Number JP19K11454.

References

1. Joy E, Clark N, Ireland ML, Martire J, Nattiv A, Varechok S (1997) Team management of the female athlete triad: part 1: what to look for, what to ask. Phys Sports Med 25:94–110 (1997)
2. Nagel M (2003) The relationship of the female athlete triad with psychocognitive functions and balance. J Exerc Sci Fit 1:116–124
3. Nattiv A, Loucks AB, Manore MM, Sanborn CF, Sundgot-Borgen J, Warren MP (2007) American college of sports medicine: American college of sports medicine position stand. The female athlete triad. Med Sci Sports Exerc 39:1867–1882
4. De Souza MJ, Nattiv A, Joy E, Misra M, Williams NI, Mallinson RJ, Gibbs JC, Olmsted M, Goolsby M, Matheson G (2014) Expert panel: female athlete triad coalition consensus statement on treatment and return to play of the female athlete triad: 1st international conference held in San Francisco, California, May 2012 and 2nd international conference held in Indianapolis, Indiana, May 2013. British J Sports Medi 48:289–308
5. Kemuriyama C, Oki J, Aamazaki M (2021) Relation between stressors and stress response in female rhythmic gymnasts. The annals of Gifu Shotoku Gakuen University. Faculty of Education. 60:37–43
6. Japan Association of Athletics Federations. What female athletes need to know, part 4: do eating disorders (anorexia, bulimia) happen to athletes?: I feel I can run better when I am thin.

Fear of eating: (2018). [Online]. [cited 2022 May 20]: Available from https://www.jaaf.or.jp/pdf/about/resist/medical/20180314-4.pdf
7. Kemuriyama C (2013) Development of the stress response scale for athletes. Ann Gifu Shotoku Gakuen Univ. Faculty of Education 52:31–38
8. Kemuriyama C, Oki J, Aamazaki M (2020) Female athlete triad, rhythmic gymnastics athletes, subjective symptoms, competitive and daily stressor depression. Ann Gifu Shotoku Gakuen Univ. Faculty of Education 59:31–38
9. Ae E (1999) Rethinking leadership in women's competitive sports: special feature: the near future of women's sports, independence of women athletes and their assistance. J Human Sci 11:20–25
10. Yamada Y, Nakajima N, Hisosawa M, Tanaka S, Mizumoto M, Susiura M (2006) A study on the similarity of stress-coping styles of athletes: focusing on sex and trait of game. J Health Sports Sci Juntendo Univ 10:21–28
11. Takenaka K, Imanaka M, Kusumoto Y, Iso S, Zaighkowsky LD, Tokoro M, Toka M, Date M (1992) Study of eating behavior for young female athletes. Descente Sports Sci 13:218–227
12. Yamatsuta K, Sato H, Sasagawa S, Yamamoto R, Nakai Y, Nomura S (2016) Development of the abnormal eating behavior scale new version for female college student. Japanese J Psychosom Med 56:737–747
13. Takahashi K (1970) Dependent behavior in female adolescents III. Jpn J Educ Young Child 18:65–75
14. Kemuriyama C (2020) Female athlete and stress: support measures for aesthetic athletes. Health Psychol 83:4
15. Takenaka K, Oka K, Oba Y (1999) Eating behavior and menstrual state for female athletes to keep drive for thinness and restriction of weight. Japan J Physical Educ Health Sport Sci 44:241–258

Application of the Health Action Process Approach to Abnormal Eating Behavior Among Female Athletes

Mitsuhiro Amazaki and Chihiro Kemuriyama

Abstract The potential cause of female athlete triad, which is considered a factor in the decline of the competitive performance of female athletes, is abnormal eating behavior. To pursue their competitive lives, addressing abnormal eating behavior without resorting to medicines is necessary for female athletes from the perspective of sports doping. Toward this end, health education interventions that promote behavior change for abnormal eating behavior are required. This study aimed to examine the of the health action process approach (HAPA) to promote behavioral change for abnormal eating behavior among female athletes. Participants included 300 Japanese female athletes, who completed a set of questionnaires on demographic variables, the abnormal eating behavior scale new version, and psychosocial factors constituting HAPA (risk perception, outcome anticipation, self-efficacy, behavioral intentions, and behavioral planning). Results revealed that the HAPA was an applicable model for explaining abnormal eating behavior (GFI = 0.987, AGFI = 0.939, CFI = 0.987, RMSEA = 0.068); however, the explanatory rate for abnormal eating behavior was small, where the only psychosocial factor reducing abnormal eating behavior was self-efficacy for eating behavior. Furthermore, there was no significant effect from negative outcome expectancy on behavioral intention. Therefore, health education interventions for improving abnormal eating behaviors in female athletes should focus on self-efficacy for eating behaviors.

Keywords Female athlete triad · Health action process approach · Self-efficacy

M. Amazaki (✉)
Aichi University, Aichi, Japan
e-mail: amazaki@vega.aichi-u.ac.jp

C. Kemuriyama
Gifu Shotoku Gakuen University, Gifu, Japan

G. Kuan et al. (eds.), *Advancing Sports and Exercise via Innovation*, Lecture Notes in Bioengineering, https://doi.org/10.1007/978-981-19-8159-3_16

1 Introduction

The female athlete triad (FAT), characterized by the three symptoms of low energy availability, athletic amenorrhea, and osteoporosis, is a significant health problem with a high incidence [1, 2]. Because FAT can afflict any female athlete regardless of athletic ability [3] sports coaches need to understand the construct.

The three symptoms of FAT affect each other, transforming the body from a healthy state to one that requires treatment over a long period [4]. Researchers have reported that female athletes not only expend a great deal of energy in daily training but also often repeat inappropriate or abnormal eating behaviors [5]. When female athletes suffer from low energy availability, a condition in which energy intake from food and other sources is insufficient relative to energy expenditure from training, female hormone secretion declines, increasing the mazarisk of amenorrhea and osteoporosis [6], and FAT is thought to start with a lack of available energy [7]. It has been reported that athletic amenorrhea and osteoporosis become more serious because of this effect [4].

Athletes have limited access to medications for treatment given anti-doping concerns. It is imperative to propose effective prevention methods for abnormal eating behavior based on behavioral science approaches that do not rely on medications. Therefore, it is important to better understand the psychological processes that lead to abnormal eating behavior to prevent female athletes from suffering from associated low energy availability. To this end, it is necessary to clarify the psychological processes that lead to abnormal eating behavior using behavioral theory.

The health action process approach [8], a behavioral theory, has been used as a predictive model for multiple health behaviors [9–11]. It is an integrated model that combines features of stage, continuum, and dual-phase social cognition models [12]. The HAPA consists of five psychological variables, risk perception, outcome expectancy, self-efficacy, behavioral intention, and behavioral planning, and the dependent variable, which is health behavior. According to the HAPA, individuals perform health behaviors in a process consisting of a motivation phase that develops behavioral intentions and a volition phase that leads to actual behavior. Because it has been reported that the HAPA can also predict healthy eating behaviors [13], we expected that it would be able to explain the psychological process leading to the occurrence of abnormal eating behaviors as well. Therefore, our aim with the present study was to examine the adaptability of the HAPA as a model for predicting abnormal eating behavior in female Japanese athletes.

2 Methods

2.1 *Survey Period and Survey Participants*

We conducted the online survey for this study from December 2021 to January 2022 among 300 female Japanese athletes aged between 20 and 39 years (mean age = 29.45 years, SD = 5.44) who were selected from among monitors registered with an Internet research firm and had participated in official sports competitions within the past 2 years. The basic attributes of the study subjects were as follows: mean height 158.05 cm (SD = 8.66), mean weight 50.34 kg (SD = 7.38), and mean years of competition 11.29 years (SD = 7.43). The study athletes were competing in 44 different types of athletic events at the time of the study including track and field (long distance). We conducted this study survey with approval from the Local Ethics Committee of Gifu Shotoku Gakuen University.

2.2 *Survey Items*

2.2.1 Demographic Variables

Participants self-reported their age in years, current sport, history of competition in current sport, highest performance in current sport, height, and weight.

2.2.2 Psychological Constructs of HAPA

To measure the psychological constructs of the HAPA, we used scales confirmed as reliable and valid by [5]. Specifically, the scales measured the female athletes' eating behaviors for risk perceptions (1 factor, 10 items), outcome expectancy (2 factors, 11 items), self-efficacy (1 factor, 9 items), behavioral intention (1 factor, 8 items), and planning (1 factor, 10 items). Respondents rated all items on 5-point Likert scales where 1 = *completely disagree* and 5 = *strongly agree* and where higher scores on each scale indicated a higher likelihood of that scale's behavior.

2.2.3 Abnormal Eating Behaviors

To measure Japanese women athletes' abnormal eating behavior, we used the Abnormal Eating Behavior Scale new version (AEBS-NV) [14]. The scale consisted of 19 items and required responses on 6-point Likert scales where 1 = *not at all true* and 6 = *completely true* and higher total scores indicated more unusual eating behavior.

2.3 Data Analysis

We conducted a structural equation model (SEM) using the total AEBS-NV score as the dependent variable and the psychological factors comprising the HAPA. We used maximum likelihood as the estimation method, and to ensure the differentiation of models, we set both the variance of each latent variable and the path from the error variable to the observed variable to 1. To confirm the suitability of the model, we calculated the following fit indices: goodness-of-fit index (GFI), adjusted goodness-of-fit index (AGFI), comparative fit index (CFI), and root-mean-square error of approximation (RMSEA). Specifically, good model fit is indicated when GFI, AGFI, and CFI exceed 0.90 and RMSEA is below 0.10 [15]. We performed all statistical analyses using IBM Amos 25.0.0.

3 Results

Tables 1 and 2 show, respectively, the mean scores for each scale used in this study and the correlation coefficients between the scales. Figure 1 presents the indices for the fit of the HAPA SEM to the data: GFI = 9.87, AGFI = 9.39, CFI = 0.987, and RMSEA = 0.068. Because these indices met sufficient values, we judged that the HAPA was fit as a model to explain the abnormal eating behavior of female Japanese athletes: The HAPA model explained 4% of the female Japanese athletes' abnormal eating behavior. The standardized coefficients between variables in the HAPA were all significant except for that from negative outcome expectancy to behavioral intention. The only direct effect of decreased abnormal eating behavior was the effect of self-efficacy on eating behavior. The variables that had the indirect effect of increasing abnormal eating behavior were risk perception, positive outcome anticipation, and self-efficacy for eating behavior.

.

Table 1 The mean scores for each scale

	Mean	SD	Min.	Max.
Risk perception for eating behaviors	40.43	7.74	10	50
Positive OE for eating behaviors	18.41	3.62	5	25
Negative OE for eating behaviors	15.88	4.71	6	30
Self-efficacy for eating behaviors	30.44	7.30	9	45
Behavioral intention for eating behaviors	30.55	5.52	10	40
Planning for eating behaviors	33.54	8.60	10	50
Abnormal eating behavior	41.00	14.54	14	84

Note OE, outcome expectancy

Table 2 The correlation coefficients between the scales

	1	2	3	4	5	6
1. Risk perception for eating behaviors	1					
2. Positive OE for eating behaviors	0.380**	1				
3. Negative OE for eating behaviors	0.016	−0.005	1			
4. Self-efficacy for eating behaviors	0.116*	0.271**	0.094	1		
5. Behavioral intention for eating behaviors	0.275**	0.337**	−0.004	0.676**	1	
6. Planning for eating behaviors	0.185**	0.336**	0.102	0.680**	0.706**	1
7. Abnormal eating behavior	−0.015	−0.011	0.510**	−0.016	−0.040	0.170**

Note OE, outcome expectancy. * $p < 0.05$, ** $p < 0.01$

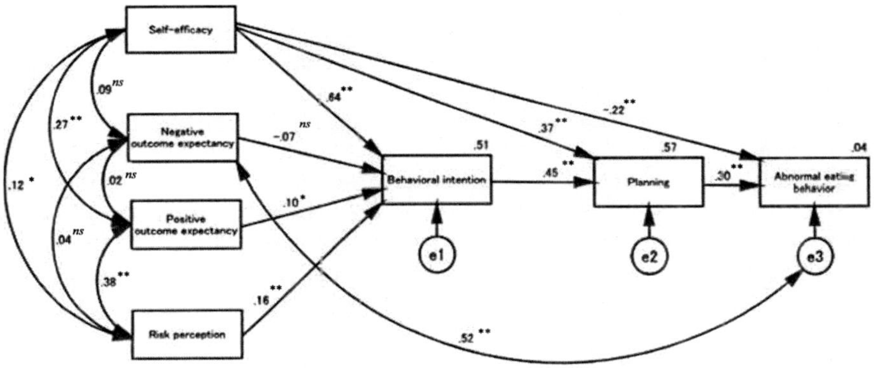

Fig. 1 HAPA model for abnormal eating behavior of Japanese women athletes. *Note* * $p < 0.05$, ** $p < 0.01$, *** $p < 0.001$

4 Conclusion

In previous attempts to explain health behaviors using the HAPA, researchers identified the psychological factors that comprise the HAPA, risk perception, outcome anticipation, self-efficacy, behavioral intention, and behavioral planning, as factors that promote behaviors considered to be good for health. However, we must keep in mind that in this study, we used a health-damaging behavior, disordered eating, as the dependent variable, and therefore, our interpretation of the variable differs from that in previous studies.

The results of this study indicated that the fit of the HAPA model for abnormal eating behavior in female Japanese athletes was adequate in that the model was able to predict abnormal eating behavior in these athletes. Specifically, female Japanese athletes who had high scores for risk perception, positive outcome anticipation, and self-efficacy related to eating behaviors were more likely to engage in abnormal eating behaviors. The results also identified the outcome expectancy and self-efficacy scales as useful instruments for screening abnormal eating behavior among female Japanese athletes. That is, although the scales included no cutoff points, scores above a certain level should require sports coaches to discuss whether female athletes have abnormal eating behaviors.

However, despite the usefulness of our findings, this study does contain limitations. First, abnormal eating is a socially undesirable behavior, and it was likely that the female Japanese athletes would hide any abnormal eating behaviors. We also recognized overeating as a potential abnormal eating behavior that could be hidden and that we would thus not know about [16, 17].

In more detail, researchers have identified the possibility of false responses on self-report questionnaires and the social desirability bias, whereby respondents answer falsely in ways they perceive will make them appear more socially desirable. In this study, the female Japanese athletes could have skewed their responses to conceal their abnormal eating behavior. Given this possibility, we anticipated that female Japanese athletes would not accurately respond to attempts to directly assess their eating behaviors, and we proposed that it would be possible to accurately screen for abnormal eating behavior among them by examining factors that indirectly influence such behavior.

Additionally, in this study, we validated the HAPA model without considering the presence or absence of subjective symptoms of FAT in a group of female Japanese athletes to better understand their overall tendencies toward abnormal eating behaviors. Future researchers should conduct multiple-group SEM analysis of the HAPA model that considers the specific FAT status of Japanese female athletes.

Acknowledgements This paper is supported by JSPS KAKENHI Grant Number 19K11454.

References

1. Joy E, Clark N, Ireland ML, Martire J, Nattiv A, Varechok S (1997) Team management of the female athlete triad: part 1: what to look for, what to ask. Phys Sportsmed 25:94–110
2. Nagel M (2003) The relationship of the female athlete triad with psychocognitive functions and balance. J Exerc Sci Fit 1:116–124
3. Koikawa N, Ogasawara E (2016) Considering sports medicine for female athletes. Japanese J Phys Fit Sport Med 65:25–26
4. Sunaga M (2018) Textbook for female athletes. Textbook for female athletes, Tokyo
5. Kemuriyama C, Oki J, Aamazaki M (2021) Relation between stressors and stress response in female rhythmic gymnasts. Ann Gifu Shotoku Gakuen Univ Fac Educ 60:37–43
6. Nattiv A, Loucks AB, Manore MM, Sanborn CF, Sundgot-Borgen J, Warren MP (2007) American college of sports medicine position stand. The female athlete triad. Med Sci Sports Exerc 39:1867–1882

7. De Souza MJ, Nattiv A, Joy E, Misra M, Williams NI, Mallinson RJ, Gibbs JC, Olmsted M, Goolsby M, Matheson G (2014) 2014 female athlete triad coalition consensus statement on treatment and return to play of the female athlete triad: 1st international conference held in San Francisco, California, may 2012 and 2nd international conference held in Indianapolis, Indiana, May 2013. Br J Sports Med 48:289
8. Schwarzer R (2008) Modeling health behavior change: how to predict and modify the adoption and maintenance of health behaviors. Appl Psychol 57:1–29
9. Amazaki M, Kemuriyama C (2013) Applicability of health action process approach to physical activity among Japanese university students. Jpn J Sport Psychol 40:125–137
10. Amazaki M, Kemuriyama C, Mori K (2014) Health action process approach for determining amount of physical exercise among Japanese workers. Japanese J Heal Psychol 27:53–62. https://doi.org/10.11560/jahp.27.1_53
11. Amazaki M, Mori K (2011) Health action process approach to condom use in university students. Japanese J Heal Psychol
12. Hamilton K, Smith SR, Keech JJ, Moyers SA, Hagger MS (2020) Application of the health action process approach to social distancing behavior during COVID-19. Appl Psychol Heal Well-Being 12:1244–1269
13. Chiu C-Y, Lynch RT, Chan F, Berven NL (2011) The health action process approach as a motivational model for physical activity self-management for people with multiple sclerosis: a path analysis. Rehabil Psychol 56:171
14. Yamatsuta K, Sato H, Sasagawa S, Yamamoto R, Nakai Y, Nomura S (2016) Development of the abnormal eating behavior scale new version for female college student. Japanese J Psychosom Med 56:737–747. https://doi.org/10.15064/jjpm.56.7_737
15. Brown TA (2015) Confirmatory factor analysis for applied research, 2nd edn. The Guilford Press
16. Yamaguchi M, Wakana N, Tanaka E, Homma K (2015) Eating disorders in Japan: a comparison with the USA. J Agric Sci Tokyo Univ Agric 60:169–177
17. Palmer B (2014) Helping people with eating disorders: a clinical guide to assessment and treatment. Wiley

Common Sports Injury in Karate Practitioner: A Bibliometric Analysis and Mini-Review (1992–2021)

Hong Yee Kek, Huiyi Tan, Keng Yinn Wong, and Garry Kuan

Abstract Karate is a Japanese martial art which emphasizes character development and self-discipline. However, noticeable karate practitioners contracted injuries during training and competition. The aim of this study is to visualize the knowledge map of sports injury in karate research through bibliometric analysis. Also, the common injury among the karate practitioner and their associated causes is included in the mini-review section. The global developments and research landscape on sports injury of karate were examined based on 139 publications (1992–2021) extracted from the Web of Science (WoS). The emerging trend in the research areas was observed in 2016. The top leading countries indicated that the USA has published the most publications related to sports injury research in karate. This manuscript reviewed the potential injuries and their associated causes. Knee injuries were the most prevalent, and they could be contracted during training or tournament. These findings would be useful for the instructor and clinicians in designing some training activities, which aim to reduce the risk of the practitioner contracting injury. In future studies, it is recommended to conduct the injury assessment on the traditional karate practitioner such as Kyokushin Kai (full-contact karate style). Such exploration could provide insight into karate-associated injuries and reduce the bias in bibliometric data analysis.

Keywords Martial art · Karate · Injury · Bibliometric analysis · Mini-review

H. Y. Kek · H. Tan · K. Y. Wong (✉)
Faculty of Engineering, Universiti Teknologi Malaysia, Johor, Malaysia
e-mail: kengyinnwong@utm.my

H. Y. Kek
e-mail: hongyee@graduate.utm.my

G. Kuan
School of Health Sciences, Universiti Sains Malaysia, Kelantan, Malaysia

G. Kuan et al. (eds.), *Advancing Sports and Exercise via Innovation*, Lecture Notes in Bioengineering, https://doi.org/10.1007/978-981-19-8159-3_17

1 Introduction

“Karate” is a martial art that was initiated in Okinawa, Japan. This martial art has been proven to have positive effects on mental and physical health [1]. Karate could be divided into two types: traditional karate and sports karate. Traditional karate mainly focuses on full-contact training, including strikes to the body with no protection for the arms and torso [1]. Also, practitioners are allowed to have a full power kick to the head, knee strikes, and kicks to the inner and outer leg. On the other hand, sports karate focuses on athletes’ performance and points scoring, which was included as one of the events in the 2020 Olympics in Japan [2]. Sports karate has two modes: sparring (Kumite) and forms (Kata) [3]. The training regime for both modes is different, as Kumite focuses on opponent-contact training, while Kata focuses on self-perform training. Both modes of practitioners were reported to suffer from different injuries.

The main objective of this study is to visualize the knowledge map of sports injury in karate research through bibliometric analysis. Also, the common injury among the karate practitioner and their associated causes is included in the mini-review section. This paper also shows the general publication trends, top-cited publications, prolific institutions, influential authors, etc. This bibliometric analysis procedure provides an overview of global trends on specific research topics to readers from different backgrounds.

2 Methodology

Web of Science (WoS) is one of the most comprehensive scientific databases, citation indexing services, and search engines when compared to Scopus, Science Direct, and Google Scholar [4]. Web of Science Core Collection (WoS) is a thorough database with the largest number of indexed journals and conference proceedings (around 12,000) and covers nearly 150 research disciplines. Therefore, WoS as one of the globally recognized databases was adopted to retrieve the publication records for this study. An electronic search was performed by inserting the search string consisting of the following keywords: (Karate AND (injur * OR hurt * OR impair *)). Next, the results are filtered by only involving research articles, review papers and conference papers written in English within the timespan from 1992 to 2021. At this stage, the inclusion criteria reduced the sample size from 195 to 153 papers. Next, the remaining papers are further screened manually to eliminate publications that are not directly related to the theme of the study. Finally, the front-page filtering method returned a result of 139 publications.

Next, the full set of downloaded publication-related data was used to generate useful statistics via Microsoft Excel. The presented tables could help visualize the growing curve of research output and identify the prolific countries, institutions, and researchers. Moreover, VOS Viewer was utilized to produce a visualization map of the collaboration network between researchers and institutions from different nations.

The size of the node denoted the importance of the representative objects in the map, while the thickness of the line connecting two nodes represented the strength of interactions between them.

Several bibliometric indicators including TP: the total amount of publications, TC: total number of citations, and h-index were employed in the current study. TP and TC could reflect the productivity of publication and its scholarly impact, respectively. The Hirsh index or h-index was known as a standard indicator to measure the cumulative impact of a scholar's output by author, institution, or nation. An author with an h-index of *N* has *N* publications each having at least *N* citations, which is a measure of the impact research work in a particular field [4].

3 Result and Discussion

3.1 Publication Trend

Figure 1 shows the chronological distribution of 139 publication outputs over the three decades, from 1992 to 2021. The publications have consisted of 118 research articles, 13 review articles, and 8 proceedings papers. As shown, the publication productivity of the research topic fluctuated over the years. It could be observed that the current topic only received little attention in the first decade (1992 to 2001), where the total production of publications did not exceed 15 in this decade. In the next decade (2002–2011), the production of articles started to increase to 34, which is contributing to approximately one-fourth of the total production. The most recent decade would be the blooming era which produced multiple folds of publications compared to the previous decade, with a peak production of 21 papers in 2019. The highest number of publications were recorded in the recent five years, indicating there is a positively growing interest of researchers in exploring the part and parcel behind the sports injuries among the Karate practitioners.

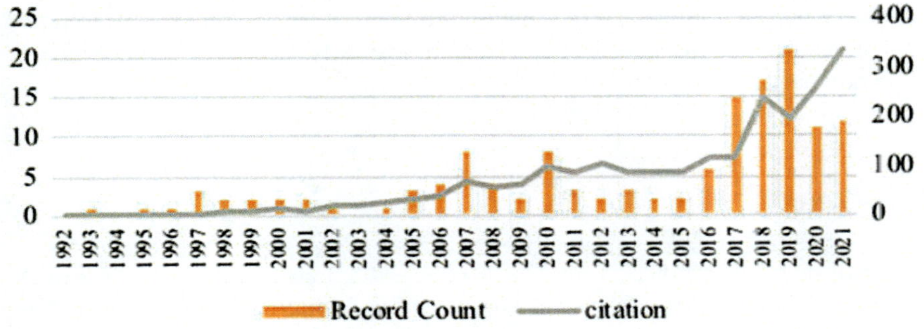

Fig. 1 Publication output distribution from 1992 to 2021

Table 1 Top leading countries in the research of sports injury in karate practitioner

No.	Countries/regions	TP	TC	h-index
1	USA	28	787	17
2	Poland	18	115	6
3	Spain	16	250	7
4	Brazil	13	108	5
5	England	11	211	6
6	Canada	10	261	8
7	Italy	10	108	5
8	Australia	8	63	5
9	Slovakia	8	45	5
10	Iran	6	63	2

3.2 *Top Leading Countries*

A total number of 41 countries Table 1 has shortlisted the top ten influential countries in the research of sports injury in Karate Practitioner with a minimum production of 6 publications. According to the analysis, it is undeniable that the USA should be recognized as the top leading country in the current research area, with its highest records of 28 publications, associated with the highest number of citations (787) and h-index (17). The distribution of research productivity from the country is highly skewed as the top ten countries had contributed over two-thirds of the total output, where the USA alone is accountable for 20 percent of the overall production. The contribution is followed by Poland (13%), Spain (11.5%), Brazil (9%), England (8%), Canada and Italy (7% each), Australia and Slovakia (6% each), and Iran (4%).

It could be also noticed that the total number of citations and h-index were not compatible with the total number of publications. For instance, Canada which produced less papers (10) scored a higher citation number of 261 compared to Poland (115) which produced a greater number of articles (18). Meanwhile, Australia and Iran which obtained similar TC (63) owned different values of h-index. Therefore, it could be deduced that the impact of the research paper may not always be proportional to the number of productions.

3.3 *Prolific Institutions*

The analysis of data indicates that 296 global institutions contributed to the total publication output. Table 2 displayed a list of prolific institutions which published at least 3 papers, together with their geographical distribution, the number of publications, number of citations and h-index. A total number of fifteen institutions complemented the list and the institutions were scattered around the globe, indicating the

Table 2 Top prolific institutions

No.	Institution	Country	TP	TC	h-index
1	Comenius University Bratislava	Slovakia	8	45	5
2	Universidade Da Coruna	Spain	6	43	4
3	Harvard University	USA	5	240	5
4	Macquarie University	Australia	5	22	3
5	University of Manitoba	Canada	5	179	4
6	Boston Children S Hospital	Canada	3	96	3
7	Gdansk University of Physical Education Sport	Poland	3	23	2
8	Instituto Medico Arriaza Asociados	Spain	3	22	3
9	Tampere University	Finland	3	278	3
10	Tehran University of Medical Sciences	Iran	3	58	2
11	University School of Physical Education	Poland	3	29	2
12	University of Calgary	Canada	3	26	3
13	University of Helsinki	Finland	3	382	3
14	University of Milan	Italy	3	28	2
15	University Of Novi Sad	Serbia	3	6	2

current research field is a global concern. Out of them, three originated from Canada, six from Spain, Finland, and Poland with two each, and the remaining six were from Slovakia, the USA, Australia, Iran, Italy, and Serbia, respectively.

Comenius University Bratislava with 8 publication records and the highest h-index of 5 was regarded as the top productive research institution in the current subject field. The production sequence was followed by Universidade Da Coruna which produced 6 research papers, and three institutions, namely Harvard University, Macquarie University, and the University of Manitoba which equally produced 5 publications. The list was continued with ten institutions which produced three publications each, including Boston Children S Hospital, Gdansk University of Physical Education Sport, Instituto Medico Arriaza Asociados, Tampere University, Tehran University of Medical Sciences, University School of Physical Education, University of Calgary, University of Helsinki, University of Milan, and University of Novi Sad.

The record of TC within the institution varied significantly ranging from 6 to 382. The significant difference in the citation count received by the institution signified that the research direction on sports injury topics has diverged and some of them were less likely to be linked with other research work. It could be noticed that only four institutions managed to obtain citations above 100. The highest number of TC (382) was documented by University of Helsinki, where the top prolific researcher, Kujala Urho was the representative of this institution. The second-highest number of TC (278) was received by Tampere University, which also originated from Finland. The citation record above 100 was then followed by Harvard University (240) and University of Manitoba (179).

3.4 Top Influential Authors

A total number of 513 authors were involved in the publications related to sports injury among Karate practitioners over the three decades, indicating that the production of a single article required three to four authors on average. Table 3 shows ten top influential authors who contributed 2 publications and above. The geographical distribution of the prolific authors is quite equal, where the authors originated from various countries. There were several pairs of authors who originated from Spain, USA, and Slovakia, while every remaining author was coming from Australia, Canada, Finland, Philippines, Poland, Scotland, and Serbia, respectively. Arriaza Rafael was crowned as the most active researcher and outnumbered other researchers with eight papers and the highest h-index of 6. The list is followed by Lystad Reidar P. who managed to produce five publications and five authors who were contributing to four publications each. Among them, Micheli Lyle J., Violan Mariona, and Zetaruk Merrilee collaborated in the four papers, thus harvesting the same number of citation counts (212) and h-index (4). Subsequently, six prolific researchers comprised of Kujala Urho, Zurakowski David, Pieter Willy, Witkowski Kazimierz, Baker Julien S., and Drid Patrik were accountable for three publication records each.

The capability of attracting citations could be closely linked to the productivity of the research team and the hot research topics. It was worth mentioning that Kujala UM was the author who received the highest number of citations (382), owning to his duo highly cited papers entitled "Acute Injuries in Soccer, Ice Hockey, Volleyball, Basketball, Judo, and Karate—Analysis of National Registry Data" [5] and "is it possible to prevent sports injuries? Review of controlled clinical trials and recommendations for future work" [6]. It is followed by the aforementioned research team who managed to score the second-highest in the total number of citations accumulated (212). The subsequent list of authors who received citations over three-digit numbers includes top productive authors Arriaza Rafael (112) and Zurakowski David who obtained 170 citations with his three publications.

Table 3 Top prolific authors

No.	Authors	Country	TP	TC	h-index
1	Arriaza R	Spain	8	112	6
2	Lystad RP	Australia	5	22	3
3	Micheli LJ	USA	4	212	4
4	Violan MA	Spain	4	212	4
5	Zetaruk MN	Canada	4	212	4
6	Cierna D	Slovakia	4	32	4
7	Augustovicova D	Slovakia	4	13	2
8	Kujala UM	Finland	3	382	3
9	Zurakowski D	USA	3	170	3
10	Pieter W	Philippines	3	43	2

3.5 Top Cited Papers

To date, the topic constituting the cause and effect of sports injury among Karate practitioners had accumulated 2248 citations, with an h-index of 27 and 16.17 citations per publication. Citation counts are frequently used in evaluating the influence of an article by determining how often it has been cited in another research. Therefore, a higher number of citations achieved could be considered an indication of greater influence, visibility, and impact. In this study, the variations in citation were normalized by adopting citation per year (CPY) as recommended by Garg and Singh [7].

Table 4 demonstrated the top-cited papers in this subject field with minimum citation counts of above 45, and the ranking of papers was based on the TC. Out of ten articles, eight of them are original research articles, while others are literature reviews. The top-cited paper (202) is a research article which discussed the acute injury profile in six sports including soccer, ice hockey, volleyball, basketball, Judo, and Karate and compared the injury rates between the sports [5]. The comprehensiveness and reliability of outcome based on 54,183 sports injuries reported by National Registry Data could best explain the considerable impact imposed by the article. Next, the review of controlled clinical trials prepared by Parkkari et al. [6] managed to score the highest CPY of 8.05 though its TC (177) was slightly lower than the top paper. This could be attributed to its various recommendations that brightened the future research path toward the prevention of sports injuries is commonly practized or high-risk sport. The last paper which obtained TC above 100 was about the comparison of injury rates among five types of martial arts [8]. The study revealed the detailed injury outcome of Shotokan Karate, Aikido, Tae kwon do, Kung Fu and Tai Chi, thus enabling the paper to become a convincing reference for the continued research from time to time.

3.6 International Collaboration Network

Figure 2 displays the network visualization map on the level of collaboration between ten countries with a minimum number of six publications. USA was the prime pioneer of the current field subject, which contributed to publication as early as 1993 [16]. In recent five years, Slovakia emerged as one of the top prolific countries in investigating the incidence and pattern of injuries in Karate, especially during Karate competitions.

As observed, Spain and Brazil held a similar number of international collaborations with five countries. However, a bigger size of the node and strong linking strength between Spain and other countries had enabling its central location in the network map. For instance, the multinational collaboration between Spain, Canada, and USA yielded a significant outcome, as their research article [8] was recognized as the top-cited papers which had attracted a great number of citation counts (177). Moreover, the thick lines forming a triangle between Spain, Slovakia, and Australia

Table 4 Top cited papers

No.	Title (Author, year)	TC	CPY
1	Acute injuries in soccer, Ice hockey, Volleyball, Basketball, Judo, and karate—analysis of national registry data [5]	202	7.21
2	Is it possible to prevent sports injuries? Review of controlled clinical trials and recommendations for future work [6]	177	8.05
3	Injuries in martial arts: a comparison of five styles [8]	116	6.44
4	Active living and injury risk [9]	98	5.16
5	No holds barred sport fighting: a 10 year review of mixed martial arts competition [10]	90	5.29
6	Injury and injury rates in Muay Thai kick boxing [11]	83	3.77
7	Pediatric martial arts injuries presenting to emergency departments, United States 1990–2003 [12]	52	3.25
8	Injury rates in Shotokan karate [13]	51	2.13
9	Effects of the new karate rules on the incidence and distribution of injuries—commentary [14]	49	2.88
10	Kinematic and electromyographic analyzes of a karate punch [15]	47	3.92

Fig. 2 Overlap visualization of the co-authorship network between nations (minimum number of publications: 6)

had indicated their joint effort in conducting prospective studies on time-loss injuries in Karate athletes [17]. Meanwhile, England and Brazilian researchers also shared a common interest in analyzing the risk of bone traumatic features depending on the engagement in sports and identifying its potential impact on healthcare costs [18].

Though Poland itself had contributed 18 publications, it demonstrated a weak linkage with the only connected country, USA. Meanwhile, Italy and Iran, which had produced ten and six publications, were coordinated at the edge of the international collaboration network, due to their less scholarly communication with other nations. Nonetheless, the active networking relationship between countries is a positive phenomenon in the current research field as it could encourage more intersections of insights between experts.

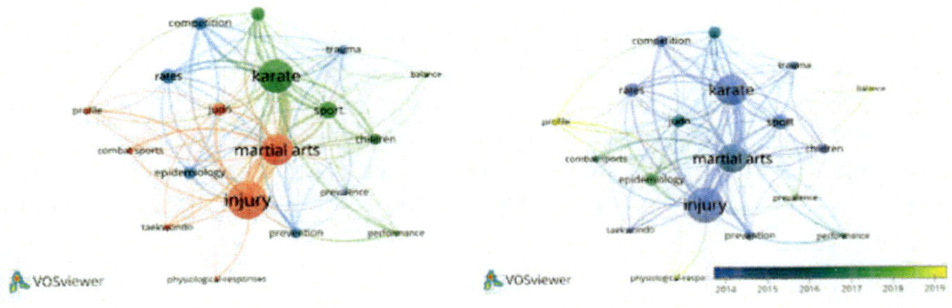

(a) (b)

Fig. 3 **a** Network visualization map and **b** overlay visualization map of the keyword co-occurrence network for research on sport karate's injury

3.7 Keyword Co-occurrence Analysis

The assessment of the co-occurrence of keywords in literature is essential to identify the research hotspots [19]. The visualization map and overlay were generated from the frequently used keywords. As shown in Fig. 3a, the terms "karate", "martial art", and "injury" are the prevalence keywords, with a total link strength of 153, 143, 178, respectively. These are followed by the keywords of "competition", "prevention", "judo", and "children". The main difference between "martial arts" and "karate" is that martial arts consist of all combat sports, i.e., karate, judo, taekwondo, mixed martial art, jiujitsu, kickboxing, etc. However, the present study indicated that most research teams focus on a specific combat sport as the case study, instead of combining several combat sports as a case study. The nineteen keywords in Fig. 3a can be categorized into three clusters, as indicated by three different colors (red, green, and blue).

"Injury" is the central mode of cluster 1 (red color), followed by "martial arts". This cluster shows the earlier focus of the researchers on the uptake of combat sports injury. Esmaeilpoor et al. [20] evaluated the experience of orofacial injuries and related factors among martial arts practitioners. They disclosed that most of the orofacial injuries were due to unsuitable or poor fitting mouthguards. On the other hand, Tulendiyeva et al. [1] reported that joint injuries were commonly suffered by Judo and Sanda martial arts, due to excessive straining and stretching of ligaments during grappling and throwing techniques. However, karate and taekwondo practitioners who emphasize striking techniques to the head region, usually experienced traumas in the head and neck areas [1].

The second cluster (green color) indicates the most recent progress in karate development among the children, which aims to promote sports performance and achieve good posture stability. Vando et al. [21] found that one week of high-intensity karate training promoted a significant improvement in static body balance among early age karate practitioners, with the age ranging from 9–12 years old. Similarly, Truszczyńska et al. [22] also conducted a similar study on non-karate practitioners aged 7–10 years. Surprisingly, they discovered that these children could develop a better mediolateral postural stability after a period of training, with 1.5 h of training sessions, twice a week. However, the dynamics balance associated with intensive karate training has not been reported so far. The third cluster (blue color) links various injuries due to martial arts during competition and training session. Various injuries have been reported associated with martial arts, i.e., knee, head, neck, limp, ankle, spine, etc., [2, 6, 20].

4 Mini-review on causes of injury

Injuries among the karate practitioners are unavoidable, especially during a sparring session, intensive training sessions, or motor consciousness weakness [23]. It is an injury associated with the nature of practitioner-to-practitioner contact inherent in the sport [2]. Injuries in sport karate could break into two main categories: contact injury and non-contact injury. Contact injury refers to the injury developed through direct contact or impact, while non-contact injury refers to the injury developed due to positioning, momentum, direction change, hyperextensions, etc., [3]. For the contact injury, the head and neck were the commonly injured body regions in karate, usually presented with mild contusions and lacerations [3]. As compared to other full-contact combat sports like boxing, kickboxing, and mixed martial arts, karate has a milder injury in these two regions [24]. One of the reasons was that sport karate's scoring system focuses on "speed" rather than the "knockdown" concept, while other combat sports focus on attacking the head region to decrease the opponent's cognitive function or perception, and subsequently losing the ability to fight with full awareness [25]. Another reason is that protective padding for hand and feet is required to be worn by the practitioner during the championship [6]. Hence, severe injuries could be prevented in most cases. The orofacial injury was also commonly encountered by the karate practitioner [26], which is due to the heavy punching or kicking technique mistakenly executed by the training partner or opponent. Most of the contact injuries occurred during the fist strikes or kicking movements, especially a combination of high-speed punching-kicking techniques was performed. Although takedown action has contributed to the injury, the reporting cases are not as frequent as punching or kicking techniques. Recently, a rare incident was encountered by a practitioner, where she sustained with spinal cord injury at the cervical and thoracic level (damaged at TH11/12) after mistakenly felling the ground during training camp [27]. Figure 4 shows the movements executed by the practitioner that could potentially cause a minor injury to the partner or opponent.

Apart from the contact injury, non-contact injury is commonly experienced by the karate practitioner. A study reported that knee injury was commonly sustained by karate practitioners [29]. Knee injury, however, could be contracted due to contact injury (accounting for 41.1% due to hitting the opponent) and non-contact injury (accounting for 32.5% due to abrupt rotation of the feet, landing from height, abrupt

Fig. 4 Karate practitioner executing a speed punch toward the opponent's head section, b takedown action during an international championship. *Source* WKF.net [28]

stopping, and falling) [29]. Approximately, 73.6% of Iranian elite karate practitioners suffered from knee injuries, i.e., anterior cruciate ligament (ACL) rupture (6.9%), articular cartilage (5.4%), and meniscus damage (3.8%). These statistics were obtained from a sample of 390 male athletes, with the age ranging from 21 to 27 years old [29]. In another study conducted in an international karate championship (a tournament used to qualify the practitioner for Tokyo Olympic Games), 24.2% of the practitioners were found to suffer from a severe knee injury [2]. The tendencies they contracted with the knee injury during training and during the competition were 54.9% and 45.1%, respectively. The researchers further highlighted that approximately 28.0% of the Kumite practitioners (sparring) and 13.9% of kata practitioners (form performance) are contracting injuries [2]. Among the karate practitioner, the upper-body injury would be lower compared to the upper body. The reason is that practitioners frequently use leg technique instead of hand technique as the scoring points awarded are two to three times higher [30, 31]. Ankle injury such as ankle sprain was also a regular injury encountered by karate practitioners [6]. The possible prevention of ankle injury would be ankle disk training, taping, and bracing [6]. A slow progression of training also helps in preventing injury, by allowing the tissues gradually adapt to the loads' increment [32]. The details of injury types associated with the elite karate practitioner in Iran are shown in Fig. 5.

5 Conclusion

The global developments and research landscape on sports injury of karate were examined based on 139 publications (1992–2021) extracted from WoS. The emerging trend in the research areas was observed from 2016. The top leading countries indicated that the USA has published the most publications related to sports injury research in karate. This manuscript reviewed the potential injuries and their associated causes. The most common injury was a knee injury, which could be contracted during training session or tournament. These findings would be useful for the

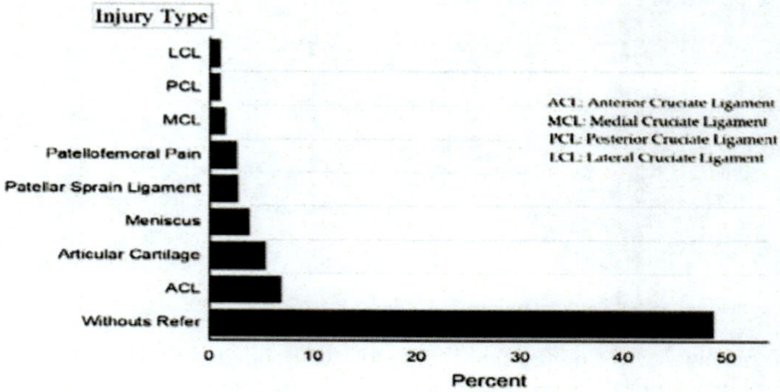

Fig. 5 Type of knee injuries suffered by the 287 elite practitioner

instructor and clinicians in designing some training activities, which aim to reduce the risk of the practitioner contracting injury. In future studies, it is highly recommended to conduct the injury assessment on the traditional karate practitioner such as Kyokushin Kai (full-contact karate style). Such exploration could provide insight into karate-associated injuries and reduce the bias in bibliometric data analysis.

Acknowledgements This study is supported by Universiti Teknologi Malaysia (UTM) Zamalah Grant, under the Vot. No. 00N04.

References

1. Tulendiyeva A, Saliev T, Andassova Z, Issabayev A, Fakhradiyev I (2021) Historical overview of injury prevention in traditional martial arts. Sport Sci Health, 1–12
2. Pérez-Martín L, Augustovicova D, Martínez-DE-Quel Ó, Tabben M, Arriaza R (2021) Severe knee injuries among karate athletes ranking towards the Tokyo olympic games: a retrospective study. J Sports Med Phys Fitness 61(8):1117–1124
3. Lystad RP, Augustovičová D, Harris G, Beskin K, Arriaza R (2020) Epidemiology of injuries in olympic-style karate competitions: systematic review and meta-analysis. Br J Sports Med 54(16):976. https://doi.org/10.1136/bjsports-2020-101990
4. Wong SL, Mong GR, Nyakuma BB, Ngadi N, Wong KY, Hernández MM, Armenise S, Chong CT (2022) Upcycling of plastic waste to carbon nanomaterials: a bibliometric analysis (2000–2019). Clean Technol Environ Policy 24(3):739–759. https://doi.org/10.1007/s10098-021-02267-w
5. Kujala UM, Taimela S, Anttipoika I, Orava S, Tuominen R, Myllynen P (1995) Acute injuries in soccer, ice hockey, volleyball, basketball, judo, and karate—analysis of national registry data. BMJ 311(7018):1465–1468. https://doi.org/10.1136/bmj.311.7018.1465
6. Parkkari J, Kujala UM, Kannus P (2001) Is it possible to prevent sports injuries? Sports Med 31(14):985–995
7. Garg KC, Singh RK (2021) A bibliometric study of papers published in library and information science research during 1994 2020. DESIDOC J Libr Inf Technol 42(1):57–63. https://doi.org/10.14429/djlit.42.1.17480

8. Zetaruk MN, Violan MA, Zurakowski D, Micheli LJ (2005) Injuries in martial arts: a comparison of five styles. Br J Sports Med 39(1):29–33. https://doi.org/10.1136/bjsm.2003.010322
9. Parkkari J, Kannus P, Natri A, Lapinleimu I, Palvanen M, Heiskanen M, Vuori I, Järvinen M (2004) Active living and injury risk. Int J Sports Med 25(3):209–216. https://doi.org/10.1055/s-2004-819935
10. Buse GJ (2006) No holds barred sport fighting: a 10 year review of mixed martial arts competition. Br J Sports Med 40(2):169–172. https://doi.org/10.1136/bjsm.2005.021295
11. Gartland S, Malik MHA, Lovell ME (2001) Injury and injury rates in Muay Thai kick boxing. Br J Sports Med 35(5):308–313. https://doi.org/10.1136/bjsm.35.5.308
12. Yard EE, Knox CL, Smith GA, Comstock RD (2007) Pediatric martial arts injuries presenting to emergency departments, United States 1990–2003. J Sci Med Sport 10(4):219–226. https://doi.org/10.1016/j.jsams.2006.06.016
13. Critchley GR, Mannion S, Meredith C (1999) Injury rates in Shotokan karate. Br J Sports Med 33(3):174–177. https://doi.org/10.1136/bjsm.33.3.174
14. Macan J, Bundalo-Vrbanac D, Romic G, Critchley G (2006) Effects of the new karate rules on the incidence and distribution of injuries—commentary. Br J Sports Med 40(4):326–330. https://doi.org/10.1136/bjsm.2005.022459
15. VencesBrito AM, Ferreira MAR, Cortes N, Fernandes O, Pezarat-Correia P (2011) Kinematic and electromyographic analyses of a karate punch. J Electromyogr Kinesiol 21(6):1023–1029. https://doi.org/10.1016/j.jelekin.2011.09.007
16. Agnew PS (1993) Taping of foot and ankle for Korean karate. J Am Podiatr Med Assoc 83(9):534–536. https://doi.org/10.7547/87507315-83-9-534
17. Augustovicova D, Lystad RP, Arriaza R (2019) Time-loss injuries in karate: a prospective cohort study of 4 consecutive world karate championships. Orthopaedic J Sports Med 7(8). https://doi.org/10.1177/2325967119865866
18. Lynch KR, Anokye NK, Vlachopoulos D, Barbieri FA, Turi-Lynch BC., Codogno JS, Fernandes RA (2020) Impact of sports participation on incidence of bone traumatic fractures and health-care costs among adolescents: ABCD—growth study. Phys Sports Med 48(3):298–303. https://doi.org/10.1080/00913847.2019.1685859
19. Wong SL, Nyakuma BB, Wong KY, Lee CT, Lee TH, Lee CH (2020) Microplastics and nanoplastics in global food webs: a bibliometric analysis (2009–2019). Mar Pollut Bull 158:111432. https://doi.org/10.1016/j.marpolbul.2020.111432
20. Esmaeilpoor A, Mohebbi SZ, Moghadam N, Ahmadian M, Razeghi S, Khami MR (2021) Self-reported experience of orofacial injury, preventive practice, and knowledge of Iranian adolescent martial art athletes towards sports-related orofacial injuries. BMC Sports Sci Med Rehabil 13(1):1–9
21. Vando S, Filingeri D, Maurino L, Chaabène H, Bianco A, Salernitano G, Foti C, Padulo J (2013) Postural adaptations in preadolescent karate athletes due to a one week karate training camp. J Human Kinet 38:45
22. Truszczyńska A, Drzał-Grabiec J, Snela S, Rachwal M (2015) Postural stability of children undergoing training in karate. Arch Budo 11:53–60
23. Skinner HB, Barrack RL (1991) Joint position sense in the normal and pathologic knee joint. J Electromyogr Kinesiol 1(3):180–190
24. Lystad RP (2015) Epidemiology of injuries in full-contact combat sports. Australas Epidemiologist 22(1):14–18
25. Tong-Iam R, Rachanavy P, Lawsirirat C (2017) Kinematic and kinetic analysis of throwing a straight punch: the role of trunk rotation in delivering a powerful straight punch. J Phys Educ Sport 17(4):2538–2543
26. Ghorpade RR, Bagal K, Karandikar K (2021) Contemporary studies on suitability of biomaterials for mouth-guards. Mater Today: Proc 47:5601–5606
27. Goraczko A, Zurek G, Lachowicz M, Kujawa K, Blach W, Zurek A (2020) Quality of life after spinal cord injury: a multiple case study examination of elite athletes. Int J Environ Res Public Health 17(20):7437

28. Federation WK (2017) Turkish celebration in day 1 of Karate European championships. https://www.wkf.net/news-center-new/turkish-celebration-in-day-1-of-karate-european-championships/475
29. Naserpour H, Baker JS, Letafatkar A, Rossettini G, Dutheil F (2021) An investigation of knee injury profiles among Iranian elite karatekas: observations from a cross-sectional study. Int J Environ Res Public Health 18(13):6888
30. Čierna D, Lystad RP (2017) Epidemiology of competition injuries in youth karate athletes: a prospective cohort study. Br J Sports Med 51(17):1285–1288
31. Rahnama N, Bambaeichi E, Bagher Nazarian A, Batavani M, Sadeghipour H (2011) Injury profile of Iranian professional male and female karate players. J Exerc Sci Med 3(1):21–37
32. Berry MJ, Renstrom P, Heinrichs KI (1995) Sports injuries: basic principles of prevention and care. The Encyclopedia of Sports Medicine vol 4. In: LWW
33. Parkkari J, Kujala UM, Kannus P (2001) Is it possible to prevent sports injuries? Review of controlled clinical trials and recommendations for future work. Sports Med 31(14):985–995. https://doi.org/10.2165/00007256-200131140-00003

Directing External Focus Instruction to Improve Overhand Volleyball Serve

Fatin Razak, Maisarah Mod Saleh, Nur Atikah Mohamed Kassim, Rozella Ab Razak, Hosni Hasan, and Cyndra Robert Budull

Abstract An explicit attentional focus is needed in performing a task. The focus of attentional cues during skill execution has received much attention in delivering the correct technique in action. However, internal attention focus that focuses on body's coordination is more complex to be followed compare to external attention focus which focusing on environmental effects. This study investigates the impact of specific attentional focus on novices in overhand volleyball serve. The participants ($N = 132$) were assigned into three groups that received different attentional focus: IF group ($n = 44$) internal focus attention, EF ($n = 44$) external focus attention, and CONF group ($n = 44$) no attentional focus attention. The measurements transpired (acquisition test and retention test). The participants performed 30 trials divided into seven blocks of five trials of overhand serve in volleyball skill using their dominant hand. North Carolina State University Volleyball Skills Test Battery and instruction scoring were used to evaluate the overhand serve performance of 7 blocks of 5 trials. ANOVA-repeated measures revealed a significant interaction between groups and measurements in a 3(groups: IF, EF, and CONF) $\times$ 2(measurements: acquisition test and retention test) design and post-hoc Bonferroni adjustment analysis revealed that the EF group demonstrated the most significant improvement in instruction scoring and serving to score. Furthermore, participants from EF understand better in executing the action compared to other groups. Conclusion: Individuals' skill-based performance can be impacted by directing their attention to internal or external stimuli. By focusing their attention externally while doing a motor task, novices can improve their serve technique. Recommendations are made to explore several criteria that may affect the participant's learning process in greater detail.

F. Razak (✉) · M. Mod Saleh · N. A. Mohamed Kassim · R. Ab Razak
Faculty of Sports Science and Recreation, Universiti Teknologi MARA Pahang, Jengka Campus, Pahang, Malaysia
e-mail: fatinaqilah@uitm.edu.my

H. Hasan
Faculty of Sports Science and Recreation, Universiti Teknologi MARA, Shah Alam, Malaysia

C. R. Budull
Faculty of Sports Science and Recreation, Universiti Teknologi MARA, Samarahan Campus, Kuching, Sarawak, Malaysia

G. Kuan et al. (eds.), *Advancing Sports and Exercise via Innovation*, Lecture Notes in Bioengineering, https://doi.org/10.1007/978-981-19-8159-3_18

Keywords Internal focus attention · External focus attention · Motor skill performance · Overhand serve · Novice

1 Introduction

According to [1], the importance of focusing on motor skill performance has influenced scientists' interest in the learning of human mobility. The precision and value of skill performance are highly dependent on where athletes direct their attention while doing a skill [2]. Hartstra et al. [3] established a link between the influence on the brain and the muscles recruited to do the task based on the attentional focus offered. Additionally, it is referred to as brain-muscle connection. Additionally, this includes assisting and mentoring new learners or beginners in attempting to apply performance. This is another level of communication; it involves communication when muscles act together. Saladin [4] revealed further that this synchronization takes position in the brain stem, a reptile brain attachment. When instructions were delivered, the person would opt to move according to the focus of attention (voluntary movement). At the same time, the theoretical basis for the efficiency of adoption and external attention resides in the constrained action hypothesis. Palmer [5] proposed this theory that the internal focus restricts the motor system's ability to perform automated movement processes and interferes with them. Comportment research that supports the CAH reveals that an external focus relates to increased job automation [6].

Many prior studies investigating the differential impacts of internal and external attentional concentration on sport performance have concentrated on skill-based activities. In sports, an athlete's external focus occurs when he or she diverts attentional resources away from the activity itself and toward its repercussions or the impact of the movement on the surrounding environment rather than the movement itself. Individuals who are athletes develop internal concentration when they direct their attentional resources inward toward the action of their bodies or their sensations about their bodies during competition. Attentional focus training, according to [7], ought should be short and precise, directing the concentration on a restricted amount of information at a time, such as a single or two items. Beginning with the most fundamental and necessary information that will aid in the execution of a new skill, it is important to maintain your attentional focus throughout the process of acquiring a new talent. For instance, in carrying out a jump, individuals can focus on how far they can perform a save landing by targeting a certain distance (i.e., external focus), or they can concentrate on pressing their feet to get off the board as strongly as possible (i.e., internal focus) [5]. Pelleck and Passmore [8] discovered that inexperienced golfers' internal emphasis on task-relevant (i.e., hands and elbows) characteristics had a less negative influence on accuracy than when they focused on fewer task-relevant internal indicators (i.e., body position). In comparison, there was no discernible variation in the accuracy of experienced golfers' focus.

When novices were instructed to pay attention to task-relevant core strength, signals that were both internal (e.g., wrist movement) or external (e.g., handle movement), the researchers discovered that they performed better on a jump-rope task than when their attention was directed toward task-relevant core strength cues that were either internal (e.g., wrist movement) or external (i.e., handle movement) (i.e., movement of the handle). Internal (foot movement) and external (handle movement) lower-body cues were shown to be less effective than external (handle movement). The attentional focus that is primarily focused on body movement coordination has been demonstrated to impede optimal cognition during the process of learning.

Uehara et al. [9] discovered that education focused on body motion may impair learning compared to no instruction. In contrast, [10] stated that it is still uncertain if an external focus continues to be beneficial after several training sessions and whether the relative detrimental influence of internal focus on performance and learning may be overcome with additional practice. In conclusion, one "dosage" of external focus attention remains as the focus on repeated practice sessions and long-term changes in motor learning are unsupported. A solid attentional focus should be emphasized and utilized in teaching or coaching individuals regardless of their skill levels. In this context, the objective of this study is to investigate the efficacy of different focus groups in volleyball improvement.

2 Methodology

2.1 Participants

Purposive sampling was used in this study, which enrolled 132 individuals (N = 132). They were selected based on the following inclusion criteria: female student, minimal knowledge of volleyball serves, age between 20 and 25, and only use of the right dominant hand. Following that, each participant was required to sign a consent form expressing their voluntary participation in the study.

2.2 Procedure

The study was divided into three sessions: the beginning phase, the acquisition test, and the retention test. The initial and acquisition tests were completed in a single day, and the retention test was completed two days later. The beginners were given two-day intervals due to the temporary influence of practice conditions [11]. Initially, the researcher invited volunteers with no prior expertise with overhand serves to practice the serve. The researcher provided fundamental stance, posture, and swing instructions and demonstrations. They were randomly separated into three groups: (n = 44) subjects were assigned into group IF (internal focus); (n = 44) subjects were assigned into group EF (external focus), and ($n = 44$) subjects were assigned into group CONF (CONTROL: no focus given).

Thirty serves were completed in the acquisition test—means acquired from 1 block of 5 trials out of a total of 6 blocks of 5 trials in the test. The maximum score for a single block of five trials was twenty points. All the participants in this study performed the same exercise, an overhand serve, but received a variety of various sorts of instruction during the process. After each of the three blocks of five trials in the acquisition test, the researcher reminded the participants of the internal and external attention instructions they had received earlier. The retention test was scheduled two days after the two-day break. There was no guidance offered throughout the retention period. Participants completed one block of five trials.

2.3 *Outcome Measures*

North Carolina State University Volleyball Skills Test Battery.

The instrument was a North Carolina State University Volleyball Skills Test Battery with coefficients of 0.81 [12]. Total points were awarded for each trial, with 4 points as a maximum score.

2.4 *Instruction Scoring*

See (Table 1).

Table 1 Internal and external focus attention order implementation for instruction scoring

Internal instruction	External instruction
• Toss the ball far enough in front of the hitting arm • While hitting the ball, snap your wrist to produce a forward rotation of the ball • Shortly before hitting the ball, move your weight from the back leg to the front leg • Arch your back and accelerate your shoulder first, then your upper arm, your lower arm, and finally your hand	• Toss the ball straight up • Imagine holding a bowl in your hand and cupping the ball with it to induce a forward rotation of the ball • Transfer weight toward the target shortly before hitting the ball • Hit the ball as if driving horses with a whip

Note Adopted from [13], principles derived from the study of simple skills do not generalize to complex skill learning

2.5 Data Analysis

This study used descriptive and inferential statistics. When evaluating the data for descriptive purposes, the mean and standard deviation were emphasized. A repeated measure ANOVA was used for the inferential statistical design to examine the differences among subjects acquired by the correctness of instruction scores and serving scores, additionally to disparities across subjects in terms of resolving the various forms of concentration attention.

2.6 Statistical Package for the Social Sciences (SPSS)

The Statistical Package for the Social Sciences was used to examine the data (SPSS. Version 24.0). The post-hoc Bonferroni correction was used to modify all comparisons. For all tests, the alpha $p > 0.05$ was set.

3 Findings

3.1 Descriptive Statistics

See (Table 2).

In the internal focus statistics table, we can see that the mean score of instruction 1 (AIS1) was 1.431 ± 1.822, which is a good starting point. One of the tests on which IF had the lowest possible score was AIS 2, with a score of 1.204 ± 1.192. It is more likely to begin with AIS 3 and progress to RIS 7. The IF received the highest possible score at RIS, which was 4.522 ± 1.284. At the AIS1 experiment, the lowest mean score for the external focus group was 6.090 ± 2.768, which was the lowest possible score. In the beginning, there was a decrease in the mean and standard deviation scores, but starting with AIS 4, RIS 7 produced greater mean and standard deviation scores, and it achieved its greatest mean score at this point in the study with 12.568 ± 1.284. Contrary to the control group, which did not get any instruction, the mean score for the CONF group improved steadily over time, starting with 0.704 ± 1.249 at AIS 1. Progressively, higher mean scores were achieved until it achieved the greatest mean score of 4.386 ± 2.442 at RIS7, which was the highest of all. As a result, the EF group received a significantly higher mean score than the other two groups, as indicated in Table 3.

The mean score of serving (ASS1), according to the internal focus statistics table, was 1.64 ± 3.111 points. One of the areas where IF's scores dropped the most was on ASS2, where the score dropped to 1.295 ± 1.456. It has been raised from ASS3 to RSS7 as a starting point. RSS7 had the highest mean score for the IF group, with a total of 5.590 ± 1.529 points. For the external focus group, the results were different:

Table 2 Instruction scores (M $\pm$ SD) for the acquisition and retention tests for each group

Group	AIS1	AIS2	AIS3	AIS4	AIS5	AIS6	RIS7
Internal	1.431 $\pm$ 1.822	1.204 $\pm$ 1.192	1.909 $\pm$ 1.639	2.250 $\pm$ 1.601	2.750 $\pm$ 1.766	2.704 $\pm$ 2.041	4.522 $\pm$ 1.284
External	6.090 $\pm$ 2.768	7.545 $\pm$ 2.609	7.340 $\pm$ 2.868	7.909 $\pm$ 3.388	9.068 $\pm$ 3.466	10.090 $\pm$ 4.141	12.568 $\pm$ 2.636
No instruction	0.704 $\pm$ 1.249	1.409 $\pm$ 1.451	2.068 $\pm$ 1.992	2.727 $\pm$ 2.670	2.977 $\pm$ 2.406	3.568 $\pm$ 2.481	4.386 $\pm$ 2.442

Table 3 Serving scores (M ± SD) on acquisition and retention tests for each group

Group	ASS1	ASS2	ASS3	ASS4	ASS5	ASS6	RSS7
Internal	1.636 ± 3.111	1.295 ± 1.456	1.500 ± 1. 691	2.236 ± 2.103	3.000 ± 2.573	3.477 ± 2.556	5.590 ± 1.529
External	4.568 ± 2.061	6.522 ± 2.183	6.386 ± 2.169	6.840 ± 2.495	8.136 ± 2.800	9.590 ± 2.545	10.522 ± 2.277
No instruction	0.545 ± 1.109	1.318 ± 1.474	1.772 ± 1.975	2.181 ± 2.127	2.704 ± 1.887	3.250 ± 2.058	4.500 ± 2.028

the ASS1 trial had the lowest mean score of 4.568 ± 2.061, whereas the ASS2 trial had the highest mean score of 4.568 ± 2.061. It had a lower mean score at ASS3 with 6.386 ± 2.169, but outcomes of mean and standard deviation score began to improve starting at ASS4 and reached its best mean score at RSS7 with 10.522 ± 2.277, which was the lowest of the seven scores. In the end, the CONF group began with an ASS1 mean score of.545 ± 1.109, but the mean scores gradually climbed until they achieved the highest mean score of 4.500 ± 2.028 at RSS7, the group with no concentration. Because of this, the table displayed the EF group as having the highest mean score when compared to the other two groups.

3.2 *Inferential Statistics*

Repeated ANOVA was used to analyze the treatment differences between two groups and one control group. Wilk's Lambda = 0.458, F (12,248) = 9.878; $p < 0.05$, partial eta2 = 0.323 demonstrated a statistically difference in significant of instruction scoring in the seven blocks of trials. Within the group, there was a difference in instruction scoring for seven blocks of trial, as measured by Wilks' Lambda = 0.122, F (6124) = 1.486; $p < 0.05$, and partial eta2 = 0.878. According to the results, there were significant variations in instruction score between treatment groups (IF, EF) and CONF. Additionally, it demonstrated that EF was the most effective treatment for increasing overhand volleyball serves among athletes, learners, and students.

In conclusion, Wilk's Lambda = 0.0.652, F (12,248) = 4.934; $p < 0.05$, partial eta2 = 0.193 suggested there was a statistically significant difference in serving scoring in the seven (blocks of five trials). Within the group, there was a statistically significant difference in serving scoring of seven (blocks of five trials), which indicated that Wilks' Lambda was in fact significant = 0.96, F(6124) = 1.954; $p < 0.05$, partial eta2 = 0.904. Additionally, it was demonstrated that EF was the most effective therapy method for verbal instruction to athletes or students.

4 Discussion

The present study aimed to examine the effectiveness of different focus attention in improving volleyball overhand serve among novices. This was achieved through the use of practice sessions with the participants in the initial phase, acquisition test, and retention test. The result of the ANOVA revealed that both training groups did show an improvement in their performance from the beginning to the end of blocks within trials. The findings confirmed that EF group attention improves tremendously and effectively serves execution toward the novices. The data analysis demonstrated a significant improvement in decision-making performance for novices trained to use an EF of attention compared to novices instructed to use an IF of attention and the control group. The EF outperformed the IF and CONF in both acquisition and retention testing in terms of the test component. Previous studies investigating object manipulation competencies have demonstrated the effectiveness of the external focus of attention. Volleyball research results [14], basketball [15], brought about the conclusion as mentioned earlier. According to Neumann et al. [1], more muscular effort rather than more coordinated or fluid motions may be responsible for the larger distance and power output with an internal concentrate rather than an external emphasis. Support for this conclusion can be seen in the findings on heart rate, which showed that heart rate was significantly higher in the internal attention condition than in the external focus condition.

Finally, there was a statistically significant difference in serving scoring between the two groups of IF and EF. The experiment's finding showed a significant improvement in serving scoring, which indicates they have improved significantly on the scoring from initial to final trials within the blocks. However, while both groups demonstrated improvement, the EF scored significantly higher than the IF and CONF. This is because the EF group practiced under movement on the environmental cues, which improved the performance of skills. The findings confirmed that EF attention could improve and is more effective in serving execution toward the novices. According to Neumann et al. [1], rather than more coordinated or fluid movements, it is possible that increased muscular effort is responsible for the greater distance and power output with internal concentration as opposed to external concentration. Palmer et al. [5], according to the teaching and learning perspective, telling participants to concentrate their attention externally on the outcome of a motor skill greatly enhanced their object control abilities when compared to when no attentional direction instructions were given.

Furthermore, this study indicated that demonstrating the correct focus attention to novices provides the right guidance for their vision, images, and action. Other research revealed a significant improvement in decision-making performance for athletes instructed to use an external focus of attention compared to athletes instructed to use an internal focus of attention and to the control group [1]. According to the research findings, an external focus appears to be helpful even after several training sessions, and ongoing practice with an internal focus does not outweigh the benefits of an external focus, according to the research findings [10]. Landers et al. [16] stated that adopting external attention training during free throws in basketball indicated greater practical implications for specialized sport performance. The current study's findings also revealed that the external teaching groups scored higher. This was due to the lack of attention and easy instruction that focuses on other factors such as equipment, surroundings, and others. External focus (i.e., focusing on the desired movement result) is consistently increase learning comparing to internal focus on physical motions or no focus concentration require (control circumstances) [17–19].

Black and Wright [20] highlighted that the necessity for explicit instruction is greater when the impact of their first fundamental of accomplishing the task is at stake. On the other side, a lack of focus or hazy attention might result in substandard output and an ineffective completion of the activity. Based on the result of the control group, [2] concluded that the athletes in the control group gained their natural internal focus through prior training that was oriented toward technique performance. Similar to finding from [21], simple attentional focus (on the surroundings or equipment) is far easier to comprehend than sophisticated attention that requires body coordination. As a result, this method is more constant in its emphasis on the external environment when instructing the new learner. It is more advantageous to educate new learners on how to perform better than it is to provide intricate concentrate attention. Even if the instruction is merely temporary, easy-to-understand instruction may provide an explicit focus of attention. Nonetheless, the possibility of capturing the correct technique can be transferred for a particular time [22].

Additionally, this external perspective may benefit other areas, such as sports coaching and training, athletic injury, and rehabilitation. It influences participants' movement patterns independent of their familiarity with the encoded activity. According to Lohse et al. [23], focusing external attention facilitates recovery and performance of motor abilities in experimental yet clinical populations.

5 Conclusion

This has brought the researcher closer conclusions and recommendations to be improved for future research purposes. For example, according to this study's findings, applying cues that guide attention externally improves decision-making in novice volleyball players. Due to this, coaches ought to be aware of learning methods, which often impact the development of technique and cognitive growth, so that players can solve difficulties during the game.

Future research should apply electromyographic recordings to directly assess muscle fiber activation in response to a variety of attentional focus commands, according to the findings. Furthermore, a comparative usage of a broader and more diverse range of participants (such as expert, novice, young, old, and gender) is advocated. Regardless of the participants' experience, it may provide a new perspective and all the effectiveness of attentional focus.

Acknowledgements A highly appreciation to the Universiti Teknologi Mara for the unconditional support. We thank you e-RTS, UiTM for the helpful comments on earlier versions of this manuscript.

References

1. Neumann DL, Walsh N, Moffitt RL, Hannan TE (2020) Specific internal and external attentional focus instructions have differential effects on rowing performance. Psychol Sport Exerc 50:101722
2. Tsetseli M, Zetou E, Vernadakis N, Michalopoulou M (2016) The effect of internal and external focus of attention on game performance in tennis. Acta Gymnica 46(4):162–173. https://doi.org/10.5507/ag.2016.021

3. Hartstra EA, Florian Waszak BC, Brass MDE (2012) The implementation of verbal instructions: Dissociating motor preparation from the formation of stimulus–response associations. Neuroimage 6(3):1143–1153
4. Saladin K (2011) Anatomy and physiology: The unity of form and function. Mc Graw Hill Companies, New York
5. Palmer KK, Matsuyama AL, Irwin JM, Porter JM, Robinson LE (2017) The effect of attentional focus cues on object control performance in elementary children. Phys Educ Sport Pedagogy 22(6):580–588. https://doi.org/10.1080/17408989.2017.1294667
6. Wulf GN, McNevin SCH (2001) The automaticity of complex motor skill learning as a function of attentional focus. Q J Exp Psychol 54(4):1143–1154
7. Edwards W (2010) Motor learning and control: From theory to practice. Cengage Learning, Wadsworth
8. Pelleck V, Passmore SR (2017) Location versus task relevance: The impact of differing internal focus of attention instructions on motor performance. Acta Physiol (Oxf) 176:23–31. https://doi.org/10.1016/j.actpsy.2017.03.007
9. Uehara LA, Button C, Davids K (2008) The effects of focus of attention instructions on novice learning soccer chip. Braz J Biomotricity 2(1):63–77
10. Diekfuss JA, Rhea CK, Schmitz RJ, Grooms DR, Wilkins RW, Slutsky AB, Raisbeck LD (2018) The influence of attentional focus on balance control over seven days of training. J Mot Behav 51(3):281–292. https://doi.org/10.1080/00222895.2018.1468312
11. Wulf G, Mcnevin NM (2003) Simply distracting learners is not enough: more evidence for the learning benefits of an external focus of attention. European. J Sports Sci 3(5):1–13
12. Singh SB (2010) Construction and standardization of volleyball skills test for men players: A usual study. Int J Educ Adm 2:139–148
13. Wulf G, Shea CH (2002) Principles derived from the study of simple skills do not generalize to complex skill learning. Psychon Bull Rev 9(2):185–211
14. Wulf G, McConnel N, Gartner M, Schwarz A (2002) Enhancing the learning of sports skills through external focus feedback. J Mot Behav 34:171–182
15. Al-Abood SA, Bennet SJ, Hernandez FM, Ashford D, Davids K (2002) Effect of verbal instructions and image size on visual search strategies in basketball free throw shooting. J Sports Sci 20:271–278
16. Landers M, Wulf G, Wallman H, Guadagnoli MA (2005) An external focus of attention attenuates balance impairment in Parkinson's disease. Physiotherapy 91:152–185
17. Lohse KR, Wulf G, Lewthwaite R (2012) In NJ Hodges, AM Williams (Eds.), Skill acquisition in sport: Research, theory & practice (2nd ed.) 40–58 London: Routlegde
18. Marchant DC (2011) Attentional focusing instructions and force production. Front Psychol 1(Article 210) doi: https://doi.org/10.3389/fpsyg.2010.00210
19. Wulf G (2013) Attentional focus and motor learning: A review of 15 years. Int Rev Sport Exerc Psychol 6:77–104
20. Black CB, Wright DL (2000) Can observational practice facilitate error recognition and movement production? Res Q Exerc Sport 71:331–339
21. Lawrence GP, Gottwald VM, Hardy J, Khan MA (2011) Internal and external focus of attention in a novel form sport. Res Q Exerc Sport 82:431–441
22. Peh SY, Chow JY, David K (2011) Focus of attention and its impact on movement behavior. J Sci Med Sport 14:70–78
23. Lohse KR, Sherwood DE, Healy AF (2011) Neuromuscular effects of shifting the focus of attention in a simple force production task. J Mot Behav 43:173–183

Effects of Physical Exercise on Post-traumatic Growth of Adolescents After Disaster Stress Events: A Conditional Process Model

Jin Chen and Hu Lou

Abstract After the disaster, stress would affect the people in the affected areas for a long time. If psychological relief and rescue were not carried out in time, it may induce a series of physical and mental diseases. Physical exercise is low cost and easy to popularize, which is in line with the characteristics of widespread disaster impact. Therefore, it has important research significance in post-disaster stress psychological rescue. The purpose of this study is to explore the comprehensive effects of adolescent physical exercise and empathy on post-traumatic growth after disaster stress events, investigate the mediating effect of empathy, the regulatory effect of gender and learning stage, and construct a conditional process model. A longitudinal follow-up design was used to investigate 1069 middle school students in Zhengzhou, which were affected by heavy rain in Henan Province. Physical exercise ($\beta = 0.75$), empathy ($\beta = 0.79$) had a significant positive impact on post-traumatic growth prosocial behavior ($P < 0.01$); when physical exercise affected post-traumatic growth, the mediating effect of empathy was significant ($P < 0.01$); in addition, the learning stage regulates the impact of physical exercise on post-traumatic growth, and gender regulates the impact of physical exercise on empathy. After disaster stress events, adolescent physical exercise cannot only directly affect post-traumatic growth, but also indirectly affect post-traumatic growth by inducing empathy; in contrast, physical exercise is more likely to trigger empathy among girls, and the body at this time, education and exercise have a greater impact on high school students' prosocial behavior.

Keywords Stress · Physical exercise · Empathy · Post-traumatic growth · Adolescent

J. Chen · H. Lou (✉)
Nantong University, Nantong 226019, Jiangsu Province, China
e-mail: 15692068@qq.com

J. Chen
e-mail: 1172113808@qq.com

G. Kuan et al. (eds.), *Advancing Sports and Exercise via Innovation*, Lecture Notes in Bioengineering, https://doi.org/10.1007/978-981-19-8159-3_19

1 Introduction

Earthquakes, floods, tsunamis, and other natural disasters are the natural enemies of human beings. Their great destruction is not only manifested in the loss of production and living facilities, materials, property, and other aspects, but also brings a strong impact on people's psychology. Even after the disaster, strong psychological stress will affect people in the affected areas for a long time. If psychological relief and rescue are not carried out in time, it is easy to produce a series of post-disaster physical and mental reactions, and even serious physical and mental diseases may be induced.

Physical exercise can antagonize acute psychological stress, improve public psychological state, relieve symptoms of post-traumatic stress, and improve individual health and behavior level [1]. In addition, physical exercise is low cost and easy to popularize, which is in line with the characteristics of the widespread impact of disasters. Therefore, it has important research significance and application potential in psychological relief of post-disaster stress. In this study, physical exercise was introduced into the post-stress intervention, and the tracking samples after the heavy rain disaster in Henan Province were used to explore the law of physical exercise and traumatic recovery of adolescents after stress events, in order to provide reference for the exercise program of psychological intervention of adolescents after stress events.

Researchers have found that individuals will not only have post-traumatic stress disorder after disaster and trauma, but also have a positive adaptive change called post-traumatic growth (PTG) [2]. Unlike post-traumatic stress disorder, which only a few people suffer from, PTG is a more common response to stress events. It can not only help the public in disaster areas recover from psychological trauma, but also has positive significance for individual coping ability, healthy development, and social adaptation in the future[3, 4]. Therefore, the concept, influencing factors, mechanism path, and promotion strategy of disaster psychology have become one of the hot topics in recent years.

Empathy is one of the most common influencing factors in PTG studies after stress events. It is an empathetic ability, which means that when disasters or mishaps occur, individuals can produce emotional responses such as sympathy, and then develop altruistic behaviors such as kindness and help, helping individuals to develop from traumatic stress to PTG [5]. Physical exercise may be a behavioral promoter of PTG through a review of physical exercise and PTG [6]. Participation in physical exercise may make post-traumatic individuals feel their physical and mental strength, relieve emotional distress, divert attention, and develop meaningful social relationships with others, which may promote the development of PTG [7–9]. Previous authors summarized the relationship between physical exercise and empathy, believing that physical exercise plays an important role in the development of empathy [10]. It should be noted that most of the previous studies on the relationship between physical exercise and PTG adopted a cross-sectional design and were generally conducted in conventional situations rather than stressful events. The cross-sectional design limits the ability to understand that physical exercise predicts changes in PTG over time, and

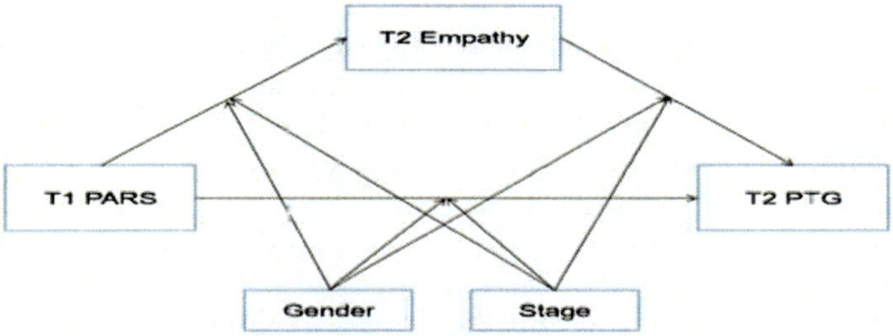

Fig. 1 Hypothetical model of post-traumatic growth affected by physical exercise after stress events

that PTG development may differ from that in the general context after a stressful event. Therefore, it is necessary to use longitudinal design to explore the effect of physical exercise on promoting PTG after stressful events.

In addition, there may be gender and stage differences in physical exercise, empathy, and post-traumatic growth. Atan conducted a survey of 400 teenagers, which showed that physical exercise had a higher impact on empathy among girls than boys [11]. The theoretical model of physical exercise and PTG proposed by Chen et al. believes that demographic variables such as age and gender may play a role in the relationship between them [6]. Based on the above considerations, a hypothetical model of conditional process is constructed (Fig. 1). Specific research hypotheses include: 1. For adolescents after stress events, there are gender and stage differences in physical exercise, empathy, and post-traumatic growth; 2. After stress events, teenagers come to physical exercise, which directly affects post-traumatic growth; 3. Empathy plays a mediating role in the influence of adolescent physical exercise on post-traumatic growth path after stress events; 4. In the intermediary model, gender and school stage have a moderating effect.

2 Methods

2.1 Study Subjects

In the summer of 2021, a severe rainstorm disaster occurred in Henan Province. According to statistics (as of 12:00 on August 2), the direct economic loss exceeded 100 billion yuan. Among them, 292 people were killed, and 47 people were missing in Zhengzhou, which not only caused huge property losses, but also brought serious psychological pain to the people of Zhengzhou. This study investigated middle school students in two schools in Zhengzhou, a serious disaster area, one month (T1) and two months (T2) after the severe rainstorm disaster in Henan Province. Cluster random sampling was used to distribute questionnaires to 200 students in each grade from junior one to senior one. A total of 2400 questionnaires were distributed twice.

Excluding 85 invalid questionnaires with missing choices, multiple choices, and consistent answers to all questions, 4 of them were the same subject, and the answers were invalid twice. 1117 subjects obtained valid data. In order to ensure that all samples are under the influence of stress events, the trauma exposure questionnaire is used for screening, and the subjects who choose "no such situation" in all items are excluded. In addition, referring to the paradigm of empathy research and using altruistic beliefs to screen the subjects, the screening standard is to exclude the subjects who choose "fully agree" for all items. In this part, 48 subjects were excluded, and 1069 effective subjects were obtained, including 487 boys and 582 girls, with an average age of 15.12 ± 1.72 years. All middle school students who participated in the questionnaire gave informed consent to the study.

2.2 *Research Tools*

2.2.1 Physical Activity Rating Scale

The Chinese version of Physical Activity Rating Scale (PARS) revised by Liang [12] was used to test the physical exercise level and the internal consistency coefficient Cronbach's $\alpha = 0.75$. The subjects were asked to recall the physical exercise in the past month and measure three aspects: the frequency of exercise every week, the duration of each exercise, and the intensity of normal exercise. The total amount of physical exercise was divided into the product of the three.

2.2.2 Interpersonal Reactivity Index

The Chinese version of Interpersonal Response Index scale (IRI) revised by Rong et al. [13] was adopted. The original scale had 28 items. According to Siu and Shek [14], 11 of them can be used to measure individual empathy. Therefore, this study uses IRI-E to measure the empathy of post-disaster adolescents, including two dimensions: point of view selection and empathic attention and its internal consistency coefficient Cronbach's $\alpha = 0.76$.

2.2.3 Post-traumatic Growth Inventory

The Post-Traumatic Growth Inventory (PTGI) revised by Zhou et al. [15] is used to reflect the post-traumatic growth status of adolescents after the disaster. The questionnaire includes three dimensions: the change of self-awareness, the change of interpersonal experience, and the change of life values. There are 20 items in total, and its internal consistency coefficient Cronbach's $\alpha = 0.93$.

2.2.4 Trauma Exposure Questionnaire

According to the trauma exposure questionnaire prepared by Wu et al. [16], the trauma exposure degree of teenagers is measured, and the first half sentence of each item is limited to "rainstorm disaster." The questionnaire includes 18 items, and its internal consistency coefficient Cronbach's $\alpha = 0.83$.

2.2.5 Philosophies of Human Nature, PHN

The altruistic belief subscale in the Chinese version of Philosophy of Human Nature (PHN) was used to measure the altruistic belief of teenagers. There were 14 items, and the internal consistency coefficient Cronbach's $\alpha = 0.78$ [17].

2.3 *Research Procedure and Data Processing*

One month (T1) after the severe rainstorm disaster in Henan, the sports activity level scale, trauma exposure questionnaire, and altruism belief scale were issued. Two months after the disaster (T2), the interpersonal response index scale and post-traumatic growth questionnaire were issued.

SPSS 23.0 was used for statistical analysis. Using Harman single factor test, the exploratory factor analysis with single factor and no rotation was carried out for all items. The variance explained by the first factor was less than 40%, indicating that there was no obvious common method deviation in this study. Then, we analyze the data based on the research hypothesis. Firstly, descriptive statistics and Mann–Whitney U test were used to investigate the gender and stage differences of various variables, and then, partial correlation analysis and regression analysis were used to investigate the direct impact of physical exercise on post-traumatic growth. Finally, bootstrap process (3.5) plug-in was used to analyze the mediating role of empathy and the regulatory role of gender and stage.

3 Result

3.1 *Group Differences in Physical Exercise, Empathy, and Post-Traumatic Growth of Adolescents After Stress Events*

Mann–Whitney U test of school stage and gender shows (see Table 1): there is significant difference in post-traumatic growth stage ($P < 0.01$). The comparison of the mean shows that the post-traumatic growth level of senior high school students

(64.51 ± 23.20) is better than that of junior high school students (58.39 ± 23.77), and there is no significant difference in other variables ($P > 0.05$). There were significant gender differences in physical exercise and empathy ($P < 0.05$). The comparison of the mean value found that the physical exercise level of boys (54.20 ± 26.53) was higher than that of girls (49.98 ± 26.13), and the empathy ability of girls (40.06 ± 18.11) was higher than that of boys (38.36 ± 16.41). However, there was no significant gender difference in post-traumatic growth ($P > 0.05$).

3.2 *Effects of Physical Exercise and Empathy on Post-traumatic Growth of Adolescents After Stress Events*

After the data were standardized, the partial correlation analysis was carried out by controlling gender and school stage. The results showed that (see Table 2): the positive correlation among physical exercise, empathy, and post-traumatic growth reached a significant level ($P < 0.05$).

The forced entry method was adopted with school stage, gender, physical exercise, and empathy as independent variables and post-traumatic growth as dependent variables. Four groups of regression analysis were conducted, respectively (see Table 3): physical exercise ($F_{(1,1067)} = 1309.60$, $\beta = 0.75$), empathy ($F_{(1,1067)} = 1703.98$, $\beta = 0.79$) had a significant effect on post-traumatic growth ($P < 0.01$).

The bootstrap mediation effect test method was adopted, and the conditional process model proposed by Hayes was tested to investigate the indirect effects of physical exercise on post-traumatic growth of adolescents after stressful events (see Tables 4 and 5, and Fig. 2). The results show that in the equation with IRI-E as dependent variable: Physical exercise positively predicts empathy ($\beta = 0.28$, $F_{(1,1067)} = 110.17$, 95% CI = [0.13, 0.33], $P < 0.01$) was statistically significant, and the interaction between gender and physical exercise had a significant positive prediction of empathy ($\beta = 0.05$, $F_{(1,1067)} = 4.23$, 95% CI = [0.01, 0.10], $P < 0.05$). In the equation with PTGI as dependent variable: physical exercise positively predicts post-traumatic growth ($\beta = 0.27$, $F_{(1,1064)} = 66.75$, 95% CI = [0.10, 0.43], $P < 0.01$) was statistically significant, and the interaction between school stage and physical exercise was a positive predictor of post-traumatic growth ($\beta = 0.08$, $F_{(1,1064)} = 4.14$, 95% CI = [0.01, 0.16], $P < 0.05$) was significant, and empathy positively predicted post-traumatic growth ($\beta = 0.86$, $F_{(1,1064)} = 6.66$, 95% CI = [0.61, 1.12], $P < 0.01$); empathy plays a conditional mediating role in the influence of post-traumatic growth.

Table 1 Group difference test of school stage and gender

	Stage			Gender		
	PARS	IRI-E	PTGI	PARS	IRI-E	PTGI
I(M ± SD)	52.06 ± 27.04	39.43 ± 18.95	58.39 ± 23.77	54.20 ± 26.53	38.36 ± 16.41	61.59 ± 23.20
II(M ± SD)	52.10 ± 25.68	38.95 ± 15.20	64.51 ± 23.20	49.98 ± 26.13	40.06 ± 18.11	60.90 ± 24.18
Mann–Whitney U	119,735.50	122,352.00	137,244.00	109,445.00	128,410.50	118,341.50
Wilcoxon W	223,475.50	226,092.00	240,984.00	229,740.00	248,705.50	238,636.50
Z	0.22	0.81	4.20	− 2.25	2.06	− 0.22
P	0.82	0.41	0.00**	0.02*	0.03*	0.82

Note In the stage difference test, I means junior high school and II means senior high school; In the gender difference test, I represents boys and II represents girls

Table 2 Correlation analysis between variables

	PARS	IRI-E	PTGI
PARS	1	0.82**	0.76**
IRI-E	0.82**	1	0.80**
PTGI	0.76**	0.80**	1

Note ** $P < 0.01$, * $P < 0.05$, the control variables are stage and gender

Table 3 Regression analysis of various variables on adolescents' post-traumatic growth

	B	*SE*	*β*	*T*	*F*	R_j^2
STAGE	6.11	1.50	0.12**	4.05	16.46	0.01
GENDER	− 0.69	1.51	− 0.01	− 0.46	0.21	0.00
PARS	0.67	0.01	0.75**	36.18	1309.60	0.57
IRI-E	1.09	0.02	0.79**	41.27	1703.98	0.63

Note ** $P < 0.01$, * $P < 0.05$

Table 4 Test of indirect effect of physical exercise on post-traumatic growth

	IRI-E				PTGI			
	β	*SE*	*t*	*P*	*β*	SE	*t*	*P*
Constant	23.05	3.60	6.39	0.00	14.17	2.21	6.39	0.00
Stage	−0.36	1.61	−0.22	0.82	−0.02	1.03	−0.02	0.98
Gender	1.75	1.62	1.07	0.28	−0.64	0.97	−0.59	0.55
PARS	0.23**	0.05	4.59	0.00	0.27**	0.08	3.20	0.00
IRI-E					0.86**	0.13	6.66	0.00
Stage × PARS	−0.01	0.02	−0.48	0.62	0.08*	0.04	2.03	0.04
Stage × IRI-E					−0.06	0.06	−1.02	0.30
Gender × PARS	0.05*	0.02	2.05	0.04*	0.03	0.03	0.78	0.43
Gender × IRI-E					−0.03	0.05	−0.59	0.55
R^2	0.34				0.33			
F	110.17				66.75			

Note ** $P < 0.01$, * $P < 0.05$

Table 5 Bootstrap test of indirect effects

Equation	Interactive item	ΔR^2	*F*	df_1	df_2	LLCI	ULCI
IRI-E	Stage × PARS	0.004	0.23	1.00	1067.00	−0.06	0.03
	Gender × PARS	0.001	4.23*	1.00	1067.00	0.01	0.10
PTGI	Stage × PARS	0.001	4.14*	1.00	1064.00	0.01	0.16
	Gender × PARS	0.002	0.618	1.00	1064.00	−0.19	0.06
	Stage × IRI-E	0.006	1.04	1.00	1064.00	−0.04	0.10
	Gender × IRI-E	0.003	0.35	1.00	1064.00	−0.14	0.07

Note ** $P < 0.01$, * $P < 0.05$ growth after stress events

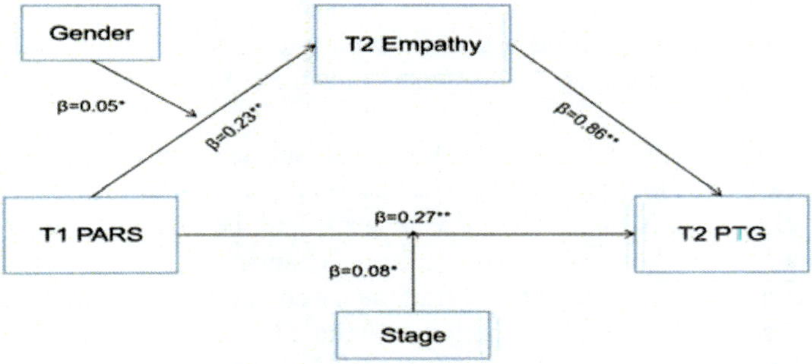

Fig. 2 Conditional process model of physical exercise affecting post-traumatic

4 Discussion

This study adopts the paradigm of follow-up longitudinal study and constructs a conditional process model to investigate the impact of adolescent physical exercise on post-traumatic growth after rainstorm stress events, as well as the mediating role of empathy and the regulatory role of gender and school stage.

5 Group Differences in Physical Exercise, Empathy, and Post-Traumatic Growth of Adolescents After Stress Events

Mann–Whitney U test showed that there were no significant differences in physical exercise and empathy in the school stage, but significant differences in post-traumatic growth in the school stage. The mean comparison found that the post-traumatic growth of senior high school students was higher than that of junior high school students, and the results were consistent with the previous views. Barakat et al. [18] found that there was a significant positive correlation between post-traumatic growth and age in adolescents aged 11–19. The theory of post-traumatic growth can explain the increase in the potential ability of post-traumatic growth of adolescents with age. Because the cognitive schema established by older adolescents is relatively less likely to be changed and have stronger tolerance for trauma, high school students may be more likely to experience post-traumatic growth [19]. Comprehensive studies have shown that post-traumatic growth may increase with the age of adolescents from junior high school to senior high school. The explanation mechanism may be related to the changes of social personal growth during adolescence. With the growth of age, teenagers show stronger psychological tolerance and social connection, which will increase the contact and negotiation between relatives, friends, and society, so as to have greater growth. Adolescence is a key period for recognizing society and developing social relations. Students from junior middle school to senior high school are more mature in mind and more stable in emotion, which may provide opportunities for senior high school students to achieve greater growth after stress events [20].

The Mann–Whitney U test of gender shows that there are significant gender differences in physical exercise and empathy. The mean comparison shows that the physical exercise level of boys is higher than that of girls, and the empathy ability of girls is higher than that of boys. The explanation mechanism of gender differences in physical exercise has the viewpoints of trait theory and gender cognition. Relatively speaking, boys are more energetic, more open-minded, and more willing to exercise, so they have a higher level of physical exercise [21]. There are gender differences in adolescents' empathy level, and girls are higher than boys, which is consistent with the previous results [22]. Colakoglu and Solak [23] also took middle school students as the survey object, and found that girls' empathy score was significantly higher than boys. The gender difference of adolescent empathy may be because girls are more understanding and sharing, easier to solve problems, and more willing to think of others than boys [11]. Role constraint theory shows that coping strategies are determined by the role played by individuals [24]. Considering that women now play multiple social roles (such as mother, worker, and housewife), compared with men, women may be more likely to integrate into positive coping strategies and consider others more after traumatic events, so as to have a higher level of personal empathy [25, 26].

6 Direct Impact of Adolescent Physical Exercise on Post-Traumatic Growth After Stress Events

The direct effect model test of physical exercise on teenagers' post-traumatic growth after stress events found that the physical exercise behavior at the previous time point can significantly and positively predict the post-traumatic growth at the later time point, which verified the research hypothesis 1. Although there are few direct studies on the prediction of adolescents' post-traumatic growth by physical exercise after stress events, physical exercise promotes pro adolescents' post-traumatic growth, and adolescents with positive behavior after stress events have more growth, which can partially support the view of this study. Love and Sabiston [27] through the investigation of young cancer patients, the results show that physical exercise is significantly positively correlated with post-traumatic growth. Physical exercise is considered to be a simple and cost-effective strategy to promote the growth of patients after trauma. Although few models of physical exercise are included in the current post-traumatic growth theory, extensive research supports that physical exercise is related to the post-traumatic growth experience of individuals who have experienced traumatic events [28, 29]. The mechanism of physical exercise affecting post-traumatic growth is not clear. However, physical exercise may change the results of mental health and social adaptation by regulating post-traumatic growth-related factors such as post-disaster emotional response, cognitive coping, and social support, so as to promote positive post-traumatic adaptation and growth [30]. However, how adolescent physical exercise affects post-traumatic growth after stress events needs to be considered in combination with other variables.

Slightly different from previous studies, the correlation coefficient of physical exercise, empathy, and post-traumatic growth found in this study is high. This may be related to the investigation period. Previous studies mainly conducted the investigation in daily situations or 1 or 2 years after the disaster, while this study investigated the early stage after the rainstorm disaster. In the early post-traumatic stage, individuals may experience major life interruptions, challenges to core beliefs, and physical and emotional pain [31]. Physical exercise can rebuild confidence and control, improve health, and provide social support at an early stage, which helps accelerate post-traumatic growth. It is also found in the study of the sick samples that the correlation between physical exercise and post-traumatic growth decreases with the delay of time. Physical exercise is conducive to regain the control of the body and cultivate confidence in the process from treatment to rehabilitation to growth, which is very important for the post-traumatic growth of patients with early diseases [32, 33]. However, individuals in the middle and late stage of the disease may have more complex physical, psychological, emotional, and social challenges, which makes the long-term effect of physical exercise on post-traumatic growth more complex [34].

7 The Mediating and Regulating Role of Empathy, Gender, and Stage in the Model

Empathy has a significant mediating effect when adolescent physical exercise affects post-traumatic growth after stress events. In the physical exercise after stress events, if it can stimulate individuals to learn other people's actions, interact with others to achieve specific goals, or predict other people's actions, identify other people's intentions, and adapt their actions to other people's behaviors, these processes will strengthen the ability of empathy [10]. From the data results, when the level of physical exercise of teenagers after stress events is not high, it is not easy to form empathy in the short term, resulting in slow growth after trauma. It can be seen that in the post-stress physical exercise situation, when individual empathy is stimulated, it is more likely to produce post-traumatic growth.

In the process model of adolescent physical exercise, empathy and post-traumatic growth after stress events, the learning stage regulates the impact of physical exercise on post-traumatic growth, and gender regulates the impact of physical exercise on empathy. High school students may have more coping resources and more flexibility to adjust the post-disaster stress environment, so that they can transfer their attention, stabilize their mentality, and rethink the positive significance of traumatic events through physical exercise in a short time, so as to have greater growth [35]. In addition, there is a more obvious longitudinal correlation between physical exercise and empathy among girls, which may be due to the differences between men and women in how to express aggression and empathy in physical exercise [36]. Stanger et al. [37] found that boys' sports attitude is more aggressive than girls, while girls' sports attitude is more empathic. Through physical exercise, girls are less likely to have aggressive behavior, and they are more likely to understand others and develop empathy through physical exercise, which is conducive to post-traumatic growth [38]. To sum up, in the group of teenagers after stress events, when considering the relationship between physical exercise, empathy, and post-traumatic growth of individuals of different stages and genders, physical exercise is more likely to stimulate girls' high-level empathy, while physical exercise in a short period after stress events has a greater impact on the post-traumatic growth of high school students.

8 Conclusion

Adolescent physical exercise after stress events can not only directly promote post-traumatic growth, but also indirectly affect post-traumatic growth by stimulating empathy. In contrast, physical exercise after stress events is easier to stimulate girls' high-level empathy, and physical exercise in a short period after stress events has a greater impact on senior high school students' post-traumatic growth. The construction of conditional process model explains the relationship between physical exercise, empathy, and post-traumatic growth after stress events to a certain extent, which can provide some reference for formulating the exercise prescription of psychological assistance for adolescents after stress events. Compared with other psychological assistance strategies after stress events, physical exercise is a low-cost and easy to implement intervention measure, which is conducive to post-traumatic psychological recovery. Therefore, it is necessary to further study and apply the precise physical exercise prescription for adolescents after stress events in the future.

Funding of Project General program of Education of the National Social Science Fund of China: "Research on sports regulation mechanism and intervention scheme of middle school students' psychological pressure." (BLA210215), Postgraduate Research and Practice Innovation Program of Jiangsu Province (Project approval No.: KYCX22-3325).

References

1. Lou H, Yan J (2020) Psychoneuroimmunity pathway and suggestion of physical exercise to cope with stress in the epidemic situation of serious infectious diseases. China Sport Sci Technol 56:35–40
2. Cordova MJ, Giese-Davis J, Golant M, Kronenwetter C, Chang V, Spiegel D (2007) Breast cancer as trauma: posttraumatic stress and posttraumatic growth. J Clin Psychol Med Settings 14:308–319

3. Diaz M, Aldridge-Gerry A, Spiegel D (2014) Posttraumatic growth and diurnal cortisol slope among women with metastatic breast cancer. Psychoneuroendocrinology 44:83–87
4. Barskova T, Oesterreich R (2009) Post-traumatic growth in people living with a serious medical condition and its relations to physical and mental health: a systematic review. Disabil Rehabil 31:1709–1733
5. Roberts W, Strayer J, Denham S (2014) Empathy, anger, guilt: emotions and prosocial behaviour. Can J Behav Sci Can des Sci du Comport 46:465
6. Chen J, Xiang X, Lee JLC, Chen C, He Y, Lou VWQ (2020) Physical activity and posttraumatic growth: a systematic review of quantitative and qualitative studies. Psychol Sport Exerc 49:101679
7. Day MC (2013) The role of initial physical activity experiences in promoting posttraumatic growth in Paralympic athletes with an acquired disability. Disabil Rehabil 35:2064–2072
8. Crawford JJ, Gayman AM, Tracey J (2014) An examination of post-traumatic growth in Canadian and American ParaSport athletes with acquired spinal cord injury. Psychol Sport Exerc 15:399–406
9. McDonough MH, Sabiston CM, Ullrich-French S (2011) The development of social relationships, social support, and posttraumatic growth in a dragon boating team for breast cancer survivors. J Sport Exerc Psychol 33:627–648
10. Sevdalis V, Raab M (2014) Empathy in sports, exercise, and the performing arts. Psychol Sport Exerc 15:173–179
11. Atan T (2017) Empathy levels of university students who do and not do sports. Univers J Educ Res 5:500–503
12. Liang DQ (1994) The stress level of college students and its relationship with physical exercise. Chin Ment Heal. J. 8:5–6
13. Rong X, Sun B, Huang X-Z, Cai M, Li W-J (2010) Reliabilities and validities of Chinese version of interpersonal reactivity index. Chin J Clin Psychol 18:158–160
14. Siu AMH, Shek DTL (2005) Validation of the interpersonal reactivity index in a Chinese context. Res Soc Work Pract 15:118–126
15. Zhou X, Wu X, An Y, Chen J (2014) The roles of rumination and social support in the associations between core belief challenge and post-traumatic growth among adolescent survivors after the Wenchuan earthquake. Acta Psychol Sin 46:1509
16. Wu X, Zhang Y, Lin C, Zang W (2013) The effect of disaster exposure on PTSD of primary and secondary students: Mediating and moderating effects. Psychol Dev Educ 29:641–648
17. Wang WC, Wu XC (2017) The bright and dark sides of empathy of adolescent survivors after earthquake. In: The 22nd national academic congress of psychology, China
18. Barakat LP, Alderfer MA, Kazak AE (2006) Posttraumatic growth in adolescent survivors of cancer and their mothers and fathers. J Pediatr Psychol 31:413–419
19. Meyerson DA, Grant KE, Carter JS, Kilmer RP (2011) Posttraumatic growth among children and adolescents: A systematic review. Clin Psychol Rev 31:949–964
20. Spenser K, Bull R, Betts L, Winder B (2020) Underpinning prosociality: age related performance in theory of mind, empathic understanding, and moral reasoning. Cogn Dev 56:100928
21. Kahlin Y, Werner S, Edman G, Raustorp A, Alricsson M (2016) Physical self-esteem and personality traits in Swedish physically inactive female high school students: an intervention study. Int J Adolesc Med Health 28:363–372
22. Durak F, Vurgun N (2006) The relation of empathy and team cohesion in terms of team sports. Sport Phys Educ Sport Sci J 4:73–77
23. Colakoglu FF, Solak N (2014) Examination of aggression levels and empathic tendency levels of secondary school students who involve or not involve in sports. Procedia-Social Behav. Sci. 152:415–420
24. Rosario M, Shinn M, Mørch H, Huckabee CB (1988) Gender differences in coping and social supports: testing socialization and role constraint theories. J Community Psychol 16:55–69
25. Cheng C, Lau H-PB, Chan M-PS (2014) Coping flexibility and psychological adjustment to stressful life changes: a meta-analytic review. Psychol Bull 140:1582

26. Mackenzie CS, Gekoski WL, Knox VJ (2006) Age, gender, and the underutilization of mental health services: The influence of help-seeking attitudes. Aging Ment Heal 10:574–582
27. Love C, Sabiston CM (2011) Exploring the links between physical activity and posttraumatic growth in young adult cancer survivors. Psychooncology 20:278–286
28. Sabiston CM, McDonough MH, Crocker PRE (2007) Psychosocial experiences of breast cancer survivors involved in a dragon boat program: exploring links to positive psychological growth. J Sport Exerc Psychol 29
29. Liu Z, Zhang L, Shi S, Xia W (2017) Objectively assessed exercise behavior in Chinese patients with early-stage cancer: A predictor of perceived benefits, communication with doctors, medical coping modes, depression and quality of life. PLoS One 12:e0169375
30. Lubans D, Richards J, Hillman C, Faulkner G, Beauchamp M, Nilsson M, Kelly P, Smith J, Raine L, Biddle S (2016) Physical activity for cognitive and mental health in youth: a systematic review of mechanisms. Pediatrics 138
31. Park CL (2010) Making sense of the meaning literature: an integrative review of meaning making and its effects on adjustment to stressful life events. Psychol Bull 136:257
32. Chen J, Liu L, Chen J, Jiang W, Wu B, Zhu J, Lou VWQ, He Y (2021) Physical activity and posttraumatic growth in patients receiving maintenance hemodialysis: a prospective study. J Health Psychol 26:2896–2907
33. Rock CL, Doyle C, Demark-Wahnefried W, Meyerhardt J, Courneya KS, Schwartz AL, Bandera EV, Hamilton KK, Grant B, McCullough M (2012) Nutrition and physical activity guidelines for cancer survivors. CA Cancer J Clin 62:242–274
34. Fitch M (2008) Supportive care framework. Can Oncol Nurs J/Revue Can soins Infirm en Oncol 18:6–14
35. Chengzhi Z (2001) Prevalence and correlated factors of PTSD in adolescents 17 months after earthquake. Chinese Ment Heal J 15:145–147
36. Moeijes J, van Busschbach JT, Bosscher RJ, Twisk JWR (2018) Sports participation and psychosocial health: a longitudinal observational study in children. BMC Public Health 18:1–11
37. Stanger N, Kavussanu M, Ring C (2017) Gender moderates the relationship between empathy and aggressiveness in sport: The mediating role of anger. J Appl Sport Psychol 29:44–58
38. Kavussanu M, Stanger N, Boardley ID (2013) The prosocial and antisocial behaviour in sport scale: further evidence for construct validity and reliability. J Sports Sci 31:1208–1221

Effect of Music and Exercise Improve Quality of Life Among Post-Stroke Patients: A Review

Wan Zhen Lee, Garry Kuan, Muhammad Hafiz Hanafi, and Yee Cheng Kueh

Abstract Stroke can result in physical impairment, cognitive or management dysfunction, emotional issues, and a decline in quality of life. Anxiety is one of the most prevalent emotional issues among stroke patients. Post-stroke anxiety may have a significant impact on the quality of life (QOL) and rehabilitation of stroke survivors. Moreover, severe anxiety symptoms are associated with an increased risk of incident stroke. Depression, cognitive impairment, fatigue, age, female sex, lesion site, and sleep disturbance are all risk factors for post-stroke anxiety. In addition to the stress caused by an acute ischemic stroke, the biological causes of post-stroke anxiety should be acknowledged. Physical activity consistently alleviates the depression symptoms in stroke patients. Regular physical activity improves the quality of life by reducing the risk of illness and disability. The therapists promote exercise and music to improve the condition of stroke patients. It also teaches them how to manage anxiety and achieve independence in daily life. Promoting music and exercise is essential for stroke patient in their daily lives. Thus, the aim of this study is to review the effect of music and exercises on reducing post stoke anxiety.

Keywords Stroke · Rehabilitation · Exercises · Music · Anxiety · Physical activity

W. Z. Lee (✉) · G. Kuan
Exercise and Sports Science Programme, School of Health Sciences, Universiti Sains Malaysia, Kubang Kerian, Kelantan, Malaysia
e-mail: leewanzhen@unisza.edu.my

W. Z. Lee
School of Rehabilitation Science, Faculty of Health Sciences, Universiti Sultan Zainal Abidin (UniSZA), Kampus Gong Badak, Terengganu, Malaysia

M. H. Hanafi
Rehabilitation Medicine Unit, School of Medical Sciences, Universiti Sains Malaysia, Kelantan, Malaysia

Y. C. Kueh
Biostatistics and Research Methodology Unit, School of Medical Sciences, Universiti Sains Malaysia, Kubang Kerian, Kelantan, Malaysia

G. Kuan et al. (eds.), *Advancing Sports and Exercise via Innovation*, Lecture Notes in Bioengineering, https://doi.org/10.1007/978-981-19-8159-3_20

1 Introduction

Stroke is indeed a prevalent chronic disease with significant mortality, morbidity, and impairment rate that poses a severe threat to human health [1]. Moreover, stroke induced by cerebral infarction or hemorrhage is the primary cause of lifelong impairment in people and the second biggest cause of mortality after heart disease and cancer [2]. Every year, more than 25 million individuals worldwide are diagnosed with a stroke, and 6.5 million people die from it, with an almost equal representation of the genders [3].

According to Oyewole and his colleagues [4], the medical developments have resulted in an increasing number of stroke survivors, but many will be disabled to varying degrees [4]. Stroke patients always experience post-stroke consequences, such as language impairment, limb hemiplegia, and other dysfunctions, and rehabilitation of these abilities might take six months or longer. In addition to physical disability, those who have had a stroke may overcome a variety of health consequences, such as mental disorders, which can compromise the rehabilitation process and have an impact on long-term recovery [5].

As a result, almost stroke survivors experience moderate to severe anxiety, which can be a natural and sometimes necessary reaction to coping with stress and life-threatening circumstances. According to Lincoln and his colleagues in the year 2013, anxiety is a common symptom of stroke onset, whether in the early period, months later, or years later. Corresponding to a 2012 systematic review, roughly, 20% of persons after stroke feel anxiety to varying degrees [6], but further research has found that more than one-third of stroke patients experience fear and anxiety post-stroke [7].

Dunton [8] was the first to describe music as a therapeutic intervention [8]. Furthermore, according to Pacchetti and his colleagues (2000), music therapy boosted physical therapy involvement among Parkinson's patients [9]. In addition, music therapy can improve neurologically injured people receiving rehabilitation therapy to enhance their mood states [10]. Music benefits for stroke inpatients at the acute stage have been reported for arm function [11], non-fluent aphasia [12], gait [13], and cognition.

The purpose of this book chapter was to summarize how music improves mood, cognitive, and reduces the physical disability among stroke patients. The investigation will contribute to the development of music training choices for stroke patients in an effort to reduce physical disability and improve their quality of life.

2 The roles of music and exercise in reducing anxiety among stroke patients

Herholz and Zatorre [14] describe that in both healthy and clinical populations, music-making activities promote neuroplastic changes in sensorimotor, auditory, and memory systems [14]. Furthermore, motor sequences are designed and implemented in a forecast moment sense, anticipating, and having to listen to sounds requires, respectively, audio feedforward and feedback systems, promoting neural communications between auditory and motor areas [15]. According to Thaut and Hoemberg[16], using music to treat stroke patients resulted in an improvement in neurological movement disorders. The researchers also demonstrate the advantages of musical activities, showing the positive sign of motor recovery in subacute and chronic stroke [16].

Besides, Maratos and his colleagues [17] stated that music therapy might indeed improve cultural and cognitive outcomes such as interpersonal communication, encouragement, emotions, anxiety, and motor improvements [17]. According to Raglio and his colleagues [18], music can enhance physical and psychological tasks. This connection will also help communication and rehabilitation by facilitating patient physiological, representation, and attenuation aspect [18].

Mental issues such as depression and anxiety are associated with lower treatment adherence, which leads to decreased social interactions and motivation. These alterations have an impact on overall health [19]. Rehabilitation intends to minimize long-term dependency and help patients to reintegrate into society. According to Nie and Yang in the year 2017 mention that the following ischemic stroke, exercise enhances movement patterns and learning function, both of which are engaged in memory and learning formation [20]. Furthermore, Mead and his colleagues [21] carry out the study that to assess the viability and effectiveness of exercise training following a sixty-six independently ambulatory stroke patients. The warm up and circuit training are included in the exercise training. Participants walked or marched in place during the middle of the circuit training. Stroke patient participated in circuit training during the first week, including shuttle walking, standing chest press, 1.4kg raising and lowering, 55cm exercise ball, and cycle ergometry. Moreover, the fourth week added the stair climbing and descending exercises. The intensity of the cycle has been raised weekly by pedal resistance. The perceived rate of exertion between 13 to 16. The duration of circuit training raised from 9 to 21 minutes by means of week 12. The result shown that the exercise training significantly greater benefits in physical function and perceived effect of physical health on daily life. In additional, exercise training was attainable for post-discharge stroke patients [21].

Moreover, aerobic exercise, strength exercise, balance exercise, and flexible exercise are all types of exercise that can help a post-stroke patient's overall function. Han and his colleagues [22] demonstrated that aerobic exercise, strength exercise, and flexibility exercise can enhance the speed of walking, endurance, strength, balance, prevent joint stiffness, and improve range of motion and motor function. It also can promote quality of life among stroke patients [22]. On the other hand, Son and his colleagues [23] studied the effects of resistance exercise training for strengthening muscles across multiple joints on the dynamic balance function of 28 stroke patients. The result showed that the exercise improved muscle strength across multiple joints, which is an effective intervention for improving the dynamic balance function of stroke patients [23]. The hydrotherapy great impact reduces level of depression and anxiety in people who have had a stroke [24]. The effect of music and exercise reduces anxiety among stroke patients is shown in Table 1.

3 Discussion

A stroke happens when the blood supply to the brain is interrupted or reduced due to a blockage or leak in the blood arteries. The brain does not receive enough oxygen or nutrients, and brain cells begin to die. It was reported that stroke patients were overall more depressed and anxiety after the post-stroke. Therefore, the discussion will focus on the effect of music and exercise reduce anxiety among stroke patients. Table 1 demonstrated numerous beneficial effects of using music and exercises on the cognitive, strength, balance, and reduce anxiety and depression among stroke patients. Exercise has a positive impact on reducing anxiety and depression and improves the quality of life among stroke patients.

Post-stroke anxiety may have a substantial impact on stroke survivors' quality of life (QOL) and recovery. As a result, high anxiety symptoms are linked to an increased risk of stroke. Moreover, several studies have looked at physical exercise during stroke rehabilitation and found that people are inactive most of the time. Severe anxiety symptoms are associated with an increased risk of incident stroke. It is essential to understand the benefits of music and exercise for stroke to the participation of the patients in the rehabilitation program actively, reduce anxiety, and improve quality of life. Sixteen weeks of cognitive rehabilitation complex exercise program (singing and dancing) and physical exercise program among 12 chronic stroke patients improve their sleep quality, cognitive, and reduce depression. In addition, three times per week, 30 minutes of music therapy in combination with rehabilitation among stroke patients to decrease of anxiety, depression, and improve muscle strength. Furthermore, six months after the stroke, the stroke patients receive music and language program showed improvement in attention, memory, and prevents negative mood. On the other hand, music combining with rehabilitation improves depression, anxiety, strength, attention, memory, and quality of life, in people who suffered a stroke.

4 Conclusion

In summary, stroke is the world's second most significant cause of mortality and a major source of disability. The most prevalent type of stroke is ischemic stroke, which is most common in underdeveloped countries. Depression and anxiety disorder are two of the most common neuropsychiatric problems following a stroke, and they commonly interact. Regular physical activity improves the quality of life by lowering the risk of sickness and disability. The therapists promote exercise and music to improve the condition of stroke patients. It also teaches them how to manage anxiety

Table 1 Effect of music and exercise reduces anxiety among stroke patients

Author (year)	Participants (N)	Intervention	Outcome measurement	Length of study and frequency per week (duration)	Result
Jun et al. [25]	$N = 30$ hospitalized stroke patients Divided in 2 groups: Experimental group ($n = 15$) Control group ($n = 15$)	Experimental group: received music-movement therapy in their wheelchairs Control group: received only routine treatment	physical outcomes: range of motion, muscle strength and activities of daily living psychological outcomes: mood states and depression	8 weeks, 3 times per week (60 minutes)	The experimental group had significantly improved in upper limb movement (shoulder flexion and elbow joint flexion), reduce level of anxiety and depression compare with control group
Doğan et al. [26]	$N = 31$ stroke patients $N = 53$ healthy volunteers	Both groups received music therapy for 50 minutes	State-Trait Anxiety Inventory (STAI)	50 minutes for one session to the groups consisting of 6–8 individuals	The result shown that their significant improvements in anxiety in both groups were observed after the music therapy. They conclude that music therapy reduces anxiety in patients with stroke and healthy individuals. This is a safe and cheap method and can support participation of the patients in the rehabilitation program actively.

(continued)

Table 1 (continued)

Author (year)	Participants (N)	Intervention	Outcome measurement	Length of study and frequency per week (duration)	Result
Aidar et al. [27]	$N = 29$ stroke patients Divided in 2 groups; Experimental group ($n = 14$) Control group ($n = 15$) There are 3 subject form Experimental group drop out ($n = 13$)	Experimental group will receive strength training such as warm-up exercises, a multi-power guided squat apparatus, and a module with eight stations, consisting of horizontal leg press, front pulley, development, bench press, bench press, and advanced	State-Trait Anxiety Inventory (STAI), Assessment of Perceived Exertion, Muscle strength (1 repetition maximum (1RM) test)	12 weeks, 3 times per week (45 to 60 minutes)	The result shown that experimental group had significantly improve in muscle strength and decrease the anxiety level compare control group. They conclude that strength training may provide an improvement in trait and state anxiety more than one year after stroke.

(continued)

Table 1 (continued)

Author (year)	Participants (N)	Intervention	Outcome measurement	Length of study and frequency per week (duration)	Result
Aidar et al. [28]	$N = 28$ stroke patients Divided in 2 groups: Experimental group ($n = 15$) Control group ($n = 13$)	Experimental group receive aquatic physical activity program (warm-up activities; walking in the swimming pool with breast level water height; pedaling work out with Spaghetti; climbing and descending of pool degrees, upper and lower limbs with educational material, breathing exercises, doing bubbles in the water, low-intensity exercise for cooling down)	Beck Depression Inventory, State-Trait Anxiety Inventory (STAI),	12 weeks, 2 times per week (45 to 60 minutes)	The result shown that experimental group had significantly decrease the anxiety and depression level compare control group for 12 weeks. They conclude that Aquatic physical activity contributes to an improvement of the levels of depression and anxiety in people who suffered a stroke.

(continued)

Table 1 (continued)

Author (year)	Participants (N)	Intervention	Outcome measurement	Length of study and frequency per week (duration)	Result
Särkämö et al. [29]	$N = 55$ stroke patients Divided in 3 groups: Music group ($n = 19$) language group ($n = 19$) control group ($N = 17$) There are 5 subjects drop out. Divided in 3 groups. Music group ($n = 18$) language group ($n = 19$) control group ($n = 17$)	Music group received portable CD players and CDs of their own favorite music in any musical genre. Language group received portable cassette players and narrated audio books on cassette selected by the patients from a collection of the Finnish Celia library. Control group received no listening material	Extensive neuropsychological assessment, which included a wide range of cognitive tests as well as mood and quality of life questionnaires	One week (baseline), 3 months, and 6 months after the stroke, daily, 1 h per day.	The result shown that recovery in the domains of verbal memory and focused attention improved significantly more in the music group than in the language and control groups. The music group also experienced less depressed and confused mood than the control group. These findings demonstrate for the first time that music listening during the early post-stroke stage can enhance cognitive recovery and prevent negative mood.

(continued)

Table 1 (continued)

Author (year)	Participants (N)	Intervention	Outcome measurement	Length of study and frequency per week (duration)	Result
Raglio et al. [30]	$N = 38$ hospitalized stroke patients Divided in 2 groups: Experimental group ($n = 19$) Control group ($n = 19$)	Experimental group received the physiotherapy and occupational therapy daily sessions and relational active music therapy treatments. Control group received standard of care.	Functional Independence Measure (FIM), Grip-Pinch Dynamometric Test, 9Hole Peg Test, Timed Up and Go Test (TUG), Aachener Aphasie Test (Italian version), Montreal Battery of Evaluation of Amusia—MBEA (rhythm and melodic contour perception)-short version, Hospital Anxiety and Depression Scale (HADS) and the Italian version of McGill Quality-of-Life Questionnaire (MQOL-It).	Music therapy intervention consisted of 20 RAMT sessions lasting 30 min each, three-weekly.	The result shown that positive significant quality of life, functional and disability levels, and gross mobility in both groups. The experimental group showed a decrease of anxiety and depression. In addition, the strength of non-dominant hand (grip) significantly increased in the experimental group.

(continued)

Table 1 (continued)

Author (year)	Participants (N)	Intervention	Outcome measurement	Length of study and frequency per week (duration)	Result
Kim and Cho [31]	$N = 12$ chronic stroke patients	The stroke patient received the cognitive rehabilitation complex exercise program (singing and dancing) and Physical exercise program (Brill Exercise).	Korean version of the Montreal Cognitive Assessment (MoCA-K), Hamilton Depression Rating Scale (K-HDRS), and Pittsburgh Sleep Quality Index (PSQI)	16 weeks, 2 times per week (60 minutes)	The result shown that significantly improve sleep quality, cognitive and reduce the depression. They conclude that the cognitive rehabilitation program used in this study had a positive effect on the cognitive function and depression in patients with chronic stroke patient.

and achieve independence in daily life. Promoting music and exercise is essential for stroke patients in their daily lives among stroke survivors in Malaysia.

Acknowledgement The study was supported by Research University Individual (RUI) grant [1001/PPSK/8012370] from Universiti Sains Malaysia.

References

1. Han Y, Liu Y, Zhang X, Tam W, Mao J, Lopez V (2017) Chinese family caregivers of stroke survivors: determinants of caregiving burden within the first six months. J Clin Nurs 26(23–24):4558–4566
2. Feigin VL, Norrving B, Mensah GA (2017) Global burden of stroke. Circ Res 120(3):439–448
3. Feigin VL, Krishnamurthi RV, Parmar P, Norrving B, Mensah GA, Bennett DA, Barker-Collo S, Moran AE, Sacco RL, Truelsen T, Davis S (2015) Update on the global burden of ischemic and hemorrhagic stroke in 1990–2013: The GBD 2013 study. Neuroepidemiology 45(3):161–176
4. Oyewole OO, Ogunlana MO, Oritogun KS, Gbiri CA (2016) Post-stroke disability and its predictors among Nigerian stroke survivors. Disabil Health J 9(4):616–623
5. Swartzman LC, Gibson MC, Armstrong TL (1998) Psychosocial considerations in adjustment to stroke. Phys Med Rehabil 12:519–542
6. Burton CAC, Murray J, Holmes J, Astin F, Greenwood D, Knapp P (2013) Frequency of anxiety after stroke: a systematic review and meta-analysis of observational studies. Int J Stroke 8(7):545–559
7. Gilworth G, Phil M, Cert AD, Sansam KAJ, Kent RM (2009) Personal experiences of returning to work following stroke: an exploratory study. Work 34(1):95–103
8. Dunton WRJr, (1946) Recreation and music therapy. Occup Ther Rehabil 25(6):247–252
9. Pacchetti C, Mancini F, Aglieri R, Fundarò C, Martignoni E, Nappi G (2000) Active music therapy in Parkinson's disease: an integrative method for motor and emotional rehabilitation. Psychosom Med 62(3):386–393
10. Magee WL, Davidson JW (2002) The effect of music therapy on mood states in neurological patients: a pilot study. J Music Ther 39(1):20–29
11. Schneider S, Schönle PW, Altenmüller E, Münte TF (2007) Using musical instruments to improve motor skill recovery following a stroke. J Neurol 254(10):1339–1346
12. Conklyn D, Novak E, Boissy A, Bethoux F, Chemali K (2012) The effects of modified melodic intonation therapy on nonfluent aphasia: a pilot study. J Speech Lang Hear Res 55(5):1463–1471
13. Thaut MH, Leins AK, Rice RR, Argstatter H, Kenyon GP, McIntosh GC, Bolay H, Fetter M (2007) Rhythmic auditory stimulation improves gait more than NDT/Bobath training in near-ambulatory patients early poststroke: a single-blind, randomized trial. Neurorehab Neural Re 21(5):455–459
14. Herholz SC, Zatorre RJ (2012) Musical training as a framework for brain plasticity: behavior, function, and structure. Neuron 76(3):486–502
15. Ross B, Barat M, Fujioka T (2017) Sound-making actions lead to immediate plastic changes of neuromagnetic evoked responses and induced β-band oscillations during perception. J Neurosci 37(24):5948–5959
16. Thaut M, Hoemberg V eds (2014) Handbook of neurologic music therapy. Oxford University Press USA
17. Maratos A, Gold C, Wang X, Crawford M (2008) Music therapy for depression. Cochrane Database Syst Rev (1)
18. Raglio A, Oasi O, Gianotti M, Rossi A, Goulene K, Stramba-Badiale M (2016) Improvement of spontaneous language in stroke patients with chronic aphasia treated with music therapy: a randomized controlled trial. Int J Neurosci 126(3):235–242

19. Rabelo DF, Néri AL (2006) Subjective wellbeing and perceived psychological adjustment among old people affected by stroke: a review. Estudos de Psicologia (Natal) 11:169–177
20. Nie J, Yang X (2017) Modulation of synaptic plasticity by exercise training as a basis for ischemic stroke rehabilitation. Cell Mol Neurobiol 37(1):5–16
21. Mead GE, Greig CA, Cunningham I, Lewis SJ, Dinan S, Saunders DH, Fitzsimons C, Young A (2007) Stroke: a randomized trial of exercise or relaxation. J Am Geriatr Soc 55(6):892–899
22. Han P, Zhang W, Kang L, Ma Y, Fu, L, Jia L, Yu H, Chen X, Hou L, Wang L, Yu X (2017) Clinical evidence of exercise benefits for stroke (eds). Exercise for cardiovascular disease prevention and treatment. Advances in Experimental Medicine and Biology 1000 Springer, Singapore. https://doi.org/10.1007/978-981-10-4304-8_9
23. Son SM, Park MK, Lee NK (2014) Influence of resistance exercise training to strengthen muscles across multiple joints of the lower limbs on dynamic balance functions of stroke patients. J Phys Ther Sci 26(8):1267–1269
24. Aidar FJ, Jaco de Oliveira R, Gama de Matos D, Chilibeck PD, de Souza RF, Carneiro AL (2017) A randomized trial of the effects of an aquatic exercise program on depression, anxiety levels, and functional capacity of people who suffered an ischemic stroke. J Sports Med Phys Fitness 58(7–8):1171–1177
25. Jun EM, Roh YH, Kim MJ (2013) The effect of music-movement therapy on physical and psychological states of stroke patients. J Clin Nurs 22(1–2):22–31
26. Doğan ŞK, Tur BS, Dilek L, Küçükdeveci A (2011) Single music therapy session reduces anxiety in patients with stroke. J Phys Med Rehabil Sci 14(1):12–15
27. Aidar FJ, de Oliveira RJ, Silva AJ, de Matos DG, Mazini Filho ML, Hickner RC, Machado Reis V (2012) The influence of resistance exercise training on the levels of anxiety in ischemic stroke. Stroke Res Treat 298375
28. Aidar FJ, Garrido ND, Silva AJ, Reis VM, Marinho DA, de Oliveira RJ (2013) Effects of aquatic exercise on depression and anxiety in ischemic stroke subjects. Health 5(02):222
29. Särkämö T, Tervaniemi M, Laitinen S, Forsblom A, Soinila S, Mikkonen M, Autti T, Silvennoinen HM, Erkkilä J, Laine M, Peretz I (2008) Music listening enhances cognitive recovery and mood after middle cerebral artery stroke. Brain 131(3):866–876
30. Raglio A, Zaliani A, Baiardi P, Bossi D, Sguazzin C, Capodaglio E, Imbriani C, Gontero G, Imbriani M (2017) Active music therapy approach for stroke patients in the post-acute rehabilitation. Neurol Sci 38(5):893–897
31. Kim S, Cho S (2022) The effect of cognitive rehabilitation program combined with physical exercise on cognitive function, depression, and sleep in chronic stroke patients. Phys Ther Rehabil Sci 11(1):32–42

Management of Physical Characteristics on Badminton Performance with Psychological Skills

Daphne Ng Chiew Yen, Lim Boon Hooi, and Teo Eng Wah

Abstract Footwork agility is one of the most important physical characteristics in badminton performance as it is the fundamental basic to execute quality technical skills. Psychological skill training (PST) is as important to the athlete as physical training. Although it has been established that PST is techniques to enhance mental skills that may improve sports performance, the interactions between PST and footwork agility in badminton have received very little attention. Therefore, this study is designed to bridge this knowledge gap of the importance of footwork agility management on badminton performance with goal setting and self-talk. There is a need to implement this PST package on footwork agility with measuring tool, Onigoe device to achieve and enhance an effective physical development on badminton performance. The participants are youth badminton players in Selangor ($N = 80$) which are randomly divided into 4 groups: goal-Setting group ($N = 20$), self-talk group ($N = 20$), combined goal-setting and self-talk group ($N = 20$), and control group ($N = 20$). This is an 8-weeks management program. All participants are required to do a pre-test before the management program. The management groups (Goal setting, self-talk, and combined goal setting and self-talk) will be given PST program before footwork agility program. The control group will not be given any program. All participants are required to do a post-test at the end of the program. The findings can be used to enhance the performance of badminton players.

Keywords Goal setting · Self-talk · Footwork agility · Onigoe · Badminton technology

1 Introduction

Badminton is one of the most popular racket sport in the world. Badminton's inclusion on the Olympic schedule didn't happen until the middle of the 1960s. In 1972, it first appeared in the Olympic Games as a demonstration sport in 1972 and as an exhibition

D. N. C. Yen (✉) · L. B. Hooi · T. E. Wah
Universiti Malaya, Kuala Lumpur, Malaysia
e-mail: daphne.ngcy@gmail.com

G. Kuan et al. (eds.), *Advancing Sports and Exercise via Innovation*, Lecture Notes in Bioengineering, https://doi.org/10.1007/978-981-19-8159-3_21

sport in 1988. At the 1992 Games, it become a full-medal Olympic sport for men singles and doubles and women singles and doubles. At the Summer Olympics in 1992 in Barcelona, Spain, the men's doubles team of Jalani and Razif Sidek won Malaysia's first-ever bronze medal [1].

In Malaysia, sports are evolving rapidly, especially badminton. Regardless of age, ethnicity, gender, groups of youngsters to adults can participate in this well-liked sport. Malaysia has produced Olympic medalists and world champions. At all levels, players are continuously looking for methods to advance and enhance their performance. Badminton is a sport which requires strong agility, endurance, speed, and intense concentration [2]. The game has become more intense which requires the highest level of physical attributes with the change of scoring system to 21 points system and no service over [3].

Peak athletic performance in sports is often the result of combination of physical condition, technical ability, psychological factors, and tactical strategies. Sports performance is based in a complex and complicated diversity of components, including physical factors (general and specific conditions), psychological (personally and motivation), and body (body morphology, anthropometry, and body composition) factors [4]. Similarly, in the sport of badminton, a number of factors, including skill and tactics, psychological training, and game strategy, as well as individuals' physical qualities, contribute to success in the sport [5]. It has been demonstrated that games effectiveness is connected to footwork agility [6]. Players that can move and change directions more quickly fare better in games. Because of this, badminton footwork is constantly stressed throughout practice and games.

The psychological component is another crucial determinant in addition to the physical, technical, and tactical ones. For the athlete, psychological skill training (PST) is just as crucial as physical training. Success in most sports is achieved by combining and employing a variety of technical, tactical, physical, and psychological skills [7]. Today's sports psychologists assist both male and female athletes in reaching their full potential [8].

Numerous earlier studies in applied sport psychology have concentrated on the psychological capabilities of athletes. Techniques for teaching or enhancing mental skills that may promote performance in competitive settings are known as psychological skills training (PST) [9].

Additionally, it was found that using the PST program improved athletic performance whether a whole strategy was implemented or just one skill [10]. Previous research showed that in the self-regulation of learning dart throwing, students who integrated self-talk with either performance or process goals did better than those in the goal only or control group condition [11].

The most crucial aspect of badminton technology, according to the most recent studies, is footwork training [12]. Children should receive instruction in good footwork; otherwise, it will negatively affect their entire technical skills in badminton [12]. This was further verified by a research, which found that good footwork is essential for hitting accurate strokes in badminton [13].

According to research on Indian badminton players, changing directions while performing a task was more closely associated to players' success on the court than other physical abilities [14]. Therefore, it would seem that managing footwork agility is something that both players and coaches should focus on.

Coaches finally resort to using tests of changes of direction like the shuttle run agility test [15] and zig-zag/slalom type tests [16] for agility evaluation due to the demand for expensive electronic equipment or the lack of a specialized test validated for a given sport [17]. Therefore, this study is designed to bridge this knowledge gap of the importance of footwork management in badminton performance as well as relying on a reliable and efficient way of training and measuring specific agility on court.

Additionally, PST programs are essential in today's sports since they train athletes of all levels to get an advantage over competitors [18]. Performance has also been discovered to be associated with improvements in PST [7]. Goal setting, imagery, self-talk, and relaxation are the most frequently used psychological techniques that help with mental toughness and performance enhancement [19]. According to research, the PST program substantially increased athletic performance whether a package approach or only one skill is applied [10].

However, previous studies stressed that most efficient psychological skills are combined PST package. There are several literature reviews showed that PST program using multimodal package has improved athletic performance in several sports such as softball [20], volleyball [21], and badminton [22].

Most studies have provided information on goal setting and self-talk multiple types of sports, but few studies have been conducted to evaluate their effectiveness on badminton performance [22]. There is no published article in this specific area of study or available data related to these characteristics in Malaysia.

Psychological preparation is crucial for successful skill learning, practice, and competition processes in the context of performance as an athlete transitions from youth sports to professional sports [23]. In accordance with the foregoing, combining goal setting with further self-use should lead to greater performance increases than each strategy used alone [24]. However, there is currently a dearth of research on the impact of self-talk and goal planning on sports [25].

Currently, to our knowledge, there is no studies on Onigoe device and psychological skills program. Onigoe device is a new instrumentation device which provides footwork agility performance feedback to participants which is significantly important in enhancing badminton performance.

Goal setting and self-talk are performance enhancement techniques in sports. Implementation of this psychological skills techniques with Onigoe device on footwork agility training will lead to new recommendations that can be adopted by coaches and players. Hence, this study will provide useful new findings in development of badminton performance.

The main objectives of this study are to examine the management of physical characteristics in badminton performance with psychological skills. The findings of the current study can be used to enhance the performance of badminton players performance and also as a reliable and evidence-based recommendations that can be adopted by sports coaches and players in their training program.

2 Methods

This study aims to examine the management of physical characteristics on badminton performance with psychological skills. Eighty participants ($N = 80$) will be recruited for this study. All players were recruited based on the following requirements. All participants had at least two years of training, have the correct badminton footwork, and are able to attend all the testing sessions. The participants for this study will be selected by using cluster sampling method where all the participants aged 12–18 years old with at least 2 years of consistent training. The participants are selected from badminton clubs in Selangor because it is the most populous state in Malaysia. Participant's parents had to sign the informed consent form before starting the test.

2.1 Procedures

This research was to examine the management of characteristics physical characteristics in badminton performance with goal setting and self-talk. The duration of this study is 8 weeks. All participants will be required to do a pre-test before the experimental program is introduced, and a post-test will be taken after eight weeks.

In this study, participants will be divided randomly to 4 groups: self-talk ($n = 20$), goal setting ($n = 20$), combined goal setting and self-talk ($n = 20$), and control ($n = 20$). All participants will be asked to do a pre-test before the program is implemented according to groups. In self-talk group, participants will be given a brief introduction and description of the technique of self-talk, presenting them with the positive cue 'I CAN DO IT'. Participants are required to familiarize using the cue by reciting their self-talk out loud or silently to themselves. In goal setting group, the program begin with an explanation of scoring system. Each participant will receive a personal score sheet and will be guided in setting personal performance goals and was asked to write down the goal for the next session. Performance goals address overall personal performance.

The program for combined goal setting and self-talk was literal combination of the treatments of combined programs in self-talk and goal-setting group with the goal-setting program being introduced first. In control group, participants will be asked to practice footwork training program with Onigoe device. In all conditions, participants will be doing footwork training program after implementation of psychological skill training programs. The measurement of the program will be evaluated by comparing between groups on pre-post test.

2.2 Instrumentation

Onigoe (Fig. 1) is an advanced developing technology instrument invented by Dr Magara Yukata from Japan. The primary function of Onigoe device is particularly for the sport of badminton with the aim to enhance players' reaction, endurance, agility, and concentration. More significantly, it is a footwork movement training tool for badminton players with in-built data collection system on reaction timing. It also improves a player's footwork on court.

Its power source is battery. The control box (170 × 85 × 60 mm, weight 290 g) and touch switch button (93 × 93 × 92 mm, 85 g). It comes with a 7 m cable, indicator lamp, and touch switch button.

This instrumentation consists of 3 main parts: the indicator lamp where there will be indication on which switch button to touch, a control box where most of the settings are set there, and a touch switch button. Both touch switch button are connected to the control box, and there are connections to the indicator lamp as well. It is all connected by cables.

The instrument provides visual cues to execute prompt agile reactions and rapid movements. LED lights, a programmable system controller, and optical sensors were used for training and measurements. The feedback/training system provided visual cues to the participants by lighting locations on the court. The players responded to these cues by moving to the lighted cylinders. The parameters and time will be recorded during the session. These measurements serve as post-training feedback, allowing coaches to modify their training programs. This instrumentation is a developing technology, but it is very effective in badminton performance.

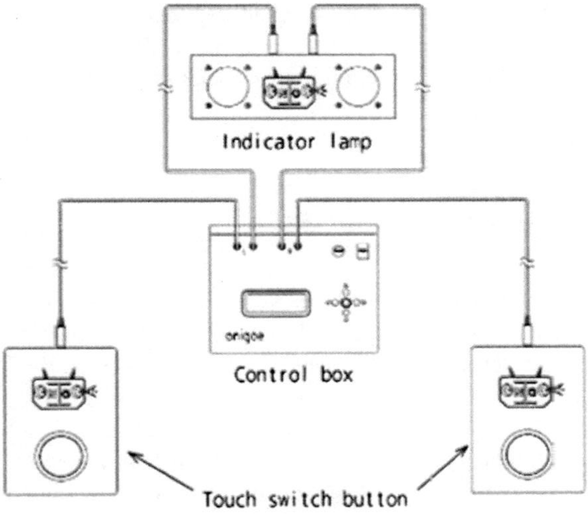

Fig. 1 Shows a diagram of how the instrumentation works

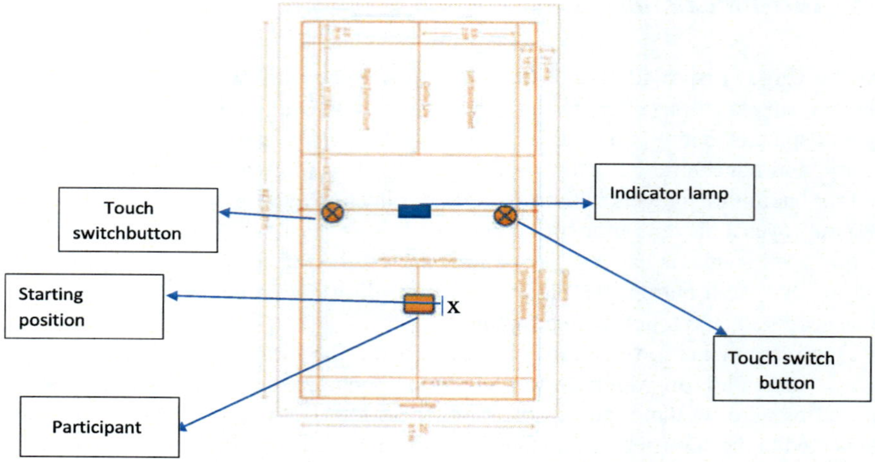

Fig. 2 Shows the two-point footwork training which will be tested (front court)

2.3 *Testing Management Setup*

The test is performed at the badminton court with 5.18 m wide approximately singles badminton court dimension—Fig. 2.

The starting position is at the center of the court which is divided into 2 squares of 0.7 m per side. The touch switch button composed of a cylinder with a rounded cover.

This instrumentation is considered low cost compared with wireless training tools with movement sensors and computers. This instrumentation can be used anywhere, easy to set up, and beneficial for coaches and players to be used as training tool and testing tool.

3 Statistical Analysis

A summary of data analyses which will be used in this study are shown in Table 1: Data analysis technique below. The reliability and validity of the Onigoe device on badminton players' footwork agility are examined in the first section. The second analysis section demonstrates that MANOVA will be used to analyze the three research goals. Multivariate analysis of variance (MANOVA) is a statistical method for comparing the means of multiple sample varieties. It is frequently used to determine whether an influence of one or more independent variables on a group of two or more dependent variables is statistically significant.

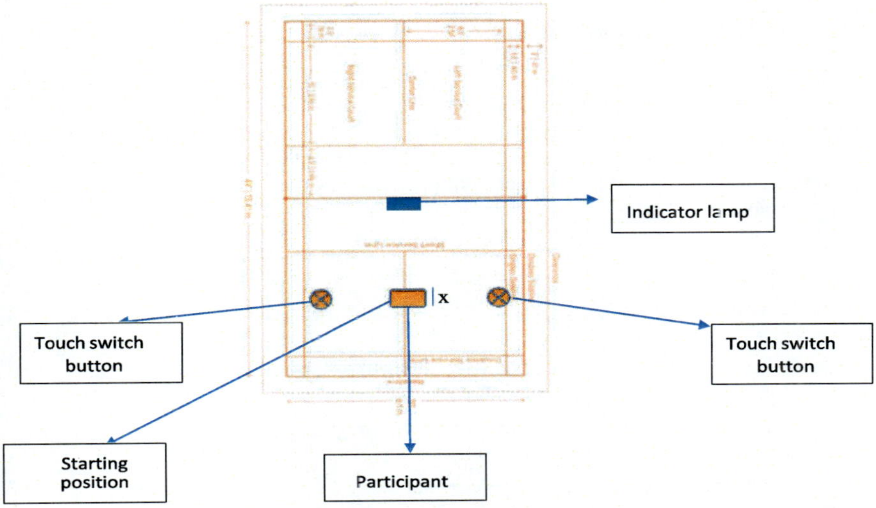

Fig. 3 Shows the testing setup of *Onigoe* device for mid-court

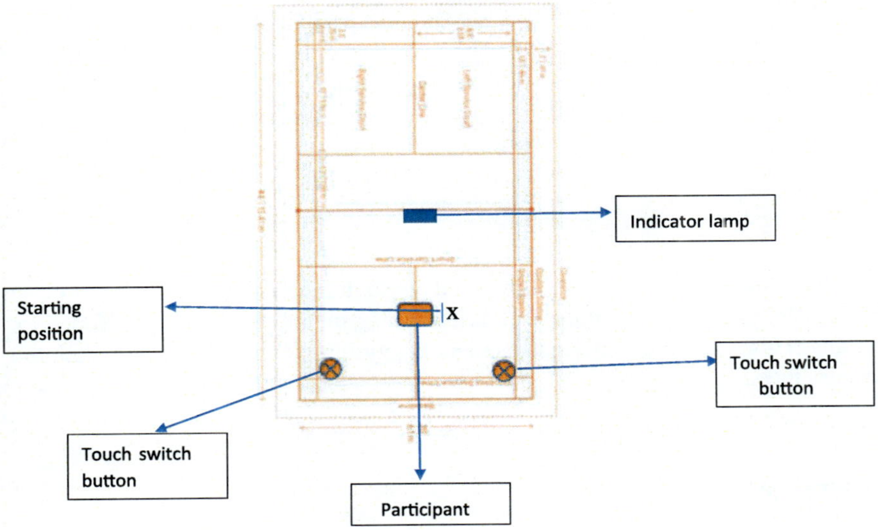

Fig. 4 shows the two-point footwork training which will be tested (rear court)

4 Conclusion

Onigoe device had been used by many international players and professional clubs across the world including national junior badminton players in Malaysia. However,

Table 1 Data analysis techniques summary

Research objective	Variable used	Statistical technique
To investigate the effects of self-talk program on physical characteristics among participants between (management and control group) and within (pre-post) groups	Dependent variable: Scores on footwork agility across two time period (pre-program and post-program)	MANOVA
	Independent variable:	
	1. Independent between-subjects variable	
	• Self-talk group	
	2. Independent within-subjects variable	
	Time period: pre-program, post-program	
3. To investigate the effects of goal-settings program on physical characteristics among participants between (management and control group) and within (pre-post) groups	Dependent variable: Scores on footwork agility across two time period (pre-program and post-program)	MANOVA
	Independent variable: 1. Independent between-subjects variable	
	• Goal-setting group	
	2. Independent within-subjects variable	
	Time period: Pre-program, Post-program	

there have not been any studied or reported research found on this device. Hence, it is important to address psychological skills training for the development of footwork training. This will enhance the management of physical characteristics on badminton performance.

References

1. Badminton World Federation (2018) The olympic journey. Available from: https://olympics.bwfbadminton.com/history/
2. Rusdiana A, Ruhayati Y, Badruzaman (2016) 3D kinematic analysis of standing and Jumping Smash Technique of Indonesian Badminton National Elite Athletes 96, pp 2525–35
3. Chen LM, Pan YH, Chen YJ (2009) A study of shuttlecock's trajectory in badminton. J Sports Sci Med 8(4):657–662
4. Campos FAD, Daros LB, Mastrascusa V, Dourado AC, Stanganelli LCR (2009) Anthropometric profile and motor performance of junior badminton players. Braz J Biomotricity 2:146–151

5. Chin M-K, Wong AS, So RC, Siu OT, Steininger K, Lo D (1995) Sport specific fitness testing of elite badminton players. British J Sports Med 29(3):153–157
6. Singh J, Raza S, Mohammad A (2011) Physical characteristics and level of performance in badminton: a relationship study. J Educ Pract 2(5):6–10
7. Weinberg RS, Gould D (2011) Foundations of sport and exercise psychology
8. Wuest DAFJL (2015) Foundations of physical education, exercise science, and sport
9. Vealey RS (1988) Future directions in psychological skills training. Sport Psychol 2(4):318–336
10. Weinberg RS, Gould D (2014) Foundations of Sport and Exercise Psychology 6th edn, Human Kinetics
11. Kolovelonis A, Goudas M, Dermitzaki I (2012) The effects of self-talk and goal setting on self-regulation of learning a new motor skill in physical education. Inter J Sport Exer Psychol 10(3):221–235
12. Hao L, Zhi J, Zhu W, Zhou L (2022) Research on badminton player's step training model based on big data and iot networks. Security and Communication Networks
13. Muthiarani A, Lismadiana RY (2021) The effect of shadow training using consecutive steps and cross steps on the agility of the footwork of badminton athletes. Jurnal Keolahragaan 9(1)
14. Tiwari L, Rai V, Srinet S (2011) Relationship of selected motor fitness components with the performance of badminton player. Asian J Phys Educ Comput Sci Sports 5(1):88–91
15. Salaj S, Markovic G (2011) Specificity of jumping, sprinting, and quick change-of-direction motor abilities. J Strength Cond Res 25(5):1249–1255
16. Sporis G, Jukic I, Milanovic L, Vucetic V (2010) Reliability and factorial validity of agility tests for soccer players. J Strength Cond Res 24(3):679–686
17. Cremasco FC, Luiz de França Bahia Loureiro Jr Mário Oliveira Costa Dias MGdS, FBdF (2017) Assessment of specificity of the badcamp agility test for badminton players
18. Birrer D, Morgan G (2010) Psychological skills training as a way to enhance an athlete's performance in high-intensity sports. Scand J Med Sci Sports 20:78–87
19. Sheard M, Golby J (2006) Effect of a psychological skills training program on swimming performance and positive psychological development. Int J Sport Exer Psychol 4(2):149–169
20. Montgomery T, Ross M, Perry J, Hansen A (2018) The mental workout: implementing phase 1 of the ten-minute toughness mental training program. J Sport Psychol in Ac 9(2):73–82
21. Shoenfelt EL, Griffith AU (2008) Evaluation of a mental skills program for serving for an intercollegiate volleyball team. Percept Mot Skills 107(1):293–306
22. Rahayu T (1997) Athletes' self confidence: Athletes' self confidence instrument arrangement and development for individual sport. Unpublished Dr. Dissertation Jakarta: State University of Jakarta
23. Ohuruogu B, Jonathan UI, Ikechukwu UJ (2016) Psychological preparation for peak performance in sports competition. J Educ Pract 7(12):47–50
24. Locke EA, Shaw KN, Saari LM, Latham GP (1981) Goal setting and task performance: 1969–1980. Psychol Bulletin 90(1):125
25. Papaioannou A, Theodorakis Y, Ballon F, Auwelle YV (2004) Combined effect of goal setting and self-talk in performance of a soccer-shooting task. Percept Mot Skills 98(1):89–99

Experienced Players and Novices in Kendo Can Discriminate a Good Preparatory Posture from Worse Ones

Motoki Okumura, Akifumi Kijima, and Yuji Yamamoto

Abstract Previous research has shown that experienced players in various sports can detect subtle differences in their opponents' movements and anticipate the outcomes of the movements accurately and rapidly. In these methods, the researchers presented opponents' movements, such as serving and striking, to their participants, and measured the precision of anticipation. Our research purpose was to examine whether experienced players and novices in kendo could discriminate between three different preparatory postures (Good-middle-bad) for striking without watching the start of the strike movements. Three preparatory postures showed very subtle differences only in trunk angle: forward tilt (Good), backward tilt (Bad), and in between these tilts (Middle). Groups of experienced players and novices watched a series of movements in which an opponent stood in a preparatory posture, stepped forward about 30 cm from right to left foot, and then stopped with one of the three preparatory postures. They watched each series of the movements 20 times on a monitor and assessed the possibility (0–100%) of the opponent striking from the preparatory postures without any preliminary movements. The results showed that both the groups could discriminate between the good preparatory posture from middle and bad ones. This indicated that the participants could detect the qualities of the next strike movements based only on watching the preparatory movements and postures, regardless of their experience levels. The perceptual abilities of vision that are used to detect kinematic patterns in sports may be more sensitive than we believe.

Keywords Visual perception · Kinematic pattern · Anticipation

M. Okumura (✉)
Tokyo Gakugei University, Tokyo 184-8501, Japan
e-mail: okumura@u-gakugei.ac.jp

A. Kijima
University of Yamanashi, Kofu 400-8510, Japan

Y. Yamamoto
Nagoya University, Nagoya 464-8601, Japan

G. Kuan et al. (eds.), *Advancing Sports and Exercise via Innovation*, Lecture Notes in Bioengineering, https://doi.org/10.1007/978-981-19-8159-3_22

1 Introduction

Many researchers have indicated that expert players can anticipate and respond more rapidly and accurately than intermediate/novice players in any sport. In these methods, the researchers asked the players to observe the actual movements of other players on screens or fields, and they were asked to anticipate the outcomes of the movements [1–3]. Thus, in past research, players anticipated the outcome of other players' movements by actually observing them. However, the question remained as to whether experienced players can perceive important information to anticipate the next movements of other players based only on the observation of the previous movements and not the next movements themselves.

It has been reported in the literature that baseball players seemed to be able to perceive invisible kinetic information about pitchers (e.g., muscle strength of lower limbs) before they executed any visible movements in anticipation to steal the base. Deriving from that notion, the following research questions are formulated in the present study, i.e., (1) Can we confirm this strange phenomenon among players in other sports, in this case, kendo or Japanese fencing? (2) If we could confirm this, what information can they perceive in anticipation? (3) Are there any differences in the perception and anticipation with respect to the players' skill levels?

2 Methods

2.1 Participants

Twenty-eight female students from Tokyo Gakugei University participated in the experiment. 14 were experienced kendo players, with an average (SD) age of 19.09 (3.99) years, athletic experience of 13.82 (2.52) years, and kendo rank 3.27 (0.45) dan. Their athletic levels ranged from the regional to national levels. 14 of the participants were novices in kendo, with an average age of 20.55 (0.89) years. Informed consent was obtained from all participants using a procedure approved by the Research Ethics Committee of Tokyo Gakugei University.

2.2 Experimental Stimuli

A female kendo player presented the stimuli as an experimental opponent. Her athletic skills were at the national level. Before the experiment, she trained sufficiently to present the same experimental stimuli consistently. In the movie clips of the stimuli, the opponent wore normal protectors and assumed a good preparatory posture with a shinai (sword) at a far distance (225 cm). From there, she approached just one step from the right to the left foot (about 30 cm) toward the participants on a monitor.

After the step, she stopped and assumed three different preparatory postures for striking. The three preparatory postures had very subtle differences only in the trunk angle: forward tilt (Good), backward tilt (Bad), and in between these tilts (Middle), as shown in Fig. 1.

In general, players would take longer to strike from the middle and bad tilts than from the good one. Because of the middle and bad tilts, the experimental opponent needed to move her posture forward before starting the strike. The length of the movie clips was approximately 3 s, including the first preparatory posture, one step, and three different postures for about a second of each movement. We recorded the movie shooting situation of the stimuli and the height of the experimental opponent, and then calculated the visual angle in the situation and reproduced the angle by adjusting the distance between the participants and a monitor in the experimental setting. The participants watched three different types of stimuli 20 times each from an opponent's view while sitting on a chair. We asked the participants to watch the opponent on the monitor in the same manner as the real matches.

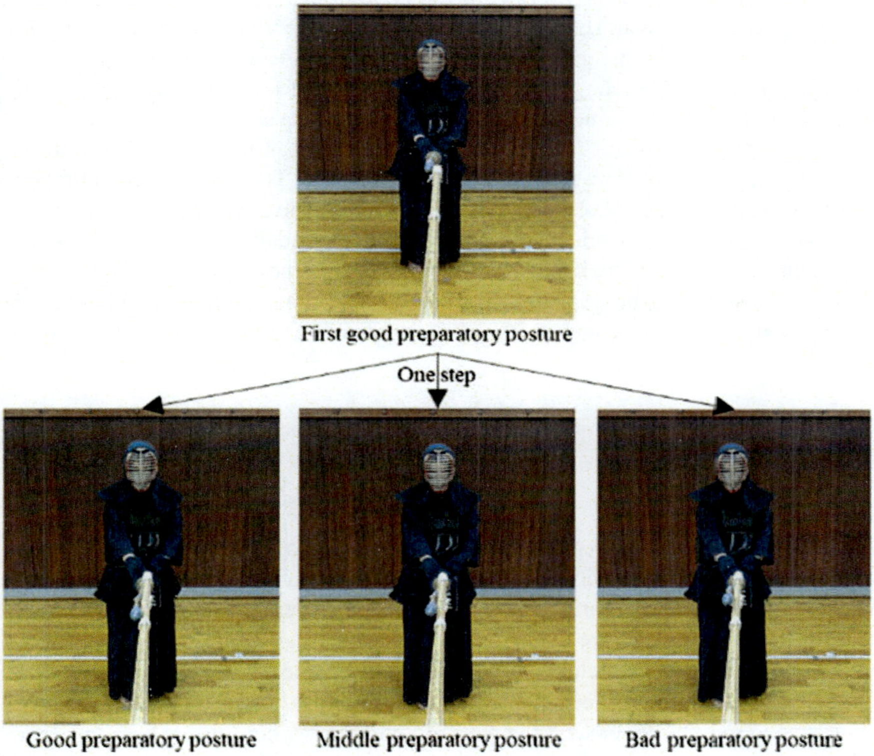

Fig. 1 Participants watched the stimuli from an opponent view. It shows the subtle differences of the stimuli in the places of shinai (sword) corresponding to forward tilt (good), backward tilt (bad), and in between these tilts (middle)

2.3 Data Collection and Analysis

After watching each stimulus, the participants were asked to answer the following question: "Do you feel the opponent could strike from the preparatory posture without any preliminary movements?" on a scale of 0 to 100%. We explained the preliminary movement in advance, which is defined as the displacement of the center of mass and flexion of the lower limbs.

We averaged each of the participants' answers for the same kind of stimuli and analyzed them using two-way ANOVA, with the independent variable being the groups, and dependent and repeated variables the average percentage of the answers.

3 Results and Discussion

The results indicated that the average of the good preparatory postures was higher than those of the middle and bad ones ($F(2, 52) = 41.19, p < 0.01$, partial $\eta^2 = 0.61$). The difference between the groups was marginally significant, and the experienced group had a higher percentage than the novice group ($F(1, 26) = 5.59$, $p < 0.10$, partial $\eta^2 = 0.12$). No significant interactions were observed.

The results depicted in Fig. 2 showed that both the groups could discriminate the good preparatory posture from middle and bad ones. The most important fact is that the participants did not observe the experimental opponent's striking movement itself, but they could discriminate the differences in the qualities of the next striking movements, regardless of their experience levels. Furthermore, as shown in Fig. 1, the experimental stimuli exhibited subtle differences. The perceptual abilities of vision that are used to detect kinematic patterns may be more sensitive than we believe [4, 5].

4 Conclusion

This study has established that the different type of kendo preparatory posture could be distinguished by both novice and experience players. In future research, we should extend the spatiotemporal range that we examine and present to players and clarify what information players are able to detect and use to anticipate their opponents' movements in real sports fields. On the other hand, although there was a marginal significance in the main effect of the group difference, it was considered that the experienced group might answer the question with higher confidence based on their athletic experience than the novice group, which made a difference.

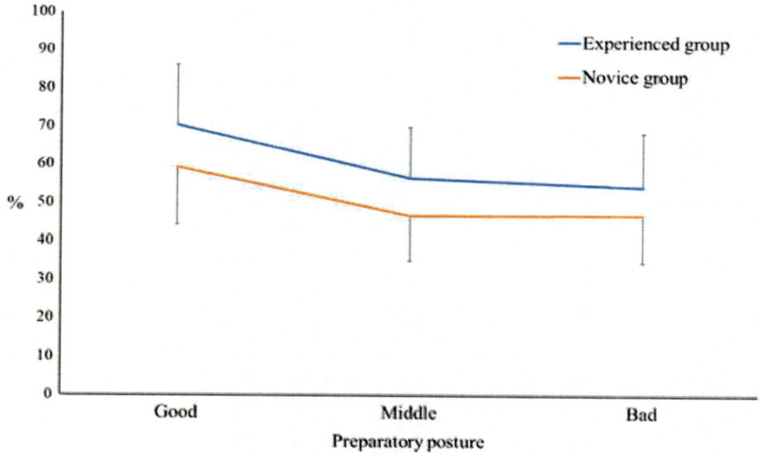

Fig. 2 Average percentage for each of the groups and the preparatory postures

Acknowledgements This study was supported by JSPS KAKENHI (grant number 20H04090).

References

1. Jackson RC, Warren S, Abernethy B (2006) Anticipation skill and susceptibility to deceptive movement. Acta Psychol (Amst) 123:355–371
2. Aglioti SM, Cesari P, Romani M, Urgesi C (2008) Action anticipation and motor resonance in elite basketball players. Nat Neurosci 11:1109–1116
3. Müller S, Abernethy B, Eid M, McBean R, Rose M (2010) Expertise and the spatio-temporal characteristics of anticipatory information pick-up from complex movement patterns. Perception 39:745–760
4. Runeson S, Frykholm G (1981) Visual perception of lifted weight. J Exp Psychol Hum Percept Perform 7:733
5. Vaziri-Pashkam M, Cormiea S, Nakayama K (2017) Predicting actions from subtle preparatory movements. Cognition 168:65–75

The Effects of Active Video Games on Children's Fundamental Movement Skills: A Systematic Review and Meta-Analysis

Zheng Su, Xuan-xi Li, Chuan-peng Hu, and Yu-cui Diao

Abstract Available evidence suggests that different types of active video games (AVGs) and interventions may have different effects on children's fundamental movement skills (FMS). Therefore, a systematic review and meta-analysis of randomized controlled trials (RCTs) using AVGs is necessary to explore the effects of different types and doses of AVGs on FMS in healthy children. The purpose of this work is to update the current evidence on the effects of AVGs on FMS in healthy children aged 3–12 years. RCTs studies up to June 2021 were retrieved from Chinese and English databases (CNKI, Wanfang database, Cochrane Library, Web of Science, Science Direct, PubMed, etc.), and the included studies were tested for bias using the Cochrane Risk Bias Assessment Tool. Review Manager 5.3 software was used to conduct meta-analysis. The study type was an RCT trial, the subjects were healthy children aged 3–12 years, and the study topic was a comparison of AVGs with other interventions and the outcome indicator was FMS. There were few high quality experimental studies, with only four of the seven included literature reaching moderate to high quality; (2) AVGs had a moderate effect on overall FMS (SMD = 0.41, 95% CI (0.15, 0.67), $p = 0.002$), where the effect on object control skills (SMD = 0.41, 95% CI (0.07, 0.75), $P = 0.02$) was slightly better than locomotor skills (SMD = 0.39, 95% CI (0.02, 0.75), $P = 0.04$); (3) For FMS, using Xbox Kinect (SMD = 0.55, 95% CI, −0.14–0.99, $P = 0.05$) in schools (SMD = 0.40, 95% CI, −0.02–0.93, $P = 0.05$) for at least 12 weeks (SMD = 0.43, 95% CI, 0.08–0.79, $P = 0.02$), 5 times/week (SMD = 0.51, 95% CI, 0.13–1.02, $P = 0.0001$), and 30 min/time (SMD = 0.49, 95% CI, 0.20–0.79, $P = 0.0009$) were more effective. For OCS, using Xbox Kinect (SMD = 0.41, 95% CI, 0.05–1.06, $P = 0.04$) at home (SMD = 0.51, 95% CI, 0.02–1.03, $P = 0.01$) for at least 12 weeks (SMD = 0.51, 95% CI, −0.00–1.15, $P = 0.01$), 5 times/week (SMD = 0.42, 95% CI, 0.05–1.09, $P = 0.05$), and 45–60 min/time (SMD = 0.47, 95% CI, −0.10–1.07, $P = 0.04$) had a more significant effect. For LOC, using Xbox Kinect (SMD = 0.47, 95% CI, −0.09–1.27, $P = 0.01$) in the home (SMD = 0.55, 95% CI, 0.04–1.25, $P = 0.05$) for at least 10 weeks (SMD = 0.48, 95% CI, 0.16–1.19, $P = 0.001$), 5–7 times/week (SMD = 0.53, 95% CI, −0.17–1.14, $P = 0.007$), and 45–60 min/time (SMD = 0.46, 95% CI, 0.07–1.04, $P = 0.002$)

Z. Su · X. Li · C. Hu · Y. Diao (✉)
Shandong Normal University, Jinan, Shandong Province, China
e-mail: diaoyucui4321@126.com

G. Kuan et al. (eds.), *Advancing Sports and Exercise via Innovation*, Lecture Notes in Bioengineering, https://doi.org/10.1007/978-981-19-8159-3_23

had a more significant effect. Only seven studies were included, the sample size was not large enough and was not comprehensive enough for possible publication bias. AVGs had a moderate facilitative effect on children's total FMS and a slightly better facilitative effect on OCS than LOC.

Keywords Active video games · Children · Fundamental movement skill · Locomotor skill · Object control skill

1 Introduction

Physical inactivity among children and adolescents had become a global epidemic, with studies indicating that three quarters of children and adolescents (aged 6–17 years) do not meet the recommended amount of physical activity guidelines set by the WHO[1], and up to 80% of children aged 13–15 years did not meet the requirement of 60 min of moderate to vigorous physical activity (MVPA) per day and engaged in prolonged sedentary behavior and electronic screen activity[2]. Physical inactivity had become a major threat to the physical and mental health of children and adolescents [3, 4]. Therefore, the promotion of physical activity (PA) in children and adolescents had become an important initiative to improve international public health [5].

FMS is considered a key influence in the promotion of PA in children and adolescents [6]. FMS are the most fundamental movements required for individuals to participate in sports, games or physical activities, and include locomotor skills (LCS), object control skills (OCS), and stability skills (SS) [7]. LCS are the ability to manipulate the body from one place to another. OCS are the ability to accomplish movement by controlling certain instruments. SS are the ability to maintain balance to control the body [8]. Multiple longitudinal studies had revealed that FMS is positively associated with moderate to vigorous PA [9–13], healthier weight and psychological in children [14, 15]. So, it is vital to create a variety of opportunities for children to practice and achieve proficiency in FMS.

Unfortunately, the development of FMS in healthy children was delayed to varying degrees both internationally [16] and in China [17]. Therefore, the use of multiple interventions to enhance the development of FMS in healthy children had become an international research hotspot in the field of motor development. Several systematic reviews of interventions for children with FMS had shown that interventions in schools [18] and with physical education teachers [19], coaches and specialists [20] are more effective. However, during the COVID-19, home-based online teaching gradually became the norm [21], which led to a shift in the implementation field of school physical education and sport from athletic fields and gymnasiums to the home and community [22], thus preventing the implementation of traditional school-based interventions. Therefore, there is a need to adopt new interventions to promote the development of FMS in children in a family-based setting.

Fortunately, in recent years, AVGs based on virtual reality technology had gradually become an integral part of international [23] and Chinese [24] home fitness distance education systems, providing a new pathway option for children's physical education reform. AVGs is a new type of video game that requires players to interact with the game environment on screen through various PA in real time and give positive feedback, aiming to improve physical health levels [25]. Studies had shown that AVGs-based interventions can significantly improve FMS in healthy children [26], however, different types and doses of AVGs may have different effects on children's FMS [27].

Despite the abundance of studies on this topic, to our knowledge, no systematic review and meta-analysis had been conducted to examine the effects of different types and doses of AVGs on FMS. Therefore, the aim of the study was to use a systematic review and meta-analysis to examine the effects of different types and doses of intervention-based AVGs on children's FMS.

2 Methods

This study strictly followed the Preferred Reporting Items for Systematic Reviews and Meta-Analyses (PRISMA) reporting guideline [28].

Search Strategy

The Chinese and English electronic search of China National Knowledge Internet (CNKI), Wanfang Database, Google Scholar, ScienceDirect, PubMed, Web of Science, and Cochrane Library was combined with manual searches of the existing literature, performed from inception to June 10, 2021. The search strategy combined the following relevant terms: *1."active video game"* OR *"electronic games"* OR *"exergame"* OR *"movement-based video game"* OR *"sedentary video games"* OR *"interactive video games"* OR *"nintendo wii"* OR *"play station" OR "xbox 360 kinect"* OR *"virtual reality"* AND *2."fundamental motor skill"* OR *"fundamental movement skill"* OR *"basic motor skill"* OR *"basic movement skill"* OR *"gross motor skill"* OR *"locomotor skill"* OR *"object control skill"* OR *"manipulative skill"* OR *"stability skill"* AND *3."children"* OR *"kid" OR "preschoolers"* OR *"kindergarten"* OR *"adolescent"* OR *"youth"* OR *"teenager"* OR *"students"*. In addition, the reference lists of the included studies were checked to find potential studies that could also be used in the review.

Eligibility Criteria and Study Selection

The criteria for study inclusion were as follows: (1) RCT published in Chinese and English, and the type of literature was journal articles; (2) apparently healthy (i.e., general population, including samples of children and adolescents with overweight or obesity but not samples of children exclusively with a diagnosed medical condition) (mean age 3–6 years) and adolescents (mean age 7–12 years); (3) The type of AVGs such as Nintendo Wii and Xbox Kinect were used as the intervention; (4) The outcome indicators included fundamental motor/movement skills, basic motor/movement skills, gross motor skill, LOC, OSC, manipulative skill, SS.

The criteria for study exclusion inclusion were as follows: (1) intervention studies for non-children and adolescents (age < 3 years or age > 12 years); (2) Meta-analyses, systematic reviews, and evaluations type of literature; (3) duplicate publications, unavailable full text, and inability to extract outcome indicators and quantitative data.

Study Selection and Data Collection Process

The entire process of study selection was carried out by two researchers who screened the 529 included papers according to the inclusion and exclusion criteria. The Cohen's Kappa coefficient for this study was 0.713, indicating a high level of consistency in the screening results [29]. For studies in which two researchers disagreed, a third researcher was invited to intervene to finalize the included studies, resulting in a total of seven publications.

For each study, data were extracted for characteristics of the study population, including: (1) first author's last name, year of publication and country; (2) purposes; (3) sample size; (4) sample characteristic (mean age and gender); (5) characteristics of intervention (type, frequency, and duration); (6) test instruments and outcome; (7) statistical methods; (8)comparison of intervention outcomes. For quantitative data, means (M) and standard deviations (SD) were extracted for each RCT at baseline as well as post-intervention. When there was insufficient information, the respective corresponding author was contacted.

Risk of Bias of Individual Studies

The risk of bias of the included studies was assessed according to the RCT risk of bias assessment tool recommended by the Cochrane systematic review manual from the following six entries: (1) random sequence generation; (2) allocation concealment; (3) blinding of participants and personnel; (4) incomplete outcome data; (5) selective reporting; (6) other bias. The maximum score for each study was 6: low risk of bias (=5 or more); moderate risk of bias (=3–4); high risk of bias (<3 or less).

3 Statistical Analysis

Meta-analysis was performed using Review Manager 5.3 software [30]. The results of this study were all continuous variables, so Standardized Mean Difference (SMD) was used to calculate the Effect Size (ES) for each study, with ES indicating the degree of difference between groups: ES < 0.2 for small effect, $0.2 \leq ES \leq 0.8$ for medium ES < 0.2 represents a small effect, $0.2 \leq ES \leq 0.8$ represents a medium effect, and ES > 0.8 represents a large effect [31]. The test for heterogeneity was assessed by two statistics, Q test and I^2: $I^2 < 50\%$ and $P \geq 0.05$ indicated that the included studies were not heterogeneous; $I^2 > 50\%$ and $P < 0.05$ indicated that the included studies were heterogeneous, and sensitivity analysis could be performed to exclude the more heterogeneous studies [32] or combined using random effect measures [33]. Funnel plots were used to test for publication bias in the included studies. Finally, subgroup analyses of the included studies were conducted to clarify the dose–effect relationship between AVGs and FMS in children as well as to uncover potential sources of heterogeneity. Statistical significance was achieved at $p \le 0.05$.

4 Results

Study Selection

A total of 529 articles were obtained by searching the databases. Among the English databases, there were 142 articles in PubMed, 95 articles in Web of Science, 43 articles in Science Direct and 172 articles in Cochrane Library. Among the Chinese databases, there were 45 articles in CNKI and 32 articles in Wanfang Database. After removing the duplicate articles by EndNote, a total of 342 articles remained. After reading the titles and abstracts, a total of 89 papers were remained. Further reading of the full text resulted in 82 papers being excluded as they did not meet the outcome indicators. A total of 7 papers were eventually included (Fig. 1).

Study Characteristics

The studies were published between 2015 and 2020, with two from USA [34, 35], one from Greece [36], one from Brazil [37], one from Australia [38], one from the UK [39], and one from Ireland [40]. For subjects, the age and gender of the children were reported in each study, with sample sizes ranging from 34 to 137 and a total of 462. The interventions in the experimental group were predominantly school-based and

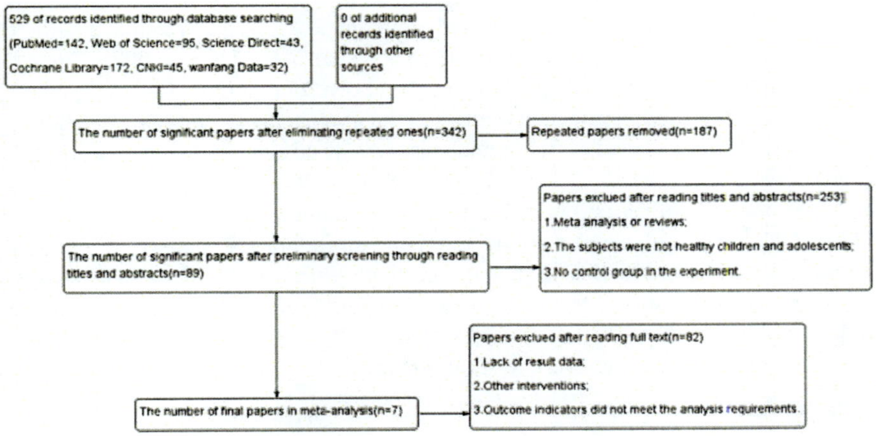

Fig. 1 PRISMA flow diagram

ranged from 8 to 18 weeks, 2–7 times/week, and 20–60 min/time. The interventions in the control group were mainly based on regular physical activities and physical education class organized by the school. Regarding statistical methods, two studies used both t-tests and ANOVA [36, 37], three studies used ANOVA [34, 35, 40], one study used t-tests [39] and one study used mixed linear models [38]. For outcome indicators, two studies included both the total FMS, OCS, and LOC [37, 39], one study included the total FMS and OCS [36], two studies included the LOC and OCS [35, 38], one study included only the total FMS [34], and one study included only the LOC [40]. All studies used the Test for Gross Motor Development (TGMD) as a test tool of FMS (Table 1).

Risk of Bias Assessment

All seven articles were low risk of bias in four areas: random sequence generation, selective reporting, incomplete outcome data, and other bias. However, in terms of allocation concealment, only three papers specified a specific concealment strategy for the allocation scheme [37–39], which achieved low risk, and the remaining four papers were of unknown risk bias, which led to a decrease in the quality of the studies. In terms of blinding of participants and personnel, only one study was blinded between subjects and assessors [37], achieving low risk bias. Given that most of the studies had children and adolescents as subjects and required informed consent or agreements from guardians prior to the experiment, this resulted in most studies not being able to use blinding between subjects and researchers and therefore most studies had a high risk of bias (Fig. 2). Based on the risk assessment tool, three articles were identified as having a low risk of bias, with one article achieving a score of 6 and the remaining four articles all achieving a moderate risk of bias (Fig. 3).

Meta-Analysis

For total FMS, heterogeneity between the four studies was not significant ($I^2 = 4\% < 50\%$, $P = 0.37$) and allowed for the selection of a fixed ES for meta-analysis. Sensitivity analyses found no effect on study outcomes. AVGs had a moderate effect on FMS in healthy children (SMD = 0.41, 95% CI, 0.15–0.67, p = 0.002) (Fig. 4). There was a low risk of bias publication for the above four studies (Fig. 5).

For OCS, there was heterogeneity between the five papers ($I^2 = 58\%$, $P = 0.05$), so a random ES was selected for meta-analysis. AVGs had a moderate effect of (SMD = 0.41, 95% CI, 0.07–0.75, $P = 0.02$) on OCS in healthy children (Fig. 6). There was a low risk of publication bias for the above five studies (Fig. 7).

For LOC, meta-analysis was conducted using a random ES because of the high heterogeneity between the five studies reporting children's LOC ($I^2 = 58\% > 50\%$, $P = 0.05$). Figure 8 shows that there was a moderate effect of AVGs on healthy children's LOC (SMD = 0.39, 95% CI, 0.02–0.75, $P = 0.04$). There was a low risk of bias publication for the above five studies (Fig. 9).

Subgroup Analyses

Because the four studies reporting total FMS had the same intervention setting (in schools) and the intervention content (using Xbox Kinect), subgroup analyses were

Table 1 Descriptive characteristics of the literature

First author, country, year of publication	Purposes	Sample size (IG, CG)	Sample characteristics (age and gender)	Interventions vs. controls	Test tools and outcomes	Statistical methods	Intervention results
Vernadakis (2015), Greece	Effects of AVGs-based training on children's FMS	66 (22, 22)	Aged 6–7 years (6.35 ± 0.73) IG(1): 11 males 11 females IG(2): 12 males 10 females CG: 13 males 9 female	8 weeks. ① IG: 2 × 30 min/week, FMS training based on AVGs; ② CG: 2 × 30 min/week, unstructured recess	TGMD-2 FMS/OCS	T-tests ANOVA	① > ②
Gao (2018), USA	Effects of school-based AVGs intervention on FMS in children	56 (20, 36)	Aged 4–5 years (4.46 ± 0.46) IG: 11 males 9 females (4.72 ± 0.34) CG:14 males 22 females (4.33 ± 0.46)	8 weeks. ① IG: 5 × 20 min/week, AVGs included in preschool curriculum, 10 min game and 10 min rest; ② CG: 5 × 20 min/week, school routine care and recess	TGMD-2 FMS	ANOVA	① > ②

(continued)

Table 1 (continued)

First author, country, year of publication	Purposes	Sample size (IG, CG)	Sample characteristics (age and gender)	Interventions vs. controls	Test tools and outcomes	Statistical methods	Intervention results
Fu (2018), USA	Effects of AVGs on physical activity, motor skill and enjoyment in preschool children	65 (36, 29)	Aged 4.9 ± 0.7 years IG:20 males 16 females CG:14 males 15 females	12 weeks. ① IG: 5 × 30 min/week, AVGs included in regular school activities and supervised by research assistants and teachers; ② CG: 5 × 30 min/week, unstructured recess	TGMD-3 FMS/OCS/LOC	ANOVA	① > ②
Medeiros (2020), Brazil	Effects of AVGs on motor skills in children aged 8–10	64 (32, 32)	Aged 8–10 years (9.09 ± 0.75) 30 males 34 females	18 weeks. ① IG: 2 × 45 min/week, AVGs included in school PE curriculum; ② CG: 2 × 45 min/ week, regular school PE curriculum	TGMD-2 FMS/OCS/LOC	T-tests ANOVA	① ≈ ②

(continued)

Table 1 (continued)

First author, country, year of publication	Purposes	Sample size (IG, CG)	Sample characteristics (age and gender)	Interventions vs. controls	Test tools and outcomes	Statistical methods	Intervention results
Trost (2020), Australia	Effects of Moovosity™ on FMS in 3–6 years old children	34 (17, 17)	Aged 3–6 years IG:8 males 9 females (5.3 ± 1.1) CG:9 males 8 females (5.3 ± 1.3)	8 weeks. ① IG: 3 times/week at home with Moovosity™; ② CG: unstructured recess	TGMD-2 OCS/LOC	Mixed linear models	① > ②
Johnstone (2019), UK	Feasibility of AVGs to improve FMS in children	137(73, 64)	IG:34 males 39 females (7.1 ± 0.3) CG: 24 males 40 females (7.0 ± 0.3)	10 weeks. ① IG: 1 × 60 min/week, Go2Play Active Play, 30 min play and 30 min recess; ② CG: unstructured recess	TGMD-2 OCS/LOC	T-tests	① ≈ ②
McGann (2019), Ireland	To examine the effects of commercial exergames and purpose-built exergames on children's locomotor skills	40 (20, 20)	Aged 5–7 years IG: 10 males 10 females CG: 6 males 4 females	8 weeks. ① IG: 7 times/week with HITL-based AVGs; ② CG: 7 times/week with commercial AVGs	TGMD-2 LOC	ANOVA	① > ②

Note ① intervention group (IG) ② control group (CG); LOC (locomotor skill); OCS (object control skill); GMS (gross motor skills)

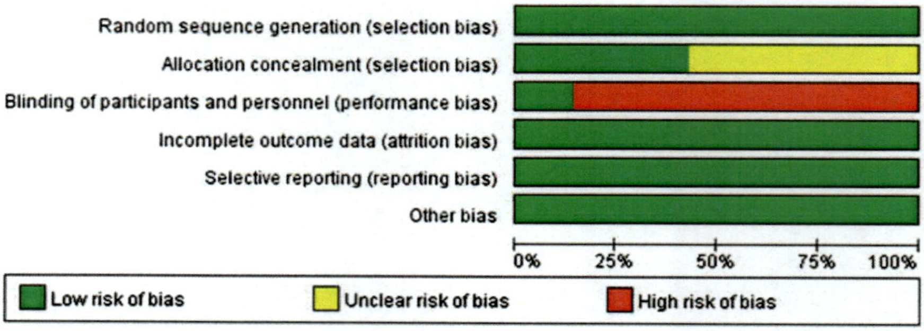

Fig. 2 The percentage of entries in the methodological assessment

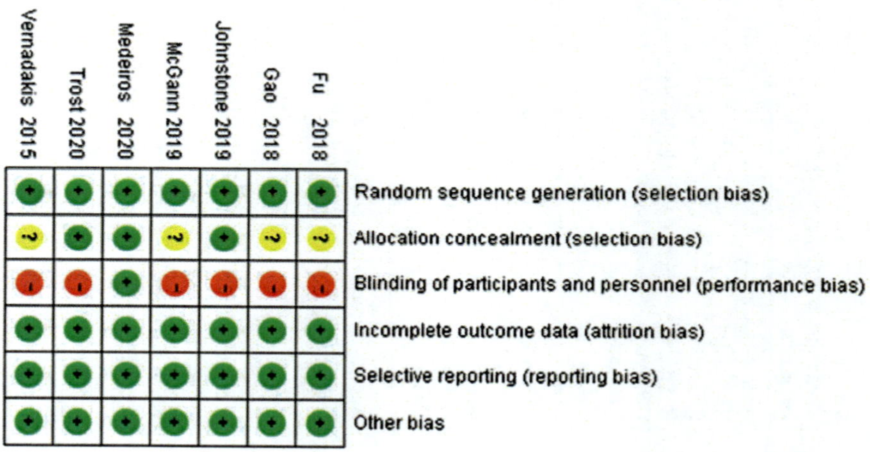

Fig. 3 Methodological quality assessment of included studies

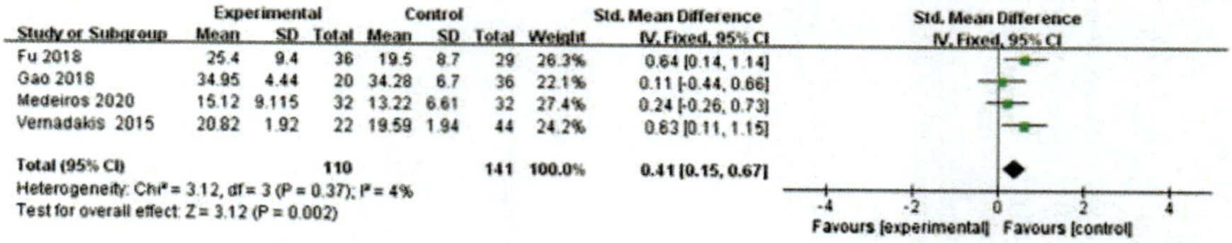

Fig. 4 Forest plot of the effects of AVGs on FMS in healthy children

not conducted for these two variables. For the combined ES of intervention weeks: although the effects of $\leq$ 8 weeks (SMD = 0.31, 95% CI, 0.00–0.68, $P = 0.03$) and 12–18 weeks (SMD = 0.43, 95% CI, 0.08–0.79, $P = 0.02$) were both moderate ES, the intervention effect of 12v18 weeks was significantly better than $\leq$ 8 weeks ($\chi^2 = 1.15$, $P = 0.04$). For the combined ES of time per intervention, there was a

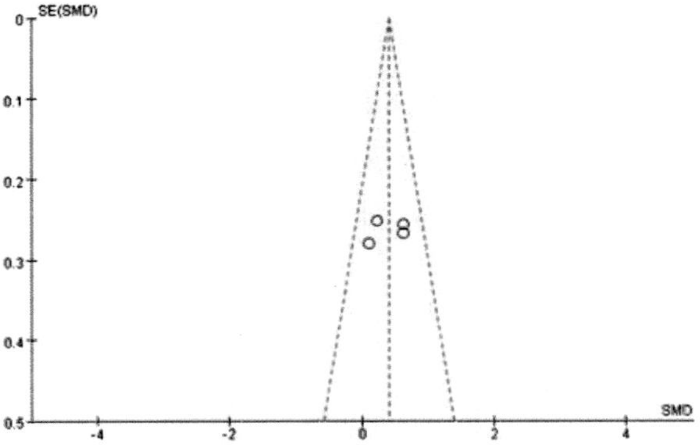

Fig. 5 Funnel plot of the effect of AVGs on FMS in healthy children

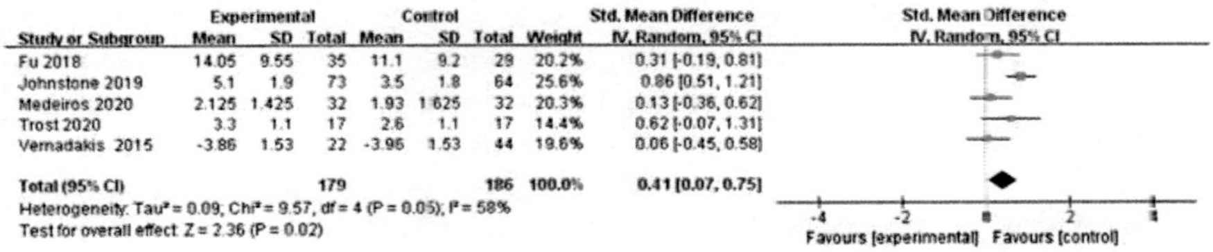

Fig. 6 Forest plot of the effects of AVGs on OCS in healthy children

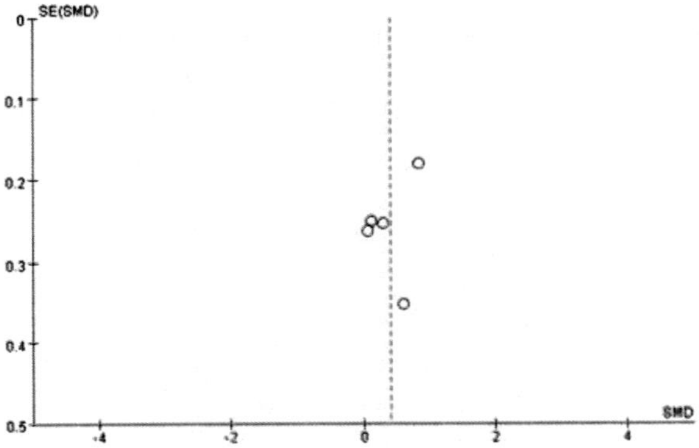

Fig. 7 Funnel plot of the effect of AVGs on OCS in healthy children

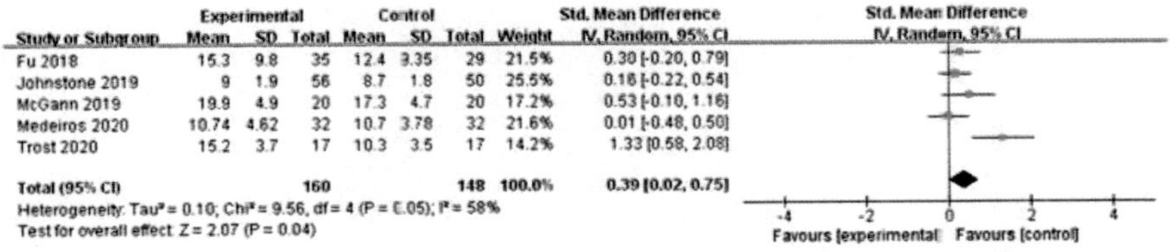

Fig. 8 Forest plot of the effects of AVGs on LOS in healthy children

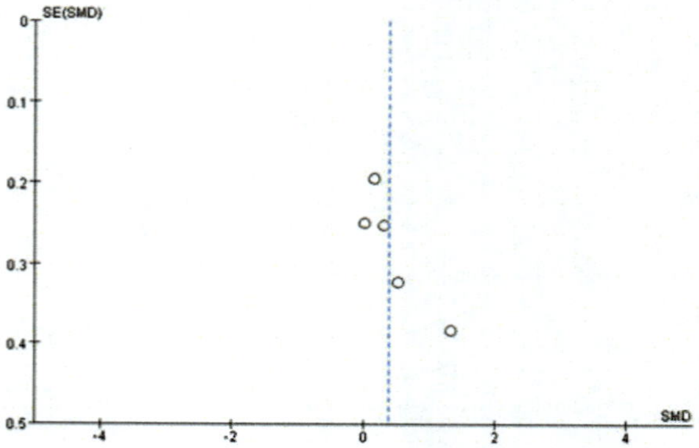

Fig. 9 Funnel plot of the effect of AVGs on LOS in healthy children

moderate effect for ≥ 30 min/time (SMD = 0.49, 95% CI, 0.20–0.79, $P = 0.0009$) and significantly better than the effect of ≤ 20 min/time ($\chi^2 = 1.24$, P = 0.05). For the ES of intervention frequency, although there was a moderate effect for both 2 times/week (SMD = 0.42, 95% CI, 0.06–0.78, $P = 0.02$) and 5 times/week (SMD = 0.51, 95% CI, 0.13–1.02, $P = 0.0001$), the effect was significantly better for 2 times/week than for 5 times/week ($\chi^2 = 2.31$, $P = 0.03$) (Table 2).

For OCS, the combined ES of intervention weeks: a moderate effect was present for 10–18 weeks intervention (SMD = 0.51, 95% CI, −0.00–1.15, $P = 0.01$) which was significantly better than for ≤ 8 weeks ($\chi^2 = 1.72$, $P = 0.03$). The combined ES of time per intervention: there was a moderate effect for 45–60 min/time (SMD = 0.47, 95% CI, −0.10–1.07, $P = 0.04$) and it was significantly better than the effect for ≤ 30 min/time ($\chi^2 = 1.66$, $P = 0.02$). The ES of intervention frequency: moderate effects were present for both 1–3 times/week (SMD = 0.39, 95% CI, −0.01–0.95, *P*

Table 2 Subgroup analysis of the effects of AVGs on total FMS

Variable	Subgroup	No. of studies	Heterogeneity	Subgroup differences	ES	95% CI	Two-tailed test (Z/P)
			$\chi^2/P/I^2$	$\chi^2/df/P/I^2$			
Intervention weeks	≤ 8 weeks	2	1.31, 0.12, 3%	1.15, 1, 0.04, 13%	0.31	(0.00, 0.68)	1.27, 0.03
	12–18 weeks	2	1.28, 0.26, 22%		0.43	(0.08, 0.79)	2.42, 0.02
Time per intervention	≤ 20 min	1	0.00, 0.00, 0	1.24, 2, 0.05, 0%	0.11	(−0.44, 0.66)	0.39, 0.69
	≥ 30 min	3	1.64, 0.44, 0%		0.49	(0.20, 0.79)	3.32, 0.0009
Intervention frequency	2 times/week	2	1.15, 0.28, 13%	2.31, 2, 0.03, 0%	0.42	(0.06, 0.78)	2.30, 0.02
	5 times/week	2	2.14, 0.02, 0%		0.51	(0.13, 1.02)	1.71, 0.0001

= 0.07) and 5 times/week (SMD = 0.42, 95% CI, 0.05–1.09, $P = 0.05$) interventions, but the effect of 5 times/week was more significant ($\chi^2 = 2.32$, $P = 0.01$). The ES of intervention setting: moderate effects were found for both interventions in school (SMD = 0.40, 95%CI, −0.02–0.93, $P = 0.05$) and in the home (SMD = 0.51, 95% CI, 0.02–1.03, $P = 0.01$), but the effect of intervention in the home were significantly better than those in school ($\chi^2 = 1.35$, $P = 0.02$). However, the inclusion studies was insufficient and the results should be viewed with caution. The combined ES of the types of AVGs: though the effect of Xbox Kinect (SMD = 0.41, 95% CI, 0.05–1.06, $P = 0.04$), Go2Play active play (SMD = 0.32, 95% CI, 0.01–0.98, $P = 0.05$) and Moovosity™ (SMD = 0.30, 95% CI, −0.03–0.81, $P = 0.07$) were all moderate, the effect of Xbox Kinect was more significant ($\chi^2 = 2.21$, $P = 0.05$) (Table 3).

For LOC, the combined ES of intervention weeks: though the effect of ≤ 8 weeks (SMD = 0.39, 95% CI, 0.12–0.65, P = 0.01) and 10–18 weeks (SMD = 0.48, 95% CI, 0.16–1.19, $P = 0.001$) were both moderate, the effect was more significant for 10–18 weeks ($\chi^2 = 2.04$, $P = 0.05$). Combined ES of time per intervention: the effect of 45–60 min/time was moderate (SMD = 0.46, 95% CI, 0.07–1.04, $P = 0.002$) which was significantly better than 30 min/time ($\chi^2 = 2.61$, $P = 0.01$). ES

Table 3 Subgroup analysis of the effects of AVGs on OCS

Variable	Subgroup	No. of studies	Heterogeneity $\chi^2/P/I^2$	Subgroup differences $\chi^2/df/P/I^2$	ES	95% CI	Two-tailed test (Z/P)
Intervention weeks	≤8 weeks	2	1.61, 0.20, 38%	1.72, 2, 0.03, 7%	0.29	(−0.24, 0.83)	1.07, 0.28
	10–18 weeks	3	2.42, 0.14, 11%		0.51	(−0.00, 1.15)	1.97, 0.01
Time per intervention	≤30 min	3	1.63, 0.44, 0%	1.66, 1, 0.02, 12%	0.28	(−0.03, 0.60)	1.75, 0.08
	45–60 min	2	1.45, 0.21, 21%		0.47	(−0.10, 1.07)	1.22, 0.04
Intervention frequency	1–3 times/week	4	1.11, 0.16, 2%	2.32, 2, 0.01, 7%	0.39	(−0.01, 0.95)	1.92, 0.07
	5 times/week	1	0.00, 0.00, 0		0.42	(0.05, 1.09)	1.75, 0.05
Intervention setting	School	4	2.35, 0.09, 23%	1.35, 3, 0.02, 18%	0.40	(−0.02, 0.93)	1.67, 0.05
	Home	1	0.00, 0.00, 0		0.51	(0.02, 1.03)	2.31, 0.01
Types of AVGs	Xbox Kinect	3	1.31, 0.07, 3%	2.21, 2, 0.05, 8%	0.41	(0.05, 1.06)	1.75, 0.04
	Go2Play Active Play	1	0.00, 0.00, 0		0.32	(0.01, 0.98)	1.22, 0.05
	Moovosity™	1	0.00, 0.00, 0		0.30	(−0.03, 0.81)	1.13, 0.07

Table 4 Subgroup analysis of the effects of AVGs on total LOC

Variable	Subgroup	No. of studies	Heterogeneity	Subgroup differences	ES	95% CI	Two-tailed test (Z/P)
			$\chi^2/P/I^2$	$\chi^2/df/P/I^2$			
Intervention week	≤8 weeks	2	0.06, 0.80, 0%	2.04, 2, 0.05, 7%	0.39	(0.12, 0.65)	2.46, 0.01
	10–18 weeks	3	1.14, 0.36, 2%		0.48	(0.16, 1.19)	3.18, 0.001
Time per intervention	≤30 min	3	0.98, 0.61, 0%	2.61, 3, 0.01, 11%	0.27	(−0.00, 0.54)	1.95, 0.05
	45–60 min	2	2.13, 0.25, 5%		0.46	(0.07, 1.04)	2.47, 0.002
Intervention frequency	1–3 times/week	3	1.18, 0.06, 4%	2.41, 2, 0.05, 4%	0.41	(−0.09, 0.97)	1.71, 0.03
	5–7 times/week	2	1.27, 0.09, 0%		0.53	(−0.17, 1.14)	1.81, 0.007
Intervention setting	School	4	1.82, 0.61, 0%	1.44, 3, 0.02, 0%	0.21	(−0.03, 0.45)	1.72, 0.08
	Home	1	0.00, 0.00, 0		0.55	(0.04, 1.25)	1.84, 0.05
Gaming platform	Xbox kinect	4	1.21, 0.07, 5%	2.14, 2, 0.001, 3%	0.47	(−0.09, 1.27)	1.71, 0.01
	Go2Play Active Play	1	0.00, 0.00, 0		0.31	(−0.10, 1.14)	1.05, 0.09

of intervention frequency: the effect of 1–3 times/week (SMD = 0.41, 95% CI, −0.09–0.97, $P = 0.03$) and 5–7 times/week (SMD = 0.53, 95% CI, −0.17–1.14, $P = 0.007$) were both moderate, but the effect of 5–7 times/week was more significant ($\chi^2 = 2.41, P = 0.05$). Combined ES of intervention setting: the effect was moderate in the home (SMD = 0.55, 95% CI, 0.04–1.25, $P = 0.05$) which was more significant than in school ($\chi^2 = 1.44, P = 0.02$). It was important to note that the inclusion study is insufficient and the results needed to be viewed with caution. Combined ES of the type of AVGs: the effects of Xbox Kinect (SMD = 0.47, 95% CI, −0.09–1.27, $P = 0.01$) and Go2Play Active Play (SMD = 0.31, 95% CI, −0.10–1.14, $P = 0.09$) were both moderate, but the effect of Xbox Kinect was more significant ($\chi^2 = 2.14, P = 0.001$) (Table 4).

5 Discussion

With the continued development of human–computer interaction technology, AVGs as a new way of sports gaming are gradually becoming a new favorite in the market.

Especially during the COVID-19 epidemic, the strength of AVGs at home are particularly highlighted. This study examined the effects of AVGs on FMS using a systematic review and meta-analysis and found a moderate facilitative effects on total FMS, OCS, and LOC.

For the total FMS, studies on the effects of AVGs on children's total FMS have been inconsistent. Such as, Vernadakis et al. [36] Gao [34] and Fu [35] found that AVGs significantly promote FMS levels in children. However, Medeiros [37] found that the facilitative effect of AVGs on FMS was not significantly different from that of the control group. It is noteworthy that there were no negative effects of AVGs in the included studies, all pointing to the positive impact on children's physical health and FMS. In line with these results, following a systematic review and meta-analysis, we also found a moderate facilitative effect of AVGs on FMS. More specifically, interventions using Xbox Kinect in schools for at least 12 weeks (5 times/week, 30 min/time) had the best impact on the promotion of FMS in healthy children. Therefore, AVGs have the potential to be a training measure for the development of FMS in healthy children.

This study found a moderate facilitative effect of AVGs on OCS in healthy children, which was consistent with the results of one systematic review [41]. In fact, studies on the effects of AVGs on OCS in healthy children are more consistent, all finding that AVGs can improve children's OCS significantly [35–39]. Furtherly, the subgroup analysis results indicated that interventions using Xbox Kinect at home for at least 12 weeks (5 times/week, 45–60 min/time) had a more significant effect on improving children's OCS. So, AVGs could be used as a supplement to school physical education [37]. Considering the low levels of OCS among international [16] and Chinese [17] children, AVGs were recommended as an effective way to promote OCS in a collaborative home-school approach.

For LOC, studies on the effects of AVGs on children's LOC have been inconsistent. McGann [40], Fu [35] and Trost [38] had shown that AVGs could significantly improve children's LOC, but a systematic review [27] shown that the facilitative effects were uncertain. In this study, the meta-analysis results suggested that AVGs have a moderate effect on children's LOC. Specifically, interventions using Xbox Kinect in the home for at least 10 weeks (5–7 times/week, 45–60 min/time) had a more significant effect on improving children's LOC. Therefore, AVGs can be an ideal intervention tool for developing children's LOC.

There were some limitations in this study. Only seven RCT trials were included and unpublished papers and non-reviewed articles were not entered. Therefore, there may be publication risk bias. The included RCTs varied widely in their intervention designs, and some of them had problems with dropouts and lost visits of study participants. The number of included RCTs was small, so subgroup analyses of age, gender, and ethnicity were not conducted.

6 Conclusions

AVGs had a moderate facilitative effect on children's total FMS and a slightly better facilitative effect on OCS than LOC.

References

1. World Health Organization (2019) Global action plan on physical activity 2018–2030: more active people for a healthier world: report of WHO scientific group. WHO, Geneva, Switzerland
2. LeBlanc AG et al (2013) Active video games and health indicators in children and youth: a systematic review. PLoS ONE 8:e65351
3. Nelson MC, Gordon-Larsen P (2006) Physical activity and sedentary behavior patterns are associated with selected adolescent health risk behaviors. Pediatrics 117:1281–1290
4. Tremblay MS et al (2011) Systematic review of sedentary behaviour and health indicators in school-aged children and youth. Int J Behav Nutr Phys Act 8:98
5. Ekelund U et al (2012) Moderate to vigorous physical activity and sedentary time and cardiometabolic risk factors in children and adolescents. JAMA 307:704–712
6. Stodden DF et al (2008) A developmental perspective on the role of motor skill competence in physical activity: an emergent relationship. Quest 60:290–306
7. Logan SW et al (2018) Fundamental motor skills: a systematic review of terminology. Sports Sci 36:781–796
8. Diao YC [刁玉翠] (2018) 3~10 岁儿童基本运动技能发展与教育促进研究. 华东师范大学博士学位论文
9. Aslankhani MA et al (2013) The effect of four motor programs on motor proficiency in 7–9 years old boys. Middle-East J Sci Res 13:1526–1532
10. Zhang L et al [张柳 等] (2020) 幼儿基本动作技能与身体素质的关联性. 中国学校卫生 (4):554–557
11. Zhang Y [张莹] (2012) 动作发展视角下的幼儿体育活动内容实证研究. 北京体育大学学报 3:133–140+145
12. Li J et al [李静 等] (2019) 3~5岁幼儿基本动作技能与体能的关系研究. 中国体育科技 6:52–58
13. Jones RA et al (2011) Promoting fundamental movement skill development and physical activity in early childhood settings: a cluster randomized controlled trial. Pediat Exer Sci 4
14. Ma R, Song H [马瑞, 宋珩] (2017) 基本运动技能发展对儿童身体活动与健康的影响. 体育科学 4:54–61+97
15. Diao YC et al [刁玉翠 等] (2017) 4~9岁儿童基本运动技能与其自我知觉的关系研究. 天津体育学院学报 4:326–331
16. Bolger LE et al (2021) Global levels of fundamental motor skills in children: a systematic review. Sports Sci 39:717–753
17. Li J, Diao YC [李静, 刁玉翠] (2013) 3~10岁儿童基本动作技能发展比较研究. 中国体育科技 3:129–132
18. Morgan PJ et al (2013) Fundamental movement skill interventions in youth: a systematic review and meta-analysis. Pediatrics 132:e1361–e1383
19. Van CA et al (2017) Interventions to improve fundamental motor skills in pre-school aged children: a systematic review and meta-analysis. Sci Med Sport 20:658–666
20. Xin F et al [辛飞 等] (2019) 国外幼儿基本动作技能干预研究系统评述. 体育科学 2:83–97
21. Dong P et al [董鹏 等] (2020) 新型冠状病毒肺炎疫情下学校体育的价值、使命与担当. 体育学研究 2:59–64

22. Chang L et al [常蕾 等] (2021) 后疫情时代学校体育与家庭体育的融合发展研究. 青少年体育 8:81–83
23. Parker K et al (2021) The use of digital platforms for adults' and adolescents' physical activity during the COVID-19 pandemic (our life at home): survey study. J Med Internet Res 2
24. Bai H et al [白桦 等] (2022) 大规模疫情时期构建中国居家健身远程教育系统探究. 河北体育学院学报 2:22–29
25. Barnett LM et al (2014) Parents' and children's views on whether active video games are a substitute for the 'real thing.' Qual Res Sport Exerc Health 1:1–16
26. Gao Z et al (2019) Effects of exergaming on motor skill competence, perceived competence, and physical activity in preschool children. Sport Health Sci 8:106–113
27. Liu W et al (2020) Effect of active video games on healthy children's fundamental motor skills and physical fitness: a systematic review. Environ Res Public Health 17:8264
28. Moher D et al (2015) Preferred reporting items for systematic review and meta-analysis protocols (PRISMA-P) 2015 statement. Syst Rev 4:1–9
29. Tang W et al (2015) Kappa coefficient: a popular measure of rater agreement. Shanghai Arch Psychiatry 27:62–67
30. Bax L et al (2007) A systematic comparison of software dedicated to meta-analysis of causal studies. BMC Med Res Methodol 1
31. Cohen J (1992) A power primer. Psychol Bull 112:155–159
32. Ren Z, Wu J (2019) The effect of virtual reality games on the gross motor skills of children with cerebral palsy: a meta-analysis of randomized controlled trials. Environ Res Public Health 16:3885
33. Guan L, Wu T (2014) Forest plot and clinical significance of meta-analysis. Evid Based Med 3:198–201
34. Gao Z, Zeng N, Pope ZC et al (2019) Effects of exergaming on motor skill competence, perceived competence, and physical activity in preschool children. Sport Health Sci 8:106–113
35. Fu Y et al (2018) Differences in step counts, motor competence, and enjoyment between an exergaming group and a non-exergaming group. Games Health 7:335–340
36. Vernadakis N et al (2015) The impact of an exergame-based intervention on children's fundamental motor skills. Comput Educ 83:90–102
37. Pâmella M et al (2020) Positive effect of a motor intervention program with exergames: a blind randomized trial. Int J Game Based Learn 10
38. Trost SG, Brookes D (2021) Effectiveness of a novel digital application to promote fundamental movement skills in 3- to 6-year-old children: a randomized controlled trial. Sports Sci 39:453–459
39. Johnstone A et al (2019) An active play intervention to improve physical activity and fundamental movement skills in children of low socio-economic status: feasibility cluster randomised controlled trial. Pilot Feasibility Stud 5:45
40. McGann J et al (2019) Hop skip jump games: the effect of "principled" exergameplay on children's locomotor skill acquisition. Educ Technol 51:798–816
41. Seth E et al (2017) The effectiveness of developing motor skills through motion-based video gaming: a review. Simul Gaming 6

The Effect of Taijiquan on Working Memory of College Students with High Obsessive–Compulsive Symptoms

Tongtong Hao

Abstract In recent years, among college students, obsessive–compulsive symptoms are more prominent than other psychological problems, and college students' working memory ability is closely related to obsessive–compulsive symptoms. Working memory plays an important role in personal learning and life. Whether college students with obsessive–compulsive problems have low working memory ability, and whether they can improve their obsessive–compulsive problems by improving their working memory ability is worth exploring. Firstly, this study takes college students as the research object and uses the method of questionnaire to study the relationship between college students' obsessive–compulsive symptoms and working memory ability; secondly, through Taijiquan exercise intervention, the working memory ability of college students with high obsessive–compulsive symptoms was intervened to explore the improvement effect of this method on college students' obsessive–compulsive symptoms.

Keywords Taijiquan · High obsessive–compulsive symptoms · Working memory · Exercise intervention

1 Introduction

Obsessive–compulsive disorder (OCD) is a common mental disease in clinic. OCD is characterized by uncontrollable obsessive–compulsive thinking or obsessive–compulsive movement, which is prone to anxiety, depression, and other adverse emotions. Previous studies have found that the cognitive control of OCD patients has defects in three aspects: inhibitory control (the ability to inhibit dominant response), working memory (the ability to maintain and update information related to needs), and cognitive flexibility (the ability to change strategies in the face of changing environmental needs) [1]. Obsessive–compulsive trait and OCD are two different

T. Hao (✉)
Chinese Wushu College of Beijing Sport University, Beijing, China
e-mail: cyndra1424@uitm.edu.my

G. Kuan et al. (eds.), *Advancing Sports and Exercise via Innovation*, Lecture Notes in Bioengineering, https://doi.org/10.1007/978-981-19-8159-3_24

concepts. Obsessive–compulsive trait is an individual who has obsessive–compulsive symptoms but fails to meet the diagnostic criteria of OCD, showing excessive responsibility and pursuing perfection [2]. Obsessive–compulsive behavior includes compulsive examination, compulsive cleansing, and compulsive counting. People with high obsessive–compulsive traits know that this is unnecessary, but in order to alleviate the negative impact of obsessive–compulsive concepts, it is difficult for them to control not to do so, which eventually leads to a vicious circle [3], which is likely to cause them to often be in a situation of contradiction and conflict. In addition, they are often in mental pain due to various bad emotions [4]. In the mental health survey of Chinese college students, it is found that obsessive–compulsive symptoms are common among college students [5]. Obsessive–compulsive symptoms rank first among college students' psychological problems.

Working memory (WM) is the ability of individuals to temporarily maintain and operate information during the execution of cognitive tasks. It is the core of human high-level cognitive activities and an important component of learning, reasoning, problem-solving, and intellectual activities [6–8]. Working memory tasks are ubiquitous in human life. For example, when reading, it is necessary to remember the characters at the beginning of a sentence and use them to understand the semantics behind. Working memory is used throughout life, so it is often associated with human intelligence, information processing ability, executive function, problem-solving ability, and learning ability [9–11]. People with high obsessive–compulsive trait have similar working memory impairment as patients with obsessive–compulsive disorder. Although they can maintain normal social life, their social function and working memory function are abnormal [12]. A recent report found that adolescents with high obsessive–compulsive traits have mild impairment in cognitive functions such as working memory [13]. The study found that compared with college students with low obsessive–compulsive traits, college students with high obsessive–compulsive traits have abnormal cognitive function indicators such as inhibition control and cognitive flexibility. Compared with obsessive–compulsive symptoms, there is still insufficient attention to the cognitive symptoms of obsessive–compulsive disorder [14].

Taijiquan has been fully proved to be one of the sports interventions for the treatment of obsessive–compulsive disorder and other mental diseases [15–17]. At this stage, there are many studies on the impact of Taijiquan on executive function [18–21]. To sum up, this study explored the effect of Taijiquan on working memory of college students with high obsessive–compulsive traits. The results are reported as follows.

2 Methods

2.1 Intervention of Taijiquan on Working Memory Ability of College Students with High Obsessive–Compulsive Symptoms

2.1.1 Research Object

From April to May 2021, two colleges and universities were selected in Beijing. A total of 816 questionnaires were distributed by cluster sampling based on class units. 762 questionnaires were recovered. 19 invalid questionnaires were eliminated. 743 valid questionnaires were obtained. A total of 56 subjects who met the inclusion criteria were divided into the experimental group (28) and the control group (28) according to the random number table method.

Inclusion criteria: score of the revised Obsessive–Compulsive Scale $\geq$ 29, normal vision or corrected vision, age over 18 years old, exclusion criteria: having mental diseases such as schizophrenia and depression, color blindness or color weakness, and serious physical diseases. In the process of Taijiquan intervention, 58 subjects completed the intervention process. No adverse events occurred in both groups within the intervention period. There was no significant difference between the experimental group and the control group in age, education level, and gender ($p > 0.05$). All the subjects participated in this study voluntarily and signed the informed consent form.

2.1.2 Research Tool

2.1.2.1 Obsessive–compulsive Inventory Revised OCI-R

FOA et al. [16] revised the Chinese version according to the simplified version in 1998, and Qinghuan et al. [17] revised the Chinese version in China. OCI-R scale has 18 questions and 6 dimensions, which are cleaning, forced thinking, hoarding, sorting, verification, psychological neutralization, and each dimension has 3 questions. According to the situation in the past month, the subjects made a self-assessment on their own obsessive–compulsive symptoms. Each topic has 5 levels. 0 means "never"; 1 means "almost never"; 2 means "sometimes"; 3 means "often"; and 4 means "almost always". The total score of the revised version of the Obsessive–Compulsive Scale is normally distributed, with the top 27% [18] scores, that is, the group with a total score $\geq$ 29 points is a high obsessive–compulsive trait group.

2.1.2.2 Neuropsychological Test

Digital span test (DST) [20] digital span is divided into back-to-back and back-to-back. The back-to-back contains 8 digital strings, and the back-to-back contains 7 digital strings. Each length has two attempts. Can correctly recall the length of the digital string and continue after adding 1; if 2 attempts fail, stop the test. The maximum length of digit string that subjects can correctly repeat is their working memory span. The larger the index, the better the working memory ability.

2.1.3 Exercise Intervention Program

The leader of this Taijiquan exercise intervention is a national level II athlete, who has received professional Taijiquan training and accepted the supervision of relevant experts. The evaluation of the two groups of subjects was completed by a graduate student who had received training in the neuropsychological-scale system. The specific intervention plan is as follows: the exercise item is 24 Style Taijiquan; the duration of exercise intervention is 12 weeks; the exercise frequency is 3 times/week, and each time is 60 min. The exercise intervention of Taijiquan experimental group was divided into three stages. The first stage was to learn the preparatory activities of related exercises for the first to fourth weeks; The second stage is to learn and practice the basic movements of relevant movements, and the time is the 5th to 8th weeks; The third stage is the study and practice of complete sets of movements, and the time is arranged in the 9th to 12th weeks. Basic requirements for sports intervention: warm up for 10 min before each activity, and then enter the learning and practice process of relevant actions. Keep your attention as much as possible during learning, and finally relax for 5 min. The control group did not receive any intervention.

3 Results

3.1 Comparison of Scores of OCI-R and Neuropsychological Tests Between the Two Groups Before and After Intervention

Before the intervention, there was no significant difference in OCI-R scale and DST between the experimental group and the control group ($t = 0.72, 0.83, p > 0.05$). After the intervention, the OCI-R score of the experimental group was lower than that of the control group ($t = 2.39, p < 0.05$). The total number of error reactions and the total number of persistent errors in the experimental group on the OCI-R scale and DST were significantly improved ($p < 0.05$), while the other differences in the control group ($p < 0.05$) were not statistically significant ($P > 0.05$). See Table 1.

Table 1 Comparison of OCI-R and neuropsychological test results of two groups of college students before and after intervention ($x \pm s$)

Group	Before and after intervention	Number of persons	Statistical value	OCI-R scale	DST	
					Anteroposterior	Inverted back
Experimental group	Before intervention	28		18.23 ± 5.08	8.25 ± 3.02	6.29 ± 1.56
	After intervention	28		11.57 ± 6.42	9.87 ± 3.17	7.93 ± 2.26
			t	3.19	−2.40	−2.53
			P	< 0.05	< 0.05	< 0.05
Control group	Before intervention	28		17.45 ± 7.06	8.91 ± 2.89	6.19 ± 1.38
	After intervention	28		16.91 ± 7.85	9.05 ± 2.11	6.54 ± 1.41
			t	0.37	−0.32	−0.92
			P	> 0.05	> 0.05	> 0.05

4 Conclusion

This study whether Taijiquan can improve the working memory function of college students with high obsessive–compulsive symptoms. According to the experimental results, the OCI-R score of the experimental group is significantly lower than that of the control group after Taijiquan intervention, that is, the obsessive–compulsive symptoms are reduced, and the working memory ability is improved after Taijiquan intervention. In general, Taijiquan can improve the working memory function of college students with high obsessive–compulsive symptoms.

The study found that the population with high obsessive–compulsive trait is one of the high-risk groups to develop obsessive–compulsive disorder. The prevalence and treatment cost of obsessive–compulsive disorder are high. Therefore, it is of great practical significance to carry out preventive screening and intervention in the general population. The reason why Taijiquan has the effect of alleviating obsessive–compulsive disorder and improving working memory function is that it is light, gentle, continuous, both internal and external, and has the characteristics of life cultivation. It can not only divert attention, but also regulate negative emotions. At the same time, in the process of practicing Taijiquan, attention will shift from anxious things to things that are willing to experience and accept, reducing obsessive–compulsive symptoms. And practicing Taijiquan is to memorize the movement sequence in the process of exercise, so it can effectively stimulate the working memory function of college students. In addition, practicing Taijiquan can also enrich after-school life, occupy the time of compulsive behavior and thought, reduce the interference of compulsive psychology, and improve learning ability and memory ability.

The results of this study showed that the obsessive–compulsive symptoms in the experimental group were significantly reduced after the intervention. Although this study is aimed at different nonclinical groups (college students with high obsessive–compulsive symptoms), Taijiquan and drug therapy can significantly improve obsessive–compulsive symptoms. The results of digit inversion in DST task showed that compared with the control group, the digit span of the experimental group increased after the intervention, indicating that Taijiquan can improve the working memory of college students with high obsessive–compulsive traits.

This study has a positive guiding significance for the early psychological intervention and mental health promotion of college students with high obsessive–compulsive trait. At the same time, this study also has some limitations: first, this study was not included in the healthy control group, so it is impossible to objectively discuss the performance of neuropsychological tests. Secondly, the subjects were not followed up, and it was not verified whether the working memory impairment in the high obsessive–compulsive trait population could predict the long-term intervention results of this kind of population. Finally, the neuropsychological tests selected in the study only have response indicators and lack objective electrophysiological indicators, which may lead to the lack of accuracy and accuracy of the research results. Objective indicators can be further adopted in future research.

References

1. Qian MT, Sui PJ, Cai Z, Chen L, Lin Y, Chen L, Zhu C (2022) Effect of cognitive behavior group counseling on cognitive control of college students with high obsessive compulsive trait. Chin Sch Health 43(04):548–552. https://doi.org/10.16835/j.cnki.1000-9817.2022.04.016
2. Deng M, Wang Y, Xiao Z (2021) The latest progress in the research on the common genetic basis of obsessive-compulsive disorder and other severe mental disorders. Chin J Psych 54(06):421–427
3. Lu D, Li P, Chen Y, Shang TH, Sun L, Jia C, Zhang G, Wang Y, Sun Z, Guo Y, Liu W (2021) Preliminary observation on brain spontaneous neural activity and functional connection in patients with OCD at rest. Chin J Psych 54(06):440–446
4. Liu ZJ, Kang YW, Li ZQ, Wu Q, Zhang ZM (2021) The effect and influencing factors of CBT combined with drugs on OCD in China: a meta-analysis. Chin J Clin Psychol 29(03):639–647. https://doi.org/10.16128/j.cnki.1005-3611.2021.03.040
5. Shen X, Zhong H, Li X, Mo D, Cao XM, Geng F, Wang A, Chen BB (2021) Correlation between childhood trauma, obsessive-compulsive symptoms, implicit and explicit memory in patients with obsessive-compulsive disorder. Chin J Behav Med Brain Sci 30(02):129–133
6. Cai Z, Yin JI, Wang N, Chen Z, Wang J, Wang X (2022) Meta analysis of the effect of physical exercise on working memory of the elderly. Chin J Ment Heal 36(02):140–146
7. Xu MC, Jiang CH (2021) A systematic review of physical activity improving cognitive neurological function in children with attention deficit hyperactivity disorder. Chin Rehab Theor Pract 27(12):1422–1429
8. Ke J, Wang B (2021) Effect of aerobic exercise on memory and its neurobiological mechanism. Progr Psychol Sci 30(01):115–128
9. Chen X, Xing F, Liu TT, Sun JM, Mo Y, Yang Q, Wang K, Hu P (2021) Correlation analysis of olfactory function, cognitive function and motor symptoms in patients with Parkinson's disease. Chin J Mod Neurol Dis 21(10):887–892

10. Guo W, Wang B, Ren J (2019) Study on the mechanism of open exercise training the advantage of visual spatial working memory in the elderly. China Sports Sci Technol 55(10):50–55+80. https://doi.org/10.16470/j.csst (two million nineteen thousand and sixty)
11. Yang Y, Tang YK, Hao XJ, Li Y (2019) Research progress of aerobic exercise in improving executive function of the elderly. Chin J Prevent Med 20(11):1110–1114. https://doi.org/10.16506/j.1009-6639.2019.11.022
12. Solution Meta analysis of the effect of different exercise intensity on working memory of children and adolescents. China Sch Health 41(03):356–360+364. https://doi.org/10.16835/j.cnki.1000-9817.2020.03.010
13. Luo J, Ouyang Y (2018) Review on the influence of fitness on working memory. J Shandong Inst Phys Educ 34(06):70–77. https://doi.org/10.14104/j.cnki.1006-2076.2018.06.013
14. Lou Y, Zhou CL, Lu YZ (2019) Study on the characteristics of biological movement working memory processing of college students with different physical activity levels. J Capital Inst Phys Educ 31(04):364–369+384. https://doi.org/10.14036/j.cnki.cn11-4513.2019.04.015
15. Ye Y, Han P (2022) Effect of integrated exercise of five elements music Taijiquan in traditional Chinese medicine on mental health and debilitating state of the elderly. Chin J Gerontol 42(03):725–728
16. Zhang J, Li Z, Xiao T (2021) Meta analysis of the effect of health preserving exercises of traditional Chinese medicine on anxiety and depression symptoms of college students. Mod Prevent Med 48(23):4324–4330
17. Zhu ZC, Li X, Li M, Wan H (2021) Meta analysis of the effect of Taijiquan on physical and psychological intervention in ATS dependent patients of different genders. China Sports Sci Technol 57(06):53–66. https://doi.org/10.16470/j.csst (two million twenty-one thousand and sixty-two)
18. Li X, Yin H, Cui L, Wang Y, Li J, Shen Q, Li Y, Wang D, Zhang J, Luo R (2020) ERP study on the effect of short-term Tai Chi (eight methods and five steps) on inhibitory function of college students. J Tianjin Inst Phys Educ 35(01):29–35. https://doi.org/10.13297/j.cnki.1005-0000.2020.01.006
19. Wang Q, Sheng Y (2016) Influence of Taijiquan on cognitive function of the elderly with mild cognitive impairment in the community. Chin Rehabil Theor Pract 22(06):645–649
20. Wang Q, Sheng Y (2016) Effect of Taijiquan on memory and executive function in the elderly with mild cognitive impairment. J Shandong Inst Phys Educ 32(03):68–72. https://doi.org/10.14104/j.cnki.1006-2076.2016.03.014
21. Liang DM, Tang WQ, Luo C, Zhang MQ (2014) A review of research on Taijiquan exercise promoting cognitive function of the elderly. J Phys Educ 21(04):61–65. https://doi.org/10.16237/j.cnki.cn44-1404/g8.2014.04.017

Validity and Reliability of a Pictorial Instrument for Assessing Movement Skill Perceived Competence in Chinese Children

Xuanxi Li, Lijun Chen, Yucui Diao, and Jing Li

Abstract This research aims to examine the reliability and validity of the Pictorial Scale of Perceived Movement Skill Competence (PMSC-TGMD-2 version) in Chinese children aged 4 to 10 years. A total of 214 children aged 4–10 years (M_{age} = 7.6 years, SD = 1.9 years, 100 girls) from four schools in Jinan City were selected for testing using the random sampling principle. Cronbach's alpha was used for internal consistency analysis; 74 children were retested at one-week intervals to determine the retest reliability by intra-group correlation coefficient (ICC), and 48 children were asked four questions to verify the face validity. Bayesian structural equation modeling (BSEM) and maximum likelihood method (ML) were used to test the structural validity, and the model competition method was used to test the discriminant validity. The internal consistency alpha coefficients for the scale overall and each dimension ranged from 0.76 to 0.86, and the retest reliability ranged from 0.82 to 0.89. (2) For the face validity, children were able to correctly identify most of the skills and had a good understanding of all skills. (3) Regarding structural validity, confirmatory factor analysis (CFA) under the Bayesian approach showed a good fit of the two-factor model of perceived locomotor skill competence and perceived object control skill competence with a posteriori predicted p value of 0.516 and 95% confidence interval of [-39.045, 36.477] after incorporating a priori information. (4) The two-factor model of locomotor and object control skill was highly supported in the model competition, demonstrating that the scale has good discriminant validity. The PMSC-TGMD-2 version has good reliability and validity in Chinese children aged 4–10 years.

Keywords Children · Fundamental movement skills · Perceived competence · Bayesian structural equation modeling

X. Li (✉) · L. Chen · Y. Diao · J. Li
College of Physical Education, Shandong Normal University, Jinan 200241, Shandong, China
e-mail: 754049014@qq.com

G. Kuan et al. (eds.), *Advancing Sports and Exercise via Innovation*, Lecture Notes in Bioengineering, https://doi.org/10.1007/978-981-19-8159-3_25

1 Introduction

Fundamental movement skill (FMS) is the proficiency of individuals to complete a variety of basic motor movements, specifically in locomotor skills, object control skills, and stable ability 3 aspects [1]. It is the basis for advanced and complex sports and physical activities [1–3]. Numerous studies have shown that FMS development not only has a critical impact on preschool children's mental health, healthy fitness, and healthy weight status but also has a significant impact on the development of FMS during childhood [4]. It has been shown in numerous studies that FMS development not only has a critical impact on preschool children's mental health, healthy fitness, and healthy weight status but also that proficiency in object control skills during childhood predicts physical activity behavior during adolescence [5]. The FMS has been shown to significantly impact preschool children's mental health, healthy physical fitness, and healthy weight status.

With the development of cognitive neuroscience and behavioral science, the importance of self-perception in behavior execution and action development has been gradually recognized. Among them, the influence of physical self-perception (PSP) on physical activity behavior is becoming a new hot topic of research. PSP is widely defined as an individual's perception of their actual motor ability [6]. PSP is widely defined as an individual's perception of their actual athletic ability. As one of the sub-dimensions of self-concept [7]. It is considered to be the factor most closely associated with physical activity level [8]. The motivational theory of ability considers PSP as the main motivating factor for voluntary participation in any sport or physical activity [9]. It has a significant impact on the persistence of physical activity participation [10]. Stodden et al. [11] in their Developmental Mechanisms Influencing Physical Activity Trajectories of Children model, they suggest that PSP plays an important mediating role between FMS and physical activity levels. The study showed that [12] the interaction between FMS and PSP and the two is the main potential mechanism to increase physical activity levels.

Given the important value of FMS and PSP for physical activity promotion in children and adolescents, Barnett et al. [13, 14] based on the TGMD-2, which is widely used to measure FMS in children aged 4–10 years, designed the Pictorial Scale of Perceived Movement Skill Competence (PMSC-TGMD-2), which has become an internationally used "self-report" form of PSP measurement tool, which has the characteristics of meeting the cognitive level of younger children, simplicity and efficiency, and high applicability [13, 14]. It has achieved acceptable reliability and validity in Australia [13, 14], Portugal [15], Brazil [16], Spain [17], and other countries. However, there are significant cultural differences between China and other countries, and the applicability of the scale among Chinese children is unknown. Therefore, this study aimed to examine the reliability and validity of the PMSC-TGMD-2 scale in Chinese children, and then, provide Chinese scholars with a valid instrument for measuring PSP in children, and enrich research on PSP in different cultural contexts, with the final aim of making a Chinese contribution to the international field of motor development and health promotion.

2 Method

2.1 Subjects

A total of 216 children aged 4–10 years (M = 7.6, SD = 1.9) were selected using the random sampling principle, and two of them did not complete the test, leaving 214 valid data. After a one-week interval, a random sample ($n = 48$) was selected to determine face validity and another group ($n = 74$) was retested according to standard procedures to determine retest reliability.

2.2 Self-Perception Measures of Fundamental Motor Skills

Using the Barnett et al. [13, 14]. The FMS was measured using the FMS Self-Perception Picture Scale (PMSC-TGMD-2) developed by Dr. Ulrich, University of Michigan, USA [18]. The FMS in this scale corresponds to those in the Test of Gross Motor Development, Second Edition (TGMD-2) developed by Dr. Ulrich et al. It includes six locomotor skills (Run, Gallop, Hop, Leap, Horizontal Jump, and Slide) and six object control skills (Striking a Stationary Ball, Stationary Dribble, Catch, kick, Overhand Throw, and Underhand Roll). The test uses a cartoon format to depict the 12 motor skills and provides two versions of the test atlas for boys and girls by gender as detailed in Fig. 1.

2.2.1 Scale Translation

The English scale was translated independently into Chinese by authors 3 and 4, and then a first draft of the Chinese scale was jointly agreed upon. Barnett optimized the scale. For example, a tester asking the question, "Which child in the picture looks more like you?" was changed to "Which child in the picture looks most like you?".

Fig. 1 Example of Pictorial Scale of Perceived Movement Skill Competence

Second, the word "good at" was difficult for younger children to understand, so it was changed to "good or bad at playing". In response to these comments, the Chinese version of the scale was refined and finally, many experts agreed to form the Chinese version of the PMSC-TGMD-2.

2.2.2 Test Procedure

Subjects were questioned one-on-one in a quiet classroom or setting, strictly according to the test manual. First, subjects were informed of the FMS that the child in the picture was doing and then asked if the subject had played with the skill. For those who had played the skill, they were further informed which of the two children in the picture was doing it well or not so well and then asked, "which child do you think is doing it the most like you". For subjects who had not played a particular FMS, they were first shown the skill and then asked to imagine which child would do the action most like you if they played it themselves. After the participant chose between a good skill and a bad skill, the tester pointed to the two circles below the chosen picture and asked, "Do you consider that you play very well or relatively well?" or "Do you think you played not well or okey?". The "large circle" below the picture of good skill performance means that you played "very well (4 points)" and the "small circle" means that you played "relatively well (3 points)"; the skill performance of the "large circle" means that you played "very well (4 points)" and the "small circle" means that you played "relatively well (3 points)". 3 points"; "large circles" under the pictures with less good skills indicate that they play "not well (1 point)", and "small circles" indicate that they play "okay (2 points)". Since younger children have difficulty in understanding the proper nouns in FMS, individual skills are replaced by realistic and colloquial nouns, for example, "hitting a ball with a bat in both hands" is replaced by "playing baseball" and "throwing a ball with the upper. For example, the expression "hitting a ball with a bat in both hands" was replaced with "playing baseball", and "throwing a ball from above" was replaced with "throwing a ball from above". The total scores of the locomotor skill and object control skill self-perceptions were 6–24, and the total scores of the whole scale were 12–48.

Four questions were administered to 48 subjects after a one-week interval to verify face validity. (1) The subjects were asked if they knew what the skill shown in the picture was, and their responses were recorded to determine the correct recognition rate. If any of the actions were incorrectly identified, such as interpreting a forward slide step as a running jump step, they were informed of the correct answer. (2) In which sports/games/activities the skill in the picture can be used, record each statement of the respondent and then classify the answers into three categories according to their answers, ① don't know ② general activities ③ specific activities. For example, if the skill of running is used in "playing in the park or running when you are in a good mood", the answer is classified as "general activity", and if the answer for overhand ball throwing is "used when serving basketballs". For example, if the answer to the question "I use it when I serve a basketball or when I hit a sandbag" was "specific activity", the answer was categorized as "specific activity". (3) Subjects were asked to identify which diagram showed the better FMS, and if they were incorrect, the tester told them the correct answer and proceeded to the next question. (4) Ask the subject why each FMS picture did well or not so well, and the tester will keep asking whether there are other differences between the two pictures to make sure the subject will say what he or she understands about the good and not so good actions. The answers can be classified into four categories based on their responses: ① don't know ② confusing understanding ③ partial understanding ④ complete understanding. For example, when a subject gives a response that has nothing to do with the quality of the action, it is classified as "confusing", when the subject can answer any one piece of information related to the good or bad action, it is classified as partial understanding, and when the subject can answer more than two pieces of information, it is classified as complete understanding.

3 Statistics Methods

(1) Descriptive statistics were conducted on the percentage of movement demonstrations required; the percentage of each FMS played to the four levels of self-perception for each skill. FMS with less than 75% of attempts using SPSS 22.0 [14] Wilcoxon rank-sum test was conducted to examine the difference in scores between children who had played with the FMS and those who had not played with the FMS.
(2) For reliability, intra-class correlation (ICC) and internal consistency were calculated to determine the reliability of PMSC-TGMD-2. The intra-class correlation ICC of the data before and after retesting the samples was examined, and ICC values less than 0.5 indicated poor reliability, between 0.5 and 0.75 indicated moderate reliability, between 0.75 and 0.9 indicated good reliability, and greater than 0.90 indicated excellent reliability [19]. Internal consistency was tested using Cronbach's alpha for six locomotor skills, six object control skills, and 12 FMS items of self-perception, and Cronbach's alpha values of 0.70 or higher were considered acceptable [20].
(3) For structural validity, confirmatory factor analysis (CFA) was conducted using MPLUS 8.0 to test its structural validity, and the model settings are shown in Fig. 2.

    Since the models of ML analysis are often poorly fitted due to the strict restriction that the zero cross-loading and zero residual correlation must be zero [21]. Therefore, Muthén and Asparouhov [22] provide a method for CFA using the Bayesian structural equation model BSEM, which allows to obtain of analytical results that converge to the real theory. Instead of the strict constraints of maximum likelihood estimation (ML), which usually leads to unrecognizable models in classical statistical methods, researchers can set a prior distribution with zero mean and very small variance for cross-loadings and residual correlations based on the theoretical conception. Unrecognizable, while borrowing

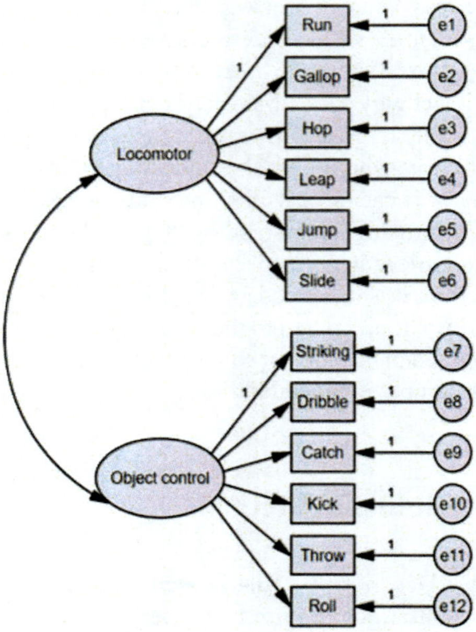

Fig. 2 Two factors structural factor analysis of locomotor and object control

based on BSEM can compensate for these problems. According to the scale developers Barnett [14], et al. reported factor load and Asparouhov et al. [23] suggested that the factor loadings be set to a priori distribution of $N(0.7, 0.02)$. According to the recommendations of Muthén and Asparouhov [24, 25], set the prior distribution of cross-loadings to $N(0, 0.02)$. Set the prior distribution of residual correlation to IW (0, 18) and the prior distribution of residual variance to IW (1, 18) for the Inverse-Wishart (IW) distribution. Since the variance of the prior information setting is set small, which will produce an increase in convergence difficulty, the number of iterations of Markov Chain Monte Carlo (MCMC) is changed from the default 50,000–100,000, and two MCMC chains are set in parallel to improve the speed of. Convergence is considered to be reached when the autocorrelation map is in a reasonable interval, and the trace map shows rapid up and down changes without long-term orientation or arbitrary drift [24, 25]. Gelman et al. [26] suggest that the fitness is best when the posterior predictive p (PPp) value is equal to 0.5 and the probability of the observed data is as high as that of the generated data, converging to 0 or 1 represents poor fitness. In practice, a p of 0.30–0.70 is an acceptable fit [27, 28].

Three different types of prior distributions are set in this study: no prior information, partial information prior, and information prior distribution. This study also uses ML for CFA, and an acceptable model fit is indicated when the fit metric CFI > 0.90 and both SRMR and RMSEA < 0.06.

(4) Regarding the discriminant validity, the model competition method was used to test the discriminant validity of the scale, and the one-factor model was set as the competition model. Bayesian information index (BIC) and abnormal information index (DIC) are commonly used to compare competing models, and according to Kass et al. [27] suggested that when the difference of BIC between two competing models $\Delta BIC > 10$, the model with a large BIC index is a good fit. According to Spiegelhalter et al. [28] suggested that the model with a small DIC is a good fit.

4 Results

4.1 Descriptive Statistics

Table 1 presents the percentage of people who required movement demonstration, the percentage of people who did each movement, and the percentage of each FMS self-perception level. Among the locomotor skills, the movement that required the highest number of demonstrations was the front slide (79.9%), followed by the straddle jump (66.4%) and the side slide (43.0), while the number of demonstrations required for the standing jump, the single-leg jump and the running was less than 5%. Among the object control skills, the most requested movement was the underhand roll (18.2%), followed by hitting the ball with a bat in both hands (15.3%) and overhand throw (7.3%), while no subjects requested demonstrations for the slap and kick.

The least number of people played with a two-handed bat to hit the ball and forward sliding, 41.1% and 66.8%, respectively. The number of those who played running, one-legged jump, standing long jump, tapping, kicking, and catching with both hands were higher than 90%. For attempt rates below 80% [14]. The Wilcoxon rank-sum test was conducted for FMS (playing baseball, forward sliding, and underhand rolling) and showed that children who played playing baseball, forward sliding, and underhand rolling rated themselves higher than those who did not play these skills [$Z_{打棒球} = -3.90$, $P_{打棒球} < 0.000$, $Z_{前滑步} = -2.92$, $P_{前滑步} = 0.003$, $Z_{下手滚球} = -5.27$, $P_{下手滚球} < 0.000$].

4.2 Reliability

4.2.1 Internal Consistency Reliability

As shown in Table 2, the internal consistency alpha values for locomotor and object control skill self-perceptions in the first test were 0.79 and 0.76, respectively, and the overall alpha value for the 12-item FMS was 0.86. The internal consistency alpha values for locomotor and object control skill self-perceptions in the second test were

Table 1 Pictorial scale of perceived movement skill competence level table ($N = 214$)

FMS	Demonstration (%)	Tried item (%)	Not too good (%)	A little good (%)	Good (%)	Very good (%)
Locomotor						
Run	0	100	4.6	17.4	31.2	46.8
Gallop	79.9	66.8	8.7	23.4	29.8	38.1
Slide	43	82.2	7.8	18.3	23.9	50
Hop	2.8	96.7	4.1	22.5	22.5	50.9
Hop	3.3	98.1	8.7	20.6	34.4	36.2
Jumping forward	66.4	86.9	7.8	23.4	19.7	49.1
Object control						
Hitting a ball with two hands	15.3	41.1	20.2	23.9	25.7	30.3
Stationary dribble	0	98.6	3.7	11.5	21.1	63.8
Catch	1.5	92.5	8.7	16.1	28.4	46.8
Kick	0	98.6	8.7	17.9	28.9	44.5
Throw	7.3	87.4	10.1	20.2	28.4	41.3
Underhand Roll	18.2	77.6	12.8	24.3	22.9	39.9

0.79 and 0.81, respectively, and the overall alpha value for the 12-item FMS was 0.84. The scale has good internal consistency reliability.

4.2.2 Retest Reliability

When the 74 children were retested one-week later according to the standard procedure, the overall retest reliability of the scale's 12-item FMS was 0.89, and the retest reliabilities of locomotor and object control skill self-perceptions were 0.88 and 0.82, respectively, indicating that the scale has high retest reliability.

4.3 Validity

4.3.1 Face Validity

Table 3 presents the face validity of the PMSC-TGMD-2 version. In terms of the correct recognition rate of the FMS, the higher recognition rate of the locomotor skills were for running and standing jump (100% and 95.8%, respectively), while the lower recognition rate was for straddle jump (39.6%), single-leg jump (29.2%), and forward slide (22.9%). All object control skills were correctly identified at a rate higher than 66%, with subjects correctly identifying kicking (97.9%), slapping

Table 2 Intra-class correlation coefficient and test–retest reliability of two tests

	First test ($n = 214$)				Second test (n = 74)				Retest Reliability
	Range	Skill mean	Total score mean	α	Range	Skill mean	Total score mean	α	ICC
Locomotor	6–24	2.8	17.1	0.79	8–24	2.9	17.3	0.79	0.88
Object Control	10–24	2.9	17.1	0.76	9–24	2.9	17.2	0.81	0.82
12 FMS	15–46	2.9	34.2	0.86	16–47	2.9	35	0.84	0.89

Note ICC is the intra-group correlation coefficient

(95.8%) and hitting the ball with a two-handed bat (91.7%), and a lower rate of correct identification of underhand rolling (66.7%).

The percentages linking FMS to daily sports/games/activities were highest for running, kicking, and tapping (97.9%, 95.8%, and 93.7%, respectively), lowest for forward skating and underhand rolling (35.4% and 47.9%, respectively), and the remaining FMS ranged from 54.2 to 89.6%, with the exception of forward skating (at 83.3%), where more than 90% of children were able to correctly distinguish between good and bad FMS for each item. More than 93% of the children were able to understand the key factors influencing the performance of each FMS partially and fully.

4.3.2 Structural Validity

The fit indices obtained from the maximum likelihood estimation were as follows: $\chi^2 = 113.7$, $\chi^2/\text{df} = 2.146$, CFI = 0.918, TLI = 0.897, RMSEA = 0.073, SRMR = 0.056, and the observed data were poorly fitted to the theoretical model. The Bayesian method was then used to test the results of the three different prior information as shown in Table 4, and the model convergence could be judged by reviewing all the posterior parameter distribution plots, autocorrelation plots, and trace plots.

$p = 0.000$ for the model without prior information and the model with partial prior information under the Bayesian approach, and the lower limits of the 95% confidence intervals are both greater than 0, suggesting that the cardinality of the observed data is greater than the cardinality of the data generated by the posterior distribution [27], indicating a poor fit. The p value for the a priori information model incorporating factor loadings, cross-loadings, and residual correlations is 0.516, which is very close to the fit criterion of 0.5. 95% confidence intervals range from −39.045 to 36.477 and are essentially centered symmetrically at 0.

Table 5 shows the model normalized factor loadings estimated by ML and BSEM, where the locomotor factor loadings obtained by ML range from 0.49 to 0.74 and the object control factor loadings range from 0.51 to 0.64. The locomotor factor loadings obtained by the information prior vary from 0.57 to 0.65 and the object control loadings range from 0.57 to 0.74. The cross-loadings range from -0.05 to 0.29, so the factor loadings and cross-loadings estimated by the Bayesian approach are within a reasonable range.

4.3.3 Differential Validity

The BIC value of the two-factor model of **locomotor** and **object control** in Table 6 is 7043.86 and the DIC value is 6588.03, while the BIC of the one-factor model = 6717.13 and DIC = 6595.98. $\Delta\text{BIC} = 326.73 > 10$, the BIC value of the two-factor model is greater than that of the one-factor model, the DIC value is smaller than that of the one-factor model, and the fit index p value is better than that of the one-factor model.

Table 3 Summary table of face validity of PMSC-TGMD-2 version ($N = 48$)

FMS	Identification	What sport/game/activity is the picture showing			Understood picture	What is it that makes one picture good and one not so good?			
	Correct %	Does not know %	Generic activity %	Specific sport/activity %	Which is good and not so good %	Did not know %	Did not know %	Partial understanding %	Good understanding %
Locomotor									
Run	100	2.1	27.1	70.8	100	0	0	14.6	85.4
Gallop	22.9	64.6	14.6	20.8	83.3	4.2	2.1	35.4	58.3
Slide	62.5	45.8	22.9	31.3	95.8	2.1	2.1	25	70.8
Hop	29.2	22.9	39.6	37.5	95.8	2.1	0	29.2	68.8
Hop	95.8	14.6	31.3	54.2	100	0	0	12.5	87.5
Jumping forward	39.6	20.8	27.1	52.1	93.8	0	0	31.3	68.8
Object control									
Hitting a ball with two hands	91.7	10.4	8.3	81.3	97.9	0	2.1	27.1	70.8
Stationary dribble	95.8	6.3	6.3	87.5	95.8	0	0	37.5	62.5
Catch	87.5	35.4	14.6	50	89.6	6.3	0	33.3	60.4
Kick	97.9	4.2	0	95.8	91.7	0	0	25	75
Throw	70.8	37.5	18.8	43.8	95.8	4.2	0	29.2	66.7
Underhand roll	66.7	52.1	33.3	14.6	97.9	0	2.1	31.3	66.7

Table 4 Bayesian CFA simulation fit index under different prior information

A priori information	Number of parameters	PPp value	95% CI	
			Lower 2.5%	Upper 2.5%
Non-informative	37	0.000	28.500	93.359
Informative (factor load, cross-load)	50	0.000	26.665	92.695
Informative (factor loadings, cross-loadings). (residual correlation)	116	0.516	−39.045	36.477

Table 5 Two estimation methods are standardized factor load and cross-load

FMS	Maximum likelihood method		Bayesian (information a priori)	
	Locomotor	Object control	Locomotor	Object control
Run	0.69	–	**0.60**	0.20
Gallop	0.49	–	**0.52**	0.06
Slide	0.54	–	**0.49**	0.26
Hop	0.64	–	**0.64**	0.19
Horizontal Jump	0.74	–	**0.56**	0.29
Leap	0.62	–	**0.69**	0.14
Stationary dribble	–	0.51	−0.05	**0.59**
Striking	–	0.54	0.18	**0.54**
Catch	–	0.64	0.04	**0.65**
Kick	–	0.62	0.05	**0.70**
Throw	–	0.60	0.03	**0.68**
Underhand roll	–	0.61	0.03	**0.70**

Note Bold indicates factor main load

Table 6 BIC value of the two-factor model

Models	Number of parameters	*p* value	95% CI		BIC	DIC
			Lower 2.5%	Upper 2.5%		
One-factor model	36	0.000	36.861	101.575	6717.13	6595.98
Two-factor model	116	0.516	−39.045	36.477	7043.86	6588.03

5 Discussion

PSP is becoming a research hotspot in cognitive neuroscience and behavioral science, and it is especially important to have a reliable and valid instrument for measuring PSP. This study examined the reliability and validity of the PMSC-TGMD-2 in children aged 4–10 years in this domestic study and demonstrated that the PMSC-TGMD-2 is equally reliable and valid in Chinese children.

5.1 FMS Exercise Situation

Most children did the FMS shown in the figure, but due to cultural differences, the skill of hitting a ball with a bat in both hands is not very popular among Chinese children, resulting in most children not having done this maneuver. The forward sliding action is often used in the mechanics of basketball and badminton, which may lead to a higher number of people who have not played this skill because the subjects are young and have not learned the mechanics. Running has many application scenarios, wide range of use and the most common in life, so all children have done it. Second, the Ministry of Education's Guidelines for Learning and Development of Children Aged 3–6 Years [29] recommended including the skill of shooting a ball to develop children's coordination and flexibility of movement, so physical activities related to shooting a ball are included in the preschool and compulsory education levels, so the number of people who have played shooting a ball is also higher. In addition, the General Office of the Ministry of Education issued the "Notice on the Creation of National Kindergartens with Football Characteristics in 2020" [30]. It is clearly required to accelerate the popularization of early childhood soccer and guide kindergartens of all levels and types to widely carry out early childhood soccer activities. Under the guidance of the policy, the popularity of soccer has greatly increased, and the data from this survey can show that the number of people who have played tap soccer reached 98.6%, reflecting the good implementation of the policy.

It was found that children who had played two-handed bat striking, front sliding, and underhand rolling had higher levels of self-perception than those who had not played these skills, so extensive exposure to various FMS can provide rich motor experience and have the effect of improving self-perception. In addition, given the large number of children who needed a demonstration of FMS such as front slide, side slide, and single-leg jump, it is recommended that these three skills should be demonstrated in the follow-up test to enhance children's knowledge and understanding of the movements.

5.2 Reliability

The scale's internal consistency reliabilities for locomotor skills, object control skills, and overall, 12-item FMS were above 0.8, indicating high consistency across the dimensions of the scale, and the ICCs for both the retest and the first test were above 0.82, which in this study is consistent with Barnett [12] reported ICCs (0.76–0.84) are generally consistent, indicating that the scale has good retest reliability.

5.3 Face Validity

Most children were able to identify the FMS shown in the picture, but the recognition rate was low for the forward slide and one-legged jump of the locomotor skill, which is consistent with Barnett [14]. This is consistent with the results of Barnett and Lopes' study. The reason for this may be that a single still picture does not reflect well the continuous dynamic process of single-leg jump and forward glide. In addition, the skill of front sliding is often used unintentionally in daily games or sports, and its limited application makes it a skill that most children lack the experience of playing, which may also lead to a low correct recognition rate. This also suggests that the researcher should increase the demonstration of these two actions during the test and describe them with names that are as easy to understand as possible. In addition, more than half of the children were unable to associate the forward slide and underhand roll with specific sports, games, or activities, which may be related to the children's little exposure to these activities, suggesting that teachers at the early elementary level should include more practice of these skills in physical education to promote the overall balanced development of children's FMS.

The vast majority of children aged 4–10 years can understand the actions presented in the pictures and can distinguish between good and not so good, proving that designing the scale in the form of pictures is effective and has good face validity.

5.4 Structural Validity

Two methods were used to explore the structural validity of the scale and it was found that the CFA with zero cross-loadings and zero correlated residuals based on ML did not produce a well-fitting model, while the BSEM with a prior of correlated cross-loadings and residuals was able to produce a well-fitting model when incorporated. Barnett et al. [12] reported a similar situation, mainly due to the difficulty of satisfying the harsh assumption of zero cross-loadings for non-target factors in empirical studies [14]. Theoretically, the various forms of running and jumping in locomotor skills all involve lower limb movements, and it is reasonable that these running and jumping skills are inherently correlated and thus cause some covariance problems. Among the object control skills, ball rolling involves almost no movement of the lower extremities and examines hand control of the ball, so it is understandable that it is not correlated with kicking, which requires lower extremity movement and foot control. The present study followed Barnett's [14] suggestion to take the above factors into account by setting cross-loadings and residual correlations that converge to, but are not equal to, zero, and showed a good fit, indicating that the PMSC-TGMD-2 has stable construct validity in a cross-cultural context.

5.5 *Differential Validity*

The comparison of the BIC and DIC results, shows that the single-factor model combining the two dimensions in a competition between the two models is a poor fit and cannot replace the two-factor model based on Ulrich's [18]. The two-factor model of locomotor and object control skills set up based on Ulrich's theory shows that there is a difference and a connection between the two dimensions of locomotor and object control skills, which proves that the scale has good discriminant validity.

6 Conclusion

The PMSC-TGMD-2 has good retest reliability, internal consistency reliability, structural validity, face validity, and discriminant validity in children aged 4–10 years and can be used to measure children's motor skill self-perceptions.

References

1. Gallahue DL, Ozmun JC, Goodway J (2012) Understanding motor development: infants, children, adolescents, adults, 7th eds. McGraw-Hill, New York, NY
2. Seefeldt V (1980) Developmental motor patterns: implications for elementary school physical education. In: Nadeau C, Holliwell W, Newell K, Roberts G (eds) Psychology of motor behavior and sport. Human Kinetics, Champaign, IL, pp 314–323
3. Clark JE, Metcalfe JS (2002) The mountain of motor development: a metaphor. In: Clark JE, Humphrey JH (eds) Motor development: research and reviews Reston. National Association of Sport and Physical Education, V A, pp 163–190
4. Barnett LM, van Beurden Eric, Morgan Philip J, Brooks Lyndon O, Beard John R (2009) Childhood motor skill proficiency as a predictor of adolescent physical activity. J Adoles Health Off Publ Soc Adoles Med 44(3)
5. Harter S (1985) Manual for the self-perception profile for children. University of Denver, Denver, CO
6. Harter S, Pike R (1969) The pictorial scale of perceived competence and social acceptance for young children. Child Develop
7. Babic M, Morgan P, Plotnikoff R, Lubans D, Lonsdale C, White R (2014) Physical activity and physical self-concept in youth: systematic review and met. J Sci Med Sport 18
8. Sollerhed AC, Apitzsch E, Rastam L et al (2008) Factors associated with young children's self-perceived physical competence and self-reported physical activity. Health Educ Res 23(1):125–136
9. Coppens E, de Meester A, Deconinck JFA et al (2021) Differences in weight status and autonomous motivation towards sports among children with various profiles of motor competence and organized sports participation. Child Basel 8(2):156
10. Stodden DF, Goodway JD, Langendorfer SJ et al (2008) Garcia. A Developmental perspective on the role of motor skill competence in physical activity: an emergent relationship. Quest 60(2)
11. Valentini NC, Rudisill ME (2004) Motivational climate, motor-skill development, and perceived competence: two studies of developmentally delayed kindergarten children. J Teach Phys Educ 23(3):216–234

12. Barnett LM, Ridgers ND, Zask A, Salmon J (2015) Face validity and reliability of a pictorial instrument for assessing fundamental movement skill perceived competence in young children. J Sci Med Sport 18(1)
13. Barnett LM, Vazou S, Abbott G et al (2016) Construct validity of the pictorial scale of perceived movement skill competence. Psychol Sport Exer 22
14. Lopes VP, Barnett LM, Saraiva L et al (2016) Validity and reliability of a pictorial instrument for assessing perceived motor competence in Portuguese children. Child Care Health Develop 42(5)
15. Valentini, Barnett, Bandeira et al (2018) The pictorial scale of perceived movement skill competence: determining content and construct validity for Brazilian children. J Motor Learn Develop 6(S2)
16. Estevan MG, Abbott et al (2018) Evidence of reliability and validity for the pictorial scale of perceived movement skill competence in Spanish children. J Motor Learn Develop 6(S2)
17. Dale A (2000) Ulrich. Test of gross motor development, 2nd edn. Examiner's manual. Pro-ed Publishers, Austin, TX
18. Nunnely JC, Bernstein IH (1994) Psychometric theory, 3rd edn. McGraw-Hill, New York
19. Burton CR (2002) Research in health care: concepts, designs, and methods. J Adv Nurs
20. Cole DA, Ciesla JA, Steiger JH (2007) The insidious effects of failing to include design-driven correlated residuals in latent-variable covariance structure analysis. Psychol Meth
21. Asparouhov T, Muthen B (2009) Exploratory structural equation modeling. Struct Equ Modeling 16(3):397–438
22. Muthén B, Asparouhov T (2017) Bayesian analysis using Mplus: technical implementation [EB/OL]. https://www.Statmodel.com/download/BayesAdvantages18.pdf
23. Muthén B, Asparouhov T (2012) Bayesian structural equation modeling: a more flexible representation of substantive theory. Psychol Methods 17(3):313–335
24. Gelman A, Carlin JB, Stern HS et al (2014) Bayesian data analysis, 3th edn. CRC Press, Boca Raton
25. Depaoli S, van de Schoot R (2017) Improving transparency and replication in Bayesian statistics: the WAMBS-checklist. Psychol Methods 22(2):240–261
26. Marsh HW (2007) Application of confirmatory factor analysis and structural equation modeling in sport and exercise psychology Hoboken. Wiley, NJ
27. Kass RE, Raftery AE (1995) Bayes factors. J Am Stat Assoc 90(430)
28. Spiegelhalter DJ, Best NG, Carlin BP, Van Der Linde A (2002) Bayesian measures of model complexity and fit. J Royal Stat Soc Ser B Stat Methodol
29. Ministry of Education of the People's Republic of China. Guidelines for the learning and development of children aged 3–6 years [EB/OL]. [2020–03–01]. http://www.moe.gov.cn/jyb_xwfb/xw_zt/moe_357/jyzt_2015nztzl/xueqianjiaoyu/yaowen/202104/t20210416_526630.html
30. Ministry of Education of the People's Republic of China. Notice the creation of national soccer characteristic kindergartens in 2020 [EB/OL]. http://www.moe.gov.cn/srcsite/A17/moe_938/s3273/202008/t20200814_478069.html

The Cognitive Effects of Physical Activity on Structural Changes in the Brain: A Review Study

Hou Yuxin, Jia Wang, and Zhao Yang

Abstract This study aims at investigating whether the changes of brain plasticity caused by exercise intervention in different groups can lead to changes in individual cognitive ability under different intervention cycles. With "Exercise", "physical activity", "cognition", "brains" and "prominent" inscriptions in PubMed, according to the literature search conducted by Web of Science, Google Scholar and other databases, 49 studies met the inclusion criteria of the initial search (sports/cognition/brain), and 16 studies were finally included after reading the abstract and full text. The plasticity of exercise intervention on brain tissue volume was mainly reflected in the changes of brain volume, hippocampus volume, gray matter volume, and thalamus volume. (2) The activation of brain network was most significant in frontal lobe and was more active in left hemisphere. (3) Exercise intervention can directly affect the changes of cerebral arterial flow velocity and pressure, as well as the concentration of oxygenated hemoglobin and brain-derived neurotrophic factor. Through the above mechanisms and pathways, exercise intervention has an impact on the plasticity of brain structure, and then promotes the cognitive change after exercise. (1) Exercise intervention can affect the level of BDNF, promote the increase of cerebral hemoglobin, the function of cerebral neural network connectivity and brain structure changes. (2) BDNF level changes in children and adolescents show correlation with the level of cerebral neural network connectivity. (3) There was a positive correlation between motor ability and cognitive level without group restriction.

Keywords Physical activity · Cognitive · Brain · Plasticity

1 Introduction

Cognitive ability is the most important psychological condition for people to perform activities successfully. For children and adolescents in the growth stage, cognitive

H. Yuxin (✉) · J. Wang · Z. Yang
Shanxi University, Shanxi 237016, China
e-mail: cyndra1424@uitm.edu.my

G. Kuan et al. (eds.), *Advancing Sports and Exercise via Innovation*, Lecture Notes in Bioengineering, https://doi.org/10.1007/978-981-19-8159-3_26

change plays a key role in their academic achievement [1]. The World Health Organization (WHO) [2], reports that the number of people living with dementia worldwide in 2015 was about 47.47 million; Alzheimer's Disease International (ADI) predicts that this number will rise to 82 million in 2030 and 139 million by 2050, with the largest increases in low and middle-income countries. Already 60% of people with dementia live in low and middle-income countries, but by 2050 that proportion will rise to 71%. Promoting cognitive change in the elderly plays an important role in alleviating social pressure in developing countries [1, 3] physical activity has been shown to promote changes in brain plasticity, the main physiological changes are the change of brain volume [4–6], the promotion of brain network connection function [7], the change of brain-derived nutritional factor (BDNF) level [8, 9]. However, it remains to be further studied which areas of the brain are represented by the changes in brain plasticity and whether the changes in brain plasticity can further lead to the changes in individual cognition. In this paper, the changes of brain plasticity in different populations under different intervention cycles and whether it will lead to individual cognitive changes are explored.

2 Method

2.1 Literature Retrieval

Keywords ("exercise" [MeSH Terms] OR "exercise" [All Fields] OR ("physical" [All Fields] AND "activity" [AllFields]) OR "physical activity" [All Fields]) AND ("promote" [All Fields] OR "promoted" [All Fields] OR "promotes" [All Fields] OR "promoting" [All Fields] OR "promotion" [All Fields] OR "promotional" [All Fields] OR "promotions" [All Fields] OR "promotive" [All Fields]) AND ("cognition" [MeSH Terms] OR "cognition" [All Fields] OR "cognitions" [All Fields] OR "cognitive" [All Fields] OR "cognitively" [All Fields] OR "cognitives" [All Fields]) AND ("change" [All Fields] OR "changed" [All Fields] OR "changes" [All Fields] OR "changing" [All Fields] OR "changings" [All Fields]) AND ("brain" [MeSH Terms] OR "brain" [All Fields] OR "brains" [All Fields] OR "brains" [All Fields]), PubMed, Web of Science, Google Scholar, and other data were used for literature retrieval. The retrieval strategy combines the main inscription with free words, and traces the reference materials of relevant literature. The retrieval year is limited to April 2022.

2.2 Inclusion and Exclusion Criteria

Exclusion criteria:

(1) The research content is inconsistent

(2) Systematic review research and statement
(3) Animal experiments
(4) Only conference abstracts or documents that cannot be obtained in full
(5) Literature with unclear research data
(6) Repeated literature.

3 Results

According to the 16 included literatures, intervention cycles were divided into 12 weeks or less, 12 weeks or more, single exercise intervention and other types.

3.1 *Intervention Period of 3 Months or Less*

There are two studies on children and adolescents. In the study design of Gerd Wagner (2016), the experimental group was determined to participate in strenuous exercise, while the control group did not participate in exercise, and the 6-week exercise intervention was conducted to study the activation of the anterior hippocampus and motor cortex. The experimental group first showed significant changes in body function after intervention. Compared with the pre-intervention group, the maximum workload P_{max} ($P < 0.001$), individual anaerobic threshold (IAnT) ($P = 0.002$), and maximum oxygen uptake (VO2max) ($P = 0.07$) all showed significant changes. Functional magnetic resonance imaging showed that exercise-induced serum BDNF concentration had a significant intergroup and time interaction ($P = 0.007$). Exercise-induced serum BDNF level was positively correlated with activation of the left anterior hippocampus ($P = 0.03$), but BDNF level had no effect on individual cognitive change. Changes in hippocampal activation and cognition remain to be further studied [10].

Riggs et al. [11] conducted neurorestorative exercise training for pediatric brain tumor survivors. In the experimental design, the experimental group was given aerobic exercise while the control group had no intervention. After 12 weeks of exercise intervention, it was found that the exercise group increased the fraction anisotropy (FA) of bilateral corpus callosum, cingulate gyrus, and superior longitudinal tract, as well as the right corticospinal tract and inferior fronto-occipital tract in pediatric brain tumor survivors ($P < 0.001$). Meanwhile, the hippocampal volume of exercise participants increased ($P = 0.001$) and tended to be healthy. Comparison with the bilateral hippocampi of healthy samples showed no significant difference ($P = 0.08$). Meanwhile, reaction time after exercise intervention was also studied in this study, and it was found that the increase of hippocampal volume was significantly correlated with the reduction of reaction time ($P = 0.01$).

Karl-jurgen et al. [12] conducted a study on hippocampal-brainstem connectivity related to vagal nerve regulation in healthy men after exercise intervention. The experimental group was required to undergo vigorous exercise for 6 weeks, while the control group received no intervention. The results showed significant changes in brain network connectivity after intervention. The resting state functional connection between right hippocampus (aHC) and frontal lobe, temporal lobe and precuneus was significantly enhanced ($P < 0.05$). However, resting state functional connectivity (RSFC) between aHC and dorsal vagal complex (DVC) was significantly decreased ($P = 0.001$), and the relationship between this change and individual cognition needs further study.

There are three studies on the elderly that studied the influence of resistance training on the cognition or related serum biomarkers of the elderly [13]. Elderly people around 90 years old were selected as research objects for 8 weeks of moderate intensity resistance training, and it was found that in this group, overall changes in BDNF, epidermal growth factor (EGF) and tum or necrosis factor (TNF) were not associated with changes in functional ability and cognitive measures after exercise intervention ($P > 0.05$). Exercise intervention focusing on resistance exercise did not affect cognitive function, nor did it affect BDNF level of 90-year-old people.

Monteblanco et al. [14] uses sports games as an intervention method to study the neuroplasticity and cognitive improvement of the elderly. The control group has no intervention measures, and BDNF is selected as the result indicator. After 6 weeks of exercise intervention, it was found that the level of BDNF in the elderly who participated in sports game training was significantly improved ($P = 0.0042$), and the cognitive level in the "language" area was significantly improved after exercise intervention ($P = 0.034$). However, the total score of Montreal Cognitive Assessment (MoCA) was not significantly different from that before the intervention ($P = 0.326$).

Takahiro et al. [15] conducted a study on the influence of exercise intervention on the plasticity of the prefrontal lobe. Elderly people were selected for moderate intensity exercise, while the control group had no intervention. After 12 weeks of experimental intervention, it was found that exercise had a significant impact on the frontal volume. Specifically, the volume of left orbitofrontal gyrus (OFG) increased ($P = 0.028$), gray matter (GM) volume of rACC ($P = 0.009$), GM volume of inferior frontal sulcus (IFS) and middle frontal sulcus (MFS) increased ($P = 0.008$). After exercise intervention, changes in brain internal volume were significantly correlated with MoCA test results ($P = 0.011$).

Under the place on put together is narrated, 12 weeks and exercise intervention volume change has a positive effect on the brain, six weeks of exercise intervention in children and adolescents, the elderly in the two groups showed the change of the serum levels of BDNF, and exercise intervention significantly correlated, but did not find the change and the direct link between individual cognitive; The 6-week exercise intervention promoted the activation of the brain and the functional connections between various parts of the brain, mainly in the activation of the right hippocampus and the resting state functional connections between the hippocampus and the frontal, temporal, and precuneus lobes. Exercise-induced changes in intracerebral volume were found to be associated with behavioral changes in children and adolescents, but the 8-week exercise intervention showed different results. In the elderly population, no changes in BDNF levels were observed after exercise intervention, and no significant differences were found in cognitive measures from before the intervention. The 12-week exercise intervention showed changes in brain volume in both children and adolescents, with significant increases in hippocampal volume in children and adolescents and in prefrontal volume in older adults, which are associated with cognitive changes in older adults.

3.2 Intervention Period is More Than 3 Months

There are three studies on the elderly. Hsu et al. [16] carried out progressive aerobic exercise in the elderly to study the relationship between aerobic exercise and frontoparietal network connectivity. After intervention, it was found that changes in frontoparietal neural network (FPN) connectivity were significantly correlated with changes in 6-min walking test results ($P = 0.05$). The change of FPN connectivity was significantly correlated with the change of Timed up-and-go test (TUG) performance ($P = 0.02$).

Shaaban et al. [17] conducted a two-year exercise intervention study, in which the elderly was also selected as research objects. The experimental group participated in moderate intensity exercise intervention, and the control group participated in health education. The results showed that the percentage change in BDNF was positively correlated with the percentage change in straight vein length ($P = 0.07$), but both intervention groups showed an increase in straight vein length; percentage change of vascular endothelial growth factor (VEGF) was not associated with percentage change of straight vein length, and no pa-related increase in peripheral blood BDNF and VEGF was observed after intervention ($P > 0.1$).

Yogev-Seligmann et al. [18] conducted exercise intervention study on elderly people with amnestic mild cognitive impairment. The experimental group participated in aerobic exercise while the control group did Balance and Toning, (BAT). The increased activity of left lower frontal lobe ($P = 0.03$) and anterior central gyrus ($P = 0.03$) in the experimental group was associated with cardiopulmonary health changes after the intervention. The change of physical strength level after exercise is correlated with the increase of "language" ability ($P = 0.02$) and tends to be correlated with the change of individual "adaptability" ($P = 0.06$).

Above all, found in the long-term exercise intervention in this group is the main focus of the scholars in the elderly group, the physical level is a factor in the effects of exercise on the brain, has also sparked concern, good participants in a long-term study found that physical level to show after aerobic exercise the brain activity and brain neural network connectivity changes significantly, this change has certain correlation with the change of individual behavior.

3.3 Single Acute Exercise Intervention

Four studies have been carried out on young people. Giles et al. [19], studied the effects of strenuous exercise on oxygenated and deoxygenated hemoglobin in the prefrontal cortex, and selected at least 30 min of moderate cardiopulmonary exercise recommended by the American College of Sports Medicine as an intervention program for college students. The results showed that oxygenated hemoglobin (O_2Hb) in prefrontal cortex increased with the increase of exercise load ($P < 0.001$).With the increase of exercise time, the change of O_2Hb was significantly higher than before exercise ($P < 0.001$). Deoxyhemoglobin dHb increased with the increase of exercise load ($P < 0.01$). With the increase of exercise time, the change of dHb was significantly higher than that before exercise ($P < 0.01$). The total hemoglobin content tHb increased with the increase of exercise load ($P < 0.001$). With the increase of exercise time, tHb change was significantly higher than that before exercise ($P < 0.001$). However, the relationship between changes in hemoglobin and cognition remains to be further studied.

Toshiaki et al. [20] studied the response of BDNF in young people under the combination of cognition and physical exercise. Moderate intensity cycling exercise was used as physical exercise intervention, while the control group only received cognitive intervention. The results showed that there was no significant correlation between cognitive intervention alone and changes in peripheral BDNF level ($P \geq 0.05$), but there was significant change in peripheral BDNF level after exercise intervention ($P = 0.006$; Cognitive intervention (CE) had no significant effect on the increase of plasma BDNF level ($P = 0.373$), while plasma BDNF level had a positive effect on physical exercise (PE) ($P = 0.002$).There was no significant correlation between PE alone and combining cognitive and physical exercise (CCPE) combined intervention on BDNF level in peripheral blood ($P = 0.861$).However, there is no evidence that exercise-induced changes in BDNF levels are related to individual cognitive ability.

Tsai et al. [21] studied the effects of acute aerobic exercise and cardiopulmonary fitness on serum BDNF level. The experimental group was divided into low level exercise group and high-level exercise group based on exercise level. The control group received non-exercise intervention, and the cognitive change was shown as the allocation of attention resources. The results showed that the EIH group with high level of exercise had a significantly greater range of attention allocation resources than the low level of exercise and the non-exercise intervention group ($P = 0.001$; Only the high-level exercise group had significantly higher amplitude of attention allocation resources after exercise than the non-exercise intervention group ($P = 0.001$). And the study of the biological indicators to determine serum BDNF, compared with before and after the exercise group compared with the change of serum BDNF levels significantly ($P = 0.031$), but the single changes of BDNF level caused by acute exercise and neural electrical performance found no significant association between resource allocations (attention), has caused the change of individual attention.

The scholars in the experimental design again two years later, using the same acute aerobic exercise on different cardiopulmonary health level of the young man's task switching scheme and studies the influence of BDNF, motion induced acute BDNF concentration change before and after the intervention group and control group of serum BDNF concentration changes with significant difference ($P = 0.011$). However, the association between BDNF concentration changes, neurophysiological (attention resource allocation) changes and acute exercise in both the intervention group and the control group did not reach the required level of significance, suggesting that BDNF cannot serve as a mediator of the association between acute exercise and neurocognitive performance [22].

There are three intervention studies on the elderly. Doi (2013) study on the effect of exercise intervention on the brain activation of the elderly with mild cognitive impairment. Dual task walking and normal walking had significant effects on O_2Hb ($P < 0.001$). The level of performance after cognitive exercise was significantly correlated with the change of O_2Hb in the frontal lobe left inferior frontal gyrus (LIFG) ($P = 0.005$), and the change of hemoglobin level also caused significant changes in individual performance ($P = 0.005$) [23].

Doi et al. [24] conducted an in-depth study using physical activity level as a reference and found that the more low intensity physical activity and medium and high intensity physical activity, the lower the atrophy rate ($P < 0.001$) in the elderly population without controlling the sample. Partial correlation analysis controlling for age, sex, and TUG showed that Lower physical activity (LPA) was not significantly associated with brain atrophy ($P = 0.062$), but moderate to high intensity physical activity was significantly associated with atrophy rate ($P < 0.001$). Consistent results were also shown in the TUG test ($P = 0.025$).

Makizako et al. [25] studied the influence of exercise ability on memory and brain volume of the elderly with MILD cognitive impairment. Taking six-minute walking distance as the test method, the results showed that higher exercise ability was related to better cognitive level. After adjusting for age and gender, gray matter density in left middle temporal gyrus, middle occipital gyrus, and hippocampus was positively correlated with 6-min walking distance ($P < 0.05$).

In conclusion, a single exercise intervention program can cause changes in BDNF level, and this change is significantly correlated with exercise intervention, but the changes in BDNF level are not associated with changes in cognitive ability. Exercise intervention has a significant correlation with the effect of hemoglobin in the young and the elderly, which is not affected by exercise intensity, and has a certain promoting effect on the change of individual executive ability.

4 Discussion

Under the difference of exercise intervention cycle, the factors that scholars focus on are mainly BDNF level, brain internal volume and brain network connection.

The results of BDNF level changes after exercise intervention were consistent in children, adolescents, and young adults with significant correlation, but the intervention cycle was less than three months. Therefore, future studies should focus on the stability of long-term exercise intervention on BDNF level changes. The level of BDNF in the elderly group showed correlation with the exercise intervention after one and a half months, but the change of BDNF level in the exercise intervention of more than two months and two years had nothing to do with the exercise intervention. Therefore, it can be preliminarily inferred that the long-term exercise intervention in the elderly group will not cause the change of BDNF level. But in the above research has not been found in the three groups of BDNF level changes cause the change of individual cognition, the study of exercise intensity relates to moderate, severe intensity, exercise types are aerobic exercise, covering a single intervention cycle, the following three months, more than three months, so you can preliminary determination of BDNF levels of change will not cause the individual cognitive change.

Studies on the changes of the internal brain volume after exercise intervention have covered two groups: children and adolescents, and the elderly. In children and adolescents, the hippocampal volume increases significantly after exercise, and this change causes the shortening of the individual reaction time of children. The prefrontal lobe volume and brain atrophy rate were significantly correlated in the elderly population, and this change caused significant cognitive changes in the elderly individuals, which were positively correlated with the TUG test results. Preliminary estimate, therefore, changes in brain volume can promote individual cognitive and behavioral changes, due to the above research cycle are in three months or more, and intensity are higher, in the future research should further explore other strength grade and different intervention cycle effect on cognitive and behavior of children, the elderly, Attention should also be paid to changes in the youth group.

Exercise intervention after brain network connections to related research become cover children, youth, the elderly three groups, and all are consistent results, namely exercise intervention can significantly improve the network connection, based on this, found in the study of the elderly, the change, and the individual "language" the improvement of the ability to have significant correlation. Therefore, it can be preliminarily inferred that brain network connectivity can promote behavior improvement in the elderly population and is not affected by exercise intervention cycle. However, attention should be paid to short-term exercise intervention in the future studies, and changes caused by different exercise intervention cycle and exercise intervention situation should be compared in the same population.

5 Conclusion

In cognitive changes caused by exercise intervention effects on the brain, most studies focused on brain source sex nutrition factor BDNF, brain network connectivity, changes in brain structure, and so on, which focus on the elderly subjects, intervention of moderate intensity, based on the combing of literature, the following conclusions: (1) Exercise intervention can affect the BDNF level; (2) Exercise intervention can promote the brain neural network connection function; (3) Exercise can promote the changes of brain structure, especially in hippocampus volume and frontal lobe; (4) Exercise intervention can promote the increase of cerebral hemoglobin; (5) Changes in BDNF levels in children and adolescents were associated with brain neural network connections; (6) There was a positive correlation between motor ability and cognitive level.

References

1. Donnelly JE, Hillman CH, Castelli D et al (2016) Physical activity, fitness, cognitive function, and academic achievement in children: a systematic review. Med Sci Sports Exerc 48(6):1197
2. WHO. https://www.who.int/zh/news/item/17-03-2015-governments-commit-to-advancements-in-dementia-research-and-care
3. ADI. Dementia facts and figures | Alzheimer's Disease International (ADI) (alzint.org)
4. Koblinsky ND, Meusel LAC, Greenwood CE et al (2021) Household physical activity is positively associated with gray matter volume in older adults. BMC Geriatr 21(1):1–10
5. McEwen SC, Hardy A, Ellingson BM et al (2015) Prefrontal and hippocampal brain volume deficits: role of low physical activity on brain plasticity in first-episode schizophrenia patients. J Int Neuropsychol Soc 21(10):868–879
6. Demirakca T, Brusniak W, Tunc-Skarka N et al (2014) Does body shaping influence brain shape? Habitual physical activity is linked to brain morphology independent of age. World J Biol Psychi 15(5):387–396
7. Chaddock-Heyman L, Weng TB, Kienzler C et al (2020) Brain network modularity predicts improvements in cognitive and scholastic performance in children involved in a physical activity intervention. Front Human Neurosci 346
8. Walsh EI, Smith L, Northey J et al (2020) Towards an understanding of the physical activity-BDNF-cognition triumvirate: a review of associations and dosage. Ageing Res Rev 60:101044
9. Huang T, Gejl AK, Tarp J et al (2017) Cross-sectional associations of objectively measured physical activity with brain-derived neurotrophic factor in adolescents. Physiol Behav 171:87–91
10. Wagner G, Herbsl Eb M, Feliberto D et al (2017) Changes in fMRI activation in anterior hippocampus and motor cortex during memory retrieval after an intense exercise intervention. Biol Psychol 124:65–78
11. Riggs L, Piscione J, Laughlin S et al (2017) Exercise training for neural recovery in a restricted sample of pediatric brain tumor survivors: a controlled clinical trial with crossover of training versus no training. Neuro Oncol 19(3):440–450
12. Karl-Jürgen B, Marco H, Andy S, et al (2016) Hippocampal-brainstem connectivity associated with vagal modulation after an intense exercise intervention in healthy men. Front Neurosci 10(278)
13. Ruiz JR, Gil-Bea F et al (2015) Resistance training does not have an effect on cognition or related serum biomarkers in nonagenarians: a randomized controlled trial. Int J Sports Med
14. Cavalcante MM, Fraga I, Dalbosco B et al (2021) Exergame training-induced neuroplasticity and cognitive improvement in institutionalized older adults: a preliminary investigation. Physiol Behav
15. Takahiro S, Michael A, Toshikazu K et al (2021) Prefrontal plasticity after a 3-month exercise intervention in older adults relates to enhanced cognitive performance. Cerebral Cortex

16. Hsu CL, Best JR, Wang S et al (2017) The impact of aerobic exercise on fronto-parietal network connectivity and its relation to mobility: an exploratory analysis of a 6-month randomized controlled trial. Front Hum Neurosci 11:344
17. Shaaban CE, Aizenstein HJ, Jorgensen DR et al (2019) Physical activity and cerebral small vein integrity in older adults. Med Sci Sports Exerc 51(8):1684
18. Yogev-Seligmann G, Eisenstein T, Ash E et al (2021) Neurocognitive plasticity is associated with cardiorespiratory fitness following physical exercise in older adults with amnestic mild cognitive impairment. J Alzheimer's Dis JAD 81(1):1–22
19. Giles GE, Brunyé TT, Eddy MD et al (2014) Acute exercise increases oxygenated and deoxygenated hemoglobin in the prefrontal cortex. Neuroreport 25(16):1320–1325
20. Toshiaki M, Saya H, Hideya Y et al (2018) Response of brain-derived neurotrophic factor to combining cognitive and physical exercise. Europ J Sport Sci 1–9
21. Tsai CL, Chen FC, Pan CY et al (2014) Impact of acute aerobic exercise and cardiorespiratory fitness on visuospatial attention performance and serum BDNF levels. Psychoneuroendocrinology 41(3):121–131
22. Tsai CL, Pan CY, Chen FC et al (2016) Effects of acute aerobic exercise on a task-switching protocol and brain-derived neurotrophic factor concentrations in young adults with different levels of cardiorespiratory fitness. Experim Physiol
23. Doi T, Makizako H, Shimada H et al (2013) Brain activation during dual-task walking and executive function among older adults with mild cognitive impairment: a fNIRS study. Aging Clin Experim Res 25(5):539–544
24. Doi T, Makizako H, Shimada H et al (2015) Objectively measured physical activity, brain atrophy, and white matter lesions in older adults with mild cognitive impairment. Experim Gerontol 62:1–6
25. Makizako H, Shimada H, Doi T et al (2013) Six-minute walking distance correlated with memory and brain volume in older adults with mild cognitive impairment: a voxel-based morphometry study. Dement Geriatric Cogn Disorders Extra 3(1):223–32

Effectiveness of Cognitive Aspects of Exercise During School Age on Adult Exercise Habits

Ryo Konno

Abstract The purpose of this study was to verify the effectiveness of cognitive aspects of exercise, after examining the effect of sports club belonging and exercise cognition during school age on the formation of exercise habits in adults. The participants were 492 adults (226 men, 262 women, average age: 50.1 ± 4.09 years old, age range 38–65 years old). The contents of the survey were the current exercise habits, sports club belonging during school age, and the cognition of exercise during school age. First, the chi-square test was conducted to examine the relationship between current exercise habits and the sports club belonging during school age. The result showed that the current exercise habits of males were related to the exercise habits during middle school age but not during other school ages, and the current exercise habits of females were not related their exercise habits during school age. Next, to examine the relationship between current exercise habits and exercise cognition during school age, two-way analysis of variance, gender $\times$ exercise habits, was performed. The result showed that males scored significantly greater than females on exercise cognition, and the scores of exercise cognition of both high and low exercise habit groups were significantly greater than those of no exercise habit group. The results of this study suggested that the cognitive aspects of exercise during school age influenced more on forming exercise habits in adulthood than behavioral aspects of exercise during school age did.

Keywords Exercise habits in adults · During school age · Cognitive aspects of exercise · Behavioral aspects of exercise

1 Introduction

Recently, people's interest in exercise and sports has been increasing due to the heightened awareness of the importance of health and various policies. What are the factors behind the formation of exercise habits? Studies investigating the factors that

R. Konno (✉)
Meiji Pharmaceutical University, Tokyo, Japan
e-mail: konno@my-pharm.ac.jp

G. Kuan et al. (eds.), *Advancing Sports and Exercise via Innovation*, Lecture Notes in Bioengineering, https://doi.org/10.1007/978-981-19-8159-3_27

influence exercise and sports activities identified many factors, which can be categorized as social, psychological, and physical [1]. With regard to the exercise habits of adults who have completed their schooling a long time ago, some studies have examined the relationship between current exercise habits and the types of exercise performed in the past, while others focused on the frequency of exercise in the past [25]. An overview of the findings of previous studies on factors influencing exercise habits shows that past exercise and sports experiences are related [1, 18, 7, 9, 13, 20, 21, 36]. Kaneko and Otsuki [11] reported that "school club activities" accounted for the majority of the "sports and exercise" that junior high school students engage in, and that students' sports and exercise tended to be dependent on school sports clubs. In addition, it is clear that in junior high schools, there are no opportunities for students to engage in sports and exercise in their free time beyond taking part in health and physical education classes and activities in the sports clubs. Yamamoto [37] states that the motives for participation in sports and the motives for participation in sports clubs overlap considerably in terms of content, as sports clubs are primarily organized to conduct sports activities. Tsuruyama et al. [36] mention that sports club activities can promote future sports activities. Many studies have concluded that joining an athletic club during the school years was associated with participation in exercise and sports at that time, and furthermore, led to the acquisition of an exercise habit in the future. Nakasuga et al. [20] suggest that an athlete's satisfaction (with their team) was important for the future continuation of exercise among athletic club participants, beyond simply belonging to an athletic club, the degree to which they were satisfied about their experience was important. Research has shown that satisfaction with the athletic team has an impact on overall satisfaction with school life [10], and that it is positively related to subsequent sports activities [2]. However, these studies were limited to investigating those who had joined an athletic club in school, and hence, the data are not applicable as factors related to the exercise habits of all adults. In other words, most of the studies that refer to the exercise habits of adults do not include data on those individuals who did not join an athletic club during school age.

Do all junior and senior high school students join and continue in an athletic club if they want to exercise and play sports? Some students who are active in athletic clubs drop out or burn out for various reasons, such as injuries, disappointment over the loss of the special status of being a regular member, and stagnation or decline in their performance [19]. Alternatively, it is possible that students may have negative attitudes toward athletic teams or have experienced an unfortunate event in the past. One such factor is corporal punishment by a coach or an individual in charge. In 2012, this topic attracted significant attention as a social problem, when the captain of the boys' basketball team of Osaka Municipal Sakuraomiya High School committed suicide due to corporal punishment by the team's former coach. In addition, "sports harassment" [5] is now regarded as a problem among coaches, which further worsens the impression of sports clubs. Sports harassment [5] is the harassment of children in sports. It is defined as "any conduct in a child's sport that causes mental or physical pain or makes the child's sporting environment worse beyond what is appropriate, against a background of superiority in the sporting arena, including role status, level of competition, human relationships, and economic status. This includes sexual harassment." In addition, the ability to adapt to strict discipline, which is a characteristic of athletic clubs, may be an issue. Hyodo [8] reported that a common reason for not wanting to join an athletic club was "there is practice every day, and it is intense and the discipline is strict." It has also been reported that teams with a high level of competition tend to impose excessive practice in order to win external games, which is not conducive to development [24]. In addition, studies have shown that interpersonal conflicts in athletic teams is one of the distinguishing reasons for leaving sports clubs in Japan [1, 3]. There is a possibility that individuals who have left an athletic club for any reason may not join an athletic club in the future despite joining an athletic club in the past. However, some of them may not have lost their initial positive feelings toward sports.

This suggests that a certain number of students exist who may have wanted to play sports or exercise during school age but did not join an athletic club because the instructor's teaching methods or conflicts with friends temporarily undermined the essence and significance of sports or exercise. Kaneko and Otsuki [11] reported that nearly 30% of junior high school students who did not play sports or exercise answered "Yes" to the question "Do you want to play sports or exercise in the future?" The report shows that it is important for schools and local governments to consider options to support children who do not belong to clubs and want to play, but are not aware of how to do so. The most common answer for the preferred sports and physical activities that non-players would like to engage with in the future was "I want to play a variety of sports." This may be due to the fact that they had few opportunities to engage in sports and exercise enthusiastically, which may in turn be the reason they were unable to choose a specific sport or exercise. In short, it is thought that how one perceives and understands exercise and sports is more important for the establishment of exercise habits among adults than membership in sports clubs in the past. Sugihara et al. [32] reported that the degree of exercise and sports experience, past pleasant as well as unpleasant experiences, and awareness and perception of exercise and sports have a strong influence on whether people did or did not engage in exercise and sports. Nishimura et al. [22] revealed that pleasant or unpleasant experiences during school age had a significant influence on subsequent participation in exercise and sports. In addition, an enthusiastic attitude toward exercise and sports has been cited as a factor that determines participation in exercise and sports [12, 20, 33]. Tadano et al. [34] reported that the factor that determines students' sports behavior is their fondness for physical education.

Considering these factors, the author has hypothesized that the cognitive aspect of past exercise may have a stronger influence on the formation of exercise habits in adults than the behavioral aspect of past exercise and sports. To date, no research has comprehensively examined both the behavioral aspect (past membership in an athletic club) and the cognitive aspect (perception of exercise) as factors in shaping exercise habit in adults. The present study was conducted to examine the validity of the cognitive aspect of exercise by comparing it with the behavioral aspect of exercise. In this study, the cognitive aspect of exercise focuses on pleasant experience in order to obtain basic data. In addition, it is possible that some of the participants have been active in local club teams for various reasons. Therefore, the behavioral aspect of past exercise in this study was defined as membership in school athletic clubs and local club teams (hereafter referred to as "athletic clubs").

Therefore, this study aims to examine the effects of past exercise club membership and perceptions of exercise on the formation of exercise habits in adults, and to test the hypotheses by investigating the influence of each of these factors.

2 Methods

2.1 Survey Targets, Survey Period

The study participants were 492 adults (226 males and 266 females, mean age 50.1 $\pm$ 4.09 years, age range 38–65 years). The participants were parents of first-year university students enrolled in the departments of sports and health sciences or medicine.

The data were collected between December 2013 and January 2014.

2.2 Survey Contents

2.2.1 Current Exercise Habits

Information on the participants' frequency of exercise, duration of exercise, and duration of current exercise habits were collected. The participants were classified into three groups: (1) habitual exercisers, defined as those who exercise for 30 min or more twice a week or more for at least one year; (2) irregular exercisers, defined as those who do not satisfy category (1) but have some exercise habit; and (3) no exercise habit, defined as those who do not have an exercise habit. The habitual exercisers (1) are considered to be those who exercise habitually by the Ministry of Health, Labor and Welfare (MHLW).

2.2.2 Past Membership in Athletic Clubs

The respondents were asked to answer "yes" or "no" to the question of whether they belonged to a school sports club or a local club team when they were (1) junior high school students, (2) high school students, or (3) university students.

2.2.3 Perception of Past Exercise

The past exercise awareness questions asked participants about their perception of exercise during school age as well as their past membership in exercise clubs. Specifically, they were asked to choose one from "1. I disliked it" to "5. I liked it" for Question (1) "Did you like exercises?" and "1. Was bad at it" to "5. Was good at it" for Question (2) "Were you good at exercise?" and the responses were marked as scores ranging from 1 to 5 on a 5-point scale. The higher the score, the more the student liked exercise (Q1) and the better the student was at exercise (Q2).

2.3 Ethical Consideration

This survey was conducted after obtaining approval from the Research Ethics Review by the Juntendo University School of Medicine (Juntendo University School of Medicine Ethics No. 2012057, Receipt number 813). The purpose of the survey, the fact that participation in the survey was voluntary, and research ethics, such as confidentiality of personal information, were explained in writing to the survey participants, and only those who consented were included in the study. Participants were also informed that they could withdraw from the study at any time, in which case, all information obtained up to that point would be discarded and would not be used for further research, and that they would not suffer any disadvantages.

2.4 Procedure

First, the author distributed the questionnaire to the university students, who were the children of the participants of the study, for a maximum of two parents each. The students then handed the questionnaires to their parents and asked them to answer the questions. The completed questionnaires were submitted by the students on behalf of their parents or directly mailed to the author by the parents. The questionnaire included the author's contact information, and participants were informed they could contact the author if they had any questions.

2.5 Analytical Method

Since various surveys have reported that the level of frequency of exercise and sports activities is higher in males than in females [4, 15, 14, 28], responses were analyzed by gender. First, a chi-square test was conducted to examine the association between current exercise habits and past exercise club membership. Next, a two-way ANOVA was conducted to examine the relationship between current exercise habits and perception of exercise in the past ((1) whether they liked exercise or (2) whether they were good at exercise) by gender (male or female) × exercise habit (habitual exerciser group, irregular exerciser group, and no exercise habit group)). The means and standard deviations were calculated, and multiple comparisons were made using Tukey's HSD test on significant main effects as sub effect tests.

The statistical packages IBM SPSS Statistics 22 and IBM SPSS Amos version 22 were used for statistical analysis.

3 Results

3.1 Relationship Between Current Exercise Habits and Past Exercise Club Membership

To examine the relationship between current exercise habits and past athletic club membership, Pearson's chi-square tests were conducted separately for males and females. The cells with an expected frequency of less than 5 were excluded from the analysis because 50.0% of both males and females had an exercise habit of less than 5 in their high school years. The relationship between current exercise habits and past membership in exercise clubs was significant only for men ($x^2(2) = 6.82, p < 0.05$), and no relationship was found for all other groups. In other words, there was little association between current exercise habits and past membership in exercise clubs. This result roughly supports the hypothesis of the present study.

3.2 Relationship Between Current Exercise Habits and Past Perceptions of Exercise

Two-way ANOVA was conducted to examine the relationship between current exercise habits and perception of exercise in the past. First, equal variances were assumed for both liking and being good at exercise by the Levene Test as a test of equal variances. Then, analysis of variance was conducted for each cognition (Tables 1 and 2). A main effect of gender was found for liking exercise, which was significantly higher in males than in females at all three time points. A significant main effect of current exercise habits was found for all three time periods, and no interaction effects were found. Multiple comparisons revealed significantly higher scores in the irregular and habitual exerciser groups than in the no exercise habit group. The results indicate that the degree of liking of exercise in the past influences the exercise habits of adults.

An analysis of variance was conducted on whether the participants were good at exercise, and on whether they liked exercise. The results showed a main effect of gender, and the score was significantly higher in males than in females at all three time points. There was a significant main effect of current exercise habits at all three

Table 1 Averages and standard deviations of perceptions (like exercise) of past exercises by exercise habit

		Exercise habits						Main effect of sex	Main effect of exercise habits	Alternating action
		High group		Low group		None				
		N	M (SD)	N	M (SD)	N	M (SD)	F	F	F
Junior high students	Male (M)	71	4.6 (0.79)	67	4.6 (0.62)	84	4.3 (1.00)	18.05** M > F	6.46**	0.53
	Female (F)	62	4.3 (0.98)	66	4.1 (1.03)	129	3.9 (1.18)			
High school students	Male (M)	71	4.5 (0.87)	67	4.6 (0.65)	84	4.2 (0.91)	17.96** M > F	8.36**	0.46
	Female (F)	62	4.2 (0.94)	66	4.1 (1.00)	129	3.8 (1.16)			
University students	Male (M)	71	4.3 (0.93)	67	4.5 (0.66)	84	4.0 (0.94)	11.06** M > F	7.58**	0.64
	Female (F)	62	4.1 (1.02)	66	4.0 (1.00)	129	3.7 (1.09)			

$^{**}p < 0.01$

Table 2 Averages and standard deviations of perceptions (exercises are good at) of past exercises by exercise habit

		Exercise habits						Main effect of sex	Main effect of exercise habits	Alternating action
		High group		Low group		None				
		N	M (SD)	N	M (SD)	N	M (SD)	F	F	F
Junior high students	Male (M)	71	4.3 (0.90)	67	4.3 (0.84)	84	3.9 (1.12)	7.33** M > F	4.11*	0.57
	Female (F)	62	4.0 (1.01)	66	3.9 (1.19)	129	3.8 (1.22)			
High school students	Male (M)	71	4.2 (0.93)	67	4.3 (0.85)	84	3.8 (0.97)	12.65** M > F	5.25**	0.52
	Female (F)	62	3.8 (1.06)	66	3.8 (1.15)	129	3.6 (1.18)			
University students	Male (M)	71	4.0 (0.97)	67	4.2 (0.88)	84	3.7 (0.91)	9.16** M > F	6.22**	0.41
	Female (F)	62	3.8 (0.99)	66	3.7 (1.19)	129	3.4 (1.13)			

$^{*}p < 0.05$; $^{**}p < 0.01$

time points, and no interaction effects were found. Multiple comparisons revealed significantly higher scores in the irregular and habitual exerciser groups than in the no exercise habit group. The degree of being good at exercise in the past was shown to influence adult exercise habits; in terms of gender, the two exercise perceptions were significantly higher in males than in females, and in terms of exercise habits, was higher in the two groups with exercise habits than in the group without exercise habits. These findings supported the results of the previous studies [4, 15, 14, 28] and the current study hypotheses.

4 Discussion

To examine the association between current exercise habits and past athletic club membership, chi-square tests were conducted, results showed that the association was significant only for males' exercise habits during middle school, but not for all other periods. Previous studies have reported that athletic club activities during school age are one of the factors that define exercise habits in middle and older ages [23] and promote future sports activities [36]. However, the findings of the present study differed from these results. The first possible reason for this is the difference in the target population. The participants in the previous study were high school and university students, and the results obtained in the previous study were used to predict the exercise behavior of adults. In other words, the previous studies did not actually examine the relationship between past membership in exercise clubs and current exercise habits among middle-aged people. On the other hand, the current study has examined this question, which has led to new results that are different from those of preceding studies.

Yamamoto [37] conducted a questionnaire survey on the motives for participation in university athletic clubs and pointed out that "avoidance" as seen in responses "because I can't quit" and "adherence" as seen in responses "because I want to continue without stopping until the end" were important factors for participation in athletic clubs. Oyama et al. [27] reported that mentally exhausting events may occur and belonging to a club itself becomes an objective, rather than continuing club activities because they are enjoyable. Nakasuga et al. [20] suggest that for the future continuation of exercise by participants in exercise club activities, it is important not only to belong to an exercise club activity but also to have a sense of satisfaction for the continuation of exercise, i.e., it depends on the extent to which an individual is satisfied with the activity. Thus, research findings so far regarding the behavioral aspect of athletic club affiliation have not shown that it determines adults' attitudes toward exercise and sports. This can be seen from the fact that there are few studies on the reasons for joining an athletic club, furthermore, the reasons for not continuing participation in an athletic club have not been examined [33]. However, school sports clubs are also a place where students can further develop the skills acquired through physical education classes and acquire the qualities and abilities necessary for lifelong sports by challenging themselves with things they cannot experience in the classroom [29]. School sports clubs have great potential to contribute to mental and physical development in young people if conducted appropriately, instead of solely aiming to improve athletic performance.

They also have the potential to teach students active and independent management and administration of sports groups and can help them experience solidarity and cooperation. Hence, it can be surmised that school sports clubs play a major role in school education [30]. Therefore, if the environment in the sports clubs during school age can be improved by resolving the problems related to coaches and conflicts in human relations, it is likely that the children will have a chance to form positive attitudes toward participating and continuing in sports and exercise in the future.

A two-way ANOVA was conducted to examine the association between current exercise habits and the cognitive aspects of past exercise, such as liking exercise or being good at exercise, and it was found that the scores of men were higher than those of women in terms of gender, and those of the two groups with exercise habits were higher than those without exercise habits, with each of these results being significant. The results for gender supported the previous studies [4, 15, 14, 28] and those for exercise habits supported the hypothesis of the present study. Although some studies on exercise continuation have focused on how people who are concerned about their health in adulthood and beyond continue to exercise [18, 6, 16, 31], the results of this study indicate that factors such as people's attitudes toward exercise and how they perceive it influence the continuation or discontinuation of exercise. The results of this study demonstrate that the cognitive aspect of exercise is more effective in forming exercise habits than the superficial behavioral aspect of exercise.

To become a lifelong sports practitioner, it is essential for children to learn the joy and pleasure unique to exercise during school age [29]. Many opportunities to engage in sports are present in school education, and academic physical education from elementary school is undeniably the most important opportunity to receive full-scale sports instruction [36]. Thus, physical education may be considered as the factor related to the cognitive aspect of exercise. In junior high school, there are no opportunities to practice sports and exercise during one's free time beyond health and physical education classes and sports club activities [11]. Therefore, it can be said that physical education in schools is important for lifelong engagement in sports. Takizawa [35] has written, "In formal physical education, children who dislike sports must be made to like sports. This is because it is necessary not only to let children experience the unique cultural world of sport, but also to develop their ability to solve problems rooted in reality." Okazawa [26] points out the importance of physical education for the development of a love of physical activity, stating that it is necessary to increase the sense of athletic competence such as "awareness of physical competence," "sense of control," and "sense of acceptance" in physical education. The above results suggest that the experience of physical education and the feelings that arise while participating in physical education may influence the subsequent practice of exercise behavior through how children perceive exercise and sports, which requires further investigation.

5 Conclusion

5.1 Effectiveness of Cognitive Aspects of Exercise During School Age

The cognitive aspects of exercise during school age, rather than the behavioral, are more effective in influencing adult exercise habits.

5.2 Future Issues

The present study was designed to examine the factors influencing the formation of exercise habits among adults in terms of past membership in exercise clubs and past perceptions of exercise. However, the study was limited to a positive approach to the perception of exercise, and therefore, a different angle of examination is required in the future. In addition, it is necessary to examine the factors that contribute to the formation of exercise habits based on the current characteristics of adults. Furthermore, new findings must be obtained by analyzing the relationship between these factors and the experiences in physical education and the emotions that arise from them.

Acknowledgements I would like to express my sincere gratitude to all the participants who cooperated in this research and to all the people concerned. This research was also supported by the Ministry of Education, Culture, Sports, Science and Technology of Japan (MEXT) under the Support Program for Strategic Research Infrastructure Formation at Private Universities (2011–2015).

References

1. Aoki K (1989) Factors affecting the continuation and withdrawal of high school athletic club members. J Phys Educ 34(1):89–100
2. Aoki K, Matsumoto K (1997) Psychosocial factors related to high school athletes adjustment to club activities. J Phys Educ 42(4):215–232
3. Ebihara O (1988) A study on dropout from organized sports. J Sociol Phys Educ Sport 7:107–129
4. Foundation for Health and Fitness (2009) National survey on active aging 2008. Foundation for Health and Fitness, Tokyo
5. Fujigo E, Ohashi E, Iume Y (2017) What is sports harassment in children's sports? Bull Tokyo Mirai Univ 12:63–73
6. Hashimoto K (2010) An attempt to construct a spiral model of exercise continuation. Health Sci 32:51–62
7. Hirata H, Aoyama S, Kikuchi Y (1989) Analysis of factors influencing sports activities of working people. Bull Phys Educ 23:39–43
8. Hyodo M (1979) The present situation and problems of sports clubs in universities. Sci Phys Educ 29(5):327–331

9. Inada T, Okada M (1998) A study on sport orientation of working people (III): relationship between sport experience during school days and current activities. Kochi Univ Facul Educ Res Rep 55:65–73
10. Kakutani S, Mutoh T (2001) Meaningfulness of junior high school club activities for the continuers of club activities: in relation to the sense of fulfillment and satisfaction with school life. J Psychol 72(2):79–86
11. Kaneko K, Otsuki K (2018) Implementation status and issues related to on- and off-campus sports and exercise of junior high school students: thinking about the support of non-performers in implementing sports and exercise. J Life Sci 41:59–69
12. Kawabata M (1998) A study on extracurricular activities of Ochanomizu university students: about athletic clubs. Bull Human Soc Sci Ochanomizu Univ 51:187–202
13. Kiku K (1991) Sport career patterns of middle-aged and older participants. Report of the 1991 Japan Sports Medicine and Science Research Report of the Japan Amateur Sports Association, "No. VII: Sociological and Psychological Studies on Sports Participation of Middle-aged and Older Adults", pp 31–60
14. Ministry of Education, Culture, Sports, Science and Technology (2013) Public Opinion Survey on Physical Fitness and Sports in 2013
15. Ministry of Health, Labour and Welfare (2012) Health Science Council, Subcommittee on Community Health and Health Promotion and Nutrition, Committee of Medical Specialists for the National Health Promotion Campaign Plan: Reference material on the promotion of Healthy Japan 21 (2nd stage)
16. Muramatsu T, Mochizuki S (2012) Meaning of exercise continuation in people's life in adulthood. Bull Univ Yamanashi Sch Nurs 14:1–10
17. Nagagahara M (2003) Research trend of physical activity in middle-aged and elderly people. J Phys Educ 48:245–268
18. Nagagahara M (2011) Sport careers of middle-aged and older athletes. Sci Phys Educ 61(9):678–683
19. Nakajima G, Ito K (2019) The impact of job choice by mid-career leavers from organized sports. Chiba Econom Rev 60:57–78
20. Nakasuga T, Sakata S, Sugiyama K (2016) Relationships among motivational atmosphere, self-disclosure, and satisfaction in college athletic club activities for exercise continuation. J Sport Perform Res 8:1–13
21. Namikoshi K, Fujii K, Tanifuji C, Izaki M (2003) Influence of the experience of sports club activities on the sports life of university students. Bull Facul Educ Chiba Univ 51:129–136
22. Nishimura K, Yamaguchi Y (2003) Processes leading to non-performance of exercise and sports: a case study of women. J Sociol Sport 11:87–101
23. Okada J, Torii T, Miyauchi T, Yanagitani T, Kato K (2003) A study on the relationship between college athletic team history and physical activity in middle-aged and older adults. Waseda Univ J Phys Educ 35:25–35
24. Okano S (2011) Growth and training in the junior period. Gendai Sports Rev 24:107–113
25. Okatsu S, Kuroda K (2016) Influence of past experience of the same sport among group ball game players in adulthood and beyond: a secondary analysis of the Sasakawa Sports Foundation "Survey on Sports Life 2012" data. Lifelong Sports Res 13(2):43–54
26. Okazawa Y (2003) Liking exercise and self-competence. Sci Phys Educ 53(12):905–909
27. Oyama Y, Oishi K (2019) A study on factors affecting the continuation of high school club activities: focusing on the difference between sports and arts. Bull Hokkaido Univ Educ 69(2):303–311
28. Sasakawa Sports Foundation (2013) Sports life data 2012
29. Shimizu N (2002) School management for lifelong sports design. Phys Educ 50(1):30–33
30. Shiromaru A, Mizuuchi H (1991) Sports club activities now. Aoki Shoten, pp 28–29
31. Sudo H (2008) A study on determinants of exercise continuation among middle-aged and older women in sports clubs. J Sports Sci 5:96–107
32. Sugihara T, Sakai S, Sasaki Y, Yamaguchi Y, Nakagome S (1995) Investigating the cause of aversion to exercise. Inform Phys Fit 12(2):2–17

33. Susaki Y, Iribe Y, Sugiyama K, Saito A (2016) A survey on the actual condition of athletic clubs in universities: reasons for joining or not continuing to join. Health Sci 38:33–41
34. Tadano H, Kanazaki R, Tokunaga M, Hashimoto K (1982) A study on the determinants of students' sports behavior (2)—Social factors. Health Sci 4:51–76
35. Takizawa F (2001) Significance of sports, especially in relation to the body. Sci Phys Educ 51(1):35–38
36. Tsuruyama H, Oyama S (2005) A study on college students' exercise experience and motivation. Bull Faculty Int Lib Stud 1:81–91
37. Yamamoto N (1990) A comparison of regular and substitute players on their motives for participation in university athletic clubs. J Phys Educ 35:109–119

Effects of Brain Breaks on Attitudes and Motivation in the Institute of Teacher Campuses

Pei-Yi Ting, Garry Kuan, Ngien Siong Chin, and Yee Cheng Kueh

Abstract Brain breaks are short mental and physical breaks incorporated periodically and can help to stimulate one's health and motivation. The study aimed to examine the effects of brain breaks on the attitudes and motives of participation of trainee teachers in Sarawak, Malaysia. The brain breaks were conducted twice a day and five times a week for a duration of 12 weeks. The participants are comprised of 48 trainee teachers (24 males and 24 females) aged 18 to 23 years ($1.48 \pm .50$) randomly divided into intervention group ($n = 24$) and control group ($n = 24$). The Attitudes towards Physical Activity Scale (APAS) and Physical Activity and Leisure Motivation Scale (PALMS-M) were utilised. For APAS, the paired-samples *t*-test revealed that the mean difference score from pre-test to post-test in intervention group was significant for self-efficacy. MD = 0.60, $t(23) = 0.32$, $p = 0.004$, and fitness. MD = 0.58, $t(23) = 2.98$, $p = 0.007$. However, there was no significant difference for all APAS variables in control group. For PALMS-M, the paired-samples t-test revealed that the mean score from pre-test to post-test in intervention group was significant for competition/ego, MD = 0.84, $t(23) = 4.54$, $p < 0.001$, appearance, MD = 0.49, $t(23) = 2.79$, $p = 0.013$, others' expectations, MD = 0.63, $t(23) = 2.29$, $p = 0.032$, psychological condition, MD = 0.49, $t(23) = 2.66$, $p = 0.014$, and enjoyment, MD = 1.08, $t(23) = 1.30$, $p < 0.001$. In the control group, the mean score from pre-test to post-test was significant for competition/ego, MD = 0.59, $t(23) = 3.01$, $p = 0.006$, others' expectations, MD = 0.75, $t(23) = 3.38$, $p = 0.003$, and enjoyment, MD = 0.55, $t(23) = 2.46$, $p = 0.022$. The results revealed that brain breaks can be implemented in the campuses in order to improve trainee teachers' attitudes and motivation towards PA.

P.-Y. Ting (✉) · G. Kuan
Exercise and Sports Science, School of Health Sciences, Universiti Sains Malaysia, Kubang Kerian, Kelantan, Malaysia
e-mail: peiyi@student.usm.my

Y. C. Kueh
Unit of Biostatistics and Research Methodology, School of Medical Sciences, Universiti Sains Malaysia, Kubang Kerian, Kelantan, Malaysia

N. S. Chin
Department of Physical Education and Health, Institute of Teacher Education Batu Lintang Campus, Kuching, Sarawak, Malaysia

G. Kuan et al. (eds.), *Advancing Sports and Exercise via Innovation*, Lecture Notes in Bioengineering, https://doi.org/10.1007/978-981-19-8159-3_28

Keywords Brain breaks · Attitudes · Motivation · Trainee teachers

1 Introduction

Physical activity (PA) is defined as a bodily movement produced by the contraction of skeletal muscles that substantially improve overall health and fitness, cognitive development as well as decrease the risk of various non-communicable chronic diseases [1]. According to the Malaysian Health Minister, Datuk Seri Dr Dzulkefly Ahmad [2], about 33 per cent of Malaysian citizens are at least overweight, while 19 per cent of the population were obese, making Malaysia as the most obese country in Asia. According to the fact sheets about PA by the World Health Organization [3], 1 in 4 adults do not meet the global recommended levels of PA and more than 80% of the world's adolescent population is insufficiently physically active. Many researches have shown that PA may improve student's cognitive abilities, attitudes as well as subsequent academic performance [4]. Interventional studies are needed to fully understand the factors that promote PA participation among young adults.

Poobalan et al. [5] revealed that levels of physical activity are low among adolescents aged 18–25 years old, which are more likely to weight gain. This transition is oftentimes accompanied by unhealthy behavioural changes or sedentary behaviours, for example, spending time on computer, playing computer games, either for pleasure or work/ study, watching television or sitting in class all day. Despite being active in school, Pooblan et al. [5] findings also showed that many teenagers did not start any new sport after moving to university. Pope, Hansen, and Harvey [6] also found that the percentage of college students who were overweight or obese had risen from 23 to 41% from freshman to senior year. In September 2015, the General Assembly adopted the 2030 Agenda for Sustainable Development that includes 17 Sustainable Development Goals (SDGs) and one of the main goals of the SDGs which are good health and well-being. A concerted shift away from what may have driven past gains—curative interventions in the case of NCDs—towards multi-sectoral, prevention-oriented policy action, and investments is required to achieve SDG goals for many health-related SDG indicators, on non-communicable diseases (NCDs), NCD-related risks, and violence-related indicators, according to current projections [7]. Wengreen and Moncur [8] study revealed that 23% of participants gained more than 5% of their body weight during their first year of college while also engaging in less physical activity than they had in high school.

Most of the countries imposed lockdown, social distancing protocols, and requirements for self-isolation during the COVID-19 pandemic. People may become more sedentary and engage in less physical activity as a result of the lockdown. This is one of the unintended outcomes of the lockdown [9]. It has also resulted in significant changes to the physical activity learning process throughout the world [10]. In order to overcome obstacles, educators must adapt to the new learning patterns brought on by the epidemic [11]. Such mobility limitations may have long-term negative health effects because physical inactivity is one of the leading risk factors for chronic illness and all-cause mortality [12]. For instance, compared to the same week last year, FitBit users' daily step counts were much lower during the week of 22 March 2020 [13].

Regarding this issue, the environment in schools and institutions of higher learning has integrated technology into their systems to enhance the learning and teaching which have also incorporated the promotion of physical activity as part of the millennium culture. Teachers or lecturers who will guide the youngsters should be prepared with an intervention which may promote physical activity to facilitate a paradigm shift towards a healthier generation. The use of video tutorials during the pandemic can complement online learning tools used by educators as discussion material and practical material which will increase students' understanding of the material presented [14]. Research studies show that providing virtual health contents (physical activity, physical education, and nutrition information) is a viable way to sustain school and family's health in times of a pandemic. Studies have shown that short physical activity breaks between lessons can improve one's cognitive and concentrations [15]. Brain breaks physical activity solutions by HopSports are one of the effective strategies which may help young adults to achieve the mentioned goals. The brain breaks physical activity is an exercise video that helps to promote health and wellness. The short videos are usually limited to three or five minutes and work best when they incorporate physical activities [16]. The brain breaks programme has given them a platform to not only be physically active but also learn new motor skills, coordination, dancing, motions of highly integrated functional muscle groups in movement activity, language, art, music, and diverse cultures [17]. Combination of some level of physical movement in brain breaks tends to be an effective way to stimulate neurological pathways and help both hemispheres of brain to work together. Previous research also indicated that brain breaks may also improve one's concentration and help to meditate their minds.

2 Methods

2.1 Study Design, Recruitment, and Sampling

This quasi-experimental design study involved pre- and post-tests. The study was conducted for a duration of 12 weeks to examine the effect of brain breaks intervention in participation motives (using PALMS) and attitudes (using APAS) among the trainee teachers in Sarawak. The participants were the trainee teachers from the Institute of Teacher Education, Batu Lintang Campus and Institute of Teacher Education Tun Abdul Razak Campus from Sarawak. The directors of the two Institute of Teacher Education campuses were approached to obtain their approval in conducting the study. The approval letter from the Ministry of Education was also given to the respective campuses as proof of approval for conducting this study. The participants who meet all inclusion criteria were given an informed consent form and signed the form before participating in the study.

The participants were assigned randomly into two groups (intervention and control group) and were gender-matched. The objective and procedures of the study were briefed to the participants. Then, the pre-test was conducted on both groups, where all participants were required to complete the PALMS and APAS. Similar procedure was repeated during the post-test. The brain break activity which was in digital platform is a video-exercise (dances) comprising structured physical activities that allows participants to follow an on-screen instructor leading an activity projected on the television and a large screen by a projector. The exercises that have moderate to high intensity were selected. The breaks were done twice per day (when trainee teachers appeared tired or distracted, brain breaks were employed) which was once in the morning and once again in mid-afternoon, five days a week, for a duration of 12 weeks. The breaks lasted for 3–5 min. But during this pandemic, the brain breaks activity was conducted through 'Google Meet' to avoid and decrease physical contact. Google classroom was created where all the intervention group's participants were added inside. A link was given from time to time by the host, and the participants would need to join the Google Meet once in the morning and once again in mid-afternoon, five days a week for a duration of 12 weeks. In addition, the participants had completed their weekly and monthly logs showing the number of brain breaks activity being followed. All participants were volunteers, and no payment was offered. However, a certificate was given to each participant at the end of the study.

The procedures of this study were summarised in Fig. 1. The Universiti Sains Malaysia (USM) Human Research Ethics Committee (USM/JEPeM/21050369) and from the Research Department of Ministry of Education (KPM.600–3/2/3-eras11814) approved the study. Before the study was conducted, a research information sheet, subject information, consent form, and the participant's material publication consent form were given to the trainee teachers to read and signed. Their participation was completely voluntary and free to withdraw at any stages of the study. Participants that had any inquiries were personally contacted. All information was kept confidential by the researchers and will not be made publicly available unless disclosure was required by law. The data presented did not identify the participants individually. There was no conflict of interest in the study.

2.2 Participants

Participants included 48 trainee teachers from 2 different Institute of Teacher Campuses in Sarawak (24 males and 24 females). The participant's age was 18–23 years old. The inclusion criteria of the study were; (1) bilingual in Bahasa Melayu and English; (2) must be in good health to participate in any physical activity; (3) agreed and signed the informed consent form. The exclusion participants were with medical conditions that may interfere with their physical ability to complete the study.

The demographic characteristics of the participants are given in Table 1. There were 48 participants involved which is comprised of 24 males (50%) and 24 females

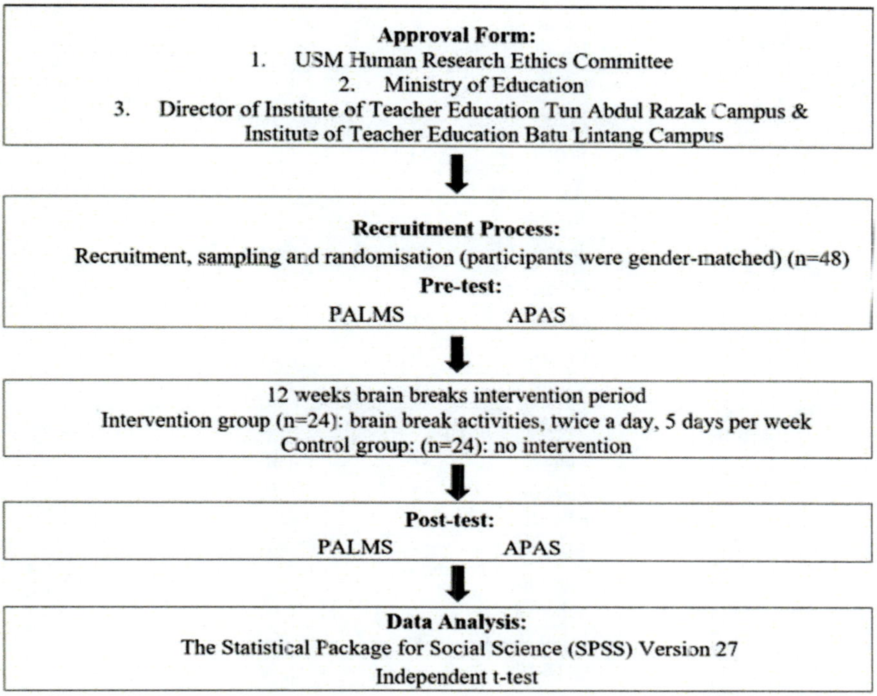

Fig. 1 Flow chart of the study procedures

(50%). The age group categories showed that 25 (52.1%) of the participants aged between 17 and 23 years. In terms of ethnicity, the Malay formed the largest percentage with 43.8%, followed by Native and Chinese who comprised 37.5% and 18.8% of the population, respectively. 43.8% of them took up Malay course, 12 (15.0%) took up physical education course, and 14 (29.2%) took up Islam course. The BMI revealed that a high percentage of the participants were in normal weight category with a total of 30 (62.5%), followed by underweight category with 8 (16.7%), overweight category with 7 (14.6%), and obese category with 3 (6.3%). Whereas, in terms of the frequency of exercise, a high percentage of the participants exercise 1–2 times (45.8%) a week averaging between 31 and 60 min (54.2%).

2.3 Research Instruments

Physical Activity and Leisure Motivation Scale (PALMS): The PALMS [18] measures the participation's motives in Physical Activity (PA). It is comprised of 40-items that assess participation in sports and PA. There are eight subscales

Table 1 Demographic characteristics of the participants

Characteristics	Frequency (F)	Percentage (%)	M (SD)
Gender			1.50 (0.51)
Male	24	50.0	
Female	24	50.0	
Age group			1.48 (0.50)
18–20	25	52.1	
21–23	23	47.9	
Ethnicity			1.94 (0.91)
Malay	21	43.8	
Chinese	9	18.8	
Native	18	37.5	
Courses			2.83 (0.86)
Malay	22	45.8	
Physical education	12	25.0	
Islam	14	29.2	
Body mass index (BMI)			2.15 (0.74)
Underweight (≤18.49)	8	16.7	
Normal (18.5–24.9)	30	62.5	
Overweight (25.0–29.9)	7	14.6	
Obese (30.0–34.9)	3	6.3	
Frequency of exercise per week			1.83 (0.93)
1–2 times	22	45.8	
3–4 times	15	31.3	
5–6 times	8	16.7	
7 times	3	6.3	
Exercise per session (minutes)			1.88 (0.84)
1–30 min	16	33.3	
31–60 min	26	54.2	
61–90 min	2	4.2	
91–120 min	4	8.3	

comprising mastery, enjoyment, psychological condition, physical condition, appearance, others' expectations, affiliation, competition/ego which are attributes of intrinsic and extrinsic motivation. There are 5 items in each of the eight subscales rated on a 5-point Likert scale ranging from 1 (strongly disagree) to 5 (strongly agree). The PALMS has shown acceptable factor structure, internal consistency (0.78–0.82), test–retest reliability (0.78–0.94), and criterion validity that can be applied in the physical activity contexts. Though it has been validated in the Malay [19], it is only validated in one population study which might not strengthen its validity, consistency, and reliability for assessment of motives of participation in another population.

Attitude towards Physical Activity Scale (APAS): The APAS [20] measures the attitudes and perceptions regarding various aspects of engagement in PA. The Cronbach's alpha ranged from 0.63 to 0.84. The 35-item APAS is comprised of seven subscales that measures benefits (6 items), importance (3 items), self-efficacy (3 items), learning (4 items), fun (7 items), self-confidence on physical fitness (7 items), and personal best (5 item). The items were rated on a 5-point Likert scale ranging from 1(strongly disagree) to 5(strongly agree).

2.4 *Statistical Analysis*

The data were analysed using the Statistical Product and Service Solution (SPSS) Version 27.0. The independent t-test or Chi-square test was used to examine whether there are any significant differences between the control group and the intervention group on the study variables. The level of the statistical significance was set as $p < 0.05$.

3 Results

Presented in Table 2 is the mean difference of APAS variables for pre- and post-intervention groups. The mean score of benefits was increased from pre (4.31 $\pm$ 0.56) to post (4.55 $\pm$ 0.50), MD = 0.24, $t(23) = 1.76, p = 0.09$. The mean score for importance was increased from pre (4.11 $\pm$ 0.73) to post (4.40 $\pm$ 0.61), MD = 0.29, $t(23) = 1.57, p = 0.13$. The mean score for learning was decreased from pre (3.47 $\pm$ 0.85) to post (3.33 $\pm$ 1.26), MD = −0.14, $t(23) = -0.50, p = 0.62$. The mean score for self-efficacy was significantly increased from pre (3.63 $\pm$ 0.74) to post (4.22 $\pm$ 0.80), MD = 0.60, $t(23) = 0.32, p = 0.004$, 95% CI: 0.21, 0.98. The mean score for fun was increased from pre (4.09 $\pm$ 0.69) to post (4.40 $\pm$ 0.65), MD = 0.32, $t(23) = 1.74$, $p = 0.10$. The mean score for fitness was significantly increased from pre (3.70 $\pm$ 0.66) to post (4.28 $\pm$ 0.59), MD = 0.58, $t(23) = 2.98, p = 0.007$, 95% CI: 0.18, 0.99. The mean score for personal best was increased from pre (4.16 $\pm$ 0.79) to post (4.50 $\pm$ 0.54), MD = 0.34, $t(23) = 1.80, p = 0.08$.

Table 2 Mean difference of APAS variables for pre- and post-intervention groups ($N = 24$)

Variables	Intervention group (N = 24)				MD (95% CI)	t-statistics (df = 23)	p-value
	Pre-test		Post-test				
	M	SD	M	SD			
APAS							
Benefits	4.31	0.56	4.55	0.50	0.24 (-0.04, 0.51)	1.76	0.09
Importance	4.11	0.73	4.40	0.61	0.29 (-0.09, 0.68)	1.57	0.13
Learning	3.47	0.85	3.33	1.26	-0.14 (-0.70, 0.43)	-0.50	0.62
Self-efficacy	3.63	0.74	4.22	0.80	0.60 (0.21, 0.98)	0.32	0.004*
Fun	4.09	0.69	4.40	0.65	0.32 (-0.06, 0.69)	1.74	0.10
Fitness	3.70	0.66	4.28	0.59	0.58 (0.18, 0.99)	2.98	0.007*
Personal best	4.16	0.79	4.50	0.54	0.34 (-0.05, 0.73)	1.80	0.08

*$p < 0.05$

Table 3 gives the mean difference of APAS variables for pre- and post-control groups. The mean score of benefits was slightly decreased from pre (4.41 ± 0.64) to post (4.40 ± 0.65), MD = −0.01, $t(23) = -0.03, p = 0.97$. However, the mean score for importance was slightly increased from pre (4.19 ± 0.58) to post (4.21 ± 0.67), MD = 0.01, $t(23) = 0.07, p = 0.95$. The mean score for learning was increased from pre (3.27 ± 0.93) to post (3.72 ± 0.83), MD = 0.45, $t(23) = 1.57$, $p = 0.13$. The mean score for self-efficacy was increased from pre (3.75 ± 0.74) to post (3.97 ± 0.77), MD = 0.22, $t(23) = 0.92$, $p = 0.37$. The mean score for fun was increased from pre (4.15 ± 0.80) to post (4.20 ± 0.70), MD = 0.04, $t(23) = 0.17, p = 0.87$. The mean score for fitness was increased from pre (3.96 ± 0.78) to post (4.02 ± 0.86), MD = 0.05, $t(23) = 0.19$, $p = 0.85$. The mean score for personal best was decreased from pre (4.36 ± 0.75) to post (4.18 ± 0.83), MD = −0.18, $t(23) = -0.66, p = 0.52$. There was no statistically significant mean difference of all the APAS subscales for the pre and post in the control group with $p > 0.05$.

Table 4 gives the mean difference of PALMS variables for pre- and post-intervention groups. The mean score of competition/ego was significantly increased from pre (3.35 ± 0.74) to post (4.19 ± 0.55), MD = 0.84, $t(23) = 4.54, p < 0.001$, 95% CI: 0.46, 1.23. The mean score for appearance was significantly increased from pre (3.98 ± 0.63) to post (4.47 ± 0.59), MD = 0.49, $t(23) = 2.79, p = 0.013$, 95% CI: 0.11, 0.87. The mean score for others' expectations was significantly increased from pre (2.64 ± 0.77) to post (3.27 ± 0.96), MD = 0.63, $t(23) = 2.29, p = 0.032$, 95% CI: 0.06, 1.19. The mean score for affiliation was increased from pre (3.99 ± 0.66) to post (4.31 ± 0.65), MD = 0.07, $t(23) = 0.50$, $p = 0.62$. The mean score for physical condition was slightly increased from pre (4.45 ± 0.53) to post (4.52 ± 0.49), MD = 0.07, $t(23) = 0.50$, $p = 0.62$. The mean score for psychological condition was significantly increased from pre (3.90 ± 0.67) to post (4.39 ± 0.54), MD = 0.49, $t(23) = 2.66, p = 0.014$, 95% CI: 0.11, 0.87. The mean score for mastery

Table 3 Mean difference of APAS variables for pre- and post-control groups ($N = 24$)

Variables	Control group ($N = 24$)				MD (95% CI)	t-statistics ($df = 23$)	p-value
	Pre-test		Post-test				
	M	SD	M	SD			
APAS							
Benefits	4.41	0.64	4.40	0.65	−0.01 (−0.44, 0.42)	−0.03	0.97
Importance	4.19	0.58	4.21	0.67	0.01 (−0.40, 0.43)	0.07	0.95
Learning	3.27	0.93	3.72	0.83	0.45 (−0.14, 1.04)	1.57	0.13
Self-efficacy	3.75	0.74	3.97	0.77	0.22 (−0.28, 0.72)	0.92	0.37
Fun	4.41	0.64	4.40	0.65	−0.01 (−0.44, 0.42)	−0.03	0.97
Fitness	4.19	0.58	4.21	0.67	0.01 (−0.40, 0.43)	0.07	0.95
Personal best	3.27	0.93	3.72	0.83	0.45 (−0.14, 1.04)	1.57	0.13

was increased from pre (4.08 ± 0.54) to post (4.38 ± 0.60), MD = 0.30, $t(23) = 1.76$, $p = 0.09$. The mean score for enjoyment was significantly increased from pre (3.73 ± 0.45) to post (4.81 ± 0.24), MD = 1.08, $t(23) = 1.30$, $p < 0.001$.

Table 5 gives the mean difference of PALMS variables for pre- and post-control groups. The mean score of competition/ego was significantly increased from pre (3.38 ± 0.74) to post (3.97 ± 0.82), MD = 0.59, $t(23) = 3.01$, $p = 0.006$, 95% CI: 0.18, 1.00. The mean score for appearance was slightly increased from pre (3.93 ± 0.61) to post (4.05 ± 0.83), MD = 0.12, $t(23) = 0.48$, $p = 0.64$. The mean score for others' expectations was significantly increased from pre (2.64 ± 0.79) to post (3.39

Table 4 Mean difference of PALMS variables for pre- and post-intervention groups ($N = 24$)

Variables	Intervention group ($N = 24$)				MD (95% CI)	t-statistics ($df = 23$)	p-value
	Pre-test		Post-test				
	M	SD	M	SD			
PALMS							
Competition/ego	3.35	0.74	4.19	0.55	0.84 (0.46, 1.23)	4.54	< 0.001*
Appearance	3.98	0.63	4.47	0.59	0.49 (0.11, 0.87)	2.79	0.013*
Others' expectations	2.64	0.77	3.27	0.96	0.63 (0.06, 1.19)	2.29	0.032*
Affiliation	3.99	0.66	4.31	0.65	0.32 (−0.09, 0.72)	1.63	0.12
Physical condition	4.45	0.53	4.52	0.49	0.07 (−0.21, 0.34)	0.50	0.62
Psychological condition	3.90	0.67	4.39	0.54	0.49 (0.11, 0.87)	2.66	0.014*
Mastery	4.08	0.54	4.38	0.60	0.30 (−0.05, 0.65)	1.76	0.09
Enjoyment	3.73	0.45	4.81	0.24	1.08 (0.87, 1.30)	1.30	< 0.001*

*$p < 0.05$

Table 5 Mean difference of PALMS variables for pre- and post-control groups ($N = 24$)

Variables	Control group ($N = 24$)				*MD* (95% CI)	*t*-statistics (*df* = 23)	*p*-value
	Pre-test		Post-test				
	M	SD	M	SD			
PALMS							
Competition/ego	3.38	0.74	3.97	0.82	0.59 (0.18, 1.00)	3.01	0.006*
Appearance	3.93	0.61	4.05	0.83	0.12 (−0.40, 0.65)	0.48	0.64
Others' expectations	2.64	0.79	3.39	0.85	0.75 (0.29, 1.21)	3.38	0.003*
Affiliation	3.96	0.65	4.05	0.86	0.10 (−0.41, 0.60)	0.40	0.70
Physical condition	4.43	0.53	4.30	0.73	−1.22 (−0.60, 0.36)	−0.53	0.60
Psychological condition	3.85	0.64	4.28	0.75	0.43 (−0.07, 0.93)	1.77	0.09
Mastery	4.04	0.52	4.18	0.81	0.14 (−0.30, 0.58)	0.66	0.52
Enjoyment	3.71	0.45	4.26	0.80	0.55 (0.09, 1.01)	2.46	0.022*

*$p < 0.05$

± 0.85), MD = 0.75, $t(23) = 3.38$, $p = 0.003$, 95% CI: 0.29, 1.21. The mean score for affiliation was increased from pre (3.96 ± 0.65) to post (4.05 ± 0.86), MD = 0.10, $t(23) = 0.40$, $p = 0.70$. The mean score for physical condition was decreased from pre (4.43 ± 0.53) to post (4.30 ± 0.73), MD = −1.22, $t(23) = -0.53$, $p = 0.60$. The mean score for psychological condition was increased from pre (3.85 ± 0.64) to post (4.28 ± 0.75), MD = 0.43, $t(23) = 1.77$, $p = 0.09$. The mean score for mastery was increased from pre (4.04 ± 0.52) to post (4.18 ± 0.81), MD = 0.14, $t(23) = 0.66$, $p = 0.52$. The mean score for enjoyment was significantly increased from pre (3.71 ± 0.45) to post (4.26 ± 0.80), MD = 0.55, $t(23) = 2.46$, $p = 0.022$.

4 Discussion

This study's goal was to ascertain how brain breaks affected trainee instructors at the Institute of Teacher Campus attitudes and motivation. During the study, brain breaks video that has moderate to high intensity was utilized on the participants. The breaks were done twice per day which was once in the morning and once again in mid-afternoon, five days a week, for a duration of 12 weeks. The breaks lasted for 3–5 min. It was shown that the significant differences for APAS were in the intervention group as compared to the control group. This showed that brain breaks have positive effects on self-efficacy and fitness which provide a source of holistic and autonomous learning among the trainee teachers. The trainee teachers could have found an improvement in their self-esteem, confidence, and ability to strive to further enhance their personal performance in their level of fitness and health. Since brain breaks were only short interval form of activities, the effects have spurred their level of motivation to further engage in other physical activities of higher intensity and durations during their free time. The finding aligns with past studies which had described self-efficacy confidence as one of the dominant factors which correlates with active life-long engagement in physical activity [21, 22].

Whereas, for PALMS, the significant differences were found mainly within the extrinsic factors of PALMS (Competition/Ego, Appearance, Others' Expectations, Psychological Conditions) and intrinsic factor of enjoyment in the intervention group as compared to the control group. This study's findings diverge from Hajar et al. [23] study that used a brain breaks intervention with elementary school pupils and found that four motives—mastery, competition, affiliation, and physical condition—had an upward tendency. Children and teens were shown to be motivated by skill development, fun, challenge, and achievement. In this study, trainee teachers possessed diverse motives in their participation according to their personal preferences. The extrinsic motives could be due to the trainee teachers participating in PA during the weekends or leisure times searching for social relatedness, appearance, health, and balanced body through competitive and challenging PA to satisfy their psychological needs. The enjoyment derived from these motivational factors had propelled the fun and eagerness to learn and participate in PA which fosters positive Attitudes towards PA through the brain breaks. Study showed that young adults are more motivated for PA which are related to maintaining fitness levels and overall health [24]. This implied that the brain breaks intervention can be just as beneficial for children as they are for adults.

The study limitation was that the trainee teachers were recruited from only two campuses in Sarawak. Therefore, for future studies, it is recommended that researchers can recruit participants from the Institute of Teacher Campus from other states or even universities students around Malaysia to solidify our understanding of the application of brain breaks which can provide some significant effect on the motives and Attitudes towards Physical Activity.

5 Conclusion

The study suggests that the integration of extrinsic and intrinsic factors of motivation would enhance and sustain the attitudes of trainee teachers to inculcate a balanced and healthy lifestyle in long run. Therefore, it is of utmost importance that tailored interventions that combined both motivational factors are necessary to improve PA and its effects on the health and well-being of the trainee teachers.

Acknowledgements The brain breaks platform was made available by Ming-Kai Chin through GCH and HOPSports® Inc., and the writers thank him for his advice and support. The Fundamental Project Grant Scheme (FRGS) of the Ministry of Higher Education of Malaysia provided funding for this research: FRGS/1/2020/SKK06/USM/03/13.

References

1. Caspersen CJ, Powell KE, Christenson GM (1985) Physical activity, exercise, and physical fitness: definitions and distinctions for health-related research. Pub Health Res 100(2):126–131
2. Ahmad D (2019) Malaysia and WHO call for more investment in primary health care the 21st century. (Y. Yang, Interviewer)
3. Physical activity (2020) World Health Organization. https://www.who.int/news-room/fact-sheets/detail/physical-activity
4. Uzunoz FS, Chin MK, Mok MMC, Edginton CR, Podnar H (2007) The effects of technology supported brain breaks on physical activity in school children. In: Passionately inclusive: towards participation and friendship in sport: Festschrift Für Gudrun Doll-Tepper. Waxmann Verlag GmbH, Münster, NY, USA
5. Poobalan AS, Aucott LS, Clarke A, Cairns S, Smith S (2012) Physical activity attitudes, intentions and behaviour among 18–25year olds: a mixed method study. BMC Pub Health 12(640). https://doi.org/10.1186/1471-2458-12-640
6. Pope L, Hansen D, Harvey J (2017) Examining the weight trajectory of college students. J Nutr Educ Behav 49(2):137–141. https://doi.org/10.1016/j.jneb.2016.10.014
7. GBD (2018) Measuring progress from 1990 to 2017 and projecting attainment to 2030 of the health related sustaintable development goals for 195 countries and territories: a systematic analysis for the global burden of disease study 2017. Lancet 392(10159):2019–2138. https://doi.org/10.1016/s0140-6736(18)32281-5
8. Wengreen HJ, Moncur C (2009) Change in diet, physical activity, and body weight among young-adults during the transition from high school to college. Nutri J 8(32):1–7. https://doi.org/10.1186/1475-2891-8-32
9. Musselwhite C, Avineri E, Susilo Y (2020) The coronavirus disease covid-19 and implications for transport and health. J Trans Health 16:1–4. https://doi.org/10.1016/j.jth.2020.100853
10. Suzuki Y, Maeda N, Hirado D, Shirakawa T, Urabe Y (2020) Physical activity changes and its risk factors among community-dwelling Japanese older adults during the covid-19 epidemic: associations with subjective well-being and health-related quality of life. Inter J Environ Res Pub Health 17(18):6591 MDPI AG. https://doi.org/10.3390/ijerph17186591
11. Jauhari MN, Sambira, Zakiah Z (2020) The impact of the covid-19 pandemic on the implementation of adaptive physical education learning in special schools. STAND: J Sports Teach Dev 1(1):63–70. https://doi.org/10.36456/j-stand.v1i1.2594
12. Lee IM, Shiroma EJ, Lobelo F, Puska P, Blair SN, Katzmarzyk PT, Lancet Physical Activity Series Working Group (2012) Effect of physical inactivity on major non-communicable diseases worldwide: an analysis of burden of disease and life expectancy. Lancet 380(9838):219–229. https://doi.org/10.1016/S0140-6736(12)61031-9
13. Fitbit S (2020) The impact of coronavirus on global activity. https://blog.fitbit.com/covid-19-global-activity/
14. Batubara HH, Batubara DS (2020) Use of video tutorials to support online learning during the corona virus pandemic. J Madrasah Ibtidayah 29:74–84
15. Willis J (2016) Using brain breaks to restore students' focus. https://www.edutopia.org/article/brain-breaks-restore-student-focus-judy-willis
16. Cox J (2019) What is brain break? ThoughtCo. https://www.thoughtco.com/what-is-a-brain-break-2081615
17. Chin MK, Edginton CR, Tang MS (2013) School physical education and health: a model of best practice - integrating local context with global trends. The Global J Health Phys Educ Pedagog 1(4):251–282. https://doi.org/10.1093/ajcn/80.3.569
18. Morris T, Rogers H (2004) Measuring motives for physical activity. Paper presented at the Int Sport Sci Congr, Seoul, Korea
19. Kueh YC, Kuan G, Morris T (2019) The physical activity and leisure motivation scale: a confirmatory study of the Malay language version. Int J Sport Exerc Psychol 17(3):250–265. https://doi.org/10.1080/1612197X.2017.1321029

20. Mok MMC, Chin MK, Chen S, Emeljanovas A, Mieziene B, Bronikowski M, Laudanska-Krzeminska I, Milanovic I, Pasic M, Balasekaran G (2015) Psychometric properties of the attitudes toward physical activity scale: A rasch analysis based on data from five locations. J Appl Meas 16(4):379–400
21. Sallis JF, Bull F, Guthold R, Health GW, Inoue S, Kelly P, Oyeyemi AL, Perez LG, Richards J, Hallal PC (2016) Progress in physical activity over the Olympic quadrennium. The Lancet 388(10051):1325–1336. https://doi.org/10.1016/S0140-6736(16)30581-5
22. Spriijt, Metz, Saelens BE (2016) Behavioral aspects in physical activity in childhood and adolescence. Handbook of Pediatric Obesity: Etiology, Pathophysiology and Prevention 227
23. Hajar MS, Rizal H, Kueh YC, Muhamad AS, Kuan G (2019) The effects of brain breaks on motives of participation in physical activity among primary school children in Malaysia. Inter J Environ Res Pub Health 16(13):2331
24. Ebben W, Brudzynski L (2008) Motivations and barriers to exercise among college students. J Exerc Physiol Online 11(5):1–11

The Effect of Brain Breaks on Health-Related Fitness Among Indigenous Primary School Children

Ting Len Lim, Garry Kuan, Ngien Siong Chin, Mohd Rahimi Che Jusoh, and Yee Cheng Kueh

Abstract The study aimed was to examine the effect of brain breaks on health-related fitness among indigenous primary school children in Sarawak, Malaysia. The participants were 70 indigenous primary school children (33 male and 37 female) aged 10–12 years old (2.03 ± .78) from two rural schools in Sarawak, Malaysia. The Malaysian National Physical Fitness Standard (SEGAK) test battery was conducted to evaluate the students' body mass index, muscle strength, muscular endurance, and flexibility. Descriptive statistics and paired-samples t-test were used to analyse the data. The paired-samples t-test based on gender revealed statistically significant difference from pre-test to post-test for both gender in the 3-min step test, push-ups test, curl-ups test, and sit and reach test with $p < 0.05$ for the intervention group. Whereas, for the control group, results revealed significant difference in push-ups test and curl-ups test for female with $p < 0.05$. Besides, the paired-samples t-test based on age groups showed statistically significant difference from pre-test to post-test for the three age groups in the 3-min step test, push-ups test, curl-ups test, and sit and reach test with $p < 0.05$ for the intervention group. Whereas, for the control group, results revealed significant difference in 3-min step test for 12 years old group, $p < 0.001$, and curl-ups test for 11 years old groups with $p = 0.001$. The study recommends school-based physical activity programmes to increase the physical activity levels of the children in order to improve their fitness and health.

Keywords Brain breaks · Health-related fitness · Indigenous school children

---

T. L. Lim (✉) · G. Kuan · M. R. Che Jusoh
Exercise and Sports Science, School of Health Sciences, Universiti Sains Malaysia, Kubang Kerian, Kelantan, Malaysia
e-mail: tinglen@student.usm.my

N. S. Chin
Department of Physical Education and Health, Institute of Teacher Education Batu Lintang Campus, Kuching, Sarawak, Malaysia

Y. C. Kueh
Unit of Biostatistics and Research Methodology, School of Medical Sciences, Universiti Sains Malaysia, Kubang Kerian, Kelantan, Malaysia

G. Kuan et al. (eds.), *Advancing Sports and Exercise via Innovation*, Lecture Notes in Bioengineering, https://doi.org/10.1007/978-981-19-8159-3_29

1 Introduction

Physical activity plays a vital role in our daily lives. Regular physical activity is recommended for all people in their reproductive years since it is associated with reduced risk of morbidity, maintaining a healthy weight, and the promotion of preconception health [1]. Despite the importance of physical activity, there is a need to increase physical activity among the Malaysian adolescent communities, especially towards indigenous children, whereby they have limited access to healthy lifestyles. To achieve this objective, this fundamental study, which is collaborating with the United Nation's Sustainable Developmental Goals, aimed to use brain breaks (BB) to transform the quality of life (QoL) among the Sarawakian indigenous primary school children. BB is an interactive web-based structured physical activity, nutrition, and hygiene (lifestyles) videos that aim to stimulate students' interest in learning and promote better health [2].

In 2019, research was conducted to see the impact of brain breaks across 600 Malay primary school children and implemented in three schools in the district of Kota Bharu, Kelantan. It is the first study conducted in Malaysia and it is important to initiate school-based physical activity policies to promote changes at decision-making levels aimed at providing children with more regular access to physical activity in school settings. However, no study was implemented for the indigenous communities, whereby there are lacking internet facilities, electricity, and proper education facilities such as the Sarawak indigenous schools.

The effect of physical activity on a holistic development of mental, proper nutrition, hygiene, and physical wellbeing can be extrapolated in school settings. This not only improves health but also can further enhance learning development and cognitive functioning more so than just emphasising on core academic subjects. In addition, physical activity can reduce diseases such as diabetes, cancer, osteoporosis, and cardiovascular diseases [3]. However, research in policies driven to improve physical activity in Malaysian schools is still lacking. A growing number of studies show a trend of decreasing physical activity level among children [4]. More worryingly, physical activity has also been found to decline with age [5, 6]. This scenario is even harder for the rural communities such as the indigenous children in Sarawak, whereby they are lacking access to clean hygiene, proper nutrition, and physical activities leading to low immune system and sometimes death due to diseases, improper cleanliness, and low immune system.

Thus, the aim of this study was to investigate the effects of using brain breaks (physical activity videos) on health-related fitness among the indigenous primary school children. We hypothesise that brain breaks (with physical activity videos) will positively impact children on health-related fitness among the indigenous primary school children.

2 Methods

2.1 Study Design, Recruitment, and Sampling

The research design used is descriptive quantitative research. The schools were selected through purposive randomisation. All indigenous schools in Sarawak were assigned a number and two schools were randomly selected (schools A and B). The headmasters were approached to obtained approval for conducting this study in their respective schools. They were informed about the objectives and procedures of the study. Letter from the education department of Sarawak was given to the respective schools as proof of approval for conducting this research. Invitation for student participation was conducted at class and/or hall with the aid of the teachers and headmaster. The headmaster and the teachers were the only to bring the students to the hall and to monitor the programme. However, the programme was conducted by the researcher. Thus, only the students that volunteered to participate in the study were included the intervention. The students were given an informed assent form to give their parents for approval. For any questions or uncertainty that arise, the parents were personally contacted to acquire verbal consent.

2.2 Participants

The study participants included 70 students from 2 different primary school in Sarawak (33 boys and 37 girls). The participants were aged between 10 and12 years old. The inclusion criteria of the study were: (1) Bilingual in Bahasa Melayu and English. (2) Those that are physically (measured by PAR-Q) and mentally healthy to participate in daily physical activity. (3) Agreed and signed the informed assent form.

Table 1 gives the demographic characteristics of the primary school students, 33 (47.1%) were males, and 37 (52.9%) were females. There were 35 students in the intervention group and 35 students in the control group. The students were classified into three age groups, 20 (28.6%) students were 10 years old, 28 (40.0%) students were 11 years old, and 22 (31.4%) students were 12 years old, respectively. Based on the pre-body mass index, the majority of the participants were in the normal weight BMI (72.9%), followed by underweight BMI (21.4%), overweight BMI (4.3%), and obese (1.4%). However, in the post-body mass index, 67.1% participants were in the normal weight BMI, 30.0% were in the underweight BMI, and 2.9% were in the overweight BMI, respectively.

Table 1 Demographic characteristics of the 70 participants

Characteristics	Frequency (F)	Percentage (%)	M (SD)
Gender			1.53 (0.50)
Male	33	47.1	
Female	37	52.9	
Age group			2.03 (0.78)
10 years old	20	28.6	
11 years old	28	40.0	
12 years old	22	31.4	
Body mass index			
Pre			1.34 (0.63)
Normal (18.50–24.99)	51	72.9	
Underweight (≤18.49)	15	21.4	
Overweight (25.00–29.99)	3	4.3	
Obese (30.00–34.99)	1	1.4	
Post			1.36 (0.54)
Normal (18.50–24.99)	47	67.1	
Underweight (≤18.49)	21	30.0	
Overweight (25.00–29.99)	2	2.9	

2.3 Research Instruments

Health-related fitness: SEGAK test for measuring the health-related fitness of an individual was used. The test comprised strength, flexibility, power, and cardiovascular endurance test (3-min step test, push-ups, curls-up, and sit and reach).

2.4 Data Analysis

The data were analysed using the Statistical Package for Social Science (SPSS) Version 27.0. The level of the statistical significance was set as $p < 0.05$.

3 Results

Table 2 gives the SEGAK tests based on gender for the intervention and control group. For the intervention group, the results revealed that the mean score for 3-min step test showed a significant decrease from pre (96.13 ± 13.01) to post (88.13 ± 12.48) for male, $t = 9.40$, $p < 0.001$, and from pre (111.32 ± 24.98) to post (103.89 ± 23.86) for female, $t = 7.09$, $p < 0.001$. The mean score for push-ups revealed a statistically significant increase from pre (13.56 ± 3.05) to post (15.56 ± 3.24) for male, $t = -8.94$, $p < 0.001$, and from pre (14.16 ± 3.47) to post (16.68 ± 2.87) for female, $t = -10.79$, $p = <0.001$. The mean score for curl-ups also showed a significant increase from pre (16.50 ± 1.90) to post (18.75 ± 2.11) for male, t = 15.59, p < 0.001, and from pre (15.47 ± 1.68) to post (17.74 ± 1.73) for female, t = 17.56, $p < 0.001$. Finally, the mean score for sit and reach test also revealed a significant increase from pre (25.56 ± 5.07) to post (28.06 ± 5.11) for male, $t = 15.81$, $p < 0.001$, and from pre (26.37 ± 6.68) to post (28.74 ± 6.79) for female, $t = 11.53$, $p < 0.001$.

Whereas, for the control group, results revealed that the mean score for 3-min step test showed no significant increase from pre (105.82 ± 16.80) to post (109.94 ± 19.54) for male, $t = 1.68$, $p = 0.11$, and from pre (105.67 ± 21.52) to post (111.89 ± 23.88) for female, $t = 1.82$, $p = 0.09$. The mean score for push-ups revealed no significant decrease from pre (11.71 ± 4.77) to post (11.65 ± 4.60) for male, $t = 0.12$, $p = 0.90$, but a significant decrease from pre (15.89 ± 5.06) to post (14.78 ± 4.57) for female, $t = 2.51$, $p = 0.022$. The mean score for curl-ups showed no significant increase from pre (14.18 ± 3.52) to post (15.00 ± 2.50) for male, $t = 1.55$, $p = 0.14$, and from pre (13.78 ± 4.31) to post (13.83 ± 3.31) for female, $t = 0.12$, $p = 0.91$. Finally, the mean score for sit and reach test also revealed a significant increase from pre (23.79 ± 5.95) to post (24.85 ± 6.00) for male, $t = 5.84$, $p < 0.001$, but no significant difference from pre (27.33 ± 4.99) to post (27.72 ± 4.19) for female, $t = -17.56$, $p = 0.38$.

Table 3 gives the SEGAK tests based on age groups for intervention and control group. For the intervention group, the results revealed that the mean score for 3-min step test showed a significant decrease from pre (94.80 ± 9.98) to post (89.40 ± 9.25) for 10 years old group, $t = 10.82$, $p < 0.001$, from pre (93.86 ± 9.54) to post (86.21 ± 8.43) for 11 years old group, $t = 7.13$, $p = 0.001$, and from pre (126.45 ± 24.22) to post (116.64 ± 25.68) for 12 years old group, $t = 6.98$, $p < 0.001$. The mean score for push-ups test revealed a significant increase from pre (10.80 ± 2.39) to post (13.20 ± 2.30) for 10 years old group, $t = 4.81$, $p = 0.001$, from pre (14.86 ± 3.28) to post (16.93 ± 2.79) for 11 years old group, $t = 10.62$, $p < 0.001$, and from pre (15.45 ± 1.81) to post (17.91 ± 1.97) for 12 years old group, $t = 6.98$, $p < 0.001$. The mean score for curl-ups tests also showed a statistically significant increase from pre (16.20 ± 2.30) to post (18.20 ± 2.35) for 10 years old group, $t = 13.42$, $p < 0.001$, from pre (15.86 ± 1.70) to post (18.00 ± 1.88) for 11 years old group, $t = 15.00$, $p < 0.001$, and from pre (15.82 ± 1.66) to post (18.45 ± 1.81) for 12 years old group, $t = 17.33$, $p < 0.001$. Finally, the mean score for sit and reach

Table 2 SEGAK test based on gender for intervention and control group

Variables	Gender	Intervention group ($N = 35$)			Control group ($N = 35$)		
		Pre	Post	p	Pre	Post	p
		M (SD)	M (SD)		M (SD)	M (SD)	
3-min step test	Male	96.13 (13.01)	88.13 (12.48)	0.001*	105.82 (16.80)	109.94 (19.54)	0.11
	Female	111.32 (24.98)	103.89 (23.86)	0.001*	105.67 (21.52)	111.89 (23.88)	0.09
Push-ups	Male	13.56 (3.05)	15.56 (3.24)	0.001*	11.71 (4.77)	11.65 (4.60)	0.90
	Female	14.16 (3.47)	16.68 (2.87)	0.001*	15.89 (5.06)	14.78 (4.57)	0.022*
Curl-ups	Male	16.50 (1.90)	18.75 (2.11)	0.001*	14.18 (3.52)	15.00 (2.50)	0.14
	Female	15.47 (1.68)	17.74 (1.73)	0.001*	13.78 (4.31)	13.83 (3.31)	0.91
Sit and reach	Male	25.56 (5.07)	26.37 (6.68)	0.001*	23.79 (5.95)	24.85 (6.00)	0.001*
	Female	28.06 (5.11)	28.74 (6.79)	0.001*	27.33 (4.99)	27.72 (4.19)	0.38

* $p < 0.05$

Table 3 SEGAK test based on age groups for intervention and control group

Variables	Age	Intervention group ($N = 35$)			Control group ($N = 35$)		
		Pre	Post	*p*	Pre	Post	*p*
		M (SD)	M (SD)		M (SD)	M (SD)	
3-min step test	10	94.80 (9.98)	89.40 (9.25)	0.001*	85.90 (7.55)	83.20 (8.90)	0.07
	11	93.86 (9.54)	86.21 (8.43)	0.001*	122.57 (15.88)	123.71 (15.20)	0.59
	12	126.45 (24.22)	116.64 (25.68)	0.001*	102.36 (7.53)	119.91 (11.25)	0.001*
Push-ups	10	10.80 (2.39)	13.20 (2.30)	0.001*	17.90 (3.51)	16.80 (3.05)	0.09
	11	14.86 (3.28)	16.93 (2.79)	0.001*	11.71 (5.25)	11.79 (5.45)	0.88
	12	15.45 (1.81)	17.91 (1.97)	0.001*	12.91 (4.95)	11.91 (3.65)	0.17
Curl-ups	10	16.20 (2.30)	18.20 (2.35)	0.001*	16.40 (2.67)	13.64 (4.14)	0.68
	11	15.86 (1.70)	18.00 (1.88)	0.001*	16.10 (2.47)	14.00 (3.09)	0.54
	12	15.82 (1.66)	18.45 (1.81)	0.001*	12.18 (3.63)	13.36 (2.77)	0.07
Sit and reach	10	25.10 (5.90)	27.40 (5.56)	0.001*	28.00 (5.29)	28.10 (4.01)	0.86
	11	24.64 (6.08)	27.00 (6.00)	0.001*	22.54 (5.89)	23.57 (5.73)	0.001*
	12	28.55 (5.43)	31.18 (5.96)	0.001*	27.36 (4.15)	28.23 (4.45)	0.09

* $p < 0.05$

test revealed a significant increase from pre (25.10 ± 5.90) to post (27.40 ± 5.56) for 10 years old group, $t = 7.67$, $p < 0.001$, from pre (24.64 ± 6.08) to post (27.00 ± 6.00) for 11 years old group, $t = 11.84$, $p < 0.001$ and from pre (28.55 ± 5.43) to post (31.18 ± 5.96) for 12 years old group, $t = 12.97$, $p < 0.001$.

Whereas, for the control group, results revealed that the mean score for 3-min step test showed no significant decrease from pre (85.90 ± 7.55) to post (83.20 ± 8.90) for 10 years old group, $t = 2.05$, $p = 0.07$, and no significant increase from pre (122.57 ± 15.88) to post (123.71 ± 15.20) for 11 years old group, $t = 0.41$, $p = 0.69$. However, there was significant increase from pre (102.36 ± 7.53) to post (119.91 ± 11.25) for 12 years old group, $t = 4.08$, $p < 0.001$. The mean score for push-ups test revealed no significant decrease from pre (17.90 ± 3.51) to post (16.80 ± 3.05) for 10 years old group, $t = 1.88$, $p = 0.09$, and from pre (12.91 ± 4.95) to post (11.91 ± 3.65) for 12 years old group, $t = 1.48$, $p = 0.17$. Push-ups test also showed no significant increase from pre (11.71 ± 5.25) to post (11.79 ± 5.45) for 11 years old group, $t = 0.15$, $p = 0.88$. The mean score for curl-ups test showed no significant decrease from pre (16.40 ± 2.67) to post (16.10 ± 2.47) for 10 years old group, $t = 0.18$, $p = 0.86$, and also no significant increase from pre (13.64 ± 4.14) to post (14.00 ± 3.09) for 11 years old group, $t = 0.64$, $p = 0.54$, and from pre (12.18 ± 3.63) to post (13.36 ± 2.77) for 12 years old group, $t = 2.02$, $p = 0.07$. Lastly, the mean score for sit and reach test revealed no significant increase from pre (28.00 ± 5.29) to post (28.10 ± 4.01) for 10 years old group, $t = 0.18$, $p = 0.86$, whereas significant increase from pre (22.54 ± 5.89) to post (23.57 ± 5.73) for 11 years old group, $t = 5.48$, $p = 0.001$. Sit and reach test showed no significant different from pre (27.36 ± 4.15) to post (28.23 ± 4.45) for 12 years old group, $t = 1.87$, $p = 0.09$.

4 Discussion

The purpose of this study was to determine the effects of brain breaks on health-related fitness among indigenous primary school children in Sarawak, Malaysia. The results revealed a significant difference for SEGAK test in the intervention group as compared to the control group.

Based on Table 2, it can be seen that male students have a higher tendency of aerobic (cardiovascular) fitness level on how fast one's heart rate returns to baseline following the workout, which revealed that male students were fitter than female students, as shown by the mean score differences. The male students recorded a higher decrease of mean score difference from pre-test and post-test, which was 8, compared to female students, who recorded only a 7.43 decrease of mean score difference. This was because male students have excellent heart rate recovery with an estimated average of 78 bpm compared to the female students' estimated average was 104 bpm. A proposed explanation is a gender variation in the mechanism by which heart rate is regulated in the autonomic nervous system. According to Ethan Sellers, female blood contains 10% less haemoglobin than male blood, and the female heart is slightly smaller than the male heart, implying that men have lower heart rates than women. However, male and female heart rates do not differ significantly until adolescence [7].

Moreover, for push-ups, the students endured themselves for a minute. However, the format contrasted according to gender, whereby the male students did the standard position with the toes as the pivot point, whereas the female students underwent customised push-ups with the knees serving as the pivot point while for curl-ups, all students concentrated on abdominal endurance, which is crucial for back support and core stability. From the table, it was shown that, statistically, female students outperformed male students since their mean scores were slightly higher. Similar studies also supported the findings reported by Dubuc et al. [8] that their findings were consistent with the findings of several other studies in children and adolescents, which also found that female students outpaced male students, with a significant relationships between cardiorespiratory fitness and muscle endurance (curl-ups and push-ups) with regard to all four academic success indicators in female students because female students had significantly higher correlation coefficient values than male students for the following correlations, indicating that the relationship between cardiorespiratory fitness and academic performance is stronger in female students [8]. Even Duckworth and Seligman [9] proposed that female students surpass male students in part because they are more conscientious.

One of the components of physical fitness is flexibility, which is the ability of a joint to move freely across its range of motion (ROM). The sit and reach test is a field test that is often used to assess trunk, hip, and hamstring flexibility [10, 11]. In the sit and reach test, the current study reported that the male students have more flexibility than the female students based on their pre- and post-mean score differences. The female students only experienced a minor upright change (from 28.06 to 28.74), whereas the male students showed a drastic shift in mean score difference (25.56 to

26.37). The more active a person is, the more probable it is that they are moving their joints through their range of motion and have a higher level of flexibility. This can be linked to the lower strength of their bodies, whereby gender is one of the factors that affect flexibility and growth rate, which consequently plays an essential role too [10]. The growth spurt in females occurs between the ages of approximately 12 and 13 years, but the growth spurt in boys occurs only between the ages of about 12 and 16 years. This means that those with short legs and/or short arms have a structural disadvantage and so score lower than those with long legs and/or long arms who offer the same degree of hip flexion [11].

As for Table 3 related to age influence, it was found that students aged 12 performed the best compared to the other two group ages (10 and 11 years old) and showed highest level of cardiovascular, muscle, and flexibility endurances as significant changes in heart and vascular structure and function related to ageing [12]. This is because 12-year-olds begin to show signs of puberty, including menstruation in girls and muscular development in boys and they become increasingly skilled in sports because of their physical development compared to 10- and 11-year-olds who are just starting to become more aware of their bodies and physical competence [13, 14].

One of the main limitations of this study is that the participants were recruited from only two indigenous primary schools in Sarawak. Therefore, for future studies, it is recommended that researchers can recruit participants from the indigenous primary school from other states around the Malaysia to solidify our understanding on the application of brain breaks which had been proven to have significant effect on the health-related fitness towards physical activity.

5 Conclusion

Physical education is taught in Malaysian primary schools twice a week for 30 min, as mandated by the Ministry of Education. The results of the study showed that adding an intervention programme to regular physical education classes helped improve their fitness and health. It indicated that an intervention programme within a physical education class had a positive effect towards their fitness and health. After 30 evenly distributed sessions, the effect of brain breaks improved students' physical fitness. It was proved to develop students' physical activity attitudes and self-esteem, as evidenced by the mean scores in Tables 2 and 3. Students learned about health-related fitness, such as cardiovascular endurance, muscular strength and endurance, flexibility, and the importance of maintaining those aspects in their lives for a healthier lifestyle and health, as well as mastering new motor skills that they never thought they could perform before, thereby unleashing their physical potential. It is strongly recommended that physical educators and curriculum planners incorporate intervention programmes into regular physical education since it is deemed vital to increase the intensity level sufficient to enhance their fitness and health among Malaysian indigenous school children. Furthermore, the current study demonstrated that both boys and girls can safely participate in the brain break exercises because there was no harm as a result of the programme.

Acknowledgements The authors would like to acknowledge Ming-Kai Chin for his guidance and for providing the Brain breaks platform, under the GCH and HOPSports® Inc. This research was supported by the Ministry of Higher Education Malaysia for Fundamental Research Grant Scheme (FRGS) with Project Code: FRGS/1/2020/SKK06/USM/03/13.

References

1. Donahue SMA, Zimmerman FJ, Starr JR, Holt VL (2010) Correlates of pre-pregnancy physical inactivity: results from the pregnancy risk assessment monitoring system. Matern Child Health J 14(2):235–244. https://doi.org/10.1007/s10995-009-0441-x
2. Glapa A, Grzesiak J, Laudanska-Krzeminska I, Chin MK, Edginton CR, Mok M, Bronikowski M (2018) The impact of brain breaks classroom-based physical activities on attitudes toward physical activity in Polish school children in third to fifth grade. Int J Environ Res Public Health 15(2):368. https://doi.org/10.3390/ijerph15020368
3. Kueh YC, Kuan G, Morris T (2017) The physical activity and leisure motivation scale: a confirmatory study of the Malay language version. Int J Sport Exerc Psychol 1–16
4. Popeska B, Jovanova-Mitkovska S, Chin MK, Edginton CR, Mok MM, Gontarev S (2018) Implementation of brain breaks in the classroom and effects on attitudes toward physical activity in a Macedonian school setting. Int J Environ Res Public Health 1127
5. Van Mechelen W, Kemper HC (1995) The Amsterdam growth study: a longitudinal analysis of health, fitness and lifestyle. J Hum Kinet 135–158
6. Schoenborn CA (1986) Health habits of U.S. adults, 1985: The "Alameda 7" revisited. PubMed Central 101(6):571–580
7. Sellers E (2009) An examination of post exercise heart rate and recovery time. https://www.lycoming.edu/schemata/pdfs/BIO_Sellers_WordPDF.pdf
8. Dubuc MM, Aubertin-Leheudre M, Karelis AD (2020) Gender differences in academic performance of high school students: the relationship with cardiorespiratory fitness, muscle endurance, and test anxiety. Int J Prev Med 201
9. Duckworth AL, Seligman ME (2006) Self-discipline gives girls the edge: gender in self-discipline, grades, and achievement test scores. J Educ Psychol 98(1):198–208
10. Akınoglu B, Pakoz B, Hasanoglu A, Kocahan T (2021) Investigation of the relationship between sit-and-reach flexibility and the height, the leg length and the trunk length in adolescent athletes. Balt J Health Phys Act 13(4):29–37
11. Graber EG (2021) Physical growth and sexual maturation of adolescents. https://www.msdmanuals.com/home/children-s-health-issues/growth-and-development/physical-growth-and-sexual-maturation-of-adolescents
12. Jakovljevic DG (2018) Physical activity and cardiovascular aging: physiological and molecular insights. Elsevier Inc 67–74
13. Vanbuskirk S (2022) 10-year-old child development milestones. https://www.verywellfamily.com/10-year-old-developmental-milestones-620710
14. Appleton N (2022) 12-year-old child development milestones. https://www.verywellfamily.com/12-year-old-developmental-milestones-4171917

Development and Prospect of Isotemporal Substitution Model in Physical Activity Research: A Narrative Review

Duan Weipeng, Garry Kuan, Lou Hu, and Yee Cheng Kueh

Abstract Human health promotion is an important research field in sports science and public health. Several studies have investigated the association of physical activity, sedentary behavior, and sleep, with various health outcomes in isolation, yet the timing of different behavioral activities within a limited 24-h day is dependent on each other. The proposal of Isotemporal Substitution Model (ISM) research method finds the best time allocation method to maximize health benefits by reallocating the time use of each activity behavior. A literature search was conducted using major computerized databases and library holdings for peer-reviewed articles. The inclusion of articles followed a three-phased approach using Preferred Reporting Items for Systematic Reviews and Meta-Analyses (PRISMA) guidelines. The studies included in this review were divided into three categories. The first category uses qualitative research to sort out the concepts related to ISM and PA, the second category uses quantitative research to review the application of ISM in PA research along the time line, and the third category describes current human 24-h activity measurement tools. Finally, the controversial points of the current ISM research and suggestions for future research development are discussed. Through literature review, scholars are suggested to improve the following research in the later research: standardize the definition and application of ISM; consider human 24-h activity behavior in a more specific manner; establish an evaluation and monitoring system for ISM research in human activities; develop convenient and scientific human activities. The evaluation tool of ISM; in-depth research and discussion on the dose–response relationship of ISM.

D. Weipeng (✉) · G. Kuan
Exercise and Sports Science Programme, School of Health Sciences, Universiti Sains Malaysia, Kubang Kerian, Kelantan, Malaysia
e-mail: duanweipeng@student.usm.my

D. Weipeng
Nantong University, Nantong, China

L. Hu · Y. C. Kueh
Biostatistics and Research Methodology Unit, School of Medical Sciences, Universiti Sains Malaysia, Kubang Kerian, Kelantan, Malaysia

G. Kuan et al. (eds.), *Advancing Sports and Exercise via Innovation*, Lecture Notes in Bioengineering, https://doi.org/10.1007/978-981-19-8159-3_30

Keywords Isotemporal substitution model · Physical activity · 24-h behavioral activity

1 Introduction

The promotion of human health is an important research direction in sports science and public health. With the development of the economy, people's living standards continue to improve, and their lifestyles have undergone tremendous changes. Due to the increased pressure from life, work, and other aspects, people's lifestyles tend to be more static and less active. At the same time, the invasion of the information age is more and more time facing electronic devices such as mobile phones and computers, and the long-term use of electronic devices threatens the physical fitness and health of young people. The amount of physical activity (PA) has different effects on the physical health and emotional state of children and adolescents. More PA time classes can improve physical health and emotional state [1–3]. In the research on the health promotion of college students' physical activity, scholars have proved that there is a correlation between college students' physical activity and anxiety. College students use moderate-to-vigorous-intensity physical activity (MVPA) for 10 min or more to replace sedentary behavior (SB) which can reduce anxiety scores, the more the anxiety score decreases with the increase of replacement time. Replacing sedentary behavior with MVPA for more than 60 min in anxious college students can relieve anxiety [4]. In studies of adults and the elderly, researchers also confirmed that PA has a certain promoting effect on human health [5], and there is a dose–response across a wide range of activity and fitness levels [1]. Further, substantial data indicate that physical inactivity also contributes to disease by being a contributor along with excess caloric intake to weight gain and obesity [2]. Via Scholar Series Studies have demonstrated the importance of physical activity in promoting human health, and encouraged national health departments and other actors to jointly develop PA promotion methods, and countries to formulate corresponding physical activity guidelines. According to statistics from the World Health Organization (WHO), insufficient physical activity is one of the main risk factors for non-communicable diseases and death, and targeted interventions are urgently needed. WHO proposes the "Global Action Plan on Physical Activity" to achieve the new goal of promoting physical activity in youth by 2030. Previous studies have mostly regarded physical activity as an isolated research factor to explore the association between physical activity and health outcomes. In fact, as an independent individual, the total activity time of a person in a day is constant, and different behaviors and activities are interdependent. The increase or decrease of one behavior time will inevitably lead to a decrease or increase in the overall time of other behaviors. Focusing on the health effects of a certain behavioral activity time in isolation is one-sided [6].

The first presentation of the Isotemporal Substitution Model (ISM) paradigm elucidates a new approach to model and study the effects of temporal engagement in different and other activities on weight change. ISM research methods have made new breakthroughs in the field of physical activity, and more and more scholars have begun to study ISM. The studies included in this review were divided into three broad categories. The first category uses qualitative research to sort out the related concepts of ISM and PA, the second category uses quantitative research to review and develop the application of ISM in PA research along the time line, and the third category describes the current human 24-h activity measurement tools. Finally, the controversial points of the current ISM research and suggestions for future research development are discussed.

2 Method

2.1 Search Strategy

A literature search was conducted using major computerized databases (e.g., PubMed, ScienceDirect, Scopus, and cnki.net) and library holdings for peer-reviewed articles. Includes systematic reviews, meta-analyses, randomized controlled, and observational studies, using the following search terms: isochronous substitution, component isochronous substitution model, 24-h behavioral activity (sedentary behavior, sleep, physical activity), etc. A manual search of the reference lists in the relevant studies found in the computerized search was also performed.

2.2 Inclusion and Exclusion Criteria

The inclusion of articles followed a three-phased approach (Fig. 1) using Preferred Reporting Items for Systematic Reviews and Meta-Analyses (PRISMA) guidelines [7]. In the first phase, 1696 records were initially collected through extensive obtained from a database search. At this stage, a total of 84 copies were identified and removed. In the second stage, the titles of 1612 records were screened and records were deleted if they did not mention the following terms (or such close variants): Isotemporal Substitution Model (ISM), Physical Activity. This process resulted in the deletion of 1303 records. In the third phase, the full text of the final 209 records was reviewed. Records that met the inclusion criteria were: (a) using ISM in research, (b) researching human activity behavior and health; (c) written in English or Chinese. A total of 64 studies met the inclusion criteria and were included in the review.

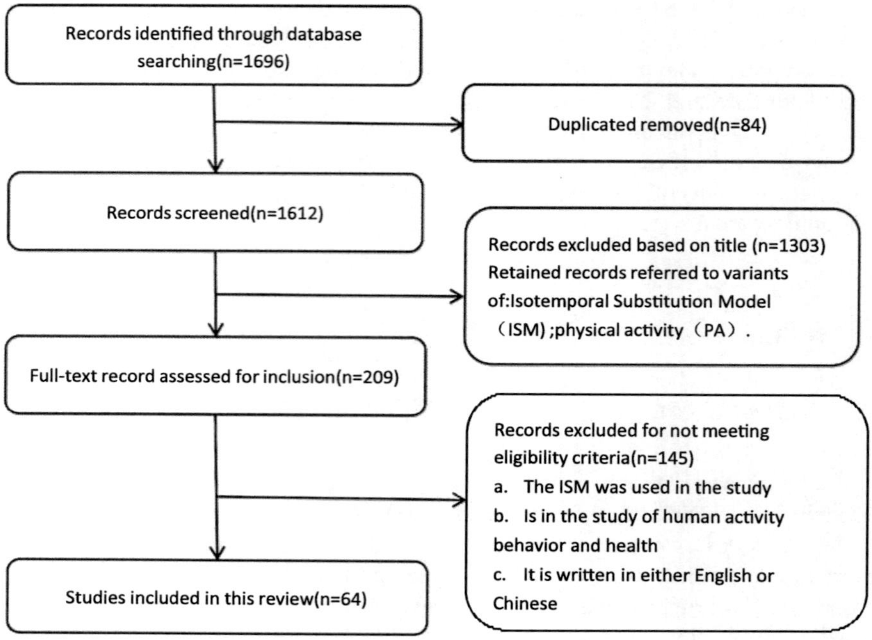

Fig. 1 Process flow diagram from identification to inclusion of studies

2.3 *Categorization of Studies*

The studies included in this review were divided into three broad categories. The first category uses qualitative research to sort out the related concepts of ISM and PA, the second category uses quantitative research to review and develop the application of ISM in PA research along the time line, and the third category describes the current human 24-h activity measurement tools. Finally, the controversial points of the current ISM research and suggestions for future research development are discussed.

3 Conceptualization of ISM in PA Research

Through qualitative research, the concepts and keywords encountered in ISM and PA related literature were sorted out and sorted out (Fig. 2). Interpret the sources, evaluation criteria, and common research directions of related vocabulary and concepts.

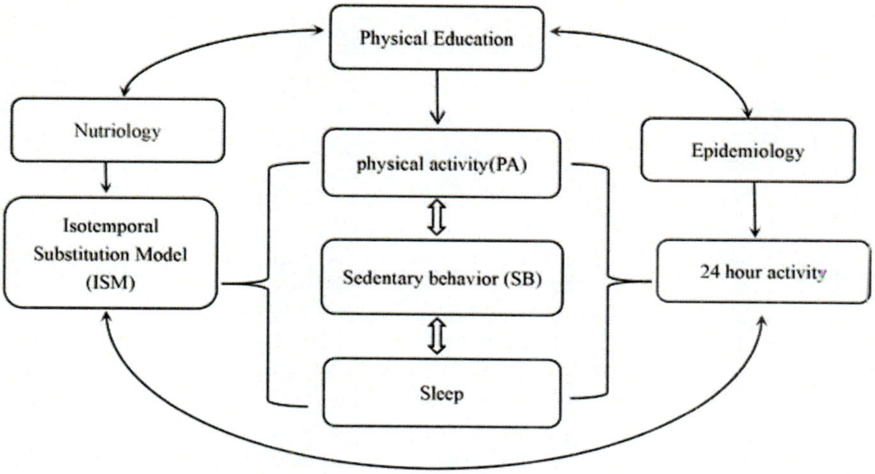

Fig. 2 Keyword combing in ISM and PA research

3.1 24-h Activity Behavior

Scholars have not given a clear definition of 24-h activity as a concept. As people pay more attention to health, a large number of studies have shown the relationship between physical activity and physical health, and countries have developed corresponding physical activity guidelines. In 2016, 24-h activity guidelines for Canadian children and adolescents: Integrating physical activity, sedentary behavior, and sleep, 24-h activity behavior was first proposed in physical education [8]. It is applied as a fixed time as a criterion for evaluating the amount of exercise. New Zealand's public health authority then followed Canada's lead and issued recommendations for physical activity, sedentary behavior, and sleep for children and adults with 24-h activity behavior. Relevant departments around the world have successively issued physical activity guidelines for the purpose of improving physical activity and promoting national health, but the amount of physical activity recommended by these guidelines can only ensure that individuals achieve basic health status. Physical Activity Guidelines for People" and Centers for Disease Control and Prevention (CDC) "Interim Report on Physical Activity Guidelines in the United States: Strategies to Increase Physical Activity in Youth", etc. Scholars summarize the evolution of the PA guideline as (Fig. 3) (Troiano et al. 2020).

3.2 ISM

The ISM concept originally originated in nutritional epidemiology, studying the total energy intake in free-living human populations and PA, body size differences and

Date	Example organisation(s)or countries	Focus	Targets	Selected features
1970s	American College of Sports Medicine(ACSM), American Heart Association 1-2	Increase fitness via exercise,minimise risk of adverse events	20min,3+times/week	Balance of endurance and muscle strength
Mid-1990s	Centres for Disease Control and Prevention (CDC)/ACSM, US Surgeon General58 59	Accumulate moderate-intensity PA to reduce non-communicable diseases (NCDs)	30min of moderate-intensity aerobic most days of week	Minimal focus on muscle strengthening
Early 2000s US	Department of Health and Human Services (HHS), WHO, Canada, Australia, other high-income countries (HICs)5 16 17 60	Accumulate moderate-intensity PA to reduce NCDs and improve quality of life (QoL)	150–300min/week moderate-intensity or equivalent aerobic, muscle strengthening 2+times/week	Increased focus on progress below target levels. 'Some is better than none'
2018–2020	US HHS, WHO, other HICs11 21	Accumulate moderate-intensity PA and reduce sedentary behaviour to reduce NCDs and improve QoL	150–300min/week moderate-intensity or equivalent aerobic, muscle strengthening 2+times/week	Increased emphasis on reducing sedentary behaviour; remove bout criterion

Fig. 3 Milestones in evolution of physical activity guideline (Troiano et al. 2020)

other factors, so if any of these factors are related to disease risk, then the total energy intake may confound associations with specific nutrients. To address such issues, the experimental design of controlled experiments evaluating the effects of a particular nutrient in nutrition usually involves substituting one nutrient for other energy sources, thereby keeping the total dietary energy intake equal across groups [9]. In the assessment of nutrient intake, the measurement error of nutrient intake is closely related to the measurement error of total energy intake. Therefore, controlling for changes in total energy intake will reduce measurement errors for specific nutrients to some extent. Adjusting total energy intake is not only more convenient but also controls external changes caused by these factors [10]. Drawing on the above models, scholars created an isochronous substitution model and introduced ISM into the field of physical activity health promotion for the first time [11].

3.3 PA

PA refers to any bodily activity that results in an increase in energy expenditure due to skeletal muscle contraction [12], including activities during work, housework, travel, and exercise-like physical activity during leisure time. Physical activity mainly includes dimensions such as frequency, duration, intensity, and activity type. According to different classification dimensions, there are different types of physical activity. Frequency refers to the number of times of participating in physical activity, such as "several times a week and several sessions per week",duration refers to the time to participate in physical activity, expressed in "minutes"; according to the intensity, it can be divided into light-intensity physical activity (LPA)) (1.5–3 METs), moderate-intensity physical activity (MPA) (3–6 METs), vigorous-intensity physical activity (VPA) (>6 METs) three types of physical activity with different intensity use energy metabolism (METs) to express its intensity. According to the type of activity, it is divided into occupation, family, transportation, housework, and leisure [13]. There is a strong relationship between physical activity and health outcomes, but the dose–response relationship between physical activity and health outcomes is still understudied at this stage.

3.4 SB

Sedentary behavior (SB) is derived from the Latin "sedere". There are two versions that are most commonly used in the definitions available. The main difference is the inclusion of the pose part. The first definition method is sedentary behavior is defined by intensity, which is commonly used in epidemiological studies to refer to any behavior or activity that consumes less than or equal to 1.5 METs of energy in the waking state [14]. The second definition adds posture to the first definition and defines it as any behavior that consumes less than or equal to 1.5 METs of energy while an individual is sitting, reclining, or leaning on the back of a chair while awake [15]. Compared with the first definition, the second definition of sedentary behavior is more complete and refined, including position and posture. With the in-depth study of physical activity by scholars, the academic community has gradually incorporated a series of physical activity-related behaviors such as LPA and SB into the field of physical activity health promotion research, and pointed out that physical activity deficiency (lack of MVPA) and SB (low energy consumption) should be clearly distinguished [16]. The Sedentary Behavior Research Network (SBRN) has published a public statement on clearly distinguishing the difference between "SB" and "insufficient physical activity" and defining them precisely [17]. Tremblay et al. [8] jointly completed a series of work on the development and specification of terms related to sedentary behavior research, including physical inactivity, immobile behavior, sedentary behavior, standing, screen time, non-screen-based sedentary time, sitting, lying, sedentary behavior patterns, etc. [8].

In the research on the promotion of human health, scholars have mostly studied the relationship between PA and health. SB has been ignored by scholars. Therefore, scholars can study the synergistic effect of PA and SB on human health from the perspective of time allocation [18–20].

3.5 Sleep

Sleep is an important physiological phenomenon of human beings. It is manifested as the disappearance of various conscious active behaviors for a certain period of time, and the ability to respond to external environmental stimuli is weakened. Problems such as insufficient sleep duration, sleep disorders, and poor sleep quality are considered to be related to human cardiovascular health. There are strong links between physical and mental health components such as metabolic health, mental health, and immune health [21–24]. At present, there is a common problem of insufficient sleep duration in the global public [21], but excessive sleep is also harmful to health [22, 23]. As well as people's habit of using electronic products before going to bed, there is a certain impact on sleep quality which will ultimately affect human health [22].

4 ISM Development and Application Research

Humans are constant 24 h a day, and the effects of variable substitution on the human body can be observed by redistributing the time occupied by various activities in the day. In the original ISM, it was the first attempt to use time variable substitution as the main research method in a series of public health-related studies such as physical activity and sedentary behavior. However, it only mentions the relationship between PA and SB on physical health, the overall health effect of assigning different specific times is not clear, and the model does not include independent variables such as sleep. At the same time, the problems of pseudo-correlation and multicollinearity will also occur in the specific data analysis process [11].

Subsequently, scholars further studied and pointed out that people's 24-h activity behavior includes sleep, SB, light-intensity physical activity (LPA), and moderate-to-vigorous-intensity physical activity (MVPA), and the sum of the time of behavioral variables is 24 h (one Fixed constants), any increase or decrease in one variable will cause corresponding changes in other variables. Therefore, statistical methods (multiple linear regression, including the ISM proposed by Mekary et al.), designed for vector-type data (continuous data) in real space, may not be suitable for temporal data [25]. According to this, the activity balance model (AB model) (Fig. 4) is proposed, in which sleep time, SB, standing, LPA, and MVPA are used as independent variables, and selected health outcomes are used as dependent variables. Potential confounders were adjusted accordingly for different groups. From another point of view, in fact time data should be regarded as compositional data and should be

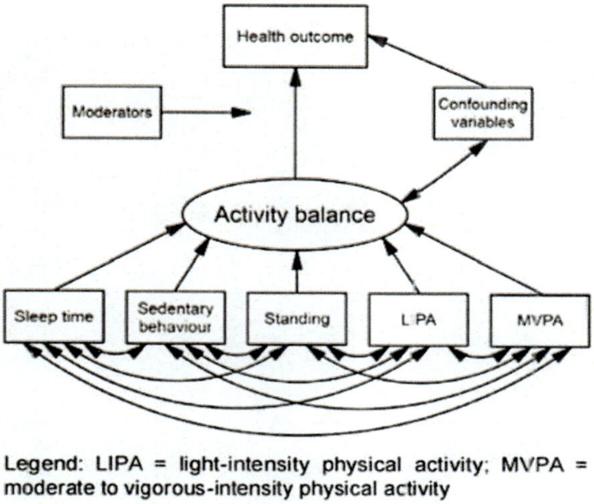

Fig. 4 Activity balance model (AB model)—a new theoretical framework for epidemiological research [25]

analyzed using isochronous substitution logic under the framework of compositional data, which can avoid multicollinearity and analyze the synthesis of all variables as a whole Influence [26].

The AB model is a comprehensive analysis of all behaviors in order to overcome the insufficient adjustment of multiple independent variables. The total time of all independent variables is 24 h. Therefore, all independent variables are regarded as component data, and the corresponding component data analysis methods are used for statistical analysis. Exploring the combined effects of independent variables such as sleep, sedentary behavior, and physical activity time on health outcomes further provides a corresponding theoretical basis.

Based on his proposed AB model, Pedišić [27] innovatively proposed a new concept of time-use epidemiology with reference to epidemiological research ideas, while expounding and describing time-use epidemiology and other epidemiological research topics relationship (Fig. 5) [27].

He defines time use epidemiology as the study of the determinants, incidence, distribution, and impact of health-related time use patterns in a population, as well as a strategy for preventing unhealthy time use patterns and achieving optimal time allocation strategies for individual health. research ideas. At the same time, he also proposed a comprehensive research framework based on the idea of time use epidemiology (framework for viable integrative research in time use epidemiology, referred to as "VIRTUE framework", Fig. 6), which incorporates social ecology key elements of the methodology, theoretical framework for behavioral epidemiology, and activity balance models [23, 25].

The VIRTUE framework starts from the perspective of finding the best balance of time use patterns and conducts comprehensive research on 5 aspects of time use determinants, time use methodologies, time use components, the relationship between time use and health outcomes, and time use interventions. Focusing on

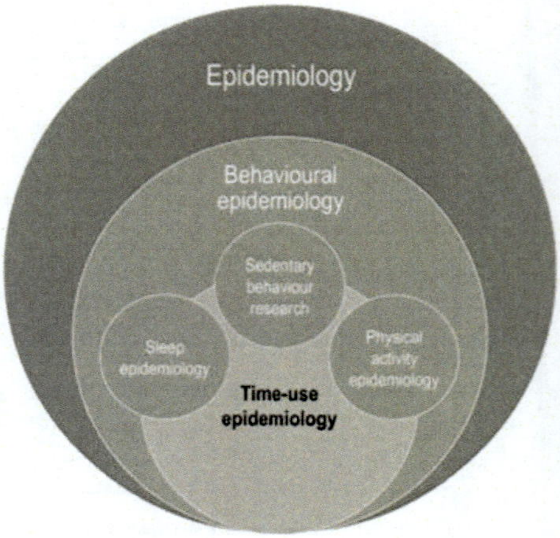

Fig. 5 Relationship between time use epidemiology and previously established epidemiological disciplines [25]

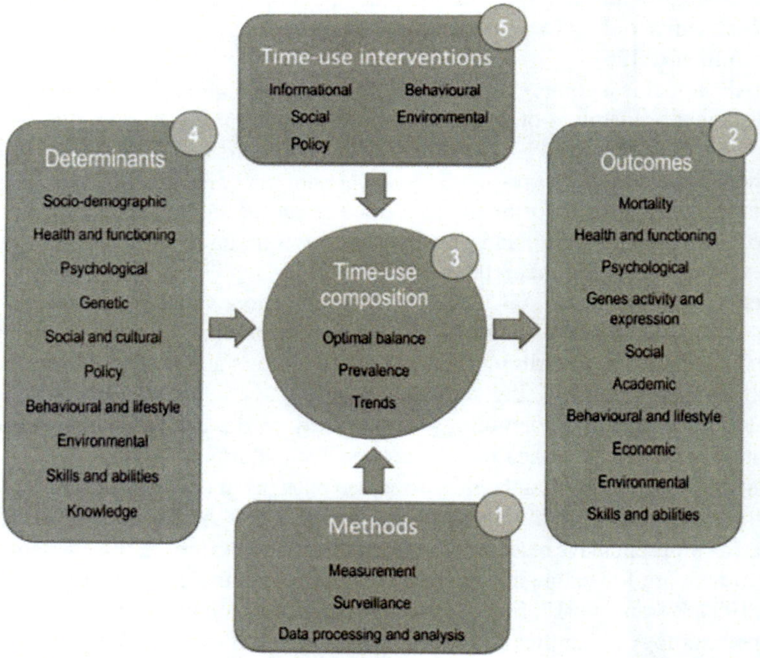

Fig. 6 Framework for viable integrative research in time use epidemiology (VIRTUE framework) [27]

individualized habit patterns of individual time use to explore the impact of time use on health.

Chastin et al. [28] used a combination method based on the Isometric Logratio (ILR) data transformation and change prediction matrix to initially process the time data, and then further test it using standard regression statistical methods, and finally convert the results back to the original unit (time unit). According to this statistical analysis idea, the time spent in all four behaviors (sleep, SB, LPA, MVPA) in a 24-h day was used as an independent variable, and backward elimination was used to add confounders as covariates to the regression model ($P < 0.2$, confounding factors were retained), and finally the linear relationship between various independent variables and health outcomes was analyzed [29].

Based on the above study, Mekary et al. [11] pointed out that component analysis is only a derivative of ISM, and there are already mathematically similar models in nutritional epidemiology. Dr. Mekary and his team believe that compositional data analysis could lead to potentially different individuals' potentially, highly inaccurate interpretations of their different free activity times, and is therefore not suitable for use in the epidemiology of physical activity.

To sum up, the advantages of ISM are: 1. It can more specifically quantify the relationship between the replacement of different physical activity behaviors and health outcomes; 2. It can enable individuals to more intuitively understand the relationship between time allocation patterns of different physical activity behaviors and health outcomes. Specific links between different health outcomes. The isochronous substitution method pioneered the substitution of different types of physical activity behaviors, which fully laid the methodological foundation for the proposal of time use epidemiology [26]. However, it should be noted that: 1. In the research on the development of ISM, scholars have not unified the concept, use method and analysis method of the existing ISM. Therefore, it is necessary to further consider the scientificity and feasibility of various alternative methods in the future research. 2. The effect of sleep behavior on health is different from that of PA and SB behavior, so it should be carefully considered when replacing it as a time component [26], 3 Human 24 h activities are complex, and more specific refinement should be ensured when SB, sleeping, and PA are considered as constituent elements to ensure scientificity.

5 Human 24-h Activity Measurement Tool

At present, human activity measurement tools are mainly divided into two categories: subjective measurement and objective measurement [30]. Scholars summarize the characteristics of all measurement tools into tables (see appendix for details) [31].

Subjective measures mainly used physical activity questionnaires, scales, or physical activity diaries. The use of questionnaire assessment can reduce the measurement cost and is convenient and quick. Questionnaire methods for assessing physical activity have remained the method of choice for epidemiological studies in later periods. Importantly, research is needed to continuously refine the questionnaire to provide more accurate physical activity information in order to test specific hypotheses about the relationship between physical activity and health [32].

Commonly used in objective measurements are the doubly labeled water (DLW) technique, indirect calorimetry, heart rate monitoring, pedometers, and accelerometers. The choice of measurement methods for human activities is affected by multiple factors, such as the number of people to be measured, the time and content of monitoring, and the cost of inputs [31]. The double-labeled water method was once considered the "gold standard" for physical activity measurement but was prohibitively expensive. The measurement data of accelerometers is relatively accurate, but if more relevant research is to be carried out from the perspective of time recording and analysis, it is necessary to make further breakthroughs in the precise measurement of sedentary behavior time and sleep time. At present, the optimal method of sedentary time measurement and assessment is controversial, and the academic community has not reached a consensus on this. At present, objective measurement instruments combined with questionnaires are commonly used, and more research is needed to examine the possibility of improving measurement accuracy by combining two or more techniques. At present, SB and sleep in different states cannot be accurately assessed and classified, and scholars are still controversial about the critical value of SB for different populations [30]. More scientific instruments and more research are needed to examine the possibility of improving measurement accuracy by combining two or more techniques.

6 Future Research Prospect of ISM in PA Application

Based on the above summary of the literature, the following characteristics can be seen:

1. At present, scholars' definitions and research methods of ISM are not unified, and there are controversial points in scholars' research;
2. The 24-h activity of the human body is divided into three sub-items: SB, PA, and sleep.

However, further research will continue to refine the sub-item classification. At the same time, studies have proved that different states of each sub-item have different effects on people. Therefore, it is necessary to further study ISM through the idea of time replacement, and to consider the mutual influence of human activities and behaviors collaboratively;

3. Most of the current research studies on PA and human health, lack of ISM analysis of other behaviors such as SB and sleep on health;
4. ISM is a virtual alternative method. Most of the current researches are cross-sectional studies and some experiments on human health using the ISM method. However, the methods of use are inconsistent and controversial. Therefore, scholars are required to accurately determine the dose–effect relationship. Deeper research on sex;
5. At present, the measurement of human activity is divided into two categories: supervisory measurement and observable measurement. However, each measurement method has certain advantages and disadvantages. Therefore, scholars need to conduct in-depth research and develop more efficient, cheap and scientific measurement tools.

Through literature review, scholars are advised to improve the following research in the later research: standardize the definition and application of ISM; consider human 24-h activity behavior in a more specific manner; establish an evaluation and monitoring system for ISM research in human activities; and develop convenient and scientific human activities.

7 Appendix

Ainslie et al. [31] (Table 1).

Table 1 Summary of key techniques

Technique	Typical no. of individuals per test	Duration of use	Cost ($US; 2002 values)[a]	Advantages	Limitations	Available models manufacturer information	References
Direct calorimetry	1	1–7 d	NA	Direct and precise measure of EE	Non-free living; only one individual can be monitored at one time; large expense of measurements	No manufacturers; normally made by specialist engineers	10, 74–77
DLW	1	1–3 wk	1000–1500	Applicable to a range of individuals and field conditions, EE is measured over long periods, is safe and does not interfere with normal physiological conditions	High costs of ^{18}O-limits large group application; requires sophisticated equipment for analysis, error introduced if FQ is not known, no information can be gained about brief or specific periods of activity	Isotec™ (USA); Marshall Isotopes™ (Israel); Cambridge Isotopes™ (Adover, MA, USA)	8, 9, 11–13,16

(continued)

Table 1 (continued)

Technique	Typical no. of individuals per test	Duration of use	Cost ($US; 2002 values)[a]	Advantages	Limitations	Available models manufacturer information	References
Indirect calorimetry	1	<9 h	20,000–60,000	Accurate in the measurement of EE and fuel utilization during rest and steady-state exercise	Cannot assess 'free living' EE; expense of systems	Oxycon Pro™ (Jaegar, Germany); Oxycon Alpha™ (Jaegar, Germany); ParvoMedics™ (Sandy, UT, USA); Pulmolab EX670™ (Morgan Medal, Kent, UK)	2, 4, 78
Indirect calorimetry-portable systems	1	< 9 h	20 000–60 000	Individual assessment of EE during a range of activities; reusable	Cost and small group usage; invalid estimation of EE during non-steady-state activities	Oxytog™ (Morgan Mecial, Kent, UK); Metamax™ (Cortex Biophsik GmbH, Germany); K4 b2™ (Cosmed, Rome, Italy)	20–23, 79

(continued)

Table 1 (continued)

Technique	Typical no. of individuals per test	Duration of use	Cost ($US; 2002 values)[a]	Advantages	Limitations	Available models manufacturer information	References
Heart-rate monitoring	1	1–3 wk	200–600	Provides information on the amount of time spent in high-intensity activity; cheap and reusable	Affected by factors other than physical activity: large potential error in estimating EE	Polar Electro™ (Oy, Kempele, Finland)	25–27, 32
Questionnaires activity and dietary recall	1	Unlimited; 1–2 wk	Cost of paper	Low cost; possible to study large individual cohort	Poor individual compliance and recoding errors; EE estimates are based predominantly on males, limited number validated against DLW		33,34,36.38, 41, 42, 51, 52, 54, 80–82
Motion sensors	1	1–2 wk	200–400	Excellent means to evaluate interventions aimed at increasing physical activity; low cost and use in a large population range	Uniaxial sensors: not sufficiently sensitive to quantify EE; triaxial; measurement of acceleration in the vertical, horizontal, and mediolateral planes		

(continued)

Table 1 (continued)

Technique	Typical no. of individuals per test	Duration of use	Cost ($US; 2002 values)[a]	Advantages	Limitations	Available models manufacturer information	References
Provides a more meaningful measurement of EE but still limited in precision	Yamax™ (New Lifestyles Inc., MO, USA); Digiwalker Caltrac™ (Torence, CA, USA); Tritrac™ (Hemokinetrics, Inc., Wl, USA); Tracmor™ (Maastricht, The Netherlands)	45,53,61–71, 83–87					
Combined systems	NA	NA	NA	Verify that elevations in heart rate are representative of responses to physical activity	Lack of validation studies; no commercial systems available on the market		88

[a] When describing cost estimates, it is important to consider the consumable and investment cost of each device. For example, the cost for a stationary or portable indirect calorimetry system may be up to $US 60,000 (2002 values): however, if maintained correctly, one would expect around 1000–2000 tests of 1–9 h over a 10-year period. As shown, the cost for the DLW dose and analysis for one individual will be approximately $US 1000–1250. Therefore, whilst the consumable cost of an indirect calorimetry set-up may be far higher than the DLW, it represents a reusable investment over time. The similar concepts hold for both heart rate monitors and motions sensors, where it is possible to do many measurements after one monitor has been purchased

References

1. Physical Activity Guidelines Advisory Committee. Report of the Physical Activity Guidelines Advisory Committee, 2008, DHHS website. http://health.gov/paguidelinesN
2. Giovannucci E, Ascherio A, Rimm EB, Colitz GA, Stampfer MJ, Willis WC (1995) Physical activity, obesity and risk of colon cancer and adenoma in men. Ann Inter Med 122:327–334
3. Chen H (2021) The isochronous substitution relationship between physical activity, sedentary behavior and body mass index in children and adolescents (Master's thesis, Shanghai Institute of Physical Education). https://kns-cnki-net-443.webvpn.caztc.edu.cn/KCMS/detail/detail.aspx?dbname=CMFD202102&filename=1021654883.nh
4. Tan J, Huang B, Huang Z, Xu H, Liu Q, Chen Q, Gao Y (2021) A component isochronous substitution model study on the effects of daily behavioral activities on anxiety of college students. J Zhengzhou Univ (Medical Edition) 2:170–175. https://doi.org/10.13705/j.issn.1671-6825.2020.09.097
5. WuZ, Wang Z, Zheng H (2022) The relationship between sedentary behavior, physical activity, body composition and grip strength in the elderly based on isochronous substitution. J Beijing Sport Univ 3:94–103. https://doi.org/10.19582/j.cnki.11-3785/g8.2022.03.010
6. Huang B, Tan J, Liu Q, Zhang D, Xu H, Huang Z, Gao Y (2022) Comparative and empirical research on the application of component and non-component isochronous surrogate models in the study of health effects of physical activity. Sports Sci 2:51–58+73. https://doi.org/10.16469/j.css.202202006
7. Moher D, Shamseer L, Clarke M, Ghersi D, Liberati A, Petticrew M, Stewart LA (2015) Preferred reporting items for systematic review and meta analysis protocols (PRISMA-P) 2015 statement. Syst Rev 4(1):1. https://doi.org/10.1186/2046-4053-4-1
8. Tremblay MS, Aubert S, Barnes JD, Saunders TJ, Carson V, Latimer-Cheung AE, Chinapaw MJ (2017) Sedentary behavior research network (SBRN)—terminology consensus project process and outcome. Int J Behav Nutr Phys Activ 14(1):1–17
9. Ravussin E, Lillioja S, Anderson TE, Christin L, Bogardus C (1986) Determinants of 24-hour energy expenditure in man. Methods and results using a respiratory chamber. J Clin Invest 78(6):1568–1578
10. Willett WC, Howe GR, Kushi LH (1997) Adjustment for total energy intake in epidemiologic studies. Am J Clin Nutr 65(4):1220S-1228S
11. Mekary RA, Willett WC, Hu FB, Ding EL (2009) Isotemporal substitution paradigm for physical activity epidemiology and weight change. Am J Epidemiol 170(4):519–527. https://doi.org/10.1093/aje/kwp163
12. Caspersen CJ, Powell KE, Christenson GM (1985) Physical activity, exercise, and physical fitness: definitions and distinctions for health-related research. Public Health Rep 100(2):126
13. World Health Organization (2017) What is moderate-intensity and vigorous-intensity physical activity?
14. Pate RR, O'neill JR, Lobelo F (2008) The evolving definition of "sedentary". Exerc Sport Sci Rev 36(4):173–178
15. Barnes J, Behrens TK, Benden ME, Biddle S, Bond D, Brassard P, Woodruff S, et al (2013) Standardized use of the terms" sedentary " and" sedentary behaviours". Ment Health Phys Activ 6(1):55–56
16. Marshall SJ, Biddle SJ, Sallis JF, McKenzie TL, Conway TL (2002) Clustering of sedentary behaviors and physical activity among youth: a cross-national study. Pediatr Exerc Sci 14(4):401–417
17. Bames J, Behrens TK, Benden ME, Biddle S, Bond D, Brassard P, Woodruff S (2012) Standardized use of the terms" sedentary " and" sedentary behaviours". Appl Physiol Nutr Metabolism-Physiologie Appliquee Nutrition Et Metabolisme 37:540–542
18. Tremblay MS, Colley RC, Saunders TJ, Healy GN, Owen N (2010) Physiological and health implications of a sedentary lifestyle. Appl Physiol Nutr Metab 35(6):725–740
19. Van der Ploeg HP, Chey T, Korda RJ, Banks E, Bauman A (2012) Sitting time and all-cause mortality risk in 222 497 Australian adults. Arch Intern Med 172(6):494–500

20. Biddle SJ, García E, Wiesner G (2017) Sedentary behaviour and adiposity in youth: a systematic review of reviews and analysis of causality. Int J Behav Nutr Phys Act 14(1):1–21
21. Pandi-Perumal SR, Verster JC, Kayumov L, Lowe AD, Santana MGD, Pires MLN, Mello MTD (2006) Sleep disorders, sleepiness and traffic safety: a public health menace. Brazilian J Med Biol Res 39(7):863–871
22. Gradisar M, Wolfson AR, Harvey AG, Hale L, Rosenberg R, Czeisler CA (2013) The sleep and technology use of Americans: findings from the National Sleep Foundation's 2011 Sleep in America poll. J Clin Sleep Med 9(12):1291–1299
23. Yin J, Jin X, Shan Z, Li S, Huang H, Li P, Liu L, et al (2017) Relationship of sleep duration with all-cause mortality and cardiovascular events: a systematic review and dose-response meta-analysis of prospective cohort studies. J Am Heart Assoc 6(9):e005947
24. Dawson D, Reid K (1997) Fatigue, alcohol and performance impairment. Nature 388(6639):235–235
25. Pedišić Ž (2014) Measurement issues and poor adjustments for physical activity and sleep undermine sedentary behaviour research—the focus should shift to the balance between sleep, sedentary behaviour, standing and activity. Kinesiology 46(1):135–146
26. Song J (2020) Exploration of the relationship between physical activity behavior time distribution and overweight/obesity (Master's thesis, Beijing Sport University). https://kns-cnki-net-443.webvpn.caztc.edu.cn/KCMS/detail/detail.aspx?dbname=CMFD202201&filename=1020027474.nh
27. Pedišić Ž, Dumuid D, S Olds T (2017) Integrating sleep, sedentary behaviour, and physical activity research in the emerging field of time-use epidemiology: definitions, concepts, statistical methods, theoretical framework, and future directions. Kinesiology 49(2):252–269
28. Chastin SF, Palarea-Albaladejo J, Dontje ML, Skelton DA (2015) Combined effects of time spent in physical activity, sedentary behaviors and sleep on obesity and cardio-metabolic health markers: a novel compositional data analysis approach. PLoS ONE 10(10):e0139984
29. Dumuid D, Stanford TE, Martin- J-A, Pedišić Ž, Maher CA, Lewis LK, Hron K, Katzmarzyk PT, Chaput J-P, Fogelholm M, Hu G, Lambert EV, Maia J, Sarmiento OL, Standage M, Barreira TV, Broyles ST, Tudor-Locke C, Tremblay MS, Olds T (2018) Compositional data analysis for physical activity, sedentary time and sleep research. Stat Methods Med Res 27(12):3726–3738. https://doi.org/10.1177/0962280217710835
30. Song J, Li H, Wang Z (2020) The application of time use epidemiology in the field of physical activity research. Sports Sci 40(1):10
31. Ainslie PN, Reilly T, Westerterp KR (2003) Estimating human energy expenditure. Sports Med 33(9):683–698
32. Washburn RA, Montoye HJ (1986) The assessment of physical activity by questionnaire. Am J Epidemiol 123(4):563–575
33. Centers for Disease Control and Prevention. Recommendations & Guidelines [EB/OL]. https://www.cdc.gov/physicalactivity/resources/recommendations.html
34. Aadahl M, Jørgensen T (2003) Validation of a new self-report instrument for measuring physical activity. Med Sci Sports Exerc 35(7):1196–1202

Validating the Chinese Version of the Social Support for Exercise Scale Among Chinese High School Students: An Exploratory Factor Analysis

Liying Yao, Garry Kuan, Ke Zhou, and Yee Cheng Kueh

Abstract While the health benefits of physical activity (PA) are well established, low levels of PA among adolescents are becoming a growing concern. The purpose of the present study is to validate the Chinese version of the social support scale for exercise (SE) using an exploratory factor analysis (EFA) among Chinese high school students. A cross-sectional study was conducted on a high school students in central China. By using the random cluster sampling method, students were recruited and given their written informed consent forms. Standard procedure of forward–backward translation method was performed to translate the English version of the SE into the Chinese version. Data were analyzed using SPSS 25. A total of 322 students (boys = 147, girls = 175) with a mean age of 17 years (SD = 1.2) participated in the study. Most of the students were engaged in regular physical activity for at least four exercise sessions per week ($M = 4.4$ sessions), with an average duration 45 min per session. EFA results suggest that the translated SE scale may best be represented through a structure that consists of two factors identified as family support and friend support. The varimax rotation was proceeded by maintaining two factors with a total variance contribution of 73.6%. For family support, items were found to have factor loading ranging from 0.561 to 0.806. While the friend support was ranging from 0.624 to 0.816. Cronbach's alpha for family support and friend support was 0.96 and 0.97, respectively, which indicated acceptable internal consistency reliability. Hence, the Chinese version of SE displayed a good model fit with good scale reliability and is

L. Yao · G. Kuan (✉)
Exercise and Sports Science Programme, School of Health Sciences, Universiti Sains Malaysia, Kubang Kerian, Kelantan, Malaysia
e-mail: garry@usm.my

L. Yao
School of Physical Education, Shangrao Normal University, Jiangxi Shangrao, China

K. Zhou
Sports Reform and Development Research Center of Henan University, School of Physical Education, Henan University, Kaifeng, China

Y. C. Kueh
Biostatistics and Research Methodology Unit, School of Medical Sciences, Universiti Sains Malaysia, Kubang Kerian, Kelantan, Malaysia

G. Kuan et al. (eds.), *Advancing Sports and Exercise via Innovation*, Lecture Notes in Bioengineering, https://doi.org/10.1007/978-981-19-8159-3_31

suitable for examining the level of social support for physical activity among Chinese high school students.

Keywords Exploratory factor analysis · Social support for exercise scale · High school students · Chinese version

1 Introduction

While the health benefits of physical activity (PA) are well established, low levels of PA among adolescents are becoming a growing concern [1]. According to WHO [2], nearly 81% of children and adolescents worldwide do not meet the recommended levels of physical activity, and about one in five children and adolescents (5–19 years old) are overweight or obese, resulting in a steady increase in adolescent health problems [3]. Therefore, strategies to effectively promote regular physical activity among adolescents are urgently needed.

Studies have suggested that improving exercise behavior requires identifying and understanding the influencing factors associated with physical activity [4]. Thus, many behavior change theories, including the Social-Ecological Model [5], Social Cognitive Theory [6], the Health Belief Model [7], and the Theory of Planned Behavior [8], emphasize the importance of social factors, such as social support (SS) and social connectedness in maintaining and/or initiating behavior change.

Social support was considered as an important mediator in changing health habits [9]. It has been demonstrated that exercisers with family support are more likely than others to continue their exercise programs [10], and social support from close friends may also be the leading intermediary of self-management among adolescents [11, 12]. Moreover, family and close friends represent a significant resource of emotional support [13, 14]. The social support for exercise (SE) scale was first proposed by Sallis et al. [15] and proved to be a valid and reliable instrument for 171 college students with a mean age of 21.3 years (SD = 6.5). The Sallis' original scale asked participants to assess the frequency of social support provided by family and friends over the past three months on a five-point scale, with the frequency of social support ranging from 1 (never) to 5 (often). The family was defined as "family members," and friends were defined as "friends, acquaintances, or coworkers." The dimensions of the family support scale were different from those of the friend support scale. The family support subscale was divided into two dimensions. The first one has named participation (retest reliability 0.77, alpha coefficient 0.91) and consists of 12 items. The second one is named reward and punishment (retest reliability 0.55, coefficient alpha 0.91) and consists of 3 items. The friends support subscale has only one dimension, exercise together, and includes five items (retest reliability 0.79, coefficient alpha 0.84). The strengths of the original scale were: (1) it is the most commonly used scale in the world. (2) The dimensions of the family and friend support sections are different and more relevant to the actual measurement needs. For example, in the family support section, it is mentioned that "taking care of my household chores gives me more time to exercise", while in the friend support section, there is no such item. The scale was then translated into Korean and confirmed as a valid and reliable measurement tool for assessing participants' perceptions of social support for exercise [16]. The validity and reliability of these translated versions have been supported by other researchers [4, 17]. However, there is no widely validated Chinese version of the scale available.

Given that, effective tools are critical for measuring factors associated with physical activity participation, the present study was designed to test the reliability and validity of the Chinese version of the SE questionnaire among high school students using an exploratory factor analysis approach.

2 Methods

2.1 Participants and Procedure

A total of 322 high school students from Shangrao City, central China, were recruited for this study. Participants were aged from 15 to 19 years old (Mage = 17 yrs old, SD = 1.2), and the sample was comprised of 175 girls (55%) and 145 boys (45%).

This study was a cross-sectional design with random cluster sampling. Ethics were approved by the Human Research Ethics Committee of Universiti Sains Malaysia and followed the Declaration of Helsinki and its subsequent amendments. Students were given written informed consent forms after acknowledging they understood the purpose of the study.

2.2 Questionnaire Translation

Standard forward–backward translation was performed to translate the English version of the SE into the Chinese version. Firstly, two bilingual scholars translated the original English SE questionnaire into Chinese and then back-translated it by two independent authorized translators from a professional translation company. Secondly, five researchers from physical education, sport psychology, exercise science, and health psychology evaluated the English translations and back-translations for meaning, accuracy, wording, and grammar. Furthermore, ten native Chinese students from different backgrounds were invited to test the draft. They were asked for feedback on language clarity and comprehension, survey functioning, time burden, and appropriateness of survey length. Based on their feedback, minor modifications were made to problematic items. After that, ten more native Chinese students were invited, and they did not raise any new issues. Therefore, the final version was finalized.

2.3 Measures

Two measures were utilized for this study, the social support for exercise scale and the demographics survey.

Social support for exercise scale. Social support is considered a valid factor in the incidence of the behavior. As an effective instrument for measuring social support, the original social support for exercise scale was proposed by Sallis et al., it consists of 20 items, including the family support scale (two factors, 15 items) and friends support scale (one factor, five items). Then, these scales were translated and revised by Korean scholars to consist of 24 items (12 reflecting family support and 12 reflecting friend support) [16]. In the present study, the Chinese version of the social support scale for exercise was based on the revised version by Yang et al. [16].

Demographics and sports activities information. The demographic survey used in this study included information about the participants' gender, age, grade, height, weight, and sports participation.

2.4 Analysis

All statistical analyses were conducted using SPSS version 25. To investigate the factor structure of the Chinese version of the SE scale, an exploratory factor analysis (EFA) was conducted. EFA allows for a preliminary analysis of how the scale measures the concepts it as designed to measure, and it was used to determine the optimal number and nature of common factors needed to explain these patterns of association separately.

Bartlett's sphericity test and Kaiser–Meyer–Olkin (KMO) were examined to ensure the assumptions associated with EFA. The principal component analysis (PCA) approach with varimax rotation was performed on the scale to determine whether all items were related to the same general construct. The number of factors was determined using the scree test and the rule of eigenvalues greater than 1 [18]. Since factors cannot be assumed to be independent in these types of data, the oblique oblivious rotation method was used to rotate the factors of EFA. Factor loadings ≥ 0.4 were considered to be meaningful and representative of the underlying constructs [19]. Psychometric properties were assessed by examining item statistics and associations between items (i.e., Cronbach's alpha).

3 Results

3.1 Demographic Characteristics of the Sample

This sample consists of 322 high school students with a 17 ± 1.2 years old mean age. Most of the participants were engaged in regular physical activity for at least four exercise sessions per week ($M = 4.4$ sessions), with an average duration of 45 min per session. Participants also reported that the most participating sports were: jogging, basketball, jump rope, badminton, and table tennis.

3.2 Exploratory Factor Analysis

Before EFA, we performed the Kaiser–Meyer–Olkin (KMO) test and Barlett's test of sphericity to assess whether the data were suitable for factor analysis. The KMO measure of sampling adequacy was 0.957, and the Barlett's test of sphericity was statistically significant (Chi-square = 8,589.606, degrees of freedom = 276, $P < 0.001$), which means that EFA can be applied to the obtained dataset.

As a result of the initial solution, Table 1 showed the principal components analysis suggested a 2-factor structure of the scale. The first two eigenvalues were 14.5 and 3.2, which explains up to 73.61% of the total variance (60.2% by Factor 1; 13.4% by Factor 2).

The scree diagram is a graphical presentation of the feature roots and is mainly used to determine the number of factors. In this study, the scree diagram (Fig. 1) also supports the Chinese version of the SE scale with a two-factor structure.

According to the reliability analysis results, the internal consistency coefficient of all subscales was excellent, and the overall Cronbach's alphas were 0.971 (0.970 for Factor 1; 0.962 for Factor 2), Cronbach's value of .70 or more was considered acceptable [20], indicating that the Chinese version of the SE could produce reliable scores. This finding is comparable to the α coefficient of the original tool. Table 2 gives the rotated factor pattern matrix and our factor interpretations.

4 Discussion

In the present study, the researchers translated the English version of the social support scale into Chinese. They then used exploratory factor analysis to determine the validity and reliability of this scale in Chinese high school students. The results obtained excellent internal consistency and convergent validity of the Chinese version of the SE and supported the use of this questionnaire in the Chinese high school student population. Given that physical inactivity and sedentary living are prevalent in the Chinese high school student population [20], the evidence of the reliability

Table 1 Total variance explained of the Chinese version of the SE

		Initial eigenvalues	Extraction sums of squared loadings			
Component	Total	% of variance	Cumulative %	Total	% of variance	Cumulative %
1	14.451	60.212	60.212	14.451	60.212	60.212
2	3.216	13.401	73.613	3.216	13.401	73.613
3	0.813	3.389	77.002			
4	0.534	2.225	79.227			
5	0.526	2.194	81.421			
6	0.498	2.077	83.498			
7	0.447	1.862	85.360			
8	0.364	1.517	86.877			
9	0.349	1.456	88.332			
10	0.304	1.267	89.600			
11	0.274	1.141	90.740			
12	0.260	1.083	91.823			
13	0.256	1.066	92.889			
14	0.244	1.017	93.906			
15	0.221	0.920	94.825			
16	0.209	0.873	95.698			
17	0.175	0.729	96.427			
18	0.167	0.697	97.124			
19	0.142	0.592	97.716			
20	0.135	0.563	98.279			
21	0.129	0.539	98.819			
22	0.112	0.466	99.285			
23	0.092	0.382	99.667			
24	0.080	0.333	100.000			

Extraction Method: Principal Component Analysis

and validity of the SE-C in this population provides a basis for further research on physical activity improvement strategies in Chinese adolescents.

Results from exploratory analysis suggested a two-factor structure: Family support (12 items) and friend support (12 items). The Cronbach's alpha values of the scale indicated excellent internal reliability (0.940 for family support and 0.936 for friend support), which seems to be consistent with previous studies(0.85 for family support and 0.88 for friend support) [21]. All item total correlation values were above 0.30, and each item contributes to measuring its core factor.

The main strengths of our study are the large sample size and the findings obtained in terms of psychometric properties. However, our study also has some limitations

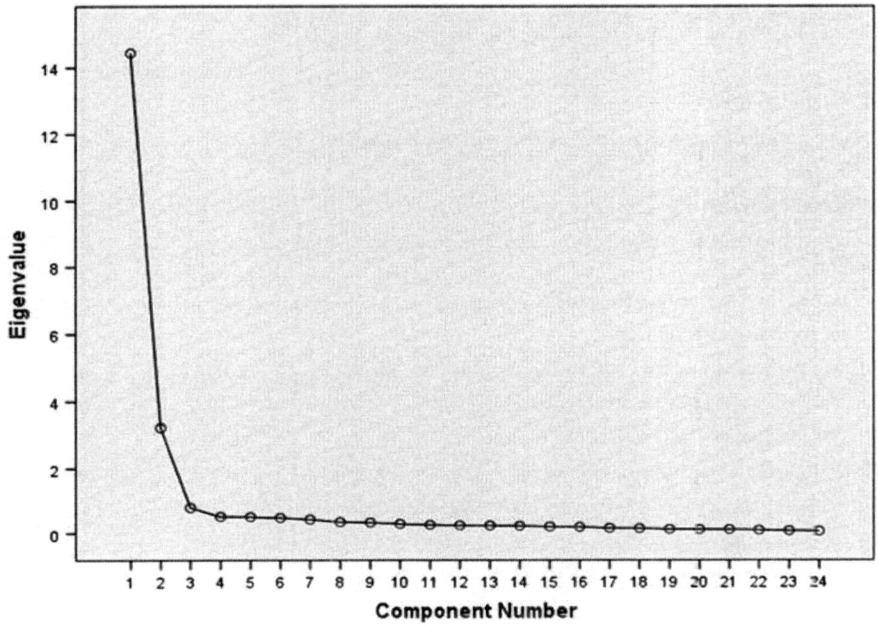

Fig. 1 Scree plot resulting from the exploratory analysis

that should be recognized appropriately. Firstly, our survey was conducted at a single high school. Therefore, the results should be generalized with caution. Further studies should attempt to replicate the results of this study using a more representative sample. Secondly, the self-report structure of the SE may be biased by social desirability as with any other self-report questionnaire. To control this intricacy, all participants are recommended to fill out the questionnaire independently and answer all items in good faith, and the confidentiality of the data is guaranteed. Nevertheless, few studies have been conducted in China on instruments for measuring social support for exercise, and the present study provides a reliable and validated tool for researchers. The researchers suggest that future studies may consider using more diverse data processing methods to further evaluate the replicability of the social support for exercise scale in populations of different age ranges, education levels, occupations, and health statuses. Also, it is crucial to study this scale in diverse Chinese populations.

5 Conclusion

Overall results demonstrate the effectiveness of using a questionnaire to assess perceived social factors associated with physical activity participation among high school students in China. The Chinese version of the social support for exercise scale

Table 2 Factor pattern matrix results for rotated factors

Item	Factors loading	
My family: 1. Exercised with me	−0.113	**0.853**
2. Gave me encouragement to stick with my exercise program	0.031	**0.729**
3. Changed their schedule so we could exercise together	−0.066	**0.908**
4. Tries to exercise with me	−0.104	**0.928**
5. Gave me helpful reminders to exercise ("Are you going to exercise tonight?")	0.024	**0.778**
6. Planned for exercise on recreational outings	0.014	**0.810**
7. Discussed exercise with me	−0.047	**0.905**
8. Talks about how much they like exercising to each other	0.098	**0.828**
9. Helped plan activities around my exercise	0.008	**0.891**
10. Talked about how much they like to exercise	0.089	**0.840**
11. Doing simple tasks for me to have more time for exercise	0.072	**0.798**
12. Praise me about the changes in my body that I got from exercising	0.161	**0.692**
My friend: 1. Exercised with me	**0.896**	−0.172
2. Gave me encouragement to stick with my exercise program	**0.912**	−0.026
3. Changed their schedule so we could exercise together	**0.881**	0.029
4. Tries to exercise with me	**0.929**	−0.128
5. Gave me helpful reminders to exercise ("Are you going to exercise tonight?")	**0.885**	0.000
6. Planned for exercise on recreational outings	**0.677**	0.220
7. Discussed exercise with me	**0.941**	−0.078
8. Talks about how much they like exercising to each other	**0.868**	0.056
9. Helped plan activities around my exercise	**0.826**	0.100
10. Talked about how much they like to exercise	**0.838**	0.077
11. Doing simple tasks for me to have more time for exercise	**0.751**	0.180
12. Praise me about the changes in my body that I got from exercising	**0.816**	0.056

Noted Extraction Method: Principal Component Analysis
[a] Rotation Method: Promax with Kaiser Normalization

is considered valid and reliable to be used among Chinese high school students. All items were identified as suitable for the sample data. It is hoped that the results of this paper will assist relevant scholars, professionals, exercise educators, and policy-makers in assessing and implementing the necessary programs to develop and build awareness of the importance of participating in regular physical activity. Researchers can use this instrument to study and explore the relationship between social variables and other psychological variables to explain the exercise behavior of Chinese adolescents.

Acknowledgements This paper is an output of the science project and is supported by the National Social Science Foundation of China (19BTY092).

References

1. Klepac Pogrmilovic B, O'Sullivan G, Milton K, Biddle SJ, Bauman A, Bull F, Pedisic Z (2018) A global systematic scoping review of studies analysing indicators, development, and content of national-level physical activity and sedentary behaviour policies. Int J Behav Phys Act 15(1):1–17. https://doi.org/10.1186/s12966-018-0742-9
2. Okuyama J, Seto S, Fukuda Y, Funakoshi S, Amae S, Onobe J, Imamura F (2021) Mental health and physical activity among children and adolescents during the COVID-19 pandemic. Tohoku J Exp Med 253(3):203–215. https://doi.org/10.1620/tjem.253.203
3. Physical activity (2020) World Health Organization. https://www.who.int/news-room/fact-sheets/detail/physical-activity
4. Sabo A, Kueh YC, Kuan G (2019) Psychometric properties of the Malay version of the self-efficacy for exercise scale. PloS One 14(5):e0215698. https://doi.org/10.1371/journal.pone.0215698
5. Bronfenbrenner U (1977) Toward an experimental ecology of human development. Am Psychol 32(7):513. https://doi.org/10.1037/0003-066X.32.7.513
6. Bandura A (1986) Social foundations of thought and action. Englewood Cliffs NJ pp 23–28
7. McLeroy KR, Bibeau D, Steckler A, Glanz K (1988) An ecological perspective on health promotion programs. Health Educ Q 15(4):351–377. https://doi.org/10.1177/109019818801500401
8. Rosenstock IM, Strecher VJ, Becker MH (1988) Social learning theory and the health belief model. Health Educ Q 15(2):175–183. https://doi.org/10.1177/109019818801500203
9. Storm V, Reinwand DA, Wienert J, Tan SL, Lippke S (2018) The mediating role of perceived social support between physical activity habit strength and depressive symptoms in people seeking to decrease their cardiovascular risk: cross-sectional study. JMIR Ment Health 5(4):e11124. https://doi.org/10.2196/11124
10. Fuller BG, Stewart Williams JA, Byles JE (2010) Active living—the perception of older people with chronic conditions. Chronic illness 6(4):294–305. https://doi.org/10.1177/1742395310377673
11. Golaszewski NM, Bartholomew JB (2019) The development of the physical activity and social support scale. J Sport Exerc Psychol 41(4):215–229. https://doi.org/10.1123/jsep.2018-0234
12. Belanger NM, Patrick JH (2018) The influence of source and type of support on college students' physical activity behavior. J Phys Act Health 15(3):183–190. https://doi.org/10.1123/jpah.2017-0069
13. Colangelo AM, Weissbrod CS (2019) The role of social support, self-efficacy, and motivation in the exercise behavior of women. Exerc Med p 3. https://doi.org/10.26644/em.2019.006
14. Van Luchene P, Delens C (2021) The influence of social support specific to physical activity on physical activity among college and university students: a systematic review. J Phys Act Health 18(6):737–747. https://doi.org/10.1123/jpah.2020-0713
15. Sallis JF, Grossman RM, Pinski RB, Patterson TL, Nader PR (1987) The development of scales to measure social support for diet and exercise behaviors. Prev med 16(6):825–836. https://doi.org/10.1016/0091-7435(87)90022-3
16. Yang JK, Lee JH, Kim YH, Hyun SK (2005) Relationship between psychological and social variables of explaining exercise behavior in adults. Korean J Sport Sci 17:119–30
17. Lee J, Kim Y (2017) Application of the social ecological constructs to explain physical activity in middle aged adults. Inter J Sport Psychol 48(2):99–110
18. Kline RB (2015) Principles and practice of structural equation modeling, Guilford publications, New York
19. Brown TA (2015) Confirmatory factor analysis for applied research, Guilford publications, New York

20. Taber KS (2018) The use of Cronbach's alpha when developing and reporting research instruments in science education. Res Sci Educ 48(6):1273–1296. https://doi.org/10.1007/s11165-016-9602-2
21. Kim Y, Kosma M (2013) Psychosocial and environmental correlates of physical activity among Korean older adults. Res Aging 35(6):750–767. https://doi.org/10.1177/0164027512462412

Research Progress on Tai Chi Intervention for Treating Depression Disorder Among the Elderly

Shiyue Wang, Garry Kuan, Yee Cheng Kueh, Ke Zhou, Yidan Wang, and Mengyuan Zhao

Abstract Depression disorder in the elderly has become one of the most prevalent mental disorders, affecting their physical and mental health. As a traditional form of physical therapy, more studies have applied Tai Chi to the prevention and treatment of depression among the elderly population. By searching CNKI and PubMed databases, the authors systematically explored literatures and data on the clinical and mechanism of Tai Chi intervention on depression disorders in the elderly, and found that 2 ~ 3 times a week, 30 ~ 60 min of Tai Chi training can significantly reduce the severity of depressive symptoms. Tai Chi training can be utilized as an effective method for preventing and treating depression disorder, and it has significant benefits for alleviating depressive symptoms and reducing loneliness among the elderly. Regular Tai Chi training has a significant effects on improving the cognitive functions, such as memory and attention of elderly patients with depression, improving sleep functions such as subjective sleep quality, and improving the quality of life, such as social ability and social participation. Practicing Tai Chi can further enhance the physical and mental health of the elderly by enhancing their subjective well-being. However, previous studies still have shortcomings in terms of study design, methods, sample size, and failure to observe the dose–response model of Tai Chi in long-term follow-up. Future studies should aim to improve the quality of evidence-based studies on Tai Chi interventions for depression disorders in the elderly.

Keywords Elderly depression disorder · Tai Chi · Traditional exercise therapy

S. Wang (✉) · G. Kuan · Y. Wang · M. Zhao
Exercise and Sports Science Programme, School of Health Sciences, Universiti Sains Malaysia, Kubang Kerian, Kelantan, Malaysia
e-mail: wangshiyue@student.usm.my

S. Wang · M. Zhao
School of Rehabilitation Medicine, Jiangsu Vocational College of Medicine, Yancheng, Jiangsu, China

Y. C. Kueh
Biostatistics and Research Methodology Unit, School of Medical Sciences, Universiti Sains Malaysia, Kubang Kerian, Kelantan, Malaysia

K. Zhou
Department of Physical Education, Henan University, Zhengzhou, Henan, China

G. Kuan et al. (eds.), *Advancing Sports and Exercise via Innovation*, Lecture Notes in Bioengineering, https://doi.org/10.1007/978-981-19-8159-3_32

1 Introduction

According to the World Federation for Mental Health [1], depression is a common disease worldwide and will be a major contributor to the global burden of disease by 2030. One of the main causes of disability, comorbidity, and impairment in the elderly, in particular, is depression [2]. It is estimated that the overall prevalence of depression among the geriatric population varies between 10 and 20% [3, 4]. Depression disorder in the elderly is a common senile mental disease [5, 6], which is accompanied by obvious indifference, high recurrence rate, and poor drug efficacy. More than 50% of patients show no significant remission after the first dose of depression [7], and about 30 to 40% of patients do not achieve complete remission after multiple doses of depression [8]. Compared with depression in youth, the etiology and clinical manifestations of depression in the elderly are more complex and varied, often accompanied by physical complications and cognitive impairment [9].

The widespread consensus is that Tai Chi is a mind–body workout that integrates physical exertion, breath inspiration and expiration, and overall mental management [10]. Due of its slow and gentle movements, Tai Chi is frequently practiced by seniors. Elderly people have also been found to benefit from Tai Chi's safety [11]. Studies have shown that Tai Chi can relieve pressure and improve depression, and has a good effect on cognitive function and sleep quality, which is conducive to the prevention and treatment of depression disorders in the elderly [12, 13]. In this paper, Chinese and foreign literatures and relevant materials were retrieved from CNKI and PubMed databases through literature review, aiming to review the current clinical studies on the effect of Tai Chi intervention on depressive symptoms, cognitive function, sleep function, and other health issues of elderly patients with depression disorder, and to understand the current status and shortcomings of the research. These will lay a foundation for better research on Tai Chi prevention and treatment of depression in the elderly.

2 Clinical Study of Tai Chi Intervention on Senile Depression Disorder

The common intervention method for elderly depression is drug therapy, but antidepressants have some shortcomings such as drug resistance, cardiovascular hazards, gastrointestinal adverse reactions, and so on [14, 15]. Many scholars have proposed non-drug treatment, such as Tai Chi. Lavretsky et al. [16] used Tai Chi combined with antidepressants to conduct interventional study among elderly patients with major depression. The participants were randomly assigned to control group or intervention group (Tai Chi with 10 mg/d of citalopram) weekly for a duration of 10 weeks (2 h each time). The Hamilton depression rating scale was used to assess the severity of depression among the study participants after the test. The results show that the treatment group had a significantly reduced severity of the depression, suggesting that Tai Chi can be used as drug therapy for a senile depressive disorder to provide an additional clinical curative effect [16].

Some scholars believe that Tai Chi alone can play a role in clinical treatment of depression in the elderly. For example, Yeung et al. [17] used a randomized controlled trial to determine the clinical efficacy of Tai Chi training for 1 h, twice a week for 12 weeks among elderly ethnic Chinese with major depression. At the evaluation of the 12th week of intervention, the response rates of the waiting exercise group, health education group, and Tai Chi intervention group were 25%, 21%, and 56%, respectively. The response rate of the Tai Chi intervention group was significantly better than those of the other two groups, and the efficacy of the Tai Chi intervention group was improved at the 24th week of follow-up. Furthermore, Chou et al. [18] conducted a randomized controlled trial with a sample of Hong Kong community, recruiting 14 elderly patients with a depressive disorder and adopting the 18-style Yang's Tai Chi training three times a week for 45 min each time for three months. Compared with the waiting exercise group, Tai Chi can effectively improve the depressive symptoms while at the same time, it can also improve physical symptoms, psychological symptoms, communication problems, and personal well-being associated with depressive symptoms. These two studies showed that Tai Chi alone was equally effective in improving depressive symptoms among elderly depressed patients.

In addition, some scholars put forward that Tai Chi combined with music therapy can reduce the symptoms of elderly patients with depression. For example, Liao et al. [19] recruited 142 elderly people from eight communities in China and applied the intervention for 50 min at a time, three times a week over three months period. The results showed that the combination of soothing music and Yang's 24-form Tai Chi training can reduce depressive symptoms in elderly patients. Rawtaer et al. [20] enrolled 100 elderly depressed patients. Tai chi, art therapy, mindfulness awareness practice, and music reminiscence therapy are a few examples of intervention groups. A single intervention phase and a combination intervention phase were each given their own portion of the program. The results showed positive improvement in the symptoms of the elderly depressed patients within a year [20]. These studies indicate that Tai Chi combined with music training can also be used as an early intervention for the prevention and treatment of depression in the elderly, and it is easy to promote and apply in the community.

Tai Chi not only has the effect of improving senile depression during the exercise period, but also has the consistency of training effect. Liu et al. [21] recruited 213 centrally obese middle-aged and elderly patients with depression in a randomized controlled trial. After 24 weeks of Tai Chi training, the severity of depression, anxiety, and stress in the Tai Chi group was significantly improved compared with the waiting control group, and the leg strength of the patients improved in the third month of the training period. Changes in mental health and leg strength were further maintained during a three-month follow-up. Cheng et al. [12] investigated how Tai Chi exercise affected older adult women's psychological well-being and sleep quality after they stopped exercising. Over the course of 24 weeks, older adult women who practiced Tai Chi reported improvements in their energy, mood, and sleep. Even four weeks after ceasing Tai Chi, the impact on lowering depression and enhancing sleep quality persisted. Investigated the effects of Tai Chi exercise on the psychological health and sleep quality. The above study indicated that the training effect of Tai Chi showed an accumulative effect in elderly patients with depression, and it could be further speculated that the maintenance effect of Tai Chi might be related to the length of training period.

In conclusion, Tai Chi not only has a therapeutic effect on depression in the elderly, but also has a certain preventive effect. Continuous Tai Chi training is positively correlated with the decrease of the incidence of depression disorder in the elderly [22]. Tai Chi training 2 ~ 3 times a week for 30 ~ 60 min each time may be an effective way to prevent and treat depression disorder in the elderly. Elderly patients can incorporate regular Tai Chi training into their daily life to prevent and decrease the rate of depression.

3 Research Progress of Tai Chi Intervention on Depression Related Dysfunction in the Elderly

Tai Chi cannot only prevent and cure the depressive symptoms of elderly patients with depressive disorder, but also improve the common concomitant disorders such as cognitive impairment, sleep disturbance, social ability, and reduced quality of life.

3.1 The Study of Tai Chi Improving Sleep Function in Elderly Patients with Depression Disorder

Sleep disturbance is one of the common dysfunctions of depression, which increases the recurrence rate of depression [23]. Studies have shown that Tai Chi training can improve sleep quality and mood improvement in patients with depressive disorder, and there is a significant correlation between subjective sleep measurement and improvement of depressive symptoms [24]. Studies have shown that Tai Chi improves sleep quality, including better subjective sleep quality, faster sleep latency, and less daytime dysfunction. The elderly participants spent their body energy during the Tai Chi training, which increased their metabolism and further promoted deep sleep. When practicing Tai Chi, the mind and breath regulate each other to relax, relieve pressure, and negative emotions. It is speculated that this is the main reason why practicing Tai Chi can improve the sleep quality and reduce the degree of depression of elderly patients with a depressive disorder [25, 26]. Tai Chi has the positive effect of regulating breath, heart beat and emotion, reducing depression, and other bad emotions, and improving sleep quality, which can positively promote the physical and mental health of the elderly [27].

3.2 *The Study of Tai Chi Improving Cognitive Function in Elderly Patients with Depression Disorder*

Cognitive impairment is often associated with depression, and there are many associations between the two [28]. A meta-analysis showed that older adults with depressive disorder had a higher risk of developing mild cognitive impairment and even dementia than those without [29]. Depression in old age is also strongly associated with dementia, and is considered a risk factor for dementia [30]. Mirza et al. [31] tracked the depressive symptom trajectories of depressed elderly people for 10 years, and found that with increasing depressive symptom trajectories, the risk of dementia also increases.

A 2014 meta-analysis of 20 clinical studies involving a total of 2553 participants showed that Tai chi can improve cognitive function in older adults [32]. Zhao et al. [33] found that the volume of the left and right hippocampus of patients with depression was smaller than that of those without depression, and it was related to memory loss in the elderly, which was speculated to be closely related to the occurrence of Senile dementia. It was found that there were changes in default network in elderly patients with subthreshold depression, which showed that the functional connections between the ventral striatum increased significantly [34]. In addition, subthreshold depression is also associated with the impairment of the resting state functional connections of the cognitive control network, and the changes of these functional connections are relevant to depression in the elderly [35]. And after 12 weeks of Tai Chi training, there was improvement in the function of the related cognitive control network connection, and obviously found in the left frontal and left dorsal anterior cingulate cortex and left beak the cognitive control network of the anterior cingulate cortex of resting state functional connectivity, the hippocampus and the resting state functional connectivity of the medial prefrontal cortex, this change was significantly related with memory function [36].

3.3 *The Study of Tai Chi Improving Social Skills and Quality of Life in Elderly Patients with Depression Disorder*

Due to the mood change of senile depressive disorder, coupled with the trouble of aging, the psychological activities and living conditions of the elderly will change, especially the persistent loneliness and social alienation of the elderly with depressive disorder [37, 38]. Studies have shown that practicing Tai Chi improve mutual communication and important therapeutic factors of depression. Tai Chi can improve the old patients with depressive disorder [39]. Liao's [40] study found that 112 elderly people in the Chinese community were randomly recruited, and the intervention of combining traditional Tai Chi with Chinese folk music, 50 min each time, 3 times a week, for 12 consecutive weeks, could improve the quality of life of elderly patients with mild to moderate depression in the Chinese community. Tai Chi can therefore improve physical and psychological functions of practitioners as well as the quality of life and social interaction connected to health and well-being of the elderly because it is a safe and simple traditional exercise therapy to learn.

4 Study on the Mechanism of Tai Chi Intervention on Senile Depression Disorder

From the perspective of synaptic plasticity, senile depression is often accompanied by abnormalities of cognitive control network (prefrontal cortex, cingulate gyrus, parietal lobe, and other key regions) and affective control network (amygdala, orbitofrontal cortex, limbic system, and other key regions) [41, 42]. In elderly patients with depression, the cognitive control network increased connectivity with the salience network and was closely associated with depression severity [43]. The study showed that after 12 weeks of Tai Chi practice, the default network connectivity between the medial frontal cortex, anterior cingulate cortex, right middle frontal gyrus, posterior cingulate cortex, and left occipital cortex improved [44]. Another study showed that 33 community elders who practiced Tai Chi independently at home three times a week for six months further improved synaptic plasticity and hippocampal neuron growth in the brain compared with the health education group [45, 46]. Therefore, Tai Chi can further regulate emotions and alleviate depression by regulating the structure and function of related brain regions of cognitive/emotional control network (dorsolateral prefrontal lobe, cingulate gyrus, orbitofrontal cortex, etc.), promoting nerve repair and functional connectivity recovery.

From the perspective of neuroendocrine, hypothalamic–pituitary–adrenal axis (HPA) was one of the main biological mediating systems regulating human stress. Its related hormones and neurotransmitters (such as glucocorticoid, cortisol, and 5-hydroxytryptamine) participate in the regulation of mood and the occurrence and development of depressive disorder. Compared with healthy people, HPA activity was significantly increased when they had a depressive disorder [47, 48]. Studies have shown that Tai Chi exercise can reduce the elevated cortisol level caused by the abnormal HPA axis in subjects [49], and help the elderly patients with depression disorder to restore the functional balance of HPA axis, so as to delay the elderly depression disorder [50].

From the perspective of immune response, it has been proposed that depressive disorder is also an immune inflammatory disease, characterized by elevated levels of inflammatory cytokines such as interleukin-1, tumor necrosis factor, and C-reactive protein [51]. Further, research has shown that inflammatory cytokines alter the production and metabolism of neurotransmitters, including dopamine, glutamate, and serotonin, which are closely involved in the regulation of mood [52]. Inflammatory factors can even affect the growth and survival of neurons. When oxidative stress occurs, inflammatory factors can further affect the brain regions related to emotion regulation, such as the prefrontal cortex and amygdala [53], and thus aggravate the severity of depression. Robins et al. [54] through Tai Chi exercise for eight weeks, the result shows that Tai Chi helps reduce inflammatory cytokines, including interferon, tumor necrosis factor, and interleukin-4, and Tai Chi exercise can further affect inflammatory reaction markers, to enhance the health level of elderly patients with depressive disorder, to alleviate the severity of the senile depression.

From the perspective of traditional Chinese medicine, Huangdi [55] believed that the incoordination of seven emotions was the main pathogenic factor of depression syndrome. Diseases caused by emotional factors that affect the circulation of qi and blood are collectively referred to as the narrow sense of "depression syndrome", which mainly emphasizes "emotional depression" and is closely related to psychological, social, and environmental factors. In traditional Chinese medicine, "stagnation syndrome" is caused by emotional discomfort and stagnation of qi. The effect of Tai Chi spirit adduction, concentration, and static qi can balance Yin and Yang, regulate the seven emotions, with body movement, stimulate Yang qi, promote qi operation, so as to achieve the effect of improving depression [56].

5 Deficiency and Solution of Tai Chi Intervention on Depression in the Elderly

At present, the research on Tai Chi intervention in senile depressive disorder has the following deficiencies: (1) Lack of long-term effect of Tai Chi intervention on senile depression; (2) The sample size of existing studies is small and most of them are single-center studies; (3) Lack of other aerobic exercise as a positive control of the study, it is difficult to clarify the role of Tai Chi non-exercise components on depression in the elderly. In the future studies, quality control of clinical trials should be strengthened to obtain evidence-based evidence of high-level Tai Chi intervention in elderly depression disorders. For example, a double blind and triple blind studies should be adopted to avoid subjective bias in the study, and multi-center trials with large samples should be carried out to observe the long-term effects with follow-up. Other aerobic exercise with the same exercise dose as Tai Chi was used as positive control to further observe the effect of its non-exercise components.

In summary, regular and continuous Tai Chi exercise can prevent and cure depression disorder in elderly patients, and improve the cognitive function, social communication, sleep quality, and other aspects.

References

1. World Federation for Mental Health (2012) Depression: a global crisis. World Federation for Mental Health, Springfield, VA
2. Alexopoulos GS (2005) Depression in the elderly. Lancet 365(9475):1961–1970. https://doi.org/10.1016/s0140-6736(05)66665-2
3. Wang S, Blazer DG (2015) Depression and cognition in the elderly. Annu Rev Clin Psychol 11:331–360. https://doi.org/10.1146/annurev-clinpsy-032814-112828
4. Cahoon CG (2012) Depression in older adults. Am J Nurs 112(11):22–30. https://doi.org/10.1097/01.NAJ.0000422251.65212.4b
5. Diniz BS, Butters MA, Albert SM, Dew MA, Reynolds CF (2013) Late-life depression and risk of vascular dementia and Alzheimer's disease: systematic review and meta-analysis of community-based cohort studies. Br J Psychiatry 202(5):329–335. https://doi.org/10.1192/bjp.bp.112.118307
6. Tedeschini E, Levkovitz Y, Iovieno N, Ameral VE, Nelson JC, Papakostas GI (2011) Efficacy of antidepressants for late-life depression: a meta-analysis and meta-regression of placebo-controlled randomized trials. J Clin Psychiatry 72(12):1660–1668. https://doi.org/10.4088/JCP.10r06531
7. Roose SP, Schatzberg AF (2005) The efficacy of antidepressants in the treatment of late-life depression. J Clin Psychopharmacol 25(4 Suppl 1):S1-7. https://doi.org/10.1097/01.jcp.0000162807.84570.6b
8. Nelson JC, Delucchi K, Schneider LS (2008) Efficacy of second generation antidepressants in late-life depression: a meta-analysis of the evidence. Am J Geriatr Psychiatry 16(7):558–567. https://doi.org/10.1097/JGP.0b013e3181693288
9. Lyness JM (2008) Depression and comorbidity: objects in the mirror are more complex than they appear. Am J Geriatr Psychiatry 16(3):181–185. https://doi.org/10.1097/JGP.0b013e318162f186
10. Wu W, Liu X, Wang L, Wang Z, Hu J, Yan J (2014) Effects of Tai Chi on exercise capacity and health-related quality of life in patients with chronic obstructive pulmonary disease: a systematic review and meta-analysis. Int J Chron Obstruct Pulmon Dis 9:1253–1263. https://doi.org/10.2147/copd.S70862
11. Wang Y, Luo B, Wu X, Li X, Liao S (2022) Comparison of the effects of Tai Chi and general aerobic exercise on weight, blood pressure and glycemic control among older persons with depressive symptoms: a randomized trial. BMC Geriatr 22(1):401. https://doi.org/10.1186/s12877-022-03084-6
12. Cheng L, Qian L, Chang S, He B (2021) Effect of Tai Chi on depression symptoms and sleep quality among older adult women after exercise cessation. Res Sports Med 29(4):395–405. https://doi.org/10.1080/15438627.2020.1870976
13. Chi I, Jordan-Marsh M, Guo M, Xie B, Bai Z (2013) Tai Chi and reduction of depressive symptoms for older adults: a meta-analysis of randomized trials. Geriatr Gerontol Int 13(1):3–12. https://doi.org/10.1111/j.1447-0594.2012.00882.x
14. Santarsieri D, Schwartz TL (2015) Antidepressant efficacy and side-effect burden: a quick guide for clinicians. Drugs Context 4:212290. https://doi.org/10.7573/dic.212290
15. Wang SM, Han C, Bahk WM, Lee SJ, Patkar AA, Masand PS, Pae CU (2018) Addressing the side effects of contemporary antidepressant drugs: a comprehensive review. Chonnam Med J 54(2):101–112. https://doi.org/10.4068/cmj.2018.54.2.101
16. Lavretsky H, Alstein LL, Olmstead RE, Ercoli LM, Riparetti-Brown M, Cyr NS, Irwin MR (2011) Complementary use of Tai Chi Chih augments escitalopram treatment of geriatric depression: a randomized controlled trial. Am J Geriatr Psychiatry 19(10):839–850. https://doi.org/10.1097/JGP.0b013e31820ee9ef
17. Yeung AS, Feng R, Kim DJH, Wayne PM, Yeh GY, Baer L, Lee OE, Denninger JW, Benson H, Fricchione GL, Alpert J, Fava M (2017) A pilot, randomized controlled study of Tai Chi with passive and active controls in the treatment of depressed Chinese Americans. J Clin Psychiatry 78(5):e522–e528. https://doi.org/10.4088/JCP.16m10772

18. Chou KL, Lee PW, Yu EC, Macfarlane D, Cheng YH, Chan SS, Chi I (2004) Effect of Tai Chi on depressive symptoms amongst Chinese older patients with depressive disorders: a randomized clinical trial. Int J Geriatr Psychiatry 19(11):1105–1107. https://doi.org/10.1002/gps.1178
19. Liao SJ, Tan MP, Chong MC, Chua YP (2018) The impact of combined music and Tai Chi on depressive symptoms among community-dwelling older persons: a cluster randomized controlled trial. Issues Ment Health Nurs 39(5):398–402. https://doi.org/10.1080/01612840.2017.1417519
20. Rawtaer I, Mahendran R, Yu J, Fam J, Feng L, Kua EH (2015) Psychosocial interventions with art, music, Tai Chi and mindfulness for subsyndromal depression and anxiety in older adults: a naturalistic study in Singapore. Asia Pac Psychiatry 7(3):240–250. https://doi.org/10.1111/appy.12201
21. Liu X, Vitetta L, Kostner K, Crompton D, Williams G, Brown WJ, Lopez A, Xue CC, Oei TP, Byrne G, Martin JH, Whiteford H (2015) The effects of Tai Chi in centrally obese adults with depression symptoms. Evid Based Complement Alternat Med 2015:879712. https://doi.org/10.1155/2015/879712
22. Li Y, Su Q, Guo H, Wu H, Du H, Yang G, Meng G, Li C, Nagatomi R, Niu K (2014) Long-term Tai Chi training is related to depressive symptoms among Tai Chi practitioners. J Affect Disord 169:36–39. https://doi.org/10.1016/j.jad.2014.07.029
23. Carney CE, Edinger JD, Kuchibhatla M, Lachowski AM, Bogouslavsky O, Krystal AD, Shapiro CM (2017) Cognitive behavioral insomnia therapy for those with insomnia and depression: a randomized controlled clinical trial. Sleep 40(4):zsx019 https://doi.org/10.1093/sleep/zsx019
24. Ma Y, Yeung A, Yang AC, Peng CK, Clain A, Alpert J, Fava M, Yeung AS (2018) The effects of Tai Chi on sleep quality in Chinese American patients with major depressive disorder: a pilot study. Behav Sleep Med 16(4):398–411. https://doi.org/10.1080/15402002.2016.1228643
25. Zheng G, Xiong Z, Zheng X, Li J, Duan T, Qi D, Ling K, Chen L (2017) Subjective perceived impact of Tai Chi training on physical and mental health among community older adults at risk for ischemic stroke: a qualitative study. BMC Complement Altern Med 17(1):221. https://doi.org/10.1186/s12906-017-1694-3
26. Chen KM, Chen MH, Chao HC, Hung HM, Lin HS, Li CH (2009) Sleep quality, depression state, and health status of older adults after silver yoga exercises: cluster randomized trial. Int J Nurs Stud 46(2):154–163. https://doi.org/10.1016/j.ijnurstu.2008.09.005
27. Sarris J, Byrne GJ (2011) A systematic review of insomnia and complementary medicine. Sleep Med Rev 15(2):99–106. https://doi.org/10.1016/j.smrv.2010.04.001
28. Byers AL, Yaffe K (2011) Depression and risk of developing dementia. Nat Rev Neurol 7(6):323–331. https://doi.org/10.1038/nrneurol.2011.60
29. Gao Y, Huang C, Zhao K, Ma L, Qiu X, Zhang L, Xiu Y, Chen L, Lu W, Huang C, Tang Y, Xiao Q (2013) Depression as a risk factor for dementia and mild cognitive impairment: a meta-analysis of longitudinal studies. Int J Geriatr Psychiatry 28(5):441–449. https://doi.org/10.1002/gps.3845
30. Cherbuin N, Kim S, Anstey KJ (2015) Dementia risk estimates associated with measures of depression: a systematic review and meta-analysis. BMJ Open 5(12):e008853. https://doi.org/10.1136/bmjopen-2015-008853
31. Mirza SS, Wolters FJ, Swanson SA, Koudstaal PJ, Hofman A, Tiemeier H, Ikram MA (2016) 10-year trajectories of depressive symptoms and risk of dementia: a population-based study. Lancet Psychiatry 3(7):628–635. https://doi.org/10.1016/s2215-0366(16)00097-3
32. Wayne PM, Walsh JN, Taylor-Piliae RE, Wells RE, Papp KV, Donovan NJ, Yeh GY (2014) Effect of Tai Chi on cognitive performance in older adults: systematic review and meta-analysis. J Am Geriatr Soc 62(1):25–39. https://doi.org/10.1111/jgs.12611
33. Zhao ZL, Fan DQ, Tang YF, Li XQ, Guo P, Xie F (2017) The value of MRI detection of hippocampus volume and morphology in depression. Int J Psychiatry 44(5):822–825 (in Chinese)
34. Hwang JW, Xin SC, Ou YM, Zhang WY, Liang YL, Chen J, Yang XQ, Chen XY, Guo TW, Yang XJ, Ma WH, Li J, Zhao BC, Tu Y, Kong J (2016) Enhanced default mode network connectivity with ventral striatum in subthreshold depression individuals. J Psychiatr Res 76:111–120. https://doi.org/10.1016/j.jpsychires.2016.02.005

35. Ely BA, Xu J, Goodman WK, Lapidus KA, Gabbay V, Stern ER (2016) Resting-state functional connectivity of the human habenula in healthy individuals: associations with subclinical depression. Hum Brain Mapp 37(7):2369–2384. https://doi.org/10.1002/hbm.23179
36. Tao J, Chen X, Egorova N, Liu J, Xue X, Wang Q, Zheng G, Li M, Hong W, Sun S, Chen L, Kong J (2017) Tai Chi Chuan and Baduanjin practice modulates functional connectivity of the cognitive control network in older adults. Sci Rep 7:41581. https://doi.org/10.1038/srep41581
37. Tanjanai PT, Moradinazar M, Najafi F (2017) Prevalence of depression and related social and physical factors amongst the Iranian elderly population in 2012. Geriatr Gerontol Int 17(1):126–131. https://doi.org/10.1111/ggi.12680
38. Domènech-Abella J, Lara E, Rubio-Valera M, Olaya B, Moneta MV, Rico-Uribe LA, Ayuso-Mateos JL, Mundó J, Haro JM (2017) Loneliness and depression in the elderly: the role of social network. Soc Psychiatry Psychiatr Epidemiol 52(4):381–390. https://doi.org/10.1007/s00127-017-1339-3
39. Yeung A, Lepoutre V, Wayne P, Yeh G, Slipp LE, Fava M, Denninger JW, Benson H, Fricchione GL (2012) Tai Chi treatment for depression in Chinese Americans: a pilot study. Am J Phys Med Rehabil 91(10):863–870. https://doi.org/10.1097/PHM.0b013e31825f1a67
40. Liao SJ, Chong MC, Tan MP, Chua YP (2019) Tai Chi with music improves quality of life among community-dwelling older persons with mild to moderate depressive symptoms: a cluster randomized controlled trial. Geriatr Nurs 40(2):154–159. https://doi.org/10.1016/j.gerinurse.2018.08.001
41. Aizenstein HJ, Butters MA, Wu M, Mazurkewicz LM, Stenger VA, Gianaros PJ, Becker JT, Reynolds CF 3rd, Carter CS (2009) Altered functioning of the executive control circuit in late-life depression: episodic and persistent phenomena. Am J Geriatr Psychiatry 17(1):30–42. https://doi.org/10.1097/JGP.0b013e31817b60af
42. Alexopoulos GS, Hoptman MJ, Kanellopoulos D, Murphy CF, Lim KO, Gunning FM (2012) Functional connectivity in the cognitive control network and the default mode network in late-life depression. J Affect Disord 139(1):56–65. https://doi.org/10.1016/j.jad.2011.12.002
43. Li W, Wang Y, Ward BD, Antuono PG, Li SJ, Goveas JS (2017) Intrinsic inter-network brain dysfunction correlates with symptom dimensions in late-life depression. J Psychiatr Res 87:71–80. https://doi.org/10.1016/j.jpsychires.2016.12.011
44. Eyre HA, Acevedo B, Yang H, Siddarth P, Van Dyk K, Ercoli L, Leaver AM, Cyr NS, Narr K, Baune BT, Khalsa DS, Lavretsky H (2016) Changes in neural connectivity and memory following a yoga intervention for older adults: a pilot study. J Alzheimers Dis 52(2):673–684. https://doi.org/10.3233/jad-150653
45. Sungkarat S, Boripuntakul S, Kumfu S, Lord SR, Chattipakorn N (2018) Tai Chi improves cognition and plasma BDNF in older adults with mild cognitive impairment: a randomized controlled trial. Neurorehabil Neural Repair 32(2):142–149. https://doi.org/10.1177/1545968317753682
46. Kong J, Wolcott E, Wang Z, Jorgenson K, Harvey WF, Tao J, Rones R, Wang C (2019) Altered resting state functional connectivity of the cognitive control network in fibromyalgia and the modulation effect of mind-body intervention. Brain Imaging Behav 13(2):482–492. https://doi.org/10.1007/s11682-018-9875-3
47. Stetler C, Miller GE (2011) Depression and hypothalamic-pituitary-adrenal activation: a quantitative summary of four decades of research. Psychosom Med 73(2):114–126. https://doi.org/10.1097/PSY.0b013e31820ad12b
48. Belvederi Murri M, Pariante C, Mondelli V, Masotti M, Atti AR, Mellacqua Z, Antonioli M, Ghio L, Menchetti M, Zanetidou S, Innamorati M, Amore M (2014) HPA axis and aging in depression: systematic review and meta-analysis. Psychoneuroendocrinology 41:46–62. https://doi.org/10.1016/j.psyneuen.2013.12.004
49. Nedeljkovic M, Ausfeld-Hafter B, Streitberger K, Seiler R, Wirtz PH (2012) Taiji practice attenuates psychobiological stress reactivity—a randomized controlled trial in healthy subjects. Psychoneuroendocrinology 37(8):1171–1180. https://doi.org/10.1016/j.psyneuen.2011.12.007

50. Sarubin N, Nothdurfter C, Schüle C, Lieb M, Uhr M, Born C, Zimmermannc R, Bühner M, Konopka K, Rupprecht R, Baghai TC (2014) The influence of Hatha yoga as an add-on treatment in major depression on hypothalamic-pituitary-adrenal-axis activity: a randomized trial. J Psychiatr Res 53:76–83. https://doi.org/10.1016/j.jpsychires.2014.02.022
51. Maes M, Berk M, Goehler L, Song C, Anderson G, Gałecki P, Leonard B (2012) Depression and sickness behavior are Janus-faced responses to shared inflammatory pathways. BMC Med 10:66. https://doi.org/10.1186/1741-7015-10-66
52. Capuron L, Miller AH (2011) Immune system to brain signaling: neuropsychopharmacological implications. Pharmacol Ther 130(2):226–238. https://doi.org/10.1016/j.pharmthera.2011.01.014
53. Leonard B, Maes M (2012) Mechanistic explanations how cell-mediated immune activation, inflammation and oxidative and nitrosative stress pathways and their sequels and concomitants play a role in the pathophysiology of unipolar depression. Neurosci Biobehav Rev 36(2):764–785. https://doi.org/10.1016/j.neubiorev.2011.12.005
54. Robins JL, Elswick RK Jr, Sturgill J, McCain NL (2016) The effects of Tai Chi on cardiovascular risk in women. Am J Health Promot 30(8):613–622. https://doi.org/10.4278/ajhp.140618-QUAN-287
55. Zhou ZH, Gao SM, Gao S, Zhou HB, Yu CQ (2015) Analysis of medication rules of ancient prescriptions for Yu Zhi based on association rules. Tianjin Tradit Chin Med 32(3):176–179 (in Chinese)
56. Yu JS, Lin DH (2017) TCM rehabilitation from Yang to treat depression. J Tradit Chin Med 58(9):748–751 (in Chinese)

Review on the Use of Pre-Task Music Prior to Exercise Performance

Hilda Anak Entering, Ngien Siong Chin, Yee Cheng Kueh, Garry Kuan, and Adam Abdul Malik

Abstract Music has been scientifically used to be an excellent complement to sports activities. Scientific research in this field has been prolific over the past four decades, with over 120 peer-reviews articles examining various aspects of the relationship between music and sports. However, little research has been conducted on the use of music prior to exercise performance. A comprehensive literature search was undertaken between January 2012 and April 2022 using online databases such as Web of Science, PubMed, and Google Scholar. The following keywords were used in combination: "pre-task music", "pre-performance music", "music and sport", and "exercise performance". A total of sixteen studies were selected for this review. The literature search was further expanded by searching for review articles on the topic, and the articles collected for any relevant references. The majority of studies have shown an ergogenic effect of pre-task music prior to exercise performance. There is clearly a need for more research on this topic before making recommendations regarding the use of pre-task music as a means of enhancing exercise performance.

Keywords Music · Psychological · Exercise performance · Sporting events

1 Introduction

In recent decades, listening to music during warm-ups has been a part of the sport and exercise activities. The combination pair of music and exercise/sports is widely used as the new developments in portable technologies (MP3 players, iPods, and

H. A. Entering (✉) · G. Kuan · A. Abdul Malik
Exercise and Sports Science Programme, School of Health Sciences, Universiti Sains Malaysia, Kubang Kerian, Kelantan, Malaysia
e-mail: hildaentering@student.usm.my

N. S. Chin
Institute of Teacher Education Batu Lintang Campus, Kuching, Sarawak, Malaysia

Y. C. Kueh
Biostatistics and Research Methodology Unit, School of Medical Sciences, Universiti Sains Malaysia, Kubang Kerian, Kelantan, Malaysia

G. Kuan et al. (eds.), *Advancing Sports and Exercise via Innovation*, Lecture Notes in Bioengineering, https://doi.org/10.1007/978-981-19-8159-3_33

smartphones) have improved accessibility and allowed for individual music selection during exercise [1]. There are numerous advantages from listening to music while exercising, but they all come down to the fact that music makes exercise feel easier and more enjoyable, allowing them to work out harder than they would if exercising alone. Music is an external source that is used as an ergogenic aid in a wide array of exercise modes. Thus, the relationship between music and sports studies has been extensively researched, as has their potential ergogenic help to improve physical performance.

Presently, pre-task music has been used successfully as a stimulant [2], and these advantages may contribute to the ergogenic effects as addressed in existing empirical studies. A chosen set of songs can have an instant and lasting effect on people's moods. Scientifically, music has extensively provided conclusive evidence that it may boost their mood, motivation, and effort while they are working out. This finding is inclined with the finding by Biagini et al. [3] that combining music and exercise appears the most common positive outcome to increase performance measures, improve mood, and improve arousal. The impact of music on performance is well-explained from a scientific perspective.

Apparently, listening to music before and during exercise has been shown as a motivation and effort booster, resulting in better performance outcomes [4, 5]. Moreover, music is known to be capable of reducing perceived exertion during intense exercise. According to Terry et al. [6], it was revealed that listening to music while engaging in a variety of physical activities can improve physiological efficiency, boost physical performance, and reduce perceived exertion. Pre-task music has been employed in many sports and is a popular approach among competitors. The majority of studies have shown an ergogenic effect of pre-task music on various physical activities. Despite its popularity, there is minimal scientific evidence that pre-task music has ergogenic benefits on exercise performance.

Thus, this study would further explore the existing studies and investigate the ergogenic effects of pre-task music prior to exercise performance. Thus, it will provide an insight for researchers before making recommendations regarding the use of pre-task music as a means of enhancing exercise performance.

2 Methods

2.1 Procedure

This review draws upon research in the field of pre-task music and exercises performance. A comprehensive literature search was undertaken using three databases (Google Scholar, PubMed, and Web of Science) between January 2012 and April 2022 through published articles, review papers, and meta-analysis. The search was limited to peer-reviewed journals written in the English language. The following keywords were used in combination: "pre-task music", "pre-performance music",

"music and sport", and "exercise performance". Articles consisting of validation studies, dissertations, conference abstracts, monographs, and brief reports were excluded to narrow down the search category. The literature search was further expanded by searching for review articles on the topic and the articles collected for any relevant references. A total of sixteen studies were selected for this review. A manual search of the reference lists in the studies found in the computerized search was also conducted. After the duplicates were removed, the comprehensive search yielded 16 studies that matched the study criteria mentioned above. The summary of the reviewed studies is presented in Table 1.

3 Evidence Synthesis

3.1 Pre-Task Music Preference

Various categorizations of music interventions (i.e., pre-task, in-task, post-task, etc.) have been explored and reported to be beneficial for sport or exercise performance. Pre-task music is commonly used to regulate a particular mindset and is particularly effective in reducing pre-competition anxiety [7]. Following this, it is also has been investigated to improve a variety of exercise mode performance such as endurance and resistance-based workouts [8]. Even though many sports governing bodies prohibit personal audio devices in the competition area, pre-task music is often used during warm-ups and as part of a routine before an event. Therefore, it may be more suitable and practical to listen to music prior to exercise to improve one's effort (i.e., pre-task) or during a warm-up session. However, methodological limitations and few systematic research in pre-task music have been highlighted and need sufficient evidence to support scientific literature on music effect on exercise performance [9].

Music preference is a listening response that reveals the degree to which a music stimulus is preferred or not preferred at a certain point in time [10]. In previous studies, music preference has been repeatedly selected to mediate music benefits [11]. According to Bishop [1], listening to music before sport or exercise can regulate their emotional state and control their mental preparation. In fact, with the right physical and mental condition, it is often beneficial to achieve optimal performance. For coaches and athletes, it is crucial attention in the consideration of selecting pre-task music to harness the power of music. In these emerging and developing countries, new developments in portable technology (i.e., smartphones, IPods, mp3 players, etc.) have allowed individuals to select their music while exercising. Some of the playlists on most music apps (i.e., Joox, Spotify, AudioFuel, etc.) today are mostly made for workouts and exercises. Meanwhile, individuals may select their most favorable array of music choices in any genre, rhythm, etc., while engaging in exercise or workouts. A study conducted by Stork et al. [12] found that selecting preferred music (music preference) can increase athletes' motivation, enjoyment,

Table 1 Reviewed studies on pre-task music preference effects on exercise performance

Study	Music conditions	Timing/condition	Task	Results/findings
Bigliassi et al. [14]	Self-selected, no music	Warm-up, during	5 km cycling time trial	↔ Performance, ↔ RPE
Chtourou et al. [15]	High tempo, no music	Warm-up	Wingate anaerobic test	↑Power, ↔ RPE
Chtourou et al. [8]	High tempo, no music	Warm-up	Wingate anaerobic test	↑Power, ↑performance, ↔ RPE
Jarraya et al. [16]	High tempo, no music	Warm-up	Wingate anaerobic test	↔ HR, ↔ RPE, ↑ power, ↑performance
Serra de Queiroz [17]	Self-selected asynchronous, no music	Pre-task	Soccer	↔ Performance, ↑ motivation, ↑ concentration
Smirmaul et al. [18]	Self-selected, no music	Pre-task	200 m freestyle swimming timed trial	↑Motivation, ↑performance ↔ RPE
Chtourou et al. [19]	Music, no music	Warm-up	Wingate anaerobic test	↑ RPE, ↑ power, ↑ performance (sprinters)
Karageorghis et al. [20]	Fast/slow tempo, loud/quiet tempo, no music	Pre-task	Grip strength	↑Enjoyment, ↑performance
Kuant et al. [21]	Relaxing music, arousing music	Pre-task	Dart throwing	↑Performance, ↑confidence
Belkhir et al. [22]	Neutral music (WUNM), self-selected motivational (WUMM), no music	Warm-up	5 m shuttle run	↑Performance, ↑enjoyment (WUNM), ↓enjoyment (WUMM)
Fox et al. [23]	Pre-selected, self-selected, white noise	Warm-up	Wingate anaerobic test	↔ Performance, ↔ RPE
Puad et al. [24]	Fast tempo, no music	Warm-up	Running anaerobic sprint test (RAST)	↔ RPE, ↔ power, ↑performance, ↑HR
Karow et al. [4]	Non-preferred, preferred, no music	Warm-up	Rowing	↑Motivation, ↑power, ↔ RPE

(continued)

Table 1 (continued)

Study	Music conditions	Timing/condition	Task	Results/findings
Belkhir et al. [25]	Synchronous, motivational	Warm-up	Continuous jump	↑RPE, ↑enjoyment, ↑performance
Meglic et al. [26]	Preferred, non-preferred	Warm-up	Cycling, Wingate anaerobic test	↔ RPE, ↑motivation, ↑performance
Ballman et al. [27]	Self-selected, no music	Pre-exercise	Bench press	↑ Motivation, ↑RTF, ↑power ↑Performance

** *Note* Music conditions (preference of music applied for intervention), task (mode of exercise), and the result of findings are stated above
↑ = Increase
↓ = Decrease
↔ = No changes
RPE = Rate of perceived exertion, RTF = Repetitions to failure

and affective state. Although the previous study reported that pre-task music would improve performance has not been supported [13], in recent decades, numerous pieces of evidence on how pre-task music preference may optimize performance on exercises are shown in Table 1.

Listening to familiar music can increase an individual's affective valence, which simply triggers happiness, memories, and reminds them of their role models or heroes. Hence, selecting the right music can positively impact an individual's emotional state and enhance self-efficacy [20]. In spite of this, musical pieces can facilitate recalling previous successful sporting experiences which boosts motivation and self-confidence [28]. As music may evoke an individual's memories, such as imaging their previous performance, music is effective in stimulating crucial tactical or technical context. For instance, young athletes can acquire proper shot-put technique by listening to the lyrics of "Push It (130 bpm)" by Salt-N-Pepa in order to acquire the perfect technique for the shot-put throw [7]. This will lead students to "push" the shot rather than attempting to "throw" it, which is the most prevalent type of technical blunder. Consequently, selecting a particular music preference in which the lyrics may convey psychological strategies should be taken into consideration for future studies. In addition, the choice of musical selection should correspond to the athlete's characteristics as well as the components of the exercise in order to determine the best musical option to improve sports and exercise performance. As suggested by Karageorghis et al. [7], Fig. 1 shows how music selection, athlete's characteristics, and sport/exercise components may be related to each other in optimizing performance.

Although pre-task music appears to have reported a higher benefit on exercise performance, the comprehensive studies pertaining to the role of pre-task music application and which selected/preferred music imposes greater benefits are still

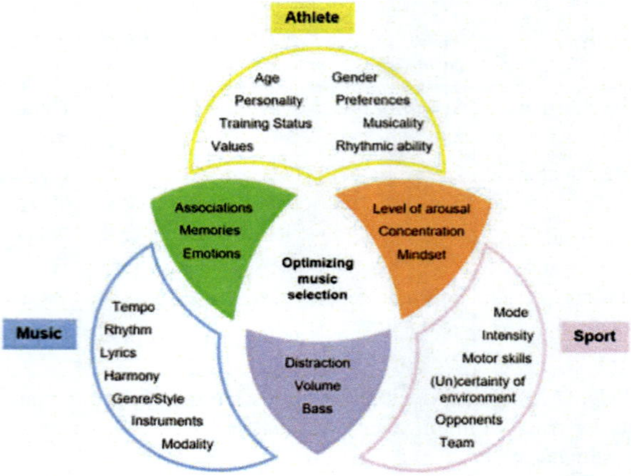

Fig. 1 Factors relevant to optimal music selection [7]

being limited. The selection of music should be done with consideration for individual requirements and tastes in order to improve affective states and provide a sense of control over their immediate surroundings [20].

4 The Role of Pre-Task Music in Exercise Performance

Pre-task music has been used successfully as an ergogenic effect. The evidence indicates that music elicits several benefits in the context of exercise and sport-related tasks, as shown in Table 1. Several music conditions have been conducted to investigate the impact on exercise performance. Five studies on the effects of high-tempo music [8, 15, 16, 20, 24] before a task (warm-up) would improve performance have been reported. As recommended by Karageorghis et al. [29], high-tempo music is defined as 120–140 bpm. A study among 12 physical education students on the diurnal variations of power output during the Wingate test showed a positive effect on peak power and mean power with music compared with no music [15]. It shows that the benefits of high-tempo music during warm-up work as an additional aid to muscle contractions, thus increasing the performance level. However, as suggested by the present authors, studies on the effects of music on diurnal fluctuations should be further examined. Other findings on the effects of high-tempo music during warm-up on short-term maximal performances showed positive effects on power, thus increasing exercise performance [8]. The Wingate test was conducted after a 10 min warm-up among nine young male sprinters and their rate of perceived exertion (RPE) was recorded at the end of the task. However, findings on the music's effect on RPE scores were unaffected.

Two studies investigating the effects of rate of perceived exertion (RPE) on anaerobic exercise using high-tempo music among 12 national and regional level sprinters [16] and 24 football players [24] also reported that RPE scores were not significant.

Nevertheless, their performance during listening to high-tempo music was reported to increase significantly. This study highlights that high-tempo music could motivate individuals during warm-up. Those coaches and practitioners interested in short-term maximal performances could examine the use of arousing music on power output during warm-up instead of high-tempo music. There has been limited research into the effects of arousing music on exercise performance.

Self-selected music conditions in six studies with different tasks; cycling [14], soccer skill test [17], swimming [18], running [22], Wingate anaerobic test [23], and bench press [27], on the other hand, were reported to have yielded diverse outcomes. The influence of self-selected music conditions on longer and predominantly aerobic tasks (5 km cycling time trial) [14], soccer skill test [17], and Wingate anaerobic test [23] resulted in no improvement in their performance compared to the other three studies [18, 22, 27]. Contradictory, Smirmaul et al. [18] reported the effects of self-selected music before performing the 200 m freestyle swimming trials were improved performance and motivation. The participants listened to the pre-task music for 5 min, and as a result, their swimming time was significantly improved with a shorter length (−1.44%) to the no music condition. The music selection selected by the participants was mainly electronic, rock, and pop genres. Ballman et al. [27] tried to understand if listening to pre-exercise music on bench press performance could have a different impact on power and motivation of participants. Ten of the participants, all male, were listening to pre-exercise music on the bench press task. Music was selected by the participants, and researchers set a minimum tempo of 120 bpm. Their findings showed significant increased on motivation with better muscle power explosiveness, and strength-endurance when listening to music before a bench press exercise. However, which selected music or songs imposes greater benefits in these six studies in improving exercise performance remains unknown.

Scientific evidence has shown different results on exercise performance with different music conditions. However, the studies on relaxing, arousing, synchronous, and motivational music are still limited even though they have revealed significant effects on exercise performance [19, 21, 25]. Therefore, further studies are strongly needed to support this type of music (e.g., relaxing, arousing, synchronous, and motivational music) as an ergogenic aid in exercise performance.

5 Conclusion

In summary, systematic reports on music as a sedative or stimulant aid in performing the exercise, particularly in pre-task music conditions are limited and infrequent [18]. There is still a lack of evidence to support the overall ergogenic effects of pre-task music on exercise performance. A comprehensive investigation pertaining to the effect of pre-task music should be conducted with diverse exercise intensities, research design and methodologies which might have influenced different outcomes. Furthermore, it would be interesting if other researchers could provide a variety of music applications and determine which selected or preferred music has the greatest impact on pre-task exercise. In fact, listening to music results in greater performance among individuals or sports practitioners. Future studies can be conducted in a more diverse populations (e.g., inactive individuals, primary school athletes).

Acknowledgements This research was supported by the Ministry of Higher Education Malaysia for Fundamental Research Grant Scheme (FRGS) with Project Code: FRGS/1/2020/SKK06/USM/03/13.

References

1. Bishop D (2010) Boom boom how: optimising performance with music. Sport Exerc Psychol Rev 6(1):35–47
2. Eliakim M, Meckel Y, Nemet D, Eliakim A (2007) The effect of music during warm-up on consecutive anaerobic performance in elite adolescent volleyball players. Int J Sports Med 28(4):321–325. https://doi.org/10.1055/s-2006-924360
3. Biagini M, Brown L, Coburn J, Judelson D, Statler T, Bottaro M, Longo N (2012) Effects of self-selected music on strength, explosiveness, and mood. J Strength Cond Res 26(7):1934–1938
4. Karow M, Rogers R, Pederson J, Williams T, Marshall M, Ballmann C (2020) Effects of preferred and nonpreferred warm-up music on exercise performance. Percept Mot Skills 127(5):912–924. https://doi.org/10.1177/0031512520928244
5. Ballmann C, McCullum M, Rogers R, Marshall M, Williams T (2021) Effects of preferred vs. nonpreferred music on resistance exercise performance. J Strength Cond Res 35(6):1650–1655. https://doi.org/10.1519/JSC.0000000000002981
6. Terry P, Karageorghis C, Curran M, Martin O, Parsons-Smith R (2019) Effects of music in exercise and sport: a meta-analytic review. Psychol Bull 146(2):91–117. https://doi.org/10.1037/bul0000216
7. Karageorghis C, Kuan G, Schiphof-Godart L (2021) Music in sport: from conceptual underpinnings to applications. Essent Exerc Sport Psychol: An Open Access Textbook 530–564. https://doi.org/10.51224/b1023
8. Chtourou H, Jarraya M, Aloui A, Hammouda O, Souissi N (2012) The effects of music during warm-up on anaerobic performances of young sprinters. Sci Sports 27(6):e85–e88. https://doi.org/10.1016/j.scispo.2012.02.006
9. Smirmaul B (2017) Effect of pre-task music on sports or exercise performance. J Sports Med Phys Fitness 57(7–8):976–984. https://doi.org/10.23736/S0022-4707.16.06411-2
10. Yeoh M, Chan CJ, Minni A, Hamid T (1998) Music preference and its intra-musical reasons. https://doi.org/10.13140/RG.2.1.1747.0161
11. Ballmann C (2021) The influence of music preference on exercise responses and performance: a review. J Funct Morphol Kinesiol 6(2). https://doi.org/10.3390/jfmk6020033
12. Stork M, Karageorghis C, Martin Ginis K (2019) Let's go: psychological, psychophysical, and physiological effects of music during sprint interval exercise. Psychol Sport Exerc 45:101547. https://doi.org/10.1016/j.psychsport.2019.101547
13. Queiroz FS De (March 2012) The effects of self-selected asynchronous pre-task music on performance in a soccer task. Dep Phys Educ Sports Sci
14. Bigliassi M, Dantas J, Carneiro J, Smirmaul B, Altimari L (2012) Influence of music and its moments of application on performance and psychophysiological parameters during a 5 km time trial. Rev Andal Med Deport 5(3):83–90
15. Chtourou H, Chaouachi A, Hammouda O, Chamari K, Souissi N (2012) Listening to music affects diurnal variation in muscle power output. Int J Sports Med 33(01):43–47

16. Jarraya M, Chtourou H, Aloui A, Hammouda O, Chamari K, Chaouachi A, Souissi N (2012) The effects of music on high-intensity short-term exercise in well-trained athletes. Asian J Sports Med 3(4):233
17. Serra de Queiroz F (2012) The effects of self-selected asynchronous pre-task music on performance in a soccer task (Master's thesis)
18. Smirmaul B, Dos Santos R, Da Silva NL (2014) Pre-task music improves swimming performance. J Sports Med Phys Fitness 55(12):1445–1451
19. Chtourou H, Hmida C, Souissi N (2017) Effect of music on short-term maximal performance: sprinters versus long-distance runners. Sport Sci Health 13(1):213–216
20. Karageorghis C, Cheek P, Simpson S, Bigliassi M (2018) Interactive effects of music tempi and intensities on grip strength and subjective affect. Scand J Med Sci Sports 28(3):1166–1175. https://doi.org/10.1111/sms.12979
21. Kuan G, Morris T, Kueh YC, Terry P (2018) Effects of relaxing and arousing music during imagery training on dart-throwing performance, physiological arousal indices, and competitive state anxiety. Front Psychol 9:14
22. Belkhir Y, Rekik G, Chtourou H, Souissi N (2019) Listening to neutral or self-selected motivational music during warm-up to improve short-term maximal performance in soccer players: effect of time of day. Physiol Behav 204:168–173
23. Fox R, Michael T, Weideman C, Hanson N (2019) Effect of listening to music during a warmup on anaerobic test performance. Sport Sci Health 15(2):369–373
24. Puad S, Ismail M, Nor M, Tumijan W, Kassim A, Amiruddin M (2020) The effects of music during warm-up on anaerobic performance of football players. Int J Hum Mov 8:477–482. https://doi.org/10.13189/saj.2020.080621
25. Belkhir Y, Rekik G, Chtourou H, Souissi N (2020) Effect of listening to synchronous versus motivational music during warm-up on the diurnal variation of short-term maximal performance and subjective experiences. Chronobiol Int 37(11):1611–1620
26. Meglic C, Orman C, Rogers R, Williams T, Ballmann C (2021) Influence of warm-up music preference on anaerobic exercise performance in division I NCAA female athletes. J Funct Morphol Kinesiol 6(3):64
27. Ballmann C, Favre M, Phillips M, Rogers R, Pederson J, Williams T (2021) Effect of pre-exercise music on bench press power, velocity, and repetition volume. Percept Mot skills 128(3):1183–1196
28. Juslin P (2013) From everyday emotions to aesthetic emotions: towards a unified theory of musical emotions. Phys Life Rev 10(3):235–266. https://doi.org/10.1016/j.plrev.2013.05.008
29. Karageorghis C, Mouzourides D, Priest D, Sasso T, Morrish D, Walley C (2009) Psychophysical and ergogenic effects of synchronous music during treadmill walking. J Sport Exerc Psychol 31(1):18–36

Online Learning Readiness Scale for Undergraduate Students in Sports and Physical Education in Malaysian Context

Teng Kie Yin, Chin Ngien Siong, and Lee Hou Yew

Abstract The initial scale of Online Learning Readiness Scale (OLRS) was developed to define the readiness towards online learning among college students. The five subscales of OLRS (Hung et al. in Comput Educ 1080–1090, [8]) were self-directed learning (SDL), motivation for learning (MFL), computer/Internet self-efficacy (CIS), learner control (LC), and online communication self-efficacy (OCS). The study's aim is to investigate the scale's cross-validation and reliability testing among Malaysian undergraduate students studying sports and physical education. Exploratory factor analysis (EFA) that has been conducted, showed that OLRS (within Malaysian context) is a multi-dimensional scale with two factors structured, namely solitary learning (SL) and computer/Internet self-efficacy (CIS). Cronbach's alpha reliability coefficient value was at 0.928. The findings revealed that the 13-item-OLRS is a valid and reliable instrument for assessing online learning readiness among Malaysian undergraduate students studying sports and physical education.

Keywords Online learning · Readiness · Exploratory factor analysis · Sports and physical education

1 Introduction

Around the world, educational institutions have been forced to close due to the COVID-19 pandemic. According to Pokhrel and Chhetri [1], more than 200 countries and roughly 1.6 billion pupils are affected, making up nearly 94% of the global student population. The traditional face-to-face learning model has been challenged, and online learning has taken centre stage instead [2]. Online education can be the answer to the COVID-19 dilemma [3]. Students had no choice but to accept and

T. K. Yin (✉) · L. H. Yew
Institute of Teacher Education, Tun Abdul Razak Campus, 94300 Kota Samarahan, Sarawak, Malaysia
e-mail: teng.kieyin@ipgm.edu.my

C. N. Siong
Institute of Teacher Education Batu Lintang Campus, 93200 Kuching, Sarawak, Malaysia

G. Kuan et al. (eds.), *Advancing Sports and Exercise via Innovation*, Lecture Notes in Bioengineering, https://doi.org/10.1007/978-981-19-8159-3_34

adjust to the crucial global incident as their classrooms were converted into online learning settings.

Numerous studies have demonstrated the benefits of online learning for students [4–6]. The learning process would be less successful without learning readiness, though. The viability of online education will benefit students' academic development [7]. The verification had been performed on the Online Learning Readiness Scale (OLRS) with five factors such as motivation to learn, self-directed learning, computer/Internet self-efficacy, learner control, and online communication self-efficacy [8]. The results of the confirmatory factor analysis (CFA) OLRS scale showed that the number of components and the loadings of measured variables on those components are consistent with expectations from underlying theories.

2 Method

The aim is to examine the ORLS measure from the Malaysian context using exploratory factor analysis. The ORLS consists of 18 items and five latent factors: computer and Internet self-efficacy, CIS (five items); self-directed learning, SDL (three items); learner control, LC (three items), motivation for learning (online context), MFL (four items); and online communication self-efficacy, OCS (three items). The OLRD five-point-Likert scale ranged from strongly disagree (1), somewhat disagree (2), neutral (3), somewhat agree (4), and strongly agree (5) [8]. A total of 210 students from four Malaysian higher education institutes volunteered to participate in the study, with males accounting for 51.43% and females with 48.57% of the total.

2.1 Cross-Cultural Adaptation Procedure

Researchers have completed a few necessary steps for scale cross-cultural adaption, which were highlighted in Bourzgui et al. [9], Silveira et al. [10], and Chae et al. [11], which included ethical considerations, forward and backward translation, and expert validation. However, subsequent paragraphs are indented.

2.2 Construct Validation

Determining the fundamental principles of the instrument's adapted scale in the context of Malaysia was also essential. Researchers used data from 210 respondents to conduct exploratory factor analysis (EFA) and obtained three underlying structures of the adapted subscales using IBM SPSS Statistics 24 and the five-step EFA technique recommended by Williams et al. [12].

2.3 Reliability Testing

Internal consistency reliability has been developed as one of the common estimators of reliability for the questionnaire's modified scales [13]. In order to determine the Cronbach's alpha for each customised subscale of the instrument used in this study, the researchers used IBM SPSS Statistics 24.

3 Results and Discussion

3.1 Exploratory Factor Analysis

A multivariate statistical technique called exploratory factor analysis (EFA) is used to show the underlying structure of a large collection of variables. The EFA objective which is a method used in factor analysis, is to discover the underlying correlations between measured variables [14]. The academics utilised it to identify a set of latent constructs that support a battery of measurable variables while developing a scale, which is a collection of questions designed to evaluate a particular research topic [15]. The researchers looked at the primary dimensions of a sizable variety of latent constructs, which are typically represented by a group of items [16–19]. This investigation used the five-step EFA process created by Williams et al. [12]

Choosing if the data are appropriate for EFA is the first step. Walker and Madden [20] state that ordinarily distributed interval or ratio data are the main presumption of factor analysis. Because ordinal data with at least four response categories are assumed to be regularly distributed when more than 200 samples are acquired, Chua [21] also thinks that ordinal data with at least four response categories are sufficient for factor analysis.

The factor extraction would be performed after the Bartlett's test of sphericity and the Kaiser-Mayer-Olkin (KMO) measure of sampling adequacy [22, 23]. Hair et al. [24] and Tabachnick and Fidell [25] recommended that the KMO index of 0.50 and above and p-value of 0.05 for the Bartlett's test of sphericity, are suitable for the test. Whereas for the EFA, it was recommended that the KMO index with a value of 0.50 or above and a p-value of 0.05 for the Bartlett's test of sphericity be utilised [24, 25]. The factor analysis was therefore allowed because the KMO index and Bartlett's test of sphericity showed that there was a relationship between the items [26].

The scale's KMO coefficient was determined to be 0.918 and satisfactory (Table 1). Bartlett's sphericity test's $p = 0.000$ result, meanwhile, indicated that there was a significant correlation between the study's variables. The extraction technique was then chosen by the researchers to carry out the EFA in the subsequent stage. The principal axis factoring (PAF) methodology was selected as the factor extraction method for this study because, throughout the factor extraction process, it applies communality estimations and examines the relationships among measured variables.

Table 1 KMO and Bartlett's test for the EFA on 13-item-OLRS

Kaiser–Meyer–Olkin measure of sampling adequacy		0.918
Bartlett's test of sphericity	Approx. chi-square	1683.199
	df	78
	sig	0.000

In step three, the EFA's quality is assessed by maintaining the proper amount of elements [27]. There were several guidelines and procedures to follow during factor extraction. For instance, the scree test, the eigenvalue, and the cumulative percent of variance all extracted at least 40% of the variance [28, 29]. Reckase [30] also advises that at least 20% of the variance in valid scales must be explained by the prime component. The test that need to be triangulated are the EV > 1 criterion and scree test with a priori hypothesis that underpins the inquiry to prevent ambiguous outcomes [31].

Table 2 displayed an eigenvalue of EV > 1 for all three components, with a cumulative percentage of variance of 54.2%. The eigenvalues of the three factors ranged from 1.204 to 7.050. The primary component explained 54.2% of the overall variation. In addition, the scree plot's third point featured a visual "elbow" (Fig. 1). The scree test's findings dictate that the data be examined for three factors.

Utilising the rotating approach enhances the interpretability of retrieved components, claim Ruscio and Roche [32]. Williams et al. [12] claim that rotation maximises high item loadings while minimising low item loadings. The elements under the adapted scales were believed to be inter-correlated, so the researchers applied the oblique Direct Oblimin (DO) rotation approach in step four. For psychological and educational studies, such as those examining human behaviours, oblique rotations result in the linked components that were thought to yield more accurate results [33, 34].

The factors' interpretation was the EFA's last stage. The researchers looked into the variables that could be attributed to a factor. The correlation matrix and factor pattern matrix were examined to ascertain the relationship of a certain factor without impacting other variables [35]. When evaluating a factor, the factor coefficients also referred to as loadings are extremely important [18]. It would be regarded appropriate to load at least 0.32 [31].

A factor normally has to have at least two or three variables applied to it in order to be relevant and interpretable [18, 36, 37]. Munro [38] contends that extraneous elements should be eliminated if they do not define the construct.

Two components were obtained via PAF extraction and DO rotation processes, with factor loadings surpassing 0.32, as predicted, according to the pattern matrix in Table 3. Five items (SDL3, LC2, OCS1, OCS2, and OCS3) that were determined to be inappropriate or cross-loaded were excluded from further investigation. However, 13 items were kept after EFA.

After the initial three components of self-directed learning (SDL), learner control (LC), and motivation for learning were integrated, the term solitary learning (SL)

Table 2 EFA total variance on 13-item-OLRS

Factor	Initial eigenvalues			Extraction sums of squared loadings			Rotation sums of squared loadings[a]
	Total	% of variance	Cumulative %	Total	% of variance	Cumulative %	Total
1	7.050	54.230	54.230	6.632	51.013	51.013	6.329
2	1.204	9.263	63.494	0.814	6.263	57.275	4.519
3	0.858	6.598	70.091				
4	0.670	5.152	75.243				
5	0.610	4.691	79.935				
6	0.491	3.774	83.709				
7	0.410	3.154	86.863				
8	0.385	2.962	89.825				
9	0.369	2.840	92.665				
10	0.299	2.302	94.968				
11	0.245	1.886	96.854				
12	0.222	1.707	98.561				
13	0.187	1.439	100.00				

Extraction method: Principal axis factoring
[a] When factors are correlated, sums of squared loadings cannot be added to obtain a total variance

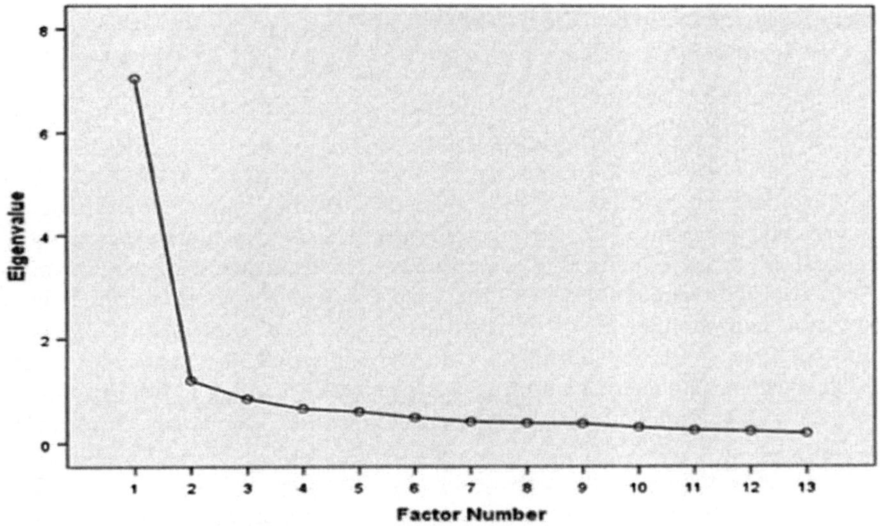

Fig. 1 Scree test criterion for the EFA on 13-item-OLRS

Table 3 Pattern matrix for the EFA on 13-item-OLRS

Item	Factor	
	1	2
CIS1		0.819
CIS2		0.723
CIS3		0.637
SDL1	0.469	
SDL2	0.455	
SDL4	0.905	
SDL5	0.888	
LC1	0.822	
LC3	0.533	
MFL1	0.674	
MFL2	0.630	
MFL3	0.661	
MFL4	0.798	

was formed (MFL). The second element, computer and Internet self-efficacy (CIS), however, was left in the original scale because it had not been deleted. SDL1, SDL2, SDL4, SDL5, LC1, LC3, MFL1, MFL2, MFL3, and MFL4 made up the ten items that made up SL. They were combined and used to gauge the pupils' capacity for independent study. The second element, CIS, had three measures about the students' computer and Internet skills (CIS1, CIS2, and CIS3).

3.2 *Reliability Coefficient of OLRS*

The reliability coefficient measures the number of real variance present in an arranged unprocessed test results [39]. The reliability coefficients, which are variance estimates, show the degree of genuine score variance. To estimate test-score reliability, at least two observations (scores) on the same group of people are required. The correlation between the sets of observations can be used to compute a dependability coefficient [40]. The item reliability for the finalised OLRS was assessed using the Cronbach's alpha internal consistency coefficient. According to Nunnally [41], Cronbach's alpha should be more than or equal to 0.70 in order to have acceptable reliability.

In this study, the reliability analysis for $N = 210$ produced satisfactory results (Table 4). The 13-item-OLRS scale as a whole was determined to be 0.928 reliable. On the other hand, the variables SL and MI have Cronbach's alpha values of 0.923 and 0.817, respectively. In conclusion, it was determined that the 13-item-OLRS was reliable.

Table 4 Reliability of the factors in the 13-item-OLRS

Factor	Number of items	Cronbach's alpha values
SL	10	0.923
CSI	3	0.817

4 Conclusion

The utility of this study hinges on its determination of the OLRS's validity and reliability. This study provided proof that the 13-item-OLRS can be a valid and reliable scale to evaluate readiness of students for online learning in sports, exercise, and physical education in Malaysia.

Acknowledgements The Institute of Teacher Education Tun Abdul Razak Campus funded this study on digital learning.

References

1. Pokhrel S, Chhetri R (2021) A literature review on impact of covid-19 pandemic on teaching and learning. High Educ Future 8(1):33–141
2. Vijayan R (2021) Teaching and learning during the covid-19 pandemic: A topic modeling study. Educ Sci 11(347):1–15
3. Dhawan S (2020) Online Learning: A panacea in the time of COVID-19. J Educ Technol Syst 49(1):5–22
4. Akkoyunlu B, Soylu MY (2006) A study on students' views on blended learning environment. Turk Online J Distance Educ 7(3):43–56
5. Alqahtani AA (2010) The effectiveness of using e-learning, blended learning and traditional learning on students. achievement and attitudes in a course on islamic culture: An experimental study (Unpublished doctoral thesis). Durham University, England
6. Chang SH, Smith RA (2008) Effectiveness of personal interaction in a learner centered paradigm distance education class based on student satisfaction. J Res Techno Educ 40(4):407–426
7. Engin M (2017) Analysis of students' online learning readiness based on their emotional intelligence level. Univers J Educ Res 5(12A):32–40
8. Hung ML, Chou C, Chen CH, Own ZY (2010) Learner readiness for online learning: Scale development and student perceptions. Comput Educ 1080–1090
9. Bourzgui F, Serhier Z, Sebbar M, Diouny S, Othmani MB, Ngom PI (2015) Adaptation and validation of the moroccan arabic version of the psychosocial impact of dental aesthetics questionnaire (PIDAQ). The Saudi Dent J 1–7
10. Silveira C, Parpinelli MA, Pacagnella RC, de Camorgo RS, Costa ML, Zanardi DM, Ferreira EC, Santos JP, Hanson L, Cecatti JG, Andreucci CB (2013) Cross-cultural adaptation of the World Health Organization Disability Assessment Schedule (WHODAS 2.0) into Portuguese. Revista da Associação Médica Brasileira 59(3):234–240
11. Chae SM, Kim CJ, Yoo H (2010) Psychometric evaluation of the Korean version of the adapted illness intrusiveness rating scale. Asian Nurs Res 4(4):194–204
12. Williams B, Brown T, Onsman A (2010) Exploratory factor analysis: A five-step guide for novices. Australas J Paramedicine 8(3):1–13

13. Nasrin P, Trisha DAM (2009) Developing and validating a questionnaire to measure spiritual: A psychometric process. Glob J Health Sci 1(1):1–11
14. Norris M, Lecavalier L (2009) "Evaluating the use of exploratory factor analysis in developmental disability psychological research. J Autism Dev Disord 40(1):8–20
15. Fabrigar LR, Wegener DT, MacCallum RC, Strahan EJ (1999) Evaluating the use of exploratory factor analysis in psychological research. Psychol Methods 4(3):272–299
16. Henson RK, Roberts JK (2006) Use of exploratory factor analysis in published research: Common errors and some comment on improved practice. Educ Psychol Meas 66(3):393–416
17. Thompson B (2004) Exploratory and confirmatory factor analysis: Understanding concepts and applications. American Psychological Association, Washington, D. C.
18. Swisher LL, Beckstead JW, Bebeau MJ (2004) Factor analysis as a tool for survey analysis using a professional role orientation inventory as an example. Phys Ther 84(9):784–799
19. Pett MA, Nancy RL, Sullivan JJ (2003) Making sense of factor analysis: The use of factor analysis for instrument development in health care research, 1st edn. Sage Publications Inc., California
20. Walker J, Madden S (2008) Chapter 14: Factor analysis, path analysis and structural equation modeling. In: Statistics in Criminology and Criminal Justice: Analysis and Interpretation, 3rd edn, Jones and Bartlett Learning
21. Chua YP (2014) "Kaedah dan statistik penyelidikan buku 5: Ujian regresi, analisis faktor dan analisis SEM". 2nd edn. McGraw-Hill Education, Selangor
22. Kaiser HF (1956) The varimax method of factor analysis. University of California, Berkeley, Doctor Philosophy
23. Bartlett MS (1950) Test of significance in factor analysis. Br J Psychol 3:77–85
24. Hair J, Anderson RE, Tatham RL, Black WC (1995) Multivariate data analysis, 4th edn. Prentice-Hall Inc., New Jersey
25. Tabachnick BG, Fidell LS (2007) Using multivariate statistics, 5th edn. Allyn & Bacon, Boston, MA
26. Burton LJ, Mazerolle SM (2011) Survey instrument validity part 1: Principles of survey instrument development and validation in Athletic Training Education Research. Athl Train Educ J 6(1):27–35
27. Conway JM, Huffcutt AIA (2003) Review and evaluation of exploratory factor analysis practices in organisational research. Organ Res Methods 6:147–168
28. Cattell RB (1966) The Scree test for the number of factors. Multivariate Behav Res 1:245–276
29. Thomas GRJr, Brad S, (2015) Exploratory factor analysis: Implications for theory, research and practice. Adv Dev Hum Resour 17(1):12–25
30. Reckase MD (1979) Unifactor latent trait models applied to multifactor tests: Results and implications. J Educ Stat 4:207–230
31. Treiblmaier H, Filzmose P (2010) Exploratory Factor Analysis revisited: How robust methods support the detection of hidden multivariate data structures in IS research. Inf Manag J 47:197–207
32. Ruscio J, Roche B (2012) Determining the number of factors to retain in an exploratory factor analysis using comparison data of known factorial structure. Psycholog Assess 24:282–292
33. Schmitt TA (2011) Current methodological considerations in exploratory and confirmatory factor analysis. J Psychoeduc Assess 29(4):304–321
34. Costello AB, Osborne JW (2005) Best practices in exploratory factor analysis: Four recommendations for getting the most from your analysis. Practical Assessment, Res Eval 10(7):1–9
35. Stevens J (1992) Applied multivariate statistics for the social sciences, 2nd edn. Lawrence Erlbaum, Hillsdale NJ
36. Raubenheimer J (2004) An item selection procedure to maximise scale reliability and validity. J Ind Psycholo 30(4):59–64
37. Isaac S, Michael WB (1997) Handbook in Research and Evaluation, 3rd ed. California: Educational and industrial testing services. Developmental stages of developmental screening: steps to implementation of successful programme. Am J Public Health 95(11):1928–1932

38. Munro BH (2005) Statistical methods for health care research. Lippincott. Williams & Wilkins, Philadelphia
39. Aron A, Aron EN (2003) Statistics for psychology, 3rd edn. Prentice Hall, Upper Saddle River
40. Piedmont RL (2014) Reliability Coefficient. In: Michalos, A. C. (eds) Encyclopedia of quality of life and well-being research. Springer, Dordrecht
41. Nunnally JC, Bernstein IH (1994) Psychometric theory. McGraw-Hill, New York

Examining Cost-Value Dimensions Among Adolescent Field Hockey Players: A Demographic Study

Kanagarajah Rarujanai, Chin Ngien Siong, and Ruebini Parama Dorai

Abstract Cost-value has been particularly difficult to assess as evidenced by the historical lack of research on it. Examining cost-value dimensions as an independent construct in the expectancy-value model could offer significant information about adolescent players' future performance and achievement, especially in field hockey. This study investigated the differences in all dimensions of cost-value in adolescent field hockey players in terms of gender, age groups and locality. Participants included 250 Malaysian adolescent field hockey players, who completed the 19-items cost scale designed by Flake et al (Contemp Educ Psychol 41:232–244, [5]) comprised of five different dimensions of cost-value. Results revealed that female players showed greater practice of cost dimension compared to male players. All five dimensions under cost-value showed no significant difference in term of age group. While, participants from northern Sarawak were more determined towards the practicing cost-value dimension compared to other participants from other parts of the state. It is suggested that incorporating cost as an important construct in the expectancy-value framework could capture best the relations between players' motivation and their performance in sports.

Keywords Cost-value · Expectancy-value · Field hockey · Adolescent

1 Introduction

Motivation science deals with numerous different theories to investigate student motivation in various fields [1]. This study focuses on the expectancy-value models

K. Rarujanai (✉)
Institute of Teacher Education, Tun Abdul Razak Campus, Sarawak, Malaysia
e-mail: kanagarajah@ipgm.edu.my

C. N. Siong
Institute of Teacher Education Batu Lintang Campus, Sarawak, Malaysia

R. P. Dorai
Sekolah Kebangsaan Landeh, Sarawak, Malaysia

G. Kuan et al. (eds.), *Advancing Sports and Exercise via Innovation*, Lecture Notes in Bioengineering, https://doi.org/10.1007/978-981-19-8159-3_35

theory [2] to understand student motivation in field hockey. Many researchers have widely employed the expectancy-value model of achievement-related choices [2] in educational and psychological researches [3, 4]. This theoretical outline explains that motivation has two different functions; expectancy belief and task value components [5].

Specifically, one element within the expectancy-value model, known as cost, explains the demand for a given task. Wigfield and Eccles [5] defined cost as "what an individual has to give up to do a task, as well as the anticipated effort one will need to put into task completion". Even though cost has been identified as one of the main components in EVM, it is the least explored component in EVM. Cost-value highlights the demand of a given task. It is said to be dealing with negative features of performing a task which includes loss of time and energy for other choices [3].

Intrinsic value, utility value, and attainment value are among the components under subjective task values. Cost is the final and most crucial element of subjective task evaluations. When someone engages in any activity, cost is seen as a crucial factor [5, 6]. They are seen as observed effects, such as the need for valuable time to participate in an activity, the dread of failure, and the loss of something significant to another person [7].

The cost of an activity is impacted by the amount of effort needed to succeed, the time lost from engaging in other activities, and the anticipated emotional states like performance anxiety and fear of failure, according to Eccles and her colleagues [2, 3]. The number of cost-value considerations that go into an activity has a significant impact on how interested people become. Even if it is their preferred or favourite activity, people are unlikely to participate in an activity if the cost is too high [6].

Athletes believe that in a sporting environment, they must restrict their access to other activities in order to participate in practices or contests. They find it difficult to participate in an activity because of the anticipated level of effort and the associated time or emotional cost [6]. For instance, a sportsperson who considers playing field hockey to be fascinating and valuable for a potential future job may feel unmotivated to participate in the scheduled training sessions due to the time and effort required to complete the sessions. The decision of the athlete to pursue the dream or decision can also be influenced by anxiety and the fear of failure [1].

Additionally, the psychological cost of potential failure is likely to affect this choice [5]. For instance, some extremely talented athletes with high accomplishment needs in sports are nevertheless unsure of their skills and qualities [8]. They frequently believe that in order to succeed in a specific sport, they will need to work incredibly hard. Failure has a significant cost for those athletes because failing to perform effectively would imply a lack of talent, which would be damaging to their perception of themselves [9].

Cost appears to be the antithesis of accomplishment, interest, and utility, which all share positive traits [10]. This kind of relationship suggests that cost is more appropriately thought of as a moderator variable of value. When there is a moderator relationship, understanding the level of one variable's influence on an outcome is crucial. Cost-value was divided into five categories by Flake et al [5]: task effort, outside effort, loss of desirable alternatives, and emotional. While outside effort focuses on the time, effort, or quantity of work put forth for tasks other than the activity of interest, task effort emphasizes the time, effort, or quantity put forward to engage in a task. What is given up as a result of engaging in the work of interest is explained next by the loss of valued alternatives. The psychological condition that develops as a result of expending effort to complete the activity is covered by emotional cost as well [8].

Considering young athletes participating actively in field hockey for various reasons, this study investigates the differences in all dimensions of cost-value in adolescent field hockey players in terms of gender, age groups, and locality. Based on the current literature, to date, there is no study on Malaysian adolescent field hockey players' cost-value dimensions had been conducted involving any specific sports especially field hockey in Malaysia. Indirectly, this study will fill in the gap of the Malaysian adolescent field hockey players' cost-value related dilemma by contributing to the knowledge body. By knowing the demands of the task for Malaysian adolescent field hockey players, better training and development programme can be designed to suit their needs and hopefully help to improve hockey performance.

For the current study, we used the 19-item cost scale designed by Flake et al [5] as a medium to achieve the objective of this study. An observed subscale score mean was created using the cost scale, which includes each cost component (task effort cost, outside effort cost, loss of valued alternatives, and emotional cost). Therefore, for the purposes of our study and upcoming research in Malaysia, we translated these surveys into Malay. No study in Malaysia has used a questionnaire to thoroughly assess beliefs, values, and sportsmanship attitudes. It is critical to comprehend how these factors affect both the performance of the individual players and the team's performance.

2 Methods

2.1 Participants

This study used a cross-sectional study design. The participants of this study came from various districts of Sarawak, Malaysia. Players in their adolescent years who play field hockey totalled 250 for the study. Participants ranged in age from 12 to 19 and were engaged in competitive field hockey in Sarawak at the club, school, and state levels. The participants for this study were 126 males (50.4%) and 124 female (49.6%). The age groups of participants are divided into 16–19 years (58.4%), and 12–15 years (41.6%). The majority of the participants were predominantly from the southern zone of Sarawak (47.2%) followed by the central zone (33.2%) and northern zone (31.7%). Participants were informed of the goal and scope of the study prior to data collection, and their informed consent was acquired.

2.2 Measures Demographic Characteristics

A number of demographic traits were added based on the literature already stated. It was specifically requested that participants provide their gender, age, birth date, race, school or institution address, and level of sport participation.

Cost-value

The 19-item cost scale created by Kanagarajah et al [11] to measure four categories of cost was translated for use with the adolescents to determine the cost values associated with their involvement in field hockey. All items on this scale had the same response scale and were presented in a random order: 7-points with 1 = "completely agree", 4 = "neither agree nor disagree", and 7 = "completely agree". Task effort cost factor was represented by five items, and emotional cost factor was denoted by six items. Outside effort cost and loss of valued alternatives factors were represented by four items each.

2.3 Procedures

The participants of this study were selected using random cluster sampling from state-level junior hockey competitions organized by state hockey associations as well as the state education department. Participants under the age of 18 gave written consent from their parents, and those above 18 signed the consent form. The study's goals and the participants' rights were explained to the participants. They were advised to only reply to any question if it makes them feel comfortable. Participants were made aware that they might leave at any time without facing any consequences.

2.4 Statistical Analysis

For the purpose of computing the descriptive analysis, correlation analysis, factorial analysis, reliability analysis, multiple comparisons (Bonferroni's test), and regression analysis, SPSS version 23 and AMOS version 23.0 software were used. Independent t-tests were used to compare gender differences independently. To analyse the variations among the age groups, a two-way ANOVA was utilized. Alpha p 0.05 was the cut-off value for significance. Later, ES was calculated and classified in accordance with Cohen's (1988) suggestions.

Table 1 Descriptive statistic results of cost-value scale dimensions ($N = 250$)

	Mean (M)	*SD*	α
Overall	4.69	0.79	0.92
Task effort cost	5.04	0.70	0.78
Outside effort cost	4.94	1.28	0.95
Loss of valued alternatives	4.89	0.51	0.68
Emotional cost	4.11	1.07	0.82

Note SD standard deviation; α Cronbach's alpha

3 Results

The descriptive statistics for the cost-value measure were presented as a whole. Gender, age group, and locality differences were presented accordingly.

3.1 *Descriptive Statistics for Cost-Value and Its Four Dimensions*

Table 1 shows the descriptive statistics of each factor included in the cost scale. Firstly, the reliability values (Cronbach's alpha coefficient) of all factors were over 0.70, with the exception of the loss of valued alternatives factor, which had a slightly lower value (0.68). Next, means and standard deviations of each factor were calculated. The overall mean value for the cost scale was $M = 4.69 \pm 0.79$. Looking at the four cost dimensions, higher values were obtained for task effort cost ($M = 5.04 \pm 0.70$) followed by outside effort cost ($M = 4.94 \pm 1.28$). Whereas emotional cost had the lowest mean ($M = 4.11 \pm 1.07$).

3.2 *Gender Differences*

Table 2 showed that there was a significant gender difference in the outside effort cost ($M = 5.17$ versus 4.70; $p = 0.004$) and emotional cost ($M = 3.81$ versus 4.43; $p = 0.000$) dimensions. Female adolescent field hockey players scored higher mean value in all dimensions (task effort cost, $M = 4.97$ versus 5.10; loss of valued alternatives, $M = 4.85$ versus 4.93, and emotional cost, $M = 3.81$ versus 4.43) except for outside effort cost ($M = 5.17$ versus 4.70). Overall, female players showed higher total scores for cost dimension compared to male players ($M = 4.62$ versus 4.77).

Table 2 Cost scale dimension according to gender

Subscales		M	*SD*	t-test	*p*
Overall	Male	4.62	0.47	−1.493	0.139
	Female	4.77	1.02		
Task effort cost	Male	4.97	0.67	−1.569	0.118
	Female	5.10	0.73		
Outside effort cost	Male	5.17	0.91	2.947	0.004*
	Female	4.70	1.54		
Loss of valued alternatives	Male	4.85	0.36	−1.158	0.250
	Female	4.93	0.62		
Emotional cost	Male	3.81	0.55	−4.830	0.000*
	Female	4.43	1.34		

Note M mean; *SD* standard deviation
* $p < 0.05$

3.3 Age Group Differences

Table 3 showed that there was no significant age group difference in terms of overall cost-value as well as all the dimensions under cost-value. The p-values for all four variables were higher than 0.05, indicating that there were no noticeable differences in the participants' ages across the age groups. Overall, men players outperformed female players in terms of overall scores across all three variables. All cost-value dimensions (task effort cost, $M = 4.99$ versus 5.10; outside effort cost, $M = 4.83$ versus 5.08; loss of valued alternatives, $M = 4.88$ versus 4.91; and emotional cost, $M = 4.11$ versus 4.13) were higher on average for younger participants (12–15 years old). Overall, 12–15 years old field hockey players showed higher total scores for cost dimension compared to players aged 16–19 years old ($M = 4.65$ versus 4.75).

Table 3 Cost scale dimension according to age group

Subscales		*M*	*SD*	t-test	*p*
Overall	16–19 years	4.65	0.97	−0.909	0.309
	12–15 years	4.75	0.43		
Task effort cost	16–19 years	4.99	0.74	−1.277	0.190
	12–15 years	5.10	0.62		
Outside effort cost	16–19 years	4.83	1.55	−1.480	0.101
	12–15 years	5.08	0.74		
Loss of valued alternatives	16–19 years	4.88	0.57	−0.427	0.651
	12–15 years	4.91	0.40		
Emotional cost	16–19 years	4.11	1.27	−0.129	0.887
	12–15 years	4.13	0.68		

Note M mean; *SD* standard deviation

3.4 Location Differences

In terms of the cost aspects, a one-way analysis of variance (ANOVA) was calculated for participants in various locations (Table 4). Participation in field hockey had a significant main effect on cost-value: $F(2, 247) = 14.49, p = 0.000$. In comparison with individuals from the central zone ($M = 4.33$, $SD = 0.58$) and southern zone ($M = 4.84$, $SD = 0.83$), participants from northern Sarawak were more committed to the entire cost-value dimension ($M = 4.95$, $SD = 0.83$). Participants from the southern zone attributed the task effort cost dimension ($M = 5.28$, $SD = 0.58$) and loss of valued alternative dimension ($M = 5.00$, $SD = 0.48$) as more valuable than participants from other two parts of Sarawak. Meanwhile, participants from northern zone attributed the outside effort cost dimension ($M = 5.39$, $SD = 1.25$) and emotional cost dimension ($M = 4.46$, $SD = 1.15$) as more valuable when compared with other participants from different locations. Participants from central zone scored the least mean score value in all the cost-value dimensions (task effort cost, $M = 4.58$, $SD = 0.64$); outside effort cost, $M = 4.60$, $SD = 1.00$); loss of valued alternative, $M = 4.70$, $SD = 0.48$); emotional cost ($M = 3.70$, $SD = 0.66$). Significant difference among variables was found in all four dimensions of cost-value with all the p-values showing < 0.05. Further, post-hoc Bonferroni test was conducted to locate those specific differences among the variables [12] (Table 5).

Bonferroni post-hoc test for multiple comparisons found that the mean value of overall cost-value was significantly different between northern and central zones ($p = 0.00$, 95% C.I. = [0.29, 0.95]) as well as central and southern zones ($p = 0.00$, 95% C.I. = [−0.77, −0.24]). There was no statistically significant difference between the northern and southern zones ($p = 1.00$). Under task effort cost dimension, northern and central zones ($p = 0.00$, 95% C.I. = [0.35, 0.89]) and central and southern zones ($p = 0.00$, 95% C.I. = [−0.91, −0.48]) showed significant differences for the mean value. Northern and southern zones showed no statistically significant difference between them ($p = 1.00$).

It was found that the mean value of outside effort cost dimension was significantly different between northern and central zones ($p = 0.00$, 95% C.I. = [0.24, 1.33]). There was no statistically significant difference between central and southern zones ($p = 0.17$) and northern and southern zones ($p = 0.11$). Under the loss of valued alternatives dimension, northern and central zones ($p = 0.02$, 95% C.I. = [0.03, 0.46]) and central and southern zones ($p = 0.00$, 95% C.I. = [−0.47, −0.13]) showed significant differences for the mean value. Northern and southern zones showed no statistically significant difference between the variables ($p = 1.000$). For the emotional cost dimension, there were no statistical differences between northern and southern zones ($p = 0.81$). Meanwhile, northern and central zones ($p = 0.02$, 95% C.I = [0.31, 1.20]) and central and southern zones ($p = 0.00$, 95% C.I. = [−0.91, −0.21]) showed significant statistical differences.

Table 4 Cost scale dimensions according to location

Subscales		*M*	*SD*	*F*	*p*
Overall	Northern	4.95	0.83	14.49	0.000*
	Central	4.33	0.58		
	Southern	4.84	0.83		
	Total	4.69	0.79		
Task effort cost	Northern	5.20	0.70	33.03	0.000*
	Central	4.58	0.64		
	Southern	5.28	0.58		
	Total	5.04	0.70		
Outside effort cost	Northern	5.39	1.25	6.16	0.002*
	Central	4.60	1.00		
	Southern	4.98	1.41		
	Total	4.94	1.28		
Loss of valued alternatives	Northern	4.94	0.53	9.68	0.000*
	Central	4.70	0.48		
	Southern	5.00	0.48		
	Total	4.89	0.51		
Emotional cost	Northern	4.46	1.15	10.63	0.000*
	Central	3.70	0.66		
	Southern	4.26	1.17		
	Total	4.11	1.07		

Note M mean; *SD* standard deviation
* $p < 0.05$

4 Discussion

To better understand the interaction between cost dimensions in predicting achievement-related outcomes, the current study examined the relationship between four cost-value components in a sample of Malaysian adolescent field hockey players. Data analyses revealed that female players showed higher total scores for cost dimension compared to male players. This result was similar to a study conducted by Vinni-Laakso et al [10] which revealed that the psychological factors involving cost dimensions of female participants were deemed greater than the male participants. This could be because girls might be more mature and ready to face the demands of the sport compared to boys [10]. In contrast, Osterlie [13] found that girls perceived significantly less practice of costs in attending their physical education or activities. Meanwhile, in the study conducted by Grasten [14] no gender differences were detected regarding interest value, utility value, or cost dimensions.

In terms of age group, this study revealed no significant difference in terms of overall cost-value as well as all the dimensions under cost-value. According to a

Table 5 Post-hoc Bonferroni test of cost-value dimension between location

Variables	μ difference	p	95% Confidence interval for difference	
			Lower bound	Upper bound
Overall				
Northern versus Central	0.62*	0.00	0.29	0.95
Northern versus Southern	0.11	1.00	−0.20	0.42
Central versus Southern	−0.51*	0.00	−0.77	−0.24
Task effort cost	0.62*	0.00	0.35	0.89
Northern versus Central	−0.08	1.00	−0.33	0.18
Northern versus Southern	−0.70*	0.00	−0.91	−0.48
Central versus Southern				
Outside effort cost	0.79*	0.00	0.24	1.33
Northern versus Central	0.41	0.17	−0.11	0.92
Northern versus Southern	−0.38	0.11	−0.81	0.06
Central versus Southern				
Loss of valued alternatives	0.25*	0.02	0.03	0.46
Northern versus Central	−0.06	1.00	−0.26	0.14
Northern versus Southern	−0.30*	0.00	−0.47	−0.13
Central versus Southern				
Emotional cost	0.75*	0.00	0.31	1.20
Northern versus Central	0.19	0.81	−0.23	0.61
Northern versus Southern	−0.56*	0.00	−0.91	−0.21
Central versus Southern				

* $p < 0.05$

study by Osterlie [13] including 338 students from six different secondary and higher secondary schools in Norway, the mean value was consistent across all of the students and did not alter noticeably from lower secondary to upper secondary classes.

Differences in regard to location were drawn to attention too in this study. The results showed that participants from northern Sarawak were more determined toward the practicing cost-value dimension compared to participants from central zone and southern zone. In Sarawak, only the northern region is equipped with field hockey stadium or turf facilities. Other regions do not have standardized field hockey playing turf/surface. This could be the reason for the participants from central and southern parts of the state to demonstrate more negative responses in form of cost-value dimensions. In their study with a total of 201,723 Korean participants, Lee et al [15] found that individuals who had easy access to sports facilities engaged in physical exercise more frequently than those who did not. Physical and mental preparedness are related to the accessibility of sports facilities [15]. Better sports facility supply is typically linked to greater sport participation and motivation, according to Eime et al [16].

5 Conclusion

Cost has been particularly difficult to assess as evidenced by the historical lack of research on it [4]. The expectancy-value model, specifically in field hockey, provided important information regarding adolescent players' future performance and achievement by looking at cost dimensions as an independent construct. Based on these findings, we propose that in order to adequately capture the relationships between players' motivation and their performance in sports, the expectancy-value framework must include cost as a significant component and evaluate expectancy, task value, and cost together. Given equal opportunity and exposure to sports, cost dimension will act as a good moderating platform for adolescent players regardless of gender or geographical differences. Knowing the weaknesses and strengthening the values through cost dimensions will surely leave a greater psychological impact on players' future participation and achievement in sports.

References

1. Pennington CG (2017) Moral development and sportsmanship in physical education and sport. J Phys Educ Recreat Dance 88:36–42
2. Eccles JS, Adler TF, Futterman R, Goff SB, Kaczala CM, Meece J, Midgley C (1983) Expectancies, values and academic behaviors. WH Freeman San Francisco
3. Simpkins SD, Fredricks JA, Eccles JS (2012) Charting the ECCLES' expectancy-value model from mothers' beliefs in childhood to youths' activities in adolescence. Dev Psychol 48:1019–1032
4. Zhu X, Chen A (2013) Adolescent expectancy-value motivation, achievement in physical education, and physical activity participation. J Teach Phys Educ 32:287–304
5. Flake JK, Barron KE, Hulleman C, McCoach BD, Welsh ME (2015) Measuring cost: the forgotten component of expectancy-value theory. Contemp Educ Psychol 41:232–244
6. Wigfield A, Eccles JS (2000) Expectancy–value theory of achievement motivation. Contemp Educ Psychol 25:68–81
7. Wigfield A, Cambria J (2010) Students' achievement values, goal orientations, and interest: definitions, development, and relations to achievement outcomes. Dev Rev 30:1–35
8. Jiang Y, Rosenzweig EQ, Gaspard H (2018) An expectancy-value-cost approach in predicting adolescent students' academic motivation and achievement. Contemp Educ Psychol 54:139–152
9. Meece JL, Eccles JS, Kaczala C, Goff SB, Futterman R (1982) Sex differences in math achievement: toward a model of academic choice. Psychol Bull 91:324–348
10. Vinni-Laakso J, Guo J, Juuti K, Loukomies A, Lavonen J, Salmela-Aro K (2019) The relations of science task values, self-concept of ability, and STEM aspirations among Finnish students from first to second grade. Front Psychol 10:1449
11. Rarujanai K, Wah TE, Ling A, Siong CN, Mansur (2020) Validation of the Malay-translated expectancy value model questionnaire and multi-dimensional sportspersonship orientations scale among Malaysian adolescence field hockey players. Cakrawala Pendidik 39:230–241
12. Lee S, Lee DK (2018) What is the proper way to apply the multiple comparison test?. Korean J Anesthesiol 71(5):353–360. https://doi.org/10.4097/kja.d.18.00242
13. Osterlie O (2018) Can flipped learning enhance adolescents' motivation in physical education? An intervention study. J Res Art Sport Educ 2:1–15

14. Grasten A (2015) Children's expectancy beliefs and subjective task values through two years of school-based program and associated links to physical education enjoyment and physical activity. J Sport Health Sci 5:500–508
15. Lee SA, Ju YJ, Lee JE, Hyun IS, Nam JY, Han K, Park E (2016) The relationship between sports facility accessibility and physical activity among Korean adults. BMC Pub Health 16:893
16. Eime RM, Harvey J, Charity MJ, Casey M, Westerbeek H, Payne WR (2017) The relationship of sport participation to provision of sports facilities and socioeconomic status: a geographical analysis. Aust N Z J Public Health 41:248–255

Profile of Sport Motivation and Sport Commitment in Malaysian Elite Badminton Players

Teo Eng Wah, Noh Young-Eun, Lim Boon Hooi, and Danaee Mahmoud

Abstract Sport-determination theory and sport commitment theory have been used to understand the motivation and commitment of athletes for the past three decades. However, no attempt has been carried to evaluate and compare these important latent variables on badminton. This study aims to evaluate the profile of sport motivation and sport commitment in elite badminton athletes. Secondly to compare gender and levels of participation on sport motivation and sport commitment profile. One hundred seventy-three elite badminton players age $M = 19.66$, $SD = 5.02$ years were recruited from entire Malaysia. The Sport Motivation Survey-II (SMS-II) and Sport Commitment Questionnaire were used for data collection. Nonparametric test was computed due to data non-normality. Analysis on level of participation (state vs. national athletes) did not showed significant difference for sport motivation except integrated regulation $H(1) = 5.756$, $p = .016$ while only sport enjoyment $H(1) = 6.638$, p = .010 was significantly different for sport commitment. Analysis on gender showed no significant different on all sport motivation constructs, while for sport commitment only desire to excel social $H(1) = 4.982$, $p = .026$ is significantly different. At elite level, the source of motivation for elite athletes is no longer intrinsic oriented but shift toward extrinsic motivated. However, the factor that push elite badminton players to continue play and trained is highly influenced by sport enjoyment (sport commitment).

Keywords Sport motivation · Sport commitment · Badminton · Elite athletes

T. E. Wah (✉) · N. Young-Eun · L. B. Hooi
Centre for Sport and Exercise Sciences, University of Malaya, Kuala Lumpur, Malaysia
e-mail: vteo2@um.edu.my

D. Mahmoud
Department of Social and Preventive Medicine, Faculty of Medicine, University of Malaya, Kuala Lumpur, Malaysia

G. Kuan et al. (eds.), *Advancing Sports and Exercise via Innovation*, Lecture Notes in Bioengineering, https://doi.org/10.1007/978-981-19-8159-3_36

1 Introduction

Self-determination theory (SDT) [1–3] is one of the popular theory used to explain sport motivation. In order to evaluate sport motivation, Brière et al [4] developed Échelle de Motivation dans les, in the same year Pelletier et al [5] translated the French version to English and named it Sport Motivation Scale (SMS). Initially, SMS consists of seven-subscale (i.e., intrinsic motivation to know, intrinsic motivation to accomplish and intrinsic motivation to experience, identified regulation, introjected regulation, extrinsic motivation, and amotivation). In 2013, Pelletier et al [6] proposed a revised six-subscale SMS-2; three intrinsic motivation subscales (from earlier version SMS) (i.e., to know, to accomplish and to experience) were collated to formed intrinsic regulation while the other subscales are integrated regulation, identified regulation, introjected regulation, external regulation, and amotivation. Based on the new revision, athlete who participates due to fun, enjoyment, and the love of sports is considered to be intrinsically motivated. This is followed by integrated regulation stage that is the assimilation of athlete's value and needs to participate. Identified regulation stage sport participation is related to the degree of importance a sport activity is to one's career for example one trained daily to qualify for the Olympics. While at introjected regulation stage, sport participation is due to feeling of guilt for instance if an individual did not play sports or go to the gym daily, there is a feel of guilt. In external regulation stage, external incentives become the main source of motivation such as rewards, money, trophy, and prestige become the source of motivation, and at the end of the continuum is amotivation, is a situation where there is lack of motivation to continue participating in any sports activities.

A closely related framework to SDT is sport commitment theory (SCT). Although both theories are commonly used to improve the understanding of sport participation but both theories look into different stages of participation. In general, SDT evaluates the reasons leading to sport participation and dropout while SCT evaluates the desire to continuously participate in sports [7]. Based on sport commitment model, Scanlan et al [7] developed Sport Commitment Questionnaire (SCQ). SCQ consists of five constructs/sources (i.e., sport enjoyment, involvement opportunities, involvement alternatives, personal investments, and social constraints). Later, Scanlan et al [8] expanded the original model to create Sport Commitment Questionnaire-2 (SCQ-2). The SCQ-2 consists of ten different sources or constructs of commitment [i.e., sport enjoyment, valuable opportunities, other priorities, personal investments (loss and quantity), social constraints, social support (emotional and informational), desire to excel (mastery achievement, and desire to excel social achievement)].

Current new SCT framework proposed to are divided sport commitment into enthusiastic commitment (functionally component or want to) and constrained commitment (obligatory component or have to). While the two newly added constructs are social support and desire to excel. According to Scanlan et al [8], social construct is posit to be the predictor of enthusiastic commitment, while desire to excel is predictor of constrained commitment. Sport enjoyment is defined as positive emotions derived through continuous sport participation; valuable opportunities can be derived through continuous participation; other opportunities refers to other non-sport activities that might derailed sport participation. Social constraint is related to "social expectations or norms which create feelings of obligation to remain in the sport activity" [8]. Sport commitment model had been used to understand competitive youth sport contexts such as softball, baseball, football, volleyball, cricket, and tennis [7, 9–11]. Beside youth, adult participants were also recruited in various sports such as exercise and fitness, tennis, windsurfing, triathlon, and swimming [12–15].

However, from our extensive literature review, no study explores sport motivation and sport commitment in badminton players especially elite players. Hence, the objective of this study is to evaluate the characteristics of sport motivation and sport commitment in elite badminton players in terms of gender and level of participation differences.

2 Subjects/Materials

2.1 Subjects

A total of 173 elite badminton players (103 males and 70 females) from 12 different states in Malaysia were recruited. Age of badminton players ranges from 13 to 39 ($M = 19.66$, $SD = 5.02$). The majority of subjects are Malaysian-Chinese (79.2%). Other Malaysian ethnic groups represented were Malay (11%), Bumiputera Sabah and Sarawak (6.9%), and Malaysian-Indian (2.9%). The main demographics of the sample are presented in Table 1.

Inclusion Criteria

The elite player in this study is defined as follows: (1) a person whose highest level of participation is at least at state level competition or (2) a player whose highest level of participation is national level tournaments or played at international tournaments representing the country. Based on Swann et al [16], the participants in this study can be defined as semi-elite or competitive-elite players.

Table 1 Demographics of the sample ($N = 173$)

	Mean	SD
Age	19.66	5.02
Weight (kg)	59.20	9.67
Height (m)	1.68	0.085
Experience in playing badminton	10.18	4.51
	Frequency	Percentage (%)
Gender		
Male	103	59.5
Female	70	40.5
Ethnicity		
Malay	19	11.0
Chinese	137	79.2
Indian	5	2.9
Bumiputera Sabah and Sarawak	12	6.9
Training period Less than 11 years 11–20 years Over 20 years	114 55 4	65.9 31.8 2.3
Highest level of sport participation		
State	144	83.2
National	29	16.8

Note Indian and Chinese is this study refers to Malaysian of Chinese or Indian origin that were born in Malaysia

2.2 Materials

The Sport Commitment Questionnaire-2 (SCQ-2) [8]. The original Sport Commitment Questionnaire (SCQ) was developed by Scanlan et al [7]. However, the original SCQ and revised the Sport Commitment Questionnaire-2 (SCQ-2) [8]. The current study used the SCQ-2. The SCQ-2 consists of ten different sources of commitment [i.e., sport enjoyment, valuable opportunities, other priorities, personal investments (loss and quantity), social constraints, social support (emotional and informational), desire to excel (mastery achievement and social achievement]. These ten sources predicted two sport commitment types (i.e., enthusiastic commitment, constrained commitment) with a total of 58 items. Each item is answered on a five-point Likert type scale ranging from 1 = strongly disagree to 5 = strongly agree. Scanlan et al [8] reported that the constructs were internally reliable and the sources explained the variance in both enthusiastic commitment (81.8%) and constrained commitment (63.9%). Spanish version of the SCQ-2 was also reported to be internally consistent and valid to assess the level of sport commitment in young athletes [17].

Sport Motivation Scale-2 (SMS-2) [6]. The original version of the Sport Motivation Scale (SMS) was developed in both French [4] and translated to English by Pelletier et al [5]. Pelletier et al [5] revised the original SMS and developed SMS-2 included the integrated subscale as suggested by SDT. The responses are based on a seven-point Likert scale, ranging from 1 (does not correspond at all) to 7 (corresponds completely). SMS-2 consisted of 18 items measuring six-subscales (i.e., intrinsic, integrated, identified, introjected, external, and amotivated regulation). The internal consistency (Cronbach's alpha) for individual subscale ranged from 0.73 to 0.86, and the discriminant and the convergent validity were supported using the average variance extracted (AVE) with 0.50 or greater.

2.3 *Procedures*

After gaining approval from Research Ethics Committee (UM.TNC2/UMREC-525), participants were recruited through the Badminton Association of Malaysia (BAM), state training centers, and private training center. The research assistants visited the selected centers in Peninsular Malaysia (pre-COVID-19) while the participants from the state of Sabah and Sarawak were collected through online survey via Google Form application with the assistance of the respective state badminton coaches. Before conducting the survey, players read the consent form regarding the purpose of the research and sign if they are interested to participate. Players were informed that there is no obligation to participate. Anonymity is guaranteed, and their data are fully confidential. On average around 30–40 min is needed to complete both questionnaires. We debriefed participants about the survey and thanked them for their involvement. For the online survey Google Form, consent statement on were indicated clearly. For example, if the player agreed to participate, he/she will check the AGREE option, likewise if he/she disagrees, the player can DECLINE or choose to close the Google Form link.

3 Data Analysis and Results

4 Discussion

Based on Table 2, sport motivation factor that scored the highest is identified regulation ($M = 4.72 \pm 0.87$), followed by integrated regulation ($M = 4.64 \pm 0.92$), amotivation ($M = 3.95 \pm 1.06$) while intrinsic motivation is at ($M = 3.85 \pm 0.95$). This results revealed that at elite level, badminton players are more motivated by

Table 2 Descriptive statistics for sport motivation and sport commitment

	Mean	SD
Sport motivation		
Intrinsic motivation	3.85	0.95
Integrated regulation	4.64	0.92
Identified regulation	4.72	0.87
Introjected regulation	3.80	0.99
External regulation	3.49	1.29
Amotivation	3.95	1.06
Sport commitment		
Enthusiastic commitment	3.97	0.88
Constraint commitment	3.18	0.80
Enjoyment	4.16	0.92
Valuable opportunities	4.00	0.92
Other priority	3.06	0.87
Personal investment loss	3.82	0.81
Personal investment quantity	4.04	0.89
Social constraint	3.31	0.94
Social support emotion	3.69	0.89
Social support information	3.86	0.86
Desire to excel mastery	3.85	0.89
Desire to excel social achievement	3.95	0.87

extrinsic motivation perhaps "degree of importance a sport activity is to one's career" or future, hence intrinsic motivation (e.g., fun) is no longer as important as compared to younger players at grassroots or beginner level.

Amotivation scored higher than intrinsic motivation, this result may be due to a group of badminton players ($n = 14$) who are older (over 25 years old), retiring from sport or at certain stage of life whereby sport might take a back seat to concentrate on their career. A more worrisome revelation if younger players are considering to quit (amotivation) badminton, hence coaches and sport administrators should look into this matter.

For sport commitment, the highest factor is enjoyment ($M = 4.16 \pm 0.92$), our finding is similar to Gómez et al [18] and Gdonteli et al [19]. Followed by personal investment quantity ($M = 4.04 \pm 0.89$), and lowest factor is other priorities ($M = 3.06 \pm 0.87$). And the lowest is other priorities ($M = 3.06 \pm 0.87$). For badminton players to continue to be committed, certain degree of enjoyment is needed. As explained earlier sport motivation (based on sport determination theory) evaluates the motivation leading to sport participation while sport commitment (sport commitment

Table 3 Sport motivation based on level of participation test statistics[a, b]

	Intrinsic motivation	Integrated regulation	Identified regulation	Introjected regulation	External regulation	Amotivation
Kruskal–Wallis H	1.319	5.756	0.318	0.059	0.291	0.007
df	1	1	1	1	1	1
Asymp. Sig	0.251	**0.016**	0.573	0.808	0.590	0.932

[a] *Kruskal–Wallis test*
[b] *Grouping variable: highest level of participation*
There is no significant difference for sport motivation in terms of level of sport participation except integrated regulation (0.02)

theory) evaluates the desire to continuously participate in sports. Hence, the concept of intrinsic motivation (SMS) and concept of enjoyment (SCQ) are slightly different but with some overlapping similarities. Personal investment quantity scored high which explained the concept of personal investment may it be on badminton equipment invested or time put into badminton that "encouraged" players to continue to be committed. Other priorities refer to other alternatives activities that are in conflict to deter continuation of sport. Hence, it can be deducted that alternatives activities were the least important factor that enticed them away from badminton.

Due to the non-normality of data, hence nonparametric test was conducted for the remaining data analysis (i.e., Kruskal–Wallis-H and Spearman's correlation). Level of participations between state versus national players (different abilities) were conducted, overall there is no significant difference for sport motivation except in integrated regulation (0.02) refer to Table 3. Integrated regulation is defined as exercise or participate in sport for its value, in our samples probably state athletes and national athletes differ in how badminton is valued.

In Table 4, there is no significant difference for sport commitment in terms of levels of sport participation (state versus national players) except enjoyment (0.01). One explanation might be certainly the component of enjoyment will be different for players at different level of expertise. National players are more mission bound during competition as they represent the honor of the country especially competing against another country, hence the element of enjoyment play less importance.

Further, analyses were conducted to evaluate gender differences in terms of sport motivation and sport commitment. In Table 5, we found no significant difference between gender for sport motivation in all factors similar to the results [20, 21] studies. While Table 6 showed no significant difference between gender for most factors in sport commitment except desire to excel social achievement (0.026). Our result differs from previous studies [22–24], for example Ling et al [23] reported significant gender difference in enthusiastic commitment for badminton, table tennis and tennis but not for squash. Desire to excel social achievement is related to the

Table 4 Sport commitment based on level of participation test statistics[a, b]

	Enthusiastic commitment	Constraint commitment	Enjoyment	Valuable opportunities	Other priorities	Personal investment loss	Personal invest quantity	Social constraints	Social support emotional	Social support information	Desire to excel mastery	Desire to excel social
Kruskal–Wallis H	1.235	0.029	6.638	0.041	0.840	0.105	0.370	0.066	0.833	0.078	1.056	0.310
df	1	1	1	1	1	1	1	1	1	1	1	1
Asymp. Sig	0.266	0.866	**0.010**	0.839	0.359	0.746	0.543	0.797	0.361	0.780	0.304	0.577

[a] *Kruskal–Wallis test*
[b] *Grouping variable: highest level of participation*
There is no significant difference for sport commitment in terms of levels of sport participation except enjoyment (0.01)

Table 5 Sport motivation between gender test statistics[a,b]

	Intrinsic motivation	Integrated regulation	Identified regulation	Introjected regulation	External regulation	Amotivation
Kruskal–Wallis H	2.115	0.423	0.002	2.306	0.117	3.231
df	1	1	1	1	1	1
Asymp. Sig	0.146	0.515	0.964	0.129	0.732	0.072

[a] *Kruskal–Wallis test*
[b] *Grouping variable: gender*
There is no significant difference between gender for sport motivation in terms for all factors

show of superiority against opponent, probably male athletes would have higher desire (higher commitment toward competitiveness) than females [25].

5 Conclusion

Our study showed that the highest source of motivation is derived from identified regulation and not intrinsic motivation among elite Malaysian badminton players. However, sport enjoyment is the preferred source of commitment, i.e., to continue to be committed to either training or to compete in tournaments. The comparison between level of participation (state versus national athletes) showed that most constructs in sport motivation did not show significant different except in integrated regulation. Similarly, only sport enjoyment was significantly different among all sport commitment constructs. There is no significant gender difference among all sport motivation constructs, while only desire to excel social achievement is significantly different among all sport commitment constructs.

Table 6 Sport commitment between gender test statistics[a,b]

	Enthusiastic commitment	Constrained commitment	Enjoyment	Valuable opportunities	Other priorities	Personal investment loss	Personal invest quantity	Social constraints	Social support emotional	Social support information	Desire to excel mastery	Desire to excel social
Kruskal–Wallis H	1.500	0.000	1.461	0.067	0.229	0.413	1.960	0.887	1.832	0.751	3.417	4.982
df	1	1	1	1	1	1	1	1	1	1	1	1
Asymp. Sig	0.221	0.991	0.227	0.796	0.632	0.520	0.161	0.346	0.176	0.386	0.065	**0.026**

[a] *Kruskal–Wallis test*
[b] *Grouping variable: gender*
There is no significant difference between gender for sport commitment except desire excel social (0.026)

References

1. Deci EL, Ryan RM (1985) Intrinsic motivation and self-determination in human behavior. Springer Science and Business Media, Berlin
2. Deci EL, Ryan RM (2000) The "what" and "why" of goal pursuits: human needs and the self-determination of behavior. Psychol Inquiry 11:227–268
3. Ryan RM, Deci EL (2007) Active human nature self-determination theory and the promotion and maintenance of sport, exercise, and health. In: Hagger M, Chatzisarantis N (eds) Intrinsic motivation and self-determination in exercise and sport 1–20 human kinetics. Champaign
4. Brière NM, Vallerand RJ, Blais MR, Pelletier LG (1995) Development and validation of a scale on intrinsic and extrinsic motivation and lack of motivation in sports: the scale on motivation in sports. Inter J Sport Psychol 26(4):465–489
5. Pelletier LG, Tuson KM, Fortier MS, Vallerand RJ, Briere NM, Blais MR (1995) Toward a new measure of intrinsic motivation, extrinsic motivation, and amotivation in sports: the sport motivation scale (SMS). J Sport Exer Psychol 17(1):35–53
6. Pelletier LG, Rocchi MA, Vallerand RJ, Deci EL, Ryan RM (2013) Validation of the revised sport motivation scale (SMS-II). Psychol Sport Exer 14(3):329–341. https://doi.org/10.1016/j.psychsport.2012.12.002
7. Scanlan TK, Simons JP, Carpenter PJ, Schmidt GW, Keeler B (1993) An introduction to the sport commitment model. J Sport Exer Psychol 15:1–15. https://doi.org/10.1123/jsep.15.1.1
8. Scanlan TK, Chow GM, Sousa C, Scanlan LA, Knifsend CA (2016) The development of the sport commitment questionnaire-2 (English version). Psychol Sport Exer 22:233–46. https://doi.org/10.1016/j.psychsport.2015.08.002
9. Carpenter PJ, Coleman RJ (1998) A longitudinal study of elite youth cricketers' commitment. Inter J Sport Psychol 29:1–16
10. Carpenter PJ, Scanlan TK, Simons JP, Lobel MA (1993) A test of the sport commitment model using structural equation modeling. J Sport Exer Psychol 15:119–133
11. Weiss MR, Kimmel LA, Smith AL (2001) Determinants of sport commitment among junior tennis players: enjoyment as a mediating variable. Ped Exer Sci 13(2):131–144
12. Alexandris K, Zahariadis P, Tsorbatzoudis C, Grouios G (2002) Testing the sport commitment model in the context of exercise and fitness participation. J Sport Beh 25(3):217–230
13. Casper JM, Gray DP, Stellino MB (2007) A sport commitment model perspective on adult tennis players' participation frequency and purchase intention. Sport Manag Rev 10(3):253–278. https://doi.org/10.1016/S1441-3523(07)70014-1
14. Crocker PRE, Augaitis L (2010) Commitment in age class adult triathletes: examining gender differences in the sport commitment model. Int J Sport Psychol 41(2):177–194
15. Jeon JH, Ridinger LL (2009) An examination of sport commitment of windsurfers. J Sport Beh 32:325–382009
16. Swann C, Moran A, Piggott D (2015) Defining elite athletes: issues in the study of expert performance in sport psychology. Psychol Sport Exer 16(1):3–14
17. Sánchez-Miguel PA, Chow GM, Sousa C, Scanlan TK, Ponseti FJ, Scanlan L, García-Mas A (2019) Adapting the sport commitment questionnaire-2 for Spanish usage. Percept Mot Skills 126(2):267–285
18. Gómez RP, Palos PA, Sánchez JCJ, Sánchez PM (2013) Sport commitment in Mexican youth. Revista Iberoamericana de Psicología del Ejercicio y el Deporte 8(2):317–330
19. Gdonteli K, Mountakis K, Gavriilidis A (2010) Self-determination and sport commitment: an evaluation by university intramural participant. Int J Fitness 6(1):41–52
20. Bayyat MM (2020) Identifying motives for sport participation from the perspective of self-determination theory: gender differences. Dirasat: Edu Sci 47:576–85
21. van Heerden CH (2014) Expressed motives of students for sport participation in a South African context. J Phy Educ Res 1–7
22. Matteo S (1986) The effect of sex and gender-schematic processing on sport participation. Sex Roles 15:417–432. https://doi.org/10.1007/BF00287981

23. Ling A, Teo EW, Chin NS (2019) Sport commitment among Malaysian racquet sports athletes. Mal J Med Sci 26(4): 1–8. https://doi.org/10.21315/mjms2019.26.4.1
24. Wigglesworth JC, Young BW, Medic N, Grove JR (2012) Examining gender differences in the determinants of Masters Swimmers' sport commitment. Inter J Sport Exer Psychol 10(3):236–250
25. Gill DL (1986) Competitiveness among females and males in physical activity classes. Sex Roles 15:233–247

Perceived Social Support, Life Enjoyment, and Satisfaction Among Polytechnic Students During COVID-19 Pandemic

Adzrul Izwan Abdullah, Ngien Siong Chin, and Boon Sian Teo

Abstract The study examined polytechnic students' social support, life enjoyment, and satisfaction among 327 polytechnic students with a mean age of 18.65 ± 0.95. The social support scale for exercise (SSSE), physical environment scale for physical activity (PESPA) and quality of life enjoyment, and satisfaction questionnaires (Q-LES-Q-M) were utilized and self-administered. Descriptive analysis, independent-samples t-test, one-way ANOVA, and two-way ANOVA were conducted. The independent-samples t-test revealed significant difference between gender in friend support, $p = 0.014$, quality, $p = 0.011$, physical health, $p = 0.007$, and leisure time activities, $p = 0.002$. The one-way ANOVA revealed significant difference between age groups in family support, $p = 0.02$. The one-way ANOVA also revealed significant difference between ethnicities in physical health, $p = 0.013$, feelings, $p < 0.001$, and homework, $p = 0.02$. Two-way ANOVA revealed the main effect of gender was significant for friend support, quality, physical health, and leisure time activities, and the main effect of age was significant for family support with $p < 0.05$. There was a significant interaction between gender × age for friend support, $p = 0.004$. Two-way ANOVA also revealed the main effect of gender was significant for friend support, availability, quality, physical health, feelings, and leisure time activities, and the main effect of ethnicity was significant for physical health, feelings, and homework with $p < 0.05$. The findings indicate the importance of significant integration of family and social supports such as peer-groups and students community and teachers in adapting to the continuous evolving learning environment.

Keywords Social support · Life enjoyment · Satisfaction · Polytechnic students

A. I. Abdullah (✉)
Politeknik Kuching Sarawak, Sarawak, Malaysia
e-mail: adzrul@poliku.edu.my

N. S. Chin · B. S. Teo
Institute of Teacher Education Batu Lintang Campus, Sarawak, Malaysia

G. Kuan et al. (eds.), *Advancing Sports and Exercise via Innovation*, Lecture Notes in Bioengineering, https://doi.org/10.1007/978-981-19-8159-3_37

1 Introduction

According to the social-ecological framework, social support, and physical environment constructs play a dominant role in influencing one's engagement towards physical activity (PA) [1]. Social support comprised of family support and friend support [2] which may synergistically function to help individuals maintain current patterns, heighten self-efficacy, and performance in executing physical activity efficiently [3]. Reference [4] study showed that social support helps professional athletes comprehend challenging issues and achieve positive outcomes through their participation in fitness cultures. Furthermore, research indicated that adolescents and women with higher perceived social support from significant others were positively associated with higher levels of PA [5, 6]. Hence, social support was a significant predictor of one's attitude towards physical activity and sport [7]. For example, perceived positive social support was found to be conducive in sports engagement which would lead to individuals achieving the recommended amounts of exercise [8]. Thus, understanding how social support can encourage students to participate in physical activity frequently is essential to prevent them from practicing sedentary lifestyles. Social support regardless of relationship and proximity is on the determinants of life satisfaction deemed important in the forms of positive therapy and happiness [9].

The physical environment included playgrounds, gymnasiums, fitness centres, parks, and schools are vital settings for PA which would likely to produce health and well-being benefits. Reference [10] found physical environment influenced the undergraduate students' participation in PA which has psychological health outcomes. Previous research has shown that these facilities available for carrying out physical activity nowadays remain grossly insufficient and inaccessible [11]. Besides, the perception of lack of serenity, safety, and security in the physical environment is an essential predictor of involvement in various physical activities as people are more reluctant to carry out physical activity like jogging and walking in environments that are considered hazardous, which may lead to decline in PA [12]. Therefore, attention should be given to the physical environment with emphasis on the availability and quality of facility infrastructure, safety, weather conditions, and community design aspects [13]. Consequently, people are able to enjoy the activities completely and become more physically active which can enhance one's quality of life and mental well-being [14]. The study aimed to examine the perceived social support, physical environment, and life enjoyment and satisfaction among polytechnic students during COVID-19 pandemic.

2 Methods

2.1 Participants

The participants were 327 polytechnic students, comprised of 183 (56.0%) males and 144 (44.0%) females aged between 18 and 20 years old (18.65 ± 0.95). In terms of ethnicity, the Native formed the largest percentage with 54.1%, followed by Malay and Chinese who comprised 37.3% and 8.6% of the population, respectively. Based on the body mass index (BMI), the majority of the participants were in normal weight category with a total of 172 (52.6%), followed by underweight category with 65 (19.9%), overweight category with 47 (14.4%), and obese category with 43 (13.1%). Participants' mean for frequency of exercise per week were 1.20 (SD = 0.40) averaging between 31 and 60 min per session. The study was carried out in one polytechnic in Kuching after the ethical approval from the Ethics Committee of the institution.

2.2 Measures

The social support scale for exercise (SSSE) was adapted and utilized to measure individual's perceived social support towards physical activity [2]. The questionnaire consists of 24 items, two subscales with 12 items each, which are as follows: family support (e.g. "discussed exercise with me") and friend support (e.g. "exercise with me"). Items are rated on a five-point Likert scale, ranging from 1 (never) to 5 (very often). All subscales and overall social support for exercise demonstrated high internal consistency with alpha reliability coefficients of 0.96, 0.95, and 0.97, respectively.

The physical environment scale for physical activity (PESPA) was adapted and used to measure the availability and quality of physical activity facilities [2]. The questionnaire consists of five items with two subscales, which are as follows: availability (three items; e.g. "My residential are offers enough facilities for me to be physically active") and quality (two items). Items are rated on a five-point Likert scale, ranging from 1 (not true at all) to 5 (definitely true). All subscales and overall social support for exercise showed high internal consistency with alpha reliability coefficients of 0.91, 0.84, and 0.84, respectively.

The quality of life enjoyment and satisfaction questionnaire (Q-LES-Q-M) was adapted to assess the degree of enjoyment and satisfaction experienced during the past week [15]. The Q-LES-Q-M consists of 43 items that comprises of four factors which are as follows: physical health (13 items; e.g. "…felt energetic?), feelings (14 items; e.g. "…felt relaxed?), homework (10 items; e.g. "…enjoyed your homework?), and leisure time activities (6 items; e.g. "How often did the leisure activities sustain your interest?). Items are rated on a five-point Likert scale, ranging from 1 (not at all) to 5 (all the time). The Q-LES-Q-M and four subscales showed high internal consistency with alpha value of 0.97, 0.93, 0.93, 0.92, and 0.90, respectively.

2.3 Statistical Analysis

The statistical package for the social science (SPSS) version 27 was utilized to analyse the data. The descriptive statistics, internal consistency, independent-samples t-test, one-way ANOVA, and two-way ANOVA were used to examine the gender, age groups, and ethnicity of social support scale for exercise, physical environment scale for physical activity, and Q-LES-Q constructs.

3 Results

Table 1 shows the demographic characteristics of the participants. A total of 327 participants were involved in this study, comprising of 183 (56.0%) males and 144 (44.0%) females. The age groups categories showed that 128 (37.3%) were 18 years old, 104 (31.8%) were 19 years old, and 95 (29.1%) were 20 years old. In terms of ethnicity, the Native formed the largest percentage with 54.1%, followed by Malay and Chinese who comprised 37.3% and 8.6% of the population, respectively. Based on the body mass index (BMI), the majority of the participants were in normal weight category with a total of 172 (52.6%), followed by underweight category with 65 (19.9%), overweight category with 47 (14.4%), and obese category with 43 (13.1%). Whereas, in terms of the frequency of exercise, a high prevalence of the participants exercises 1–3 times (80.4%) a week averaging between 31 and 0 min (41.3%).

Table 2 shows the mean, standard deviation, and reliability of SSSE, PESPA, and Q-LES-Q. For SSSE, friend support (3.27 $\pm$ 1.09) was the most crucial factor that influence students' participation towards physical activity, followed by family support (2.86 $\pm$ 0.96). The overall SSSE showed a good internal consistency of 0.96 while subscales alpha coefficients for family support and friend support were 0.95 and 0.97, respectively. For PESPA, availability (3.40 $\pm$ 1.04) was the most essential construct that influence students' engagement in physical activity, followed by quality (3.37 $\pm$ 1.10). The overall PESPA showed a good internal consistency of 0.91, while subscales alpha coefficients for availability and quality were 0.84. For Q-LES-Q, physical health (3.77 $\pm$ 0.70) was the most crucial construct that influenced the students' degree of enjoyment and satisfaction during the past week, followed by feelings (3.77 $\pm$ 0.69), leisure time activities (3.71 $\pm$ 0.71) and homework (3.48 $\pm$ 0.70). The overall Q-LES-Q showed a good internal consistency of 0.97 while subscales alpha coefficients for physical health, feelings, homework, and leisure time activities demonstrated high reliability of 0.93, 0.93, 0.92, and 0.90, respectively.

Table 3 shows the independent sample *t*-test of SSSE, PESPA, and Q-LES-Q based on gender. For SSSE, results revealed no significant difference in family

Table 1 Demographic characteristics of 327 participants

Characteristics	Frequency (F)	Percentage (%)
Gender		
Male	183	56.0
Female	144	44.0
Age		
18	128	39.1
19	104	31.8
20	95	29.1
Ethnicity		
Malay	122	37.3
Chinese	28	8.6
Native	177	54.1
Body mass index (BMI)		
Underweight (≤ 18.49)	65	19.9
Normal (18.5–24.9)	172	52.6
Overweight (25.0–29.9)	47	14.4
Obese (30.0–34.9)	43	13.1
Frequency of exercise per week		
1–3 times	263	80.4
4–6 times	64	19.6
Minutes per session		
0–30 min	90	27.5
31–60 min	135	41.3
61–90 min	49	15.0
91–120 min	53	16.2

Table 2 Mean, standard deviation, and reliability for SSSE, PESPA, and Q-LES-Q

Variables	M	SD	A
Overall SSSE			0.96
Family support	2.86	0.96	0.95
Friend support	3.27	1.09	0.97
Overall PESPA			0.91
Availability	3.40	1.04	0.84
Quality	3.37	1.10	0.84
Overall Q-LES-Q			0.97
Physical health	3.77	0.70	0.93
Feelings	3.77	0.69	0.93
Homework	3.48	0.70	0.92
Leisure time activities	3.71	0.71	0.90

Table 3 Independent t-test of SSSE, PESPA, and Q-LES-Q based on gender

Variables	Male		Female		*t*-test	*p*
	M	*SD*	*M*	*SD*		
SSSE						
Family support	2.84	0.99	2.87	0.91	−0.30	0.76
Friend support	3.40	1.05	3.10	1.11	2.47	0.014*
PESPA						
Availability	3.49	1.04	3.29	1.02	1.69	0.09
Quality	3.50	1.10	3.19	1.08	2.56	0.011*
Q-LES-Q						
Physical health	3.86	0.70	3.65	0.69	2.73	0.007*
Feelings	3.81	0.71	3.72	0.67	1.16	0.25
Homework	3.48	0.71	3.48	0.69	0.02	0.99
Leisure time activities	3.81	0.75	3.57	0.63	3.16	0.002*

* $p < 0.05$

support scores between gender, $p = 0.76$. Conversely, friend support was significant difference between gender, whereby males rated 0.30 point higher than female, $p = 0.014$. For PESPA, results revealed no significant difference in availability scores between gender, $p = 0.09$. However, quality was significant difference between gender, whereby males rated 0.31 point higher than females, $p = 0.011$. For Q-LES-Q, results revealed statistically significant difference in physical health scores between gender, whereby males rated 0.21 point higher than females, $p = 0.007$. However, there were no significant difference between gender in feelings, $p = 0.25$, and homework, $p = 0.99$. Furthermore, leisure time activities were significant difference between gender, whereby males rated 0.24 point higher than females, $p = 0.002$.

Table 4 shows the one-way ANOVA of SSSE, PESPA, and Q-LES-Q based on age groups. For SSSE, results revealed that there was a statistically significant difference between the three age groups for family support, $p = 0.02$. Post-hoc Tukey HSD adjusted comparisons for family support indicated that the mean score for 18 years old was significantly lower than the 19 years old, $\mu = -0.35, p = 0.02$. However, there was no significant difference between the three age groups for friend support, $p = 0.52$. For PESPA, results revealed that there was no statistically significant difference between the three age groups for availability, $p = 0.28$, and quality, $p = 0.11$. For Q-LES-Q, results revealed that there was no statistically significant difference between the three age groups for physical health, $p = 0.67$, feelings, $p = 0.90$, homework, $p = 0.97$, and leisure time activities, $p = 0.41$.

Table 5 shows the one-way ANOVA of SSSE, PESPA, and Q-LES-Q based on ethnicity. For SSSE, results revealed there was no statistically significant difference between the three ethnicities for family support, $p = 0.06$, and friend support, $p = 0.13$. For PESPA, results revealed there was no statistically significant difference

Table 4 One-way ANOVA of SSSE, PESPA, and Q-LES-Q based on age groups

Variables	18		19		20		$F(2, 324)$	p
	M	SD	M	SD	M	SD		
SSSE								
Family support	2.70	0.87	3.05	1.05	2.84	0.92	3.94	0.02*
Friend support	3.22	1.06	3.37	1.19	3.23	1.00	0.66	0.52
PESPA								
Availability	3.33	1.07	3.54	1.08	3.35	0.94	1.26	0.28
Quality	3.21	1.15	3.51	1.09	3.41	1.04	2.19	0.11
Q-LES-Q								
Physical health	3.75	0.70	3.82	0.73	3.74	0.69	0.41	0.67
Feelings	3.75	0.66	3.78	0.76	3.79	0.67	0.10	0.90
Homework	3.47	0.65	3.50	0.78	3.49	0.69	0.04	0.97
Leisure time activities	3.71	0.73	3.77	0.71	3.64	0.68	0.89	0.41

* $p < 0.05$

between the three ethnicities for availability, $p = 0.44$, and quality, $p = 0.27$. For Q-LES-Q, results revealed that there was a statistically significant difference between the three ethnicities for physical health, $p = 0.013$. Post-hoc Tukey HSD adjusted comparisons for physical health indicated that the mean score for Malay was significantly lower than Chinese, μ = -0.43, $p = 0.009$. Besides, there was significant different between the three ethnicities for feelings, $p < 0.001$. Post-hoc Tukey HSD adjusted comparisons for feelings indicated that the mean score for Malay was significantly lower than the Chinese, $\mu = -0.60$, $p < 0.001$. Post-hoc Tukey HSD also adjusted comparisons for feelings indicated that the mean score for Chinese was significantly higher than the Native, $\mu = 0.45$, $p = 0.003$. Moreover, there was also significant different between the three ethnicities for homework, $p = 0.02$. Post-hoc Tukey HSD adjusted comparisons for homework indicated that the mean score for Malay was significantly lower than the Chinese, $\mu = -0.36$, $p = 0.04$, but significantly higher than the Native, $\mu = 0.40$, $p = 0.014$. However, there was no significant difference between the three ethnicities for leisure time activities, $p = 0.23$.

Table 6 shows the two-way ANOVA for gender (male versus female) and age groups (18 years old versus 19 years old versus 20 years old) on SSSE, PESPA, and Q-LES-Q. For SSSE, the between-subjects ANOVA for family support showed that there was no significant main effect of gender, $p = 0.67$, but significant main effect on age groups, $p = 0.029$. Besides, there was no significant interaction between gender × age on family support, $p = 0.72$. Post-hoc Bonferroni adjusted comparisons for family support indicated that 18 years old rated 0.34 point lower than 19 years old, $p = 0.008$. The between-subjects ANOVA for friend support revealed that the main effect of gender, $p = 0.006$ was significant, whereas no significant main effect of age groups, $p = 0.87$. Besides, there was a significant interaction between gender × age on friend support, $p = 0.004$. Post-hoc Bonferroni adjusted comparisons for

Table 5 One-way ANOVA of SSSE, PESPA, and Q-LES-Q based on ethnicity

Variables	Malay		Chinese		Native		$F(2, 324)$	p
	M	SD	M	SD	M	SD		
SSSE								
Family support	2.80	1.00	3.27	1.10	2.83	0.89	2.92	0.06
Friend support	3.12	1.11	3.22	1.17	3.38	1.05	2.08	0.13
PESPA								
Availability	3.39	0.94	3.64	0.98	3.38	1.11	0.82	0.44
Quality	3.37	1.01	3.68	1.02	3.31	1.17	1.33	0.27
Q-LES-Q								
Physical health	3.68	0.72	4.11	0.57	3.78	0.70	4.43	0.013*
Feelings	3.64	0.72	4.24	0.55	3.79	0.67	9.07	< 0.001*
Homework	3.48	0.69	3.83	0.70	3.43	0.70	3.97	0.02*
Leisure time activities	3.63	0.74	3.86	0.64	3.74	0.70	1.46	0.23

* $p < 0.05$

quality indicated that male rated 0.31 point higher than female, $p = 0.013$. Post-hoc Bonferroni adjusted comparisons for friend support indicated that male rated 0.33 point higher than female, $p = 0.006$. Post-hoc Bonferroni adjusted comparisons for quality indicated that male rated 0.31 point higher than female, $p = 0.013$. Pair-wise comparison between gender and age group revealed that post-hoc Bonferroni adjusted comparisons for friend support indicated that 18 years old male rated 0.53 point lower than 19 years old male, $p = 0.005$. For PESPA, the between-subjects ANOVA for availability revealed that there was no significant main effect of gender, $p = 0.11$, and age groups, $p = 0.48$. Besides, there was also no significant interaction between gender × age on availability, $p = 0.46$. The between-subjects ANOVA for quality revealed that there was significant main effect of gender, $p = 0.013$, whereas no significant main effect of age groups, $p = 0.23$. Moreover, there was no significant interaction between gender × age on quality, $p = 0.24$. For Q-LES-Q, the between-subjects ANOVA for physical health revealed that there was significant main effect of gender, $p = 0.006$, whereas no significant main effect of age groups, $p = 0.92$. Post-hoc Bonferroni adjusted comparisons for physical health indicated that male rated 0.22 point higher than female, $p = 0.006$. Besides, there was no significant interaction between gender × age on physical health, $p = 0.25$. The between-subjects ANOVA for feelings revealed that there was no significant main effect of gender, $p = 0.22$, and age groups, $p = 0.91$. Likewise, there was no significant interaction between gender × age on feelings, $p = 0.61$. The between-subjects ANOVA for homework revealed that there was no significant main effect of gender, $p = 0.95$, and age groups, $p = 0.98$. Similarly, there was no significant interaction between gender × age on homework, $p = 0.15$. The between-subjects ANOVA for leisure time activities revealed that there was significant main effect of gender, $p = 0.002$, whereas no significant main effect of age groups, $p = 0.59$. Moreover, there was

Table 6 Main-effect of gender versus age groups for SSSE, PESPA, and Q-LES-Q

Variables	Gender (G)				Age groups (AG)					G versus AG	
	M	F	*F*	*p*	18	19	20	*F*	*p*	*F*	*p*
SSSE											
Family support	2.84	2.89	0.18	0.67	2.71	3.05	2.85	3.58	0.029*	0.33	0.72
Friend support	3.41	3.07	7.75	0.006*	3.22	3.28	3.21	0.14	0.87	5.57	0.004*
PESPA											
Availability	3.48	3.30	2.58	0.11	3.33	3.49	3.35	0.75	0.48	0.77	0.46
Quality	3.50	3.20	6.18	0.013*	3.21	3.44	3.40	1.48	0.23	1.45	0.24
Q-LES-Q											
Physical health	3.86	3.64	7.51	0.006*	3.75	3.77	3.73	0.08	0.92	1.40	0.25
Feelings	3.81	3.72	1.52	0.22	3.75	3.75	3.79	0.10	0.91	0.50	0.61
Homework	3.48	3.48	0.0004	0.95	3.47	3.47	3.49	0.03	0.98	1.93	0.15
Leisure time activities	3.81	3.57	9.55	0.002*	3.71	3.73	3.63	0.52	0.59	0.88	0.42

* $p < 0.05$

no significant interaction between gender × age on leisure time activities, $p = 42$. Post-hoc Bonferroni adjusted comparisons for leisure time activities indicated that male rated 0.24 point higher than female, $p = 0.002$.

Table 7 shows the two-way ANOVA for gender (male versus female) and ethnicity (Malay versus Chinese versus Native) on SSSE, PESPA, and Q-LES-Q. For SSSE, the between-subjects ANOVA for family support showed that there was no significant main effect of gender, $p = 0.99$, and ethnicity, $p = 0.06$. Besides, there also was no significant interaction between gender × ethnicity on family support, $p = 0.90$. The between-subjects ANOVA for friend support revealed that there was significant main effect of gender, $p = 0.017$, whereas no significant main effect of ethnicity, $p = 0.06$. Furthermore, no significant interaction between gender × ethnicity on friend support, $p = 0.18$. Post-hoc Bonferroni adjusted comparisons for friend support indicated that male rated 0.38 point higher than female, $p = 0.017$. For PESPA, the between-subjects ANOVA for availability revealed that there was significant main effect of gender, $p = 0.021$, but no significant main effect of ethnicity, $p = 0.34$. There was no significant interaction between gender × ethnicity on availability, $p = 0.32$. Post-hoc Bonferroni adjusted comparisons for availability indicated that male rated 0.36 point higher than female, $p = 0.021$. The between-subjects ANOVA for quality revealed that there was significant main effect of gender, $p = 0.032$, yet no significant main effect of ethnicity, $p = 0.20$. Results also revealed no significant interaction between gender × ethnicity on quality, $p = 0.84$. Post-hoc Bonferroni adjusted comparisons for quality indicated that male rated 0.35 point higher than female, $p = 0.032$. For Q-LES-Q, the between-subjects ANOVA for physical health revealed that there was significant main effect of gender, $p = 0.005$, and ethnicity, $p = 0.006$. However, there was no significant interaction between gender × ethnicity on physical health, $p = 0.68$. Post-hoc Bonferroni adjusted comparisons for physical health indicated that male rated 0.29 point higher than female, $p = 0.005$. Post-hoc Bonferroni adjusted comparisons for physical health indicated that Malay rated 0.47 point lower than the Chinese, $p = 0.001$, and Chinese rated 0.36 point higher than the Native, $p = 0.011$. The between-subjects ANOVA for feelings revealed that there was significant main effect of gender, $p = 0.047$, and ethnicity, $p < 0.001$. Conversely, there was no significant interaction between gender × ethnicity on feelings, $p = 0.38$. Post-hoc Bonferroni adjusted comparisons for feelings indicated that male rated 0.20 point higher than female, $p = 0.047$, 95% CI [0.003, 0.40]. Post-hoc Bonferroni adjusted comparisons for feelings indicated that Malay rated 0.62 point lower than the Chinese, $p < 0.001$, and Chinese rated 0.48 point higher than the Native, $p = 0.001$. The between-subjects ANOVA for homework revealed that there was no significant main effect of gender, $p = 0.30$, but significant main effect of ethnicity, $p = 0.017$. Besides, there was no significant interaction between gender × ethnicity on homework, $p = 0.38$. Post-hoc Bonferroni adjusted comparisons for homework indicated that Malay rated 0.37 point lower than the Chinese, $p = 0.012$, and Chinese rated 0.41 point higher than the Native, $p = 0.004$. The between-subjects ANOVA for leisure time activities revealed that there was significant main effect of gender, $p = 0.001$, but no significant main effect of ethnicity, $p = 0.11$. Moreover, there was no significant interaction between gender × ethnicity on leisure time activities, $p = 0.22$. Post-hoc Bonferroni adjusted comparisons for leisure time activities indicated that male rated 0.36 point higher than female, $p = 0.001$.

4 Discussion

The male displayed higher social support in order to manage and handle academic and social stressors as they are displaying some form of difficulties in their adaptation and transitional process from adolescence to early adulthood in search of their identity and shaping of their characteristics socially [16]. Therefore, the social support will lower the rates of stressors and improve physical health and well-being of the students in polytechnic. Indirectly, the quality of life enjoyment and satisfaction will more likely be stable from the emotional support from friends as the degree of social support will affect the quality of life. The Chinese who have higher quality of life enjoyment could have derived support from their parents and friends though they tended to hold a competitive-learning environment whereby they are expected to perform well academically. This revealed that life satisfaction was associated with mental well-being which contributed to greater forms of enjoyment and gratification [17].

The study enriches further literature on the relevance of social support in relation to physical environment to quality of life in polytechnic schools. It is of importance

Table 7 Main-effect of gender versus ethnicity for SSSE, PESPA, and Q-LES-Q

Variables	Gender (G)				Ethnicity (E)					G versus E	
	M	F	*F*	*p*	Malay	Chinese	Native	*F*	*p*	*F*	*p*
SSSE											
Family support	2.97	2.97	< 0.001	0.99	2.80	3.27	2.83	2.90	0.06	0.11	0.90
Friend support	3.42	3.04	5.78	0.017*	3.07	3.24	3.37	2.81	0.06	1.70	0.18
PESPA											
Availability	3.65	3.29	5.37	0.021*	3.37	3.67	3.37	1.08	0.34	1.13	0.32
Quality	3.62	3.27	4.63	0.032*	3.35	3.70	3.39	1.62	0.20	0.17	0.84
Q-LES-Q											
Physical health	3.99	3.71	7.84	0.005*	3.66	4.13	3.77	5.22	0.006*	0.38	0.68
Feelings	3.99	3.79	3.98	0.047*	3.64	4.26	3.78	9.64	< 0.001*	0.96	0.38
Homework	3.64	3.53	1.09	0.30	3.48	3.85	3.44	4.16	0.017*	0.97	0.38
Leisure time activities	3.92	3.56	11.80	0.001*	3.60	3.88	3.73	2.18	0.11	1.50	0.22

* $p < 0.05$

to initiate relevant programmes according to their potential and make adjustment psychologically to help the students to enhance their satisfaction in engaging in exercise and PA. The programmes need to be implemented during the early years of their polytechnic life which would lead to the desired life satisfaction. The limitation was that the participants were only obtained from one polytechnic due to the challenges in recruiting participants since it was conducted during the COVID-19 pandemic period. Therefore, the findings could only describe the students from one state instead of the majority of the states in Malaysia. Future studies could include different institution of higher education covering a broader number of populations in order to generate a more convincing finding. The finding highlighted the importance of different social supports attuned with better physical environment that can help satisfy and advocate the students' well-being.

5 Conclusion

The study showed that the environment does not necessarily support an increase in the psychological well-being of students. The findings recommend interventions that can enhance the environment for the well-being of the students. The study would recommend relevant and practical awareness of social support and interventions for the students that would support their psychological well-being.

References

1. Van KDH, Paw MJ, Twisk JW, Van WM (2007) A brief review on correlates of physical activity and sedentariness in youth. Med and Sci Sports Exerc 39(8):1241–1250. https://doi.org/10.1249/mss.0b013e318059bf35
2. Sabo A, Kueh YC, Arifin WN, Kim Y, Kuan G (2020) The validity and reliability of the Malay version of the social support for exercise and physical environment for physical activity scales. PLoS ONE 15(9):1–15. https://doi.org/10.1371/journal.pone.0239725
3. Duncan TE, McAuley E (1993) Social support and efficacy cognitions in exercise adherence: a latent growth curve analysis. J Behavioral Med 16(2):199–218. https://doi.org/10.1007/BF00844893
4. Anderson PJ, Bovard RS, Wang Z, Beebe TJ, Murad MH (2016) A survey of social support for exercise and its relationship to health behaviours and health status among endurance Nordic skiers. BMJ Open 6(6):1–6. https://doi.org/10.1136/bmjopen-2015-010259
5. Eyler AA, Brownson RC, Donatelle RJ, King AC, Brown D, Sallis JF (1999) Physical activity social support and middle- and older-aged minority women: results from a US survey. Soc Sci Med 49(6):781–789. https://doi.org/10.1016/s0277-9536(99)00137-9
6. Mendonca G, Cheng LA, Melo EN, de Farias Junior JC (2014) Physical activity and social support in adolescents: a systematic review. Health Educ Res 29(5):822–839. https://doi.org/10.1093/her/cyu017
7. Okun MA, Ruehlman L, Karoly P, Lutz R, Fairholme C, Schaub R (2003) Social support and social norms: do both contribute to predicting leisure-time exercise? Am J Health Behav 27:493–507

8. Giles-Corti B, Donovan RJ (2002) The relative influence of individual, social and physical environment determinants of physical activity. Soc Sci Med 54(12):1793–1812. https://doi.org/10.1016/s0277-9536(01)00150-2
9. Kasprzak E (2010) Perceived social support and life-satisfaction. Pol Psychol Bull 41(4):144–154
10. Ajibua MA, Onanuga OA (2020) Supportiveness of university physical environment on undergraduate students participation in leisure-time physical activity in South West Nigeria. Cent Eur J Sport Sci Med 32(4):85–98. https://doi.org/10.18276/cej.2020.4-09
11. Sallis JF, Cain KL, Conway TL, Gavand KA, Millstein RA, Geremia CM, Frank LD, Saelens BE, Glanz K, King AC (2015) Is your neighbourhood designed to support physical activity? A brief streetscape audit tool. Prev Chronic Dis 12:1–11. https://doi.org/10.5888/pcd12.150098
12. Corbin CB, Welk GJ, Corbin WR, Welk KA (2004) Concept of physical activity: active lifestyles for wellness, 12 edn. McGraw-Hill
13. Cutts BB, Dirby KJ, Boone CG, Brewis A (2009) City structure, obesity and environmental justice: an integrated analysis of physical and social barriers to walkable streets and park access. Soc Sci Med 69:1314–1322
14. Cooper SL (2020) Promoting physical activity for mental well-being. ACSM's Health Fit J 24(3):12–16
15. Ibrahim MF, Kuan G, Hashim HA, Hamzah NA, Kueh YC (2021) Psychometric properties of the Malay-language quality of life enjoyment and satisfaction scale: a confirmatory study on Malaysian children. Inter J Environ Res Public Health 18(1):1–10. https://doi.org/10.3390/ijerph18020622
16. Alsubaie MM, Stain HJ, Webster LAD, Wadman R (2019) The role of sources of social support on depression and quality of life for university students. Int J Adolesc Youth 24(4):484–496
17. Lombardo P, Jones W, Wang L, Shen X, Goldner EM (2018) The fundamental association between mental health and life satisfaction: results from successive waves of a Canadian national survey. BMC Public Health 18(1):342. https://doi.org/10.1186/s12889-018-5235-x

Perceived Motivational Climate and Mental Well-Being Among Secondary School Students in Physical Education

Matheus Chin Yan Yu, Teo Boon Sian, Liew Guo Chen, Chin Ngien Siong, Jeswenny Fresshila John, and Charissa Ann Ho Xin Hui

Abstract The purpose of the study was to examine the perceived motivational climate and mental well-being of secondary school students in physical education (PE). The participants were 230 secondary school students (127 male, 103 female) aged 15 to 17 years old (2.03 ± 0.80) in Kuching, Sarawak. The Teacher-Initiated Motivational Climate in Physical Education Questionnaire (TIMCPEQ; Papaioannou in Res Q Exerc Sport 69:267–275, [25]) and Warwick Edinburgh Mental Well-Being Scale (WEMWBS; Tennant et al. in Health Qual Life Outcomes 5(63):1–13, [26]) were used. Descriptive analysis, independent-samples *t*-test, one-way ANOVA, and two-way ANOVA were utilized. The independent-samples *t*-test revealed significant difference between gender for teacher-initiated performance orientation, $p = 0.009$. The one-way ANOVA revealed no significant difference between age groups and ethnicities for all variables in TIMCPEQ and WEMWBS with $p > 0.05$. The between subject's ANOVA between gender and age revealed the main effect of gender was significant for teacher-initiated performance orientation, $F = 5.98$, $p = 0.015$. The between subject's ANOVA between gender and ethnicity revealed the main effect of gender was significant for eudaimonic, $F = 4.12$, $p = 0.043$, and the main effect of age was significant for teacher-initiated mastery orientation, $F = 3.26$, $p = 0.04$. The findings suggest schools create mastery motivational climate intervention that can facilitate the mental well-being of students in and beyond PE.

M. C. Y. Yu
SMK Sungai Maong, Kuching, Sarawak, Malaysia

T. B. Sian (✉) · C. N. Siong
Department of Physical Education and Health, Institute of Teacher Education Batu Lintang Campus, Kuching, Sarawak, Malaysia
e-mail: boonsian1209@gmail.com

L. G. Chen
Sports Unit, Sarawak Education Department, Kuching, Malaysia

J. F. John
School of Health Science, Universiti Sains Malaysia, Kubang Kerian, Kelantan, Malaysia

C. A. H. X. Hui
SJK (C) Chung Hua Bako, Samarahan, Sarawak, Malaysia

G. Kuan et al. (eds.), *Advancing Sports and Exercise via Innovation*, Lecture Notes in Bioengineering, https://doi.org/10.1007/978-981-19-8159-3_38

Keywords Motivational climate · Mental well-being · Secondary school students

1 Introduction

Research in motivation has been addressed in numerous disciplines which are related to physical activity, sports, and physical education [1–3]. Hence, motivation is one of the fundamental processes in influencing one's participation or engagement to accomplish a goal which is deemed important. The type of motivation is stimulated by the motivational climate that is embedded within the Achievement Goal Theory (AGT; [4]) contemplating on psychological environment which is associated with individual's ability to demonstrate competence towards physical activity [4, 5]. The PE educators can revolve on either mastery-oriented or performance-oriented climate in demonstrating one's competence during the PE classes [6–9]. Mastery-oriented climate emphasizes on individual progress, skill improvement, prioritizing effort, and encourages social relations with others in the team. Conversely, performance-oriented climate capitalized on normative-based evaluation, competition, and social comparison [10, 11]. Understanding factors that motivate the individuals is important for the schools' PE teachers to adapt their teaching methodologies and pedagogies to create an environment that can lead to positive effects towards learning PE [7, 12, 13]. Therefore, PE teachers are encouraged to envisage a heartening motivational climate efficiently to motivate the students to have better engagement and sustain their learning process in PE.

Consistency engagement in PE is fundamental as it leads to positive physical and psychological health, such as evaluating one's overall health and mental well-being [14–16]. Mental well-being comprises hedonic which are pertain to life satisfaction and happiness and eudaimonic which are related to positive functioning, self-acceptance, and endeavour [17]. Research have shown lower risk of depression, anxiety, and stress in individuals who were involved in higher PA as compared with lower levels of PA [18–20]. These findings highlight the significance of being involved in physical activity actively to have better mental well-being and prevent signs of mental health disorder among the adolescents [21]. Studies [22, 23] had shown that university students' mental well-being, and health had grown for the past decades and expected to continue in future as they transit from adolescence to young adulthood. The phase of life requires greater maturity, responsibility, autonomous, and independence which cause additional stress along a continuum form normal to severe problems with sustaining of mental well-being [24].

Thus, with the changing climate of COVID-19 towards the traditional delivery of PE whereby PE is considered to be a more practical approach have resulted in shifting of PE pedagogy to more student-centred technological-based learning strategies. In addition, the mental well-being of the students need to be prioritized due to the rise in mental health challenges in which the study aimed to investigate the motivational climate and mental well-being of secondary school students in PE.

2 Methods

2.1 Participants

The number of secondary school students involved in the study was 230 (Form 3, 4, and 5) that comprised of 127 males (55.2%) and 103 females (44.8%) aged 15 to 17 years old (2.03 ± 0.80). In terms of ethnicity, a higher proportion of participants were Native (47.0%), followed by Malay (36.1%) and Chinese (17.0%). The BMI categories showed that the majority of the participants were in normal BMI (57.0%) followed by underweight (26.1%), overweight (12.2%), and obese (4.8%). Participants' mean for frequency of exercise per week were 1.72 (SD = 0.86) averaging between 1 and 30 min per session. All participants voluntarily completed the questionnaire via Google Forms and submitted their parental consent with assurance of their anonymity. The inclusion criteria included participants who participated in physical activities, whereas exclusion criteria included participants who had not acquired parental consent. The administering institution's ethics committee approved the study.

2.2 Measures

The Teacher-Initiated Motivational Climate in Physical Education Questionnaire (TIMCPEQ; [25]) was used to measure the students' perceptions of the motivational climate in PE. The TIMCPEQ was formed by 12 items, 2 subscales with 6 items each, which were as follows: Teacher-initiated Mastery Orientation (TIMO) (e.g. "The PE teacher paid special attention to whether my skills were improving") and Teacher-initiated Performance Orientation (TIPO) (e.g. "The PE teacher encouraged competition between classmate"). The 5-point Likert scale for all the items ranged from 1 (Strongly disagree) to 5 (Strongly agree). The TIMCPEQ and two subscales demonstrated high internal consistency of 0.74, 0.80, and 0.76, respectively.

Whereas, the Warwick Edinburgh Mental Well-Being Scale (WEMWBS; [26]) was used to evaluate the students' mental well-being in physical education. The WEMWBS consists of 14 items that comprises 2 subscales which are eudaimonic (10 items; e.g. "I've been feeling relaxed") and hedonic (4 items; e.g. "I've had energy to spare"). The 5-point Likert scale for all items ranged from 1 (None of the time) to 5 (All of the time). The WEMWBS and two subscales showed high internal consistency with alpha reliability coefficients of 0.93, 0.89, and 0.82, respectively.

2.3 Statistical Analysis/Data Analyses

The Statistical Package for the Social Science (SPSS) version 27 was conducted to analyse the data. The descriptive statistics, internal consistency, independent-samples t-test, one-way ANOVA, and two-way ANOVA were utilized.

3 Results

Table 1 shows the demographic characteristics of the participants. A total of 230 participants were involved in the study, comprising of 127 (55.2%) males and 103 (44.8%) females. The age groups categories showed that 69 (30.0%) of the participants aged 15 years old, 85 (37.0%) aged 16 years old, and 76 (33.0%) aged 17 years old. In terms of ethnicity, Native formed the largest percentage with 47.0%, followed by Malay and Chinese who comprised of 36.1% and 17.0% of the population, respectively. Based on the Body Mass Index, the majority of the participants were in normal weight category with a total of 131 (57.0%), followed by underweight category with 60 (26.1%), overweight category with 28 (12.2%), and overweight category with 11 (4.8%). Whereas, in terms of the frequency of exercise, a high percentage of the participants exercise 1–2 times (49.6%) a week averaging between 1 and 30 min (57.8%).

Table 2 shows the mean, standard deviation, and reliability of TIMCPEQ and WEMWBS. For TIMCPEQ, teacher-initiated mastery orientation (3.97 ± 0.66) was the most essential factor that influenced the students' perceptions of the motivational climate in PE, followed by teacher-initiated performance orientation (2.84 ± 0.71). The overall TIMCPEQ showed a good internal consistency of 0.74, while subscales alpha coefficients for teacher-initiated mastery orientation and teacher-initiated performance orientation demonstrated high reliability of 0.70, and 0.76, respectively. For WEMWBS, hedonic (3.54 ± 0.84) is the most vital factor that influence the students' mental well-being, followed by eudaimonic (3.53 ± 0.73). The overall WEMWBS and two subscales showed good internal consistency with alpha coefficient of 0.93, 0.89, and 0.82, respectively.

Table 3 shows the independent-samples t-test of TIMCPEQ and WEMWBS based on gender. For TIMCPEQ, the results revealed no statistically significant difference between gender for teacher-initiated mastery orientation scores, $p = 0.89$. However, there was significant difference between gender for teacher-initiated performance orientation scores, $p = 0.009$. The males (2.95 ± 0.68) rated 0.25 point higher than the females (2.70 ± 0.73). For WEMWBS, the results revealed no significant difference between gender for eudaimonic, $p = 0.11$, and hedonic, $p = 0.10$.

Table 4 shows the one-way ANOVA of TIMCPEQ and WEMWBS based on age groups. For TIMCPEQ, the results revealed that there was no statistically significant difference between the three age groups for teacher-initiated mastery orientation, $p = 0.47$, and teacher-initiated performance orientation, $p = 0.08$. For WEMWBS,

Table 1 Socio-demographic characteristics of the 230 participants

Characteristics	Frequency (F)	Percentage (%)	M (SD)
Gender			1.45 (0.50)
Male	127	55.2	
Female	103	44.8	
Age groups			2.03 (0.80)
15	69	30.0	
16	85	37.0	
17	76	33.0	
Ethnicity			2.58 (1.38)
Malay	83	36.1	
Chinese	39	17.0	
Native	108	47.0	
BMI			1.96 (0.76)
Normal (18.50–24.99)	131	57.0	
Underweight (≤ 18.49)	60	26.1	
Overweight (25.00–29.99)	28	12.2	
Obese (30.00–34.99)	11	4.8	
Times doing exercise (week)			1.72 (0.86)
1–2 times	114	49.6	
3–4 times	78	33.9	
5–6 times	26	11.3	
7 times	12	5.2	
Exercise per session (average)			1.60 (0.81)
1–30 min	133	57.8	
31–60 min	64	27.8	
61–90 min	26	11.3	
91–120 min	7	3.0	

the results revealed that there was no statistically significant difference between the three age groups for eudaimonic, $p = 0.49$, and hedonic, $p = 0.54$.

Table 5 shows the one-way ANOVA of *TIMCPEQ* and *WEMWBS* based on age groups. For TIMCPEQ, the results revealed that there was no statistically significant difference between the three age groups for teacher-initiated mastery orientation, $p = 0.14$, and teacher-initiated performance orientation, $p = 0.38$. For WEMWBS,

Table 2 Mean, standard deviation, and reliability for TIMCPEQ and WEMWBS

Variables	M	SD	α
TIMCPEQ			0.74
TIMO	3.97	0.66	0.80
TIPO	2.84	0.71	0.76
WEMWBS			0.93
Eudaimonic	3.53	0.73	0.89
Hedonic	3.54	0.84	0.82

Table 3 Independent-samples t-test of TIMCPEQ and WEMWBS based on gender

Variables	Male		Female		*t*-test	*p*
	M	SD	M	SD	*t* statistics (df = 228)	
TIMCPEQ						
TIMO	3.97	0.06	0.68	0.07	-0.14	0.89
TIPO	2.95	0.68	2.70	0.73	2.64	0.009*
WEMWBS						
Eudaimonic	3.60	0.75	3.44	0.69	1.60	0.11
Hedonic	3.63	0.84	3.44	0.83	1.63	0.10

* $p < 0.05$

Table 4 One-way ANOVA of TIMCPEQ and WEMWBS based on age groups

Variables	Age Groups			*F*(2, 227)	*p*
	15	16	17		
TIMCPEQ					
TIMO	3.94 (0.64)	3.93 (0.64)	4.05 (0.71)	0.76	0.47
TIPO	2.79 (0.59)	2.74 (0.70)	2.98 (0.81)	2.58	0.08
WEMWBS					
Eudaimonic	3.44 (0.66)	3.58 (0.74)	3.55 (0.77)	0.72	0.49
Hedonic	3.45 (0.85)	3.57 (0.83)	3.60 (0.85)	0.61	0.54

* $p < 0.05$

the results revealed that there was no statistically significant difference between the three age groups for eudaimonic, $p = 0.85$, and hedonic, $p = 0.91$.

Table 6 shows the two-way ANOVA for *TIMCPEQ* and *WEMWBS* based on gender and age groups. For TIMCPEQ, the between-subjects ANOVA for teacher-initiated mastery orientation revealed that there was no significant main effect of gender, $F(1, 224) = 0.12, p = 0.73, \eta_p{}^2 = 0.001$, and age groups, $F(2, 224) = 0.91$, $p = 0.40, \eta_p{}^2 = 0.008$. Besides, there was no significant interaction between Gender × Age on teacher-initiated mastery orientation, $F(2, 224) = 0.46, p = 0.63, \eta_p{}^2 =$

Table 5 One-way ANOVA of TIMCPEQ and WEMWBS based on ethnicity

Variables	Ethnicity			F	p
	Malay	Chinese	Native		
TIMCPEQ					
TIMO	3.96 (0.60)	3.76 (0.61)	4.04 (0.72)	1.98	0.14
TIPO	2.76 (0.67)	2.98 (0.67)	2.85 (0.78)	0.96	0.33
WEMWBS					
Eudaimonic	3.54 (0.76)	3.54 (0.64)	3.48 (0.75)	0.16	0.85
Hedonic	3.49 (0.80)	3.57 (0.63)	3.52 (0.91)	0.09	0.91

* $p < 0.05$

0.004. The between-subjects ANOVA for teacher-initiated performance orientation revealed that there was significant main effect of gender, $F(1, 224) = 5.98$, $p = 0.015$, ${\eta_p}^2 = 0.026$, but no significant of age groups, $F(2, 224) = 2.02$, $p = 0.14$, ${\eta_p}^2 = 0.018$. Moreover, there was no significant interaction between Gender × Age on teacher-initiated performance orientation, $F(2, 224) = 0.57$, $p = 0.57$, ${\eta_p}^2 = 0.005$. For WEMWBS, the between-subjects ANOVA for eudaimonic revealed that there was no significant main effect of gender, $F(1, 224) = 1.52$, $p = 0.22$, ${\eta_p}^2 = 0.007$, and age groups, $F(2, 224) = 0.53$, $p = 0.59$, ${\eta_p}^2 = 0.005$. Furthermore, there was no significant interaction between Gender × Age on eudaimonic, $F(2, 224) = 2.70$, $p = 0.07$, ${\eta_p}^2 = 0.024$. The between-subjects ANOVA for hedonic revealed that there was no significant main effect of gender, $F(1, 224) = 1.59$, $p = 0.21$, ${\eta_p}^2 = 0.007$, and age groups, $F(2, 224) = 0.59$, $p = 0.55$, ${\eta_p}^2 = 0.005$. Similarly, there was no significant interaction between Gender × Age on hedonic, $F(2, 224) = 2.75$, $p = 0.07$, ${\eta_p}^2 = 0.024$.

Results (Table 7) revealed that post-hoc Bonferroni adjusted comparisons for teacher-initiated performance orientation indicated that the males rated 0.23 point higher than the females, $p = 0.015$, 95% CI [0.05, 0.42].

Table 6 Main effect of gender and age for TIMCPEQ and WEMWBS

Variables	Gender (G)			Age Groups (AG)				G versus AG
	M	F	p	15	16	17	p	p
TIMCPEQ								
TIMO	3.96	3.99	0.73	3.93	3.93	4.06	0.40	0.63
TIPO	2.94	2.70	0.015*	2.80	2.71	2.94	0.14	0.57
WEMWBS								
Eudaimonic	3.57	3.45	0.22	3.44	3.54	3.56	0.59	0.07
Hedonic	3.60	3.45	0.21	3.45	3.52	3.60	0.55	0.07

* $p < 0.05$

Table 7 Pairwise comparisons of gender for TIMCPEQ and WEMWBS

Variables	μ	p	95% CI	
			LB	UB
TIMCPEQ				
TIMO				
Male versus Female	−0.03	0.73	−0.21	0.15
TIPO				
Male versus Female	0.23	0.015*	0.05	0.42
WEMWBS				
Eudaimonic				
Male versus Female	0.12	0.22	−0.07	0.31
Hedonic				
Male versus Female	0.14	0.21	−0.08	0.37

* $p < 0.05$

Table 8 shows the two-way ANOVA for *TIMCPEQ* and *WEMWBS* based on gender and ethnicity. For TIMCPEQ, the between-subjects ANOVA for teacher-initiated mastery orientation revealed that there was no significant main effect of gender, $F(1, 224) = 0.07, p = 0.79, \eta_p{}^2 < 0.001$, but significant main effect of age groups, $F(2, 224) = 3.26, p = 0.04, \eta_p{}^2 = 0.028$. Besides, there was no significant interaction between Gender × Age on teacher-initiated mastery orientation, $F(2, 224) = 0.64, p = 0.53, \eta_p{}^2 = 0.006$. The between-subjects ANOVA for teacher-initiated performance orientation revealed that there was no significant main effect of gender, $F(1, 224) = 3.00, p = 0.09, \eta_p{}^2 = 0.013$, and age groups, $F(2, 224) = 0.96, p = 0.39, \eta_p{}^2 = 0.008$. Moreover, there was no significant interaction between Gender × Age on teacher-initiated performance orientation, $F(2, 224) = 1.68, p = 0.19, \eta_p{}^2 = 0.015$. For WEMWBS, the between-subjects ANOVA for eudaimonic revealed that there was significant main effect of gender, $F(1, 224) = 4.12, p = 0.043, \eta_p{}^2 = 0.018$, but no significant main effect of age groups, $F(2, 224) = 0.10, p = 0.91, \eta_p{}^2 = 0.001$. Furthermore, there was no significant interaction between Gender × Age on eudaimonic, $F(2, 224) = 3.04, p = 0.05, \eta_p{}^2 = 0.026$. The between-subjects ANOVA for hedonic revealed that there was no significant main effect of gender, $F(1, 224) = 3.32, p = 0.07, \eta_p{}^2 = 0.015$, and age groups, $F(2, 224) = 0.12, p = 0.89, \eta_p{}^2 = 0.001$. Likewise, there was no significant interaction between Gender × Age on hedonic, $F(2, 224) = 2.80, p = 0.06, \eta_p{}^2 = 0.024$.

Results (Table 9) revealed that post-hoc Bonferroni adjusted comparisons for eudaimonic indicated that the males rated 0.21 point higher than the females, $p = 0.043$, 95% CI [0.01, 0.42].

Results (Table 10) revealed that post-hoc Bonferroni adjusted comparisons for teacher-initiated mastery orientation indicated that the Chinese rated 0.32 point lower than the Native, $p = 0.011$, 95% CI [−0.56, −0.07].

Table 8 Main effect of gender and ethnicity for TIMCPEQ and WEMWBS

Variables	Gender (G)			Ethnicity (E)				G versus E
	M	F	*p*	Malay	Chinese	Native	*p*	*p*
TIMCPEQ								
TIMO	3.92	3.94	0.79	3.97	3.75	4.07	0.04*	0.53
TIPO	2.93	2.76	0.09	2.75	2.94	2.84	0.39	0.19
WEMWBS								
Eudaimonic	3.63	3.42	0.043*	3.56	3.51	3.51	0.91	0.05
Hedonic	3.65	3.43	0.07	3.51	3.54	3.57	0.89	0.06

* $p < 0.05$

Table 9 Pairwise comparisons of gender for TIMCPEQ and WEMWBS

Variables	*u*	*p*	95% CI	
			LB	UB
TIMCPEQ				
TIMO				
Male versus Female	−0.03	0.79	−0.21	0.16
TIPO				
Male versus Female	0.18	0.09	−0.02	0.38
WEMWBS				
Eudaimonic				
Male versus Female	0.21	0.043*	0.01	0.42
Hedonic				
Male versus Female	0.22	0.07	−0.02	0.46

* $p < 0.05$

4 Discussion

The study examined the secondary school students' perceptions of motivational climate and mental well-being in PE whether the PE school environment would provide a specific climate that is perceived as conducive to learning PE. The results revealed male have higher perceived performance orientation than female implied that male have the tendency to strive for positive competency. This implied that the male basic needs and support of autonomy, competence, and relatedness needs related to their lessons were lower which could also slightly lower their level of motivation. The male also tends to compare in terms of performance or socially and the teachers might not plan sufficiently in developing the PE activities independently.

The non-significant difference between age groups revealed that all age groups of students were experiencing a higher degree and feeling of autonomy and freedom in schools which lead them adopting and adapting a more mastery orientation in learning

Table 10 Pairwise comparisons of ethnicity for TIMCPEQ and WEMWBS

Variables	μ	p	95% CI	
			LB	UB
TIMCPEQ				
TIMO				
Malay versus Chinese	0.22	0.09	−0.04	0.47
Malay versus Native	−0.10	0.31	−0.29	0.09
Chinese versus Native	−0.32	0.011*	−0.56	−0.07
TIPO				
Malay versus Chinese	−0.19	0.18	−0.46	0.08
Malay versus Native	−0.08	0.43	−0.29	0.12
Chinese versus Native	0.10	0.43	−0.16	0.36
WEMWBS				
Eudaimonic				
Malay versus Chinese	0.05	0.75	−0.23	0.32
Malay versus Native	0.04	0.68	−0.17	0.25
Chinese versus Native	−0.002	0.99	−0.27	0.27
Hedonic				
Malay versus Chinese	−0.03	0.85	−0.35	0.29
Malay versus Native	−0.06	0.62	−0.30	0.18
Chinese versus Native	−0.03	0.85	−0.34	0.28

* $p < 0.05$

PE. The PE educators could have provided more choices in learning resources for the students in learning their tasks without any forms of competing with each other. The active and strategic engagement of students through equal opportunities in learning through practising and doing and reflection on their experience on PE activities empowers them to improve their attitudes towards PE.

The importance of students' mental health and lessen any negative mental well-being need to be emphasised due to the current pandemic environment which is necessary for better quality of life and well-being of the students. The students who perceived teacher-initiated mastery motivational climate mediates personal satisfaction and enjoyment and mental well-being that facilitate future vigour. It is recommended that PE teachers would design interesting and meaningful activities and tasks that can energise the learning processes.

5 Conclusion

The study expanded the current review on perceptions of motivational climate and mental well-being of secondary school students which has practical value for PE teachers and educational organizations. Therefore, a climate that facilitates supportive attributes for students to experience psychological satisfaction and needs for volition, relatedness, and competence would lead to long-term sustainability of students' mental well-being in PE.

Acknowledgements We would like to acknowledge the participants for participating in the study.

References

1. Clancy RB, Herring MP, Campbell MJ (2017) Motivation measures in sport: a critical review and bibliometric analysis. Front Psychol 8:348. https://doi.org/10.3389/fpsyg.2017.00348
2. Ntoumanis N, Stenling A, Thøgersen-Ntoumani C, Vlachopoulos S, Lindwall M, Gucciardi DF, Tsakonitis C (2017) Longitudinal associations between exercise identity and exercise motivation: a multilevel growth curve model approach. Scand J Med Sci Sports 28(2):746–753. https://doi.org/10.1111/sms.12951
3. Venhorst A, Micklewright D, Noakes TD (2017) Towards a three-dimensional framework of centrally regulated and goal-directed exercise behaviour: a narrative review. Br J Sports Med 52(15):957–966. https://doi.org/10.1136/bjsports-2016-096907
4. Nicholls JG (1989) The competitive ethos and democratic education. Harvard University Press
5. Perlman DJ, Goc Karp G (2007) Using TARGET to enhance preservice classroom teachers' ability to identify and develop a motivational climate in physical education. Phys Ed 64(2):102–113
6. Ames C (1992) Achievement goals, motivational climate and motivational processes. In: G Roberts (ed) Motivational in sport and exercise, human kinetics
7. Chin NS, Teo BS, Wee EH (2021) Trainee teachers' situational motivation and motivational climate towards physical lessons during COVID-19 pandemic. Coll Antropl 45(3):201–207
8. Walling MD, Duda JL (1995) Goals and their associations with beliefs about success in and perceptions of the purposes of physical education. J Teach Phys Educ 14(2):140–156
9. Xiang P, Lee A (2002) Achievement goals, perceived motivational climate, and students self-reported mastery behaviors. Res Q Exerc Sport 73(1):58–65. https://doi.org/10.1080/02701367.2002.10608992
10. Castro-Sanchez M, Zurita-Ortega F, Garcia-Mamol E, Chacon-Cuberos R (2019) Motivational climate in sport is associated with life stress levels, academic performance and physical activity engagement of adolescents. Int J Environ Res Public Health 16:1–15 https://doi.org/10.3390/ijerph1607119
11. Ruiz MC, Robazza C, Tolvanen A, Haapanan S, Duda JL (2019) Coach-created motivational climate and athletes' adaptation to psychological stress: Temporal motivation-emotion interplay. Front Psychol 10(617):1–11. https://doi.org/10.3389/fpsyg.2019.00617
12. Cury F, Biddle S, Famose J, Goudas M, Sarrazin P, Durand M (1996) Personal and situational factors influencing intrinsic interest of adolescent girls in school physical education: a structural equation modelling analysis. Educ Psychol 16(3):305–314
13. Lacerda A, Filgueiras A, Campos M, Keegan R, Landeira-Fernandez J (2021) Motivational climate measures in sport: a systematic review. Span J Psychol 24(e27):1–10. https://doi.org/10.1017/SJP.2021.13

14. Cooper SL (2020) Promoting physical activity for mental well-being. ACSM's Health Fit J 24(3):12–16
15. Loureiro A, Veloso S (2017) Green exercise, health and well-being. In: Handbook of environmental psychology and quality of life research. Springer
16. Syed Ali SK (2018) The important of physical activities in our life. Int J Phys Med Rehabil 3(4):308–310
17. Pollard EL, Lee PD (2003) Child well-being: a systematic review of the literature. Soc Indic Res 61(1):59–78
18. Kleppang AL, Hartz I, Thurston M, Hagquist C (2018) The association between physical activity and symptoms of depression in different contexts—a cross-sectional study of Norwegian adolescents. BMC Public Health 18(1):1368. https://doi.org/10.1186/s12889-018-6257-0
19. Schuch FB, Vancampfort D, Firth J, Rosenbaum S, Ward PB, Silva ES, Hallgren M, Ponce De Leon A, Dunn AL, Deslandes AC, Fleck MP, Carvalho AF, Stubbs B (2018) Physical activity and incident depression: a meta-analysis of prospective cohort studies. Am J Psychiatry 175(7):631–648. https://doi.org/10.1176/appi.ajp.2018.17111194
20. Schuch FB, Stubbs B, Meyer J, Heissel A, Zech P, Vancampfort D, Rosenbaum S, Deenik J, Firth J, Ward PB, Carvalho AF, Hiles SA (2019) Physical activity protects from incident anxiety: a meta-analysis of prospective cohort studies. Depress Anxiety 36(9):846–858. https://doi.org/10.1002/da.22915
21. Bell S, Audrey S, Gunnell D, Cooper AR, Campbell R (2019) The relationship between physical activity, mental wellbeing and symptoms of mental health disorder in adolescents: a cohort study. Int J Behav Nutr Phys Act 16(138):1–12. https://doi.org/10.1186/s12966-019-0901-7
22. Hernández-Torrano D, Ibrayeva L, Sparks J, Lim N, Clementi A, Almukhambetova A, Nurtayev Y, Muratkyzy A (2020) Mental health and well-being of university students: a bibliometric mapping of the literature. Front Psychol 11:1226. https://doi.org/10.3389/fpsyg.2020.01226
23. Hunt J, Eisenberg D (2010) Mental health problems and help-seeking behavior among college students. J Adolesc Health 46(1):3–10. https://doi.org/10.1016/j.jadohealth.2009.08.008
24. Park SY, Andalibi N, Zou Y, Ambulkar S, Huh-Yoo J (2020) Understanding students mental well-being challenges on a university campus: interview study. JMIR Form Res 4(3):e15962. https://doi.org/10.2196/15962
25. Papaioannou A (1998) Students perceptions of the physical education class environment for boys and girls and the perceived motivational climate. Res Q Exerc Sport 69:267–275
26. Tennant R, Hiller L, Fishwick R, Platt S, Joseph S, Weich S, Parkinson J, Secker J, Stewart-Brown S (2007) The Warwick-Edinburgh mental well-being scale (WEMWBS): development and UK validation. Health Qual Life Outcomes 5(63):1–13. https://doi.org/10.1186/1477-7525-5-63

Comparing the Category of Rating Scale of the CVIS Instrument of Knowledge, Attitude, and Practices of School Heads and Teachers Towards COVID-19 Pandemic

Mui Yee Voon and Ngien Siong Chin

Abstract The COVID-19 pandemic has had a profound impact on the education systems and poses unprecedented challenges to the schools in terms of the health and well-being of the students, which is a crucial concern globally. The aim of this study was to develop the Coronavirus in School (CVIS) Instrument to measure the level of knowledge, attitude, and practices of primary school heads and teachers towards the COVID-19 pandemic. The instrument is comprised of three constructs: knowledge, which consists of 14 items; attitudes, with 9 items; and practice, with 8 items, respectively. The instrument was rated on a 4-point Likert scale, ranging from strongly disagree to strongly agree. The Rasch Measurement Model was utilized to verify the validity and reliability of the instrument items. The Four Building Blocks model was applied in the construction of the items, and three pilot tests were conducted through the modification and improvement of the items. The result of rating scale analysis of the tests met the assumption of the Rasch rating scale. The results revealed a significant difference between the three categories of rating scale and the four categories of rating scale, especially those involving the middle category of rating scale, the neutral category. The findings demonstrated the instrument with four categories of rating scale managed to meet all the important assumptions of the Rasch Measurement Model, especially for the rating scale analysis. This indicates that the four categories of rating scale performed better than the three categories of rating scale. Moreover, this finding also proves that a rating scale without a middle category (neutral) functions better than a rating scale with a middle category.

Keywords COVID-19 · Rasch measurement model · Rating scale

M. Y. Voon (✉)
Institute of Teacher Education, Tun Abdul Razak Campus, Samarahan, Sarawak, Malaysia
e-mail: voonmui.yee1512@gmail.com

N. S. Chin
Institute of Teacher Education, Batu Lintang Campus, Kuching, Sarawak, Malaysia

G. Kuan et al. (eds.), *Advancing Sports and Exercise via Innovation*, Lecture Notes in Bioengineering, https://doi.org/10.1007/978-981-19-8159-3_39

1 Introduction

The coronavirus, which first appeared in Wuhan, China, affected the entire world at the end of 2019. The virus has been widely spread in Malaysia since the first case was detected on 25 January 2020 by three visitors from China [1]. According to the World Health Organization [1], there are two ways of COVID-19 transmission: direct or indirect contact of infected people. Direct contact happens when individuals have close contact with an infected person when the person coughs, sneezes, or speaks. While indirect contact transmission happens when individuals touch their eyes, nose, or mouth after touching contaminated objects or surfaces, all the transmission of viruses involves droplets from the respiratory system: saliva or secretion droplets. There are 176 million confirmed cases of COVID-19 and 3.81 million deaths reported globally, updated on 14 June 2021 [1]. On the other hand, in Malaysia, there are a reported 658 thousand confirmed cases of COVID-19 with 3908 deaths from 3 January 2020 until 12 June 2021 [1]. Concerns have arisen since the first. The Movement Control Order (MCO) was announced by the Prime Minister on 16 March 2020, due to the drastic increase in the number of cases from 190 to 553 ("14-day Movement Control Order", 2020) [2]. In 2020, there were a total of four phases of movement control orders (MCO) from 18 March 2020 to 31 March 2020. Besides that, the Movement Control Order continued with the Conditional Movement Control Order, which consists of two phases from 4 May 2020 to 9 June 2020. On account of decreasing infected cases nationwide, our government implemented the Recovery Movement Control Order, which involved three phases from 10 June 2020 to 31 March 2021. At the same time, a Movement Control Order was implemented in certain districts in Selangor that have high cases of COVID-19 from 11 January to 31 March 2021. On 10 May 2021, Prime Minister Muhyiddin Yassin declared a nationwide Movement Control Order from 12 May to 7 June 2021. Furthermore, the Prime Minister announced a total lockdown called "FMCO" since the infected cases and deaths reported are increasing drastically within the period of MCO; the period of FMCO is from 1 to 28 June 2021.

During the period of Movement Control Order (MCO), the Ministry of Education came up with a plan to overcome students' learning process obstacles. Therefore, on 27 March 2020, the Ministry of Education implemented e-learning for all the students in our country, and teachers have to work from home. Online learning, or e-learning, involved all the primary school students and secondary school students through Google Classroom, which was organized by the teachers. Students from rural schools fail to access online learning compared to urban and suburban students because, in Malaysia, different states have different access to the Internet, such as Sarawak, Sabah, and Pahang, which do not have sufficient Internet access [3]. As COVID-19 cases began to decline, the Ministry of Education announced the actual date of school reopening was on 3 October 2021. To ensure students' safety and health after the reopening of schools, the Ministry of Education and the Ministry of Health, along with the National Security Council, have started establishing guidelines for Standard Operating Procedures (SOPs) for schools [4]. This guideline includes all MOE schools as well as private MOE-registered educational institutions. School heads and decision-makers have to ensure the existence of productive safety measures in schools such as physical distancing, hand hygiene, and sterilizing school buildings [5]. Besides that, school principals are responsible for giving orders and instructing all the staff members to produce a safe, highly preventable infection of coronavirus [6]. According to the European Agency for Safety and Health at Work (EU-OSHA), [7], school heads play the role as employers and are responsible for producing a safe and healthy environment in schools by following all the safety guidance documents to prevent transmission of COVID-19.

Concerns have arisen over the impact of many COVID-19 cases which involved primary school students as they are still young, and it is difficult for them to comprehend the concept of each SOP implemented in school. Therefore, school educators play the utmost role in educating students on precautions and prevention from getting infected. Teachers are responsible for ensuring students' safety and well-being in school [8]. School educators have to share exact information, control measures, and science facts about COVID-19 with different ages of students to enhance their ability to cope with this pandemic [9]. Thus, the very first step in educating students is to make sure school educators acquire sufficient knowledge of COVID-19, present positive attitudes and behaviour, and practice the preventive measures in schools for students to observe and imitate. Teachers are dynamic role models for students who can affect them in a way to be a better person [10]. The interaction between a teacher and students can influence a student's attitudes and behaviour [11]. Additionally, many studies emphasized the importance of KAP survey in finding the level of knowledge, believed, and what had practiced by a particular population [12] because untrustworthy information and negative attitude can cause panic and stress in a community [13]. As a result, the KAP survey is an extremely important survey that should be conducted to determine the community's level of awareness of knowledge, attitude, and practice regarding the COVID-19 among the community [14]. It is important to develop an instrument, specifically the KAP survey that can be used to measure knowledge, attitudes, and practices of individuals towards the COVID-19 pandemic, especially in the school community, which mainly involved educators. This is to make sure they are qualified to guide and protect students in the school compound. People's constancy in control measures of COVID-19 was influenced by their knowledge, attitudes, and practices [14, 15]. Lessons from the previous SARS outbreak in 2003 and the Ebola outbreak in 2013 demonstrated that a lack of knowledge about the diseases and a lack of perception can cause panic emotion, making infectious disease prevention more difficult [16] and will make the prevention of infectious disease more complicated [17, 18].

This study has succeeded in developing an instrument, the CVIS instrument, to measure the knowledge, attitude, and practice of school heads and teachers at school. The researcher utilized the Rasch Measurement Model as the fundamental measurement model in the study to ensure the validity, reliability, and good quality of instrument items. All the instrument items have been tested to ensure the item fit, item dimensionality, rating scale analysis, item reliability, item difficulty level, item separation, and differential item functioning (DIF) have fulfilled the assumptions of the Rasch model.

2 Methods

2.1 *Instrumentation*

The Coronavirus in School (CVIS) Instrument is an instrument that designed to measure the level of knowledge, attitudes, and practices of primary school educators towards the COVID-19. This instrument consists of three fundamental constructs as mentioned which utilized to measure primary school educators' knowledge, attitudes, and practices towards the COVID-19. In details, the instrument consists of nine themes related to the pandemic: transmission of COVID-19, general knowledge of COVID-19, misconception of COVID-19, symptoms of COVID-19, face mask, hand hygiene, preventive measures, physical distancing and Standard Operating Procedures (SOPs).

Consequently, the two sets of instruments consist of 53 items but with different categories of rating scales being tested in order to test the effectiveness of rating scales. As mentioned, the first set of instrument consists of three rating scales (1-Disagree/2-Neutral/3-Agree), while the second set of instrument consists of four rating scales (1-Strongly Disagree/2-Disagree/3-Agree/4-Strongly Agree). According to the Rasch Measurement Model, it is important to meet the requirements of rating scales in order to verify the validity of data [19]. It is also able to prove whether respondents can distinguish each of the rating scales utilized in a survey. The Rasch model can determine the zero setting and calibrate the rating scale. The range of acceptable rating scale index is between 1.4 and 5.0 [20]. As a consequence, this study has been carried out to identify the effectiveness and precision of the rating scale in the instrument, either the 3 rating scale or the 4 rating scale.

2.2 *Respondents*

The respondents involved in this study were primary school educators from Malaysian government primary schools, and the age range of respondents was between 24 and 65 years old. The respondents from 11 districts in Sarawak have been invited to participate in the study by getting permission from the Department of Education Sarawak initially. Next, the researcher proceeded with getting permission from the particular district education office. The instrument (questionnaire) distributed through Google Form to all the schools from each of the districts. Convenience sampling method implemented in the study by only involved respondents who are willing to contribute in the study. As result, there were 52 respondents involved in the Instrument 1 with 3 rating scales, while 71 respondents involved in the Instrument 2 with 4 rating scales. The respondents involved in both sets of instruments were mainly from urban and suburban areas. All the respondents were required to answer all the items in the instrument which can be divided into two fundamental sections: demographics and the items measuring the level of knowledge, attitudes, and practice towards the COVID-19. The respondents needed to rate each statement of an item based on their degree of agreement. This study has been conducted in Sarawak because this state has all types of school, with regard to urban and rural, as well as a high variation of public school qualities. Sarawak is also an excellent example of Malaysian unity, as many researchers prefer to study on the Peninsula.

2.3 Measurement Model

The Item Response Theory and the Rasch model were utilized in this research. The Rasch Measurement Model is the chosen measurement model rather than Classical Test Theory or other mathematical models due to its specific advantages. The Rasch Measurement Model enables interchangeability between item and respondent due to both having the same interval scale [21]. The equal interval scale called logit is utilized to connect the relationship between the capability of the test taker and the severity level of the item [22]. These properties of Rasch directly allow researchers to use the difficulty level of an item to represent or figure out the actual ability level of the respondent. Moreover, the Rasch Measurement Model fulfilled all the five measurement principles of human sciences [23] which allowed researchers to obtain more precise and meaningful inferences from the data collected [22]. The Rasch Measurement Model can detect any outliers or misfit data that may occur in the research. The Rasch model software called Winsteps is used to detect the outliers and misfit data through these three psychometric attributes: (1) outfit mean square value in between 0.5 and 1.5, (2) Z-standard outfit value in between −2.0 and + 2.0, and (3) value of point measure correlation value in between 0.4 and 0.85 [22]. Any items that are not in the range of these three psychometric attributes are considered misfit items. The process of reconsidering or re-test has to be done for misfit items before getting rid of them. Moreover, the Rasch Measurement Model can detect any occurrences of item bias, which is important for this research as it involved respondents with different demographic backgrounds. The issue of bias can occur when there are items in the research instrument that are beneficial to a certain group of respondents.

The Andrich Rasch Rating Scale model is implemented in the study due to its ability to be conveyed as a linear logistic model [24]. Comprehensively, it involved the process of calibration, which involves converting the raw scores of polytomous data acquired from Likert scales to a logit scale (log odds unit). The logit scale of the Rasch model enables it to accomplish all five concepts of measurement for human sciences, which consist of an equal interval on its linear ruler, solving the problem of missing data, estimation of preciseness, detecting misfits or outliers, and allowing repeatability [22, 23]. The Rasch analyses focus on the calibration of respondent ability, item difficulty, item fit, item discrimination, and reliability for both item and respondent [19, 25].

In contrast, the traditional measurement model (Classical Test Theory) has some limitations when analysing ordinal scales, such as the fact that the obtained scores are test- and sample-dependent, and total scores can affect item difficulty level, test reliability, and item discrimination [26–28]. Further, there is no equal interval between scores and no absolute zero as an origin. This indicates that the scores from the Likert scales are only able to supply incomprehensible scores in ordinal form [29]. Therefore, it is pointless to analyse the total score of the rating scale when counting the frequencies of raw data is easy and effortless. Indeed, misconstruing of the analysis results is able to mislead the result of a study.

As highlighted earlier, the Rasch Measurement Model enables interchangeability between item and respondent due to both having the same interval scale [21]. The equal interval scale called logit is utilized to connect the relationship between the capability of the respondents and the severity level of the item [22]. These properties of Rasch directly allow researchers to use the difficulty level of an item to represent or figure out the actual ability level of the respondent.

This paper will illustrate how the researcher selects the most appropriate and effective number of rating scale category by utilizing the Rasch analysis. First, the value of "Observed Average" is used to show the pattern of response and if the values are increasing from negative to positive, it shows the normal pattern. Second, the value of "Sample Expected" should approximately the same as the value of the "Observed Average". Third, the values of "Infit and Outfit Mean Squares" supposed to be equal to 1.0 logit. The most important assumption of the Rasch model, which must be fulfilled to ensure that respondents can distinguish the categories of rating scales, is the range of difference between rating each scale. In the Rasch model, the acceptable range of the difference between the scale should be between 1.4 and 5.0 logit [20, 30]. The difference between the scale which is less than 1.4 will be collapsed, while the difference between the scale which is bigger than 5.0 will form a new scale [30]. Additionally, another way of detecting the practicability of the rating scale is by referring to the probability curves [31]. Each curve of the graph represented a category of the rating scale, and the intersection of the points between adjacent curves represented the Andrich Threshold.

2.4 Findings (Validity of the CVIS Instrument)

The face and content validity of this study was determined by the process of expert judgement through the application of the Many Facets Measurement Model. Correspondingly, the construct validity of this study was verified by the unidimensionality of the instrument. The Many-Facet Rasch Measurement is an extension measurement model of the basic Rasch model by including more aspects (facets), for instance, raters and tasks [32, 33]. This measurement model is used to demonstrate how to measure the severity and leniency of raters and to determine raters' consistency [32]. There are two essential dimensions emphasized in this research: the severity of raters and raters' internal consistency. The higher the logit value of the measure, the more severe the expert judges; the lower the logit value of the measure, the more lenient the expert is [34]. Furthermore, the Many-Facet analysis also provided rater fit statistics which were used to indicate the internally self-consistent across raters [20, 23].

The Many-Facet Rasch Measurement analysis also comes up with an analysis for the quality of instrument items. The quality of items is determined by two aspects, which are item logits and item fit statistics. Item logit values are measured according to the total experts rating based on the content of the items. The fit statistics value for the items is useful for the validity purpose of the instrument [20]. The Many-Facet Rasch Measurement analysis indicates the good quality items which are suitable for the instrument and determines the low-quality items which have to be eliminated from the instrument. Besides that, the suggestions and comments from the expert judges are essential for the face validity of the instrument. There were a few instrument items were modified based on the suggestions of expert.

In Rasch, unidimensionality is the essential criteria that must be fulfilled by any instrument as an indication of construct validity. A unidimensionality instrument has a minimum of 40% of "raw variance explained by measures" and better at 60%, and the "unexplained variance in the first contrast" is below 15% [30]. Table 1 shows the "Raw Variance Explained by Measures" for both instruments are Instrument 1 = 61.0%; Instrument 2 = 41.0%, which is more than 40% and approximately matches the expected model, which is 61.7% and 40.8% separately. Besides that, the "Unexplained Variance in 1st Contrast" for Instrument 1 is at 3.0% and Instrument 2 is at 6.7%, which is below 15%. The minimum ratio of "Raw Variance Explained by Measures" and "Unexplained Variance in 1st Contrast" must be at least 3:1. Table 1 shows the ratio of "Raw Variance Explained by Measures" and the "Unexplained Variance in 1st Contrast" for both set instruments are 20.3: 1 (Instrument 1) and 6.1: 1 (Instrument 2), respectively, so both the ratios are larger than the minimum ratio of 3:1. Thus, the instrument items managed to fulfil both the criteria for unidimensionality and construct validity.

2.5 *Reliability of the CVIS Instrument*

The good item reliability is between 0.80 and 0.90, and the ideal reliability is greater than 0.94 [21]. Table 1 shows the item reliability for Instrument 1 is 0.91, while Instrument 2 is 0.97. Item separation is used to verify the item hierarchy. The greater value of item separation indicated the instrument has good quality in differentiating items into a few different difficulty levels [21]. The acceptable item separation is at least 3, able to distinguish 3 groups of difficulty levels of items [30]. Table 1 shows the value of item separation for Instrument 1 = 4.37 and Instrument 2 = 5.45. Both indices manifested the Instrument 2 (4 rating scale) had better item reliability and item separation than the Instrument 1 (3 rating scale).

Table 1 Item dimensionality and item reliability analysis of Instrument 1 and Instrument 2

	Acceptable range	Observed	
		Instrument 1	Instrument 2
Raw variance explained by measures	> 40%	61.0% Expected = 61.7%	41.0% Expected = 40.8%
Unexplained variance in 1st contrast	< 15%	3.0%	6.7%
Ratio between raw variance explained by measures and unexplained variance in 1st contrast	3:1	20.3:1	6.1:1
Item reliability		0.91	0.97
Item separation		4.37	5.45

2.6 *Rating Scale Analysis*

Table 2 shows that the value of the "Observed Average" for Instrument 1 and Instrument 2 is an increase from a negative value to a positive value, proving that both instruments have normal patterns. Besides that, the value of "Sample Expected" for both instruments are nearly the same as the value of "Observed Average". The values of "Infit and Outfit Mean Square" for these two instruments are approximately equal to 1.0 logit and within the acceptable range of the Rasch model (Refer Table 2). As highlighted, the most important assumption of the Rasch rating scale is the difference of each rating scale must between 1.4 and 5.0 logit. Table 2 shows the difference between each rating scales for Instrument 1 is not in the acceptable range (Scale 1 & 2 = 0.07; Scale 2 & 3 = 0.14 logit). This indicates that the respondents involved in the instrument failed to distinguish the three rating scales that utilized in the survey. Conversely, the difference between each rating scale for Instrument 2 are Scale 1 & 2 = 3.31; Scale 2 & 3 = 2.49; and Scale 3 & 4 = 4.96. These values clarified that each of the rating scales is functioning well, and respondents are able to differentiate them and choose the best rating scale to describe their degree of agreement towards the items.

Moreover, Figs. 1 and 2 show the graphics of the rating scales analysis for Instrument 1 and Instrument 2, which are called category probability curve graphs. Figure 1 shows there are two distinct peaks in the graph which represent the rating scale 1 (Disagree category) and rating scale 3 (Agree category). There is no distinct peak for rating scale 2 (neutral category). This indicates that the respondents were confused with the three categories of rating scales and especially failed to differentiate the differences between the neutral category and the other two rating scales.

On the contrary, Fig. 2 shows there are four distinct peaks on the graph which represent each of the rating scale categories utilized in the survey (1 = "Strongly Disagree"; 2 = "Disagree"; 3 = "Agree"; and 4 = "Strongly Agree"). The distinct peaks indicate that the four categories of the rating scale were understood by the

Table 2 Rating scale analysis of Instrument 1 and Instrument 2

Condition	Category	Observed average	Sample expected	Infit MNSQ	Outfit MNSQ	Structure calibration	Difference between categories
Instrument 1	1	−1.26	−1.24	1.02	0.96	NONE	1 & 2 = 0.07
	2	0.49	0.94	0.94	85	0.07	2 & 3 = 0.14
	3	1.96	1.06	1.06	1.07	−0.07	
Instrument 2	1	−1.96	−2.00	1.05	1.16	NONE	1 & 2 = 3.31
	2	−0.81	−0.75	0.94	0.92	−3.31	2 & 3 = 2.49
	3	1.19	1.16	0.96	0.96	−0.82	3 & 4 = 4.96
	4	2.23	2.34	1.07	1.07	4.14	

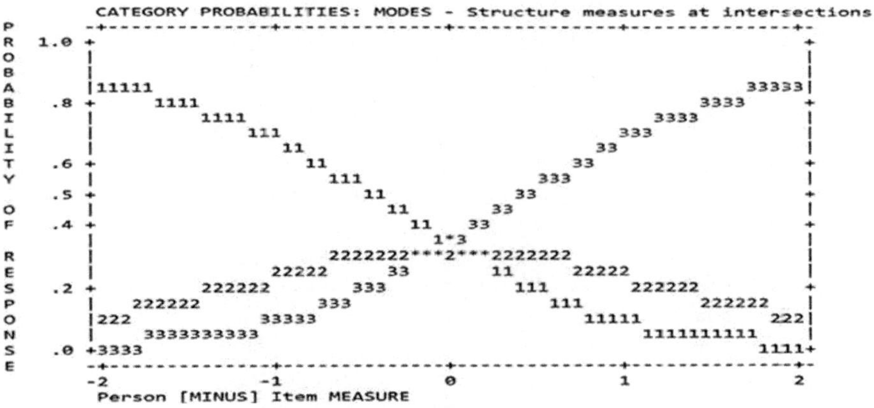

Fig. 1 Category probability curve graph of Instrument 1

respondents. As a result, the respondents were able to distinguish the differences among the four categories of the rating scales and managed to choose the most probable responses for each instrument item.

3 Discussion

In Rasch analysis, researchers managed to make a meaningful comparison on the variable intended to be measured by referring to the rating scale selected by respondents. The concept of linearity of the Rasch rating scale allowed researchers to figure

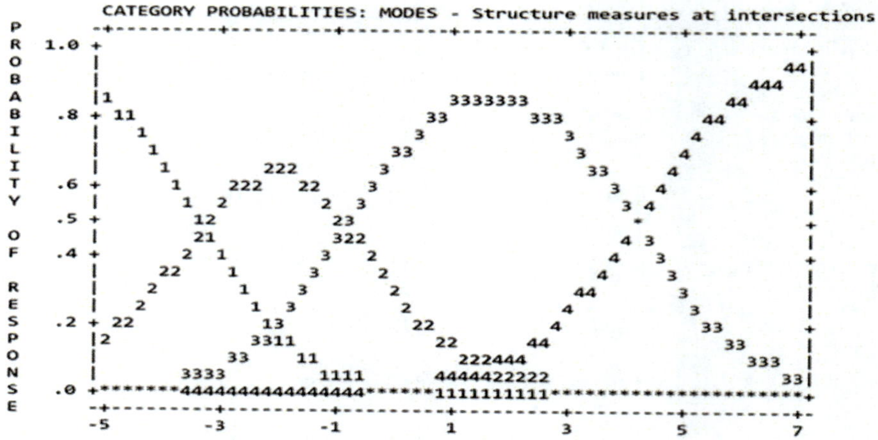

Fig. 2 Category probability curve graph of Instrument 2

out the level of agreement of the targeted population's knowledge, attitude, and practices towards the COVID-19 pandemic. The Rasch model succeeded in overcoming the shortcomings of Classical Test Theory, which utilized the raw data scores in measuring the ability level and agreeability of respondents. Moreover, utilization of impropriate rating scales can affect item fit statistics [31].

By detecting and improving the impropriate rating scale, a researcher can avoid eliminating some misfit items that might be "fit" items with the suitable rating scale. For instance, the findings of this study showed that both instruments contained the same instrument items but had different results in item fit statistics. Instrument 2, which utilized 3 rating scales, exhibited 8 outliers (misfit items), while Instrument 2, with 4 rating scales, verified all the instrument items were fit. As mentioned above, all the important assumptions in the Rasch Measurement Model have to be fulfilled to ensure the functionality of the rating scale before the researcher conducts the real survey on a particular group of samples.

References

1. Abdi M (2020) Coronavirus disease 2019 (COVID-19) outbreak in Iran: actions and problems. Infection control and hospital epidemiology. Cambridge University Press, America
2. Shah AUM, Safri SNA, Thevadas R, Noordin NK, Abd Rahman A, Sekawi Z (2020) COVID-19 outbreak in Malaysia: actions taken by the Malaysian government. Int J Infect 97:108–116
3. Ating R (2020) Education senior minister explains SOPs for schools' reopening
4. Arumugam (2020) Education Senior Minister explains SOPs for schools' reopening.
5. Shi Z, Mehrotra A, Gidengil CA, Poon SJ, Uscher-Pines RKN (2018) Quality of care for acute respiratory infections during direct-to-consumer telemedicine visits for adults. Health Aff 37(12):2014–2023

6. Stone JH, Frigault MJ, Serling-Boyd NJ, Fernandes AD, Harvey L, Foulkes AS, Woolley AE (2020) Efficacy of tocilizumab in patients hospitalized with Covid-19. N Engl J Med
7. Walters D, Wadsworth E (2020) Participation in safety and health in European workplaces: framing the capture of representation. Eur J Ind 26(1):75–90
8. Duraku ZH, Hoxha L (2020) The impact of COVID-19 on education and on the well-being of teachers, parents, and students: challenges related to remote (online) learning and opportunities for advancing the quality of education
9. Saxena R, Saxena SK (2020) Preparing children for pandemics. In: Coronavirus disease 2019 (COVID-19), pp 187–198
10. Bashir S, Bajwa M, Rana S (2014) Teacher as a role model and its impact on the life of female students. Int J Res Granthaalayah 1(1):9–20
11. Hughes JN, Chen Q (2011) Reciprocal effects of student–teacher and student–peer relatedness: effects on academic self-efficacy. J Appl Dev Psychol 32(5):278–287
12. ul Haq N, Hassali MA, Shafie AA, Saleem F, Farooqui M, Aljadhey H (2012) A cross sectional assessment of knowledge, attitude and practice towards Hepatitis B among healthy population of Quetta, Pakistan. BMC Public Health 12(1):692
13. Lin Y, Huang L, Nie S, Liu Z, Yu H, Yan W, Xu Y (2011) Knowledge, attitudes and practices (KAP) related to the pandemic (H1N1) 2009 among Chinese general population: a telephone survey. BMC Infect Dis 11(1):1–9
14. Zhong BL, Luo W, Li HM, Zhang QQ, Liu XG, Li WT, Li Y (2020) Knowledge, attitudes, and practices towards COVID-19 among Chinese residents during the rapid rise period of the COVID-19 outbreak: a quick online cross-sectional survey. Int J Biol Sci 16(10):1745
15. Geda NR, Legesse B, Kebede SW, Fekadu B (2020) Knowledge and attitude towards covid-19 in Ethiopia
16. Tao N (2003) An analysis on reasons of SARS-induced psychological panic among students. J Anhuin Agric 21:78–79
17. Fraser C, Riley S, Anderson RM, Ferguson NM (2004) Factors that make an infectious disease outbreak controllable. Proc Natl Acad Sci 101(16):6146–6151
18. Ajilore K, Atakiti I, Onyenankey K (2017) College students' knowledge, attitudes, and adherence to public service announcements on Ebola in Nigeria: suggestions for improving future Ebola prevention education programs. Health Educ J 76:648–660
19. Baghaei P (2008) The Rasch model as a construct validation tool. Rasch Measur Trans 22(1):1145–1146
20. Bond TG, Fox CM, Lacey H (2007) Applying the Rasch model: fundamental measurement. In: The social sciences, 2nd edn
21. Sumintonon B, Widhiarso W (2015) Aplikasi pemodelan rasch pada assessment pendidikan. Trim komunikata
22. Sumintono B (2017) Science education in Malaysia: challenges in the 21st century. Jurnal Cakrawala Pendidikan 35(3)
23. Wright BD, Mok M (2004) An overview of the family of Rasch measurement models. Introduction Rasch Meas 1–24
24. Andrich D (1978) A rating formulation for ordered response categories. Psychometrika 43(4):561
25. Zubairi AM, Kassim NLA (2016) Classical and Rasch analyses of dichotomously scored reading comprehension test items. MaJER 2(1):20
26. Bradley KD, Peabody MR, Akers KS, Knutson N (2015) Rating scales in survey research: using the Rasch model to illustrate the middle category measurement flaw. Surv Pract 8(2)
27. Magno C (2009) Demonstrating the difference between classical test theory and item response theory using derived test data. Int J Educ Psychol 1(1):1–11
28. Impara JC, Plake BS (1998) Teachers' ability to estimate item difficulty: a test of the assumptions in the Angoff standard setting method. J Educ Meas 35:69–81
29. Baumgartner TA, Jackson AS, Mahar MT, Rowe DA (2003) Measurement in physical education and exercise science, No. 7. Realibility and objectivity, 7th edn. Mcgraw-hill, New Work

30. Aziz AA, Masodi MS, Zaharim A (2013) Asas model pengukuran Rasch: pembentukan skala dan struktur pengukuran. Penerbit Universiti Kebangsaan Malaysia
31. Green KE, Frantom CG (2002) Survey development and validation with the Rasch model. In: International conference on questionnaire development, evaluation, and testing. Charleston, SC
32. Eckes T (2014) Reference supplement to the manual for relating language examinations to the common European framework of reference for languages: learning. Teaching, assessment, Section H: Many-Facet Rasch Measurement
33. Tavakol M, Pinner G (2019) Using the many-facet Rasch model to analyse and evaluate the quality of objective structured clinical examination: a non-experimental cross-sectional design. BMJ Open 9(9):e029208
34. Sahin MG, Teker GT, Güler N (2016) An analysis of peer assessment through many facet Rasch model. J Educ Pract 7(32):172–181

Undergraduate Students' Achievement Emotions, Life Enjoyment, Satisfaction, and Enjoyment Processes in Physical Education

Teo Boon Sian, Chin Ngien Siong, Teo Eng Wah, Arthur Ling Wei, and Ruilin Xu

Abstract One's sense of achievement and well-being can be dependent on emotions, enjoyment, and satisfaction in life. The study aimed to examine achievement emotions, life enjoyment, satisfaction, and enjoyment processes of undergraduates in physical education (PE). The participants were 260 undergraduate students (133 male, 127 female) aged 18 to 23 years old (2.00 ± 0.83) from the universities. The Achievement Emotions Questionnaire-Physical Education (AEQ-PE-M; Ibrahim et al. in BMC Public Health 21(1):1–8 [9]), Quality of Life Enjoyment and Satisfaction Questionnaire (Q-LES-Q-M; Ibrahim et al. in Int J Environ Res Public Health 18:1–10, [10]), and Physical Education Enjoyment Processes Questionnaire (PEEPQ; Hashim et al. in Res Q Exerc Sport 79(2):183–194, [8]) were used. Descriptive analysis, one-way ANOVA, and two-way ANOVA were conducted. The one-way ANOVA revealed significant difference between age groups for pride, boredom, physical health, feelings, homework, leisure time activities, self-referent competency, activity-generated excitement, teacher-generated excitement, and PE enjoyment. Results also revealed significant difference between ethnicities for boredom, physical health, and all PEEPQ subscales. The between-subjects ANOVA between gender and age revealed the main effect of age was significant for pride, boredom, physical health, feelings, homework, leisure time activities, self-referent competency, activity-generated excitement, teacher-generated excitement, peer interaction, parental encouragement, and PE enjoyment. There was significant interaction between gender × age for pride, enjoyment, and all PEEPQ subscales. The between subject's ANOVA between gender and ethnicity revealed the main effect of gender was significant for anxiety, whereas ethnicity was significant for physical health and all PEEPQ subscales. There was significant interaction between gender × ethnicity for self-referent competency. The findings concluded that well-functional environments would improve the students' emotions, life enjoyment, satisfaction, and processes in universities.

T. B. Sian (✉) · C. N. Siong · A. L. Wei
Department of Physical Education and Health, Institute of Teacher Education Batu Lintang Campus, Kuching, Sarawak, Malaysia
e-mail: boonsian1209@gmail.com

T. Eng Wah · R. Xu
Centre for Sports & Exercise Science, University of Malaya, Kuala Lumpur, Malaysia

G. Kuan et al. (eds.), *Advancing Sports and Exercise via Innovation*, Lecture Notes in Bioengineering, https://doi.org/10.1007/978-981-19-8159-3_40

Keywords Achievement emotions · Life enjoyment · Satisfaction · Life processes · Physical education

1 Introduction

One of the factors which can influence achievement, engagement, academic interest, motivation, and well-being among the undergraduates are achievement emotions and feelings during their time in the universities [1]. Achievement emotions are always involved in student learning because they are biological and psychological conditions associated with learning activities. Therefore, their performance is bounded by emotional states and be motivated to learn, discover, and overcome challenges to acquire a sense of achievement. The emotional factor is related to life satisfaction which is one's self-assessment on account of the criteria he/her the criteria being determined [2]. As one's ages, life satisfaction tends to heighten at a certain age and decreases which tend to affect learning and achievement outcomes. Achievement emotions are one's experienced emotions in relation to any achievement outcomes and activities [3]. Though there are educational literature on emotions, studies on achievement emotions on physical education (PE) are still sparse. Predominantly, positive emotions such as interest, contentment, happiness, and enjoyment are related with attention and engagement with increased motivation, effort, and achievement. Conversely, emotions that are deactivating such as anxiety can be detrimental for learning and achievement as it exhilarates disengagement from learning and challenging tasks and achievements [4].

The interactions of emotions with other attributes such as intellectual, physical social needs, and spiritual needs can also lead to a higher level of self-regulation. The elements of self-regulation could be accomplished by participation through physical activity, exercise, and games in PE. The enhancement of interpersonal skills with regards to teamwork, sharing as well as the enjoyment and satisfaction in learning and exploring can stimulate and inspire creativity in PE. In addition, the play and games forms in PE enable students to develop their transversal skills (emotional, communicative, methodological processes and social competence and solve cognitive, motor and social issues) which can aid in developing their intellectual capabilities. As regular engagement in physical activities through PE can help in improving the mental well-being and prevention of non-communicable diseases, the students would be able to enjoy and value a healthy life [5]. Positive emotions such as enjoyment and negative emotions such as frustration and anxiety are based on the degree of activation and valence in response to the PE environment [6]. Among the fundamental objectives of PE is to present an opportunity for the students to enjoy themselves providing fun, liking, and pleasure [7] in the forms of regular recreational and physical activities which can initiate active-related behaviour [8].

Though PE is a central mechanism for enjoyment, it is mandatory to examine achievement emotions, life enjoyment, satisfaction, and processes that can promote the enjoyment of students in PE which can affect the learning of PE in universities.

2 Methods

2.1 Participants

A total of two hundred and sixty participants from several institutions of higher learning were selected randomly to participate in the study. There were one hundred and twenty-seven males and one hundred and thirty females aged between 18 and 23 (1.90 + 0.87) years old. The self-administrated online Google Form questionnaires were utilized upon obtaining the participants' consent, whereby their confidentiality and anonymity were also assured. Ethics approval was obtained from the Institute of Teacher Education Campus' research committed prior to the study.

2.2 Data Analysis

The IBM Statistical Package for the Social Science (SPSS) version 27 was used to analyse the data. The descriptive statistics, independent-samples *t*-test, one-way ANOVA, and two-way ANOVA were conducted to examine the gender, age groups, and ethnicity among undergraduate students.

2.3 Measures

The Achievement Emotions Questionnaire-Physical Education (AEQ-PE-M) was adapted and validated by Ibrahim et al. [9] was utilized to measure achievement emotions experienced by undergraduate students. The AEQ-PE-M is formed by 24 items, 6 subscales with 4 items each, which are as follows: pride (e.g. "I am proud to be able to keep up with the physical education class"), enjoyment (e.g. "I am motivated to go to the physical education class because it is exciting"), anger (e.g. "I feel anger welling up in me during the physical education class"), anxiety (e.g. "I feel nervous in the physical education class"), hopelessness (e.g. "I have lost all hope of doing physical education activities effectively"), and boredom (e.g. "I get bored during the physical education class"). Participants responded to all measures on a 5-point Likert scale ranging from 1 (Strongly disagree) to 5 (Strongly agree). All subscales and overall AEQ-PE-M demonstrated adequate internal consistency with alpha reliability coefficients ranging from 0.56 to 0.72.

The Quality of Life Enjoyment and Satisfaction Questionnaire (Q-LES-Q-M) was adapted and validated by Ibrahim et al. [10]. The questionnaire is designed to help assess the degree of enjoyment and satisfaction experienced during the past week. The Q-LES-Q consists of 43 items that comprises four factors which are as follows: physical health/activities (13 items; e.g. "…felt rested?), feelings (14 items; e.g. "…felt good about your appearance?), school/course work (10 items; e.g. "…been interested in your course/class work?), leisure time activities (6 items; e.g. "How often did you enjoy the leisure activities?). Participants responded to all measures on a 5-point Likert scale ranging from 1 (not at all) to 5 (all the time). The Q-LES-Q and four subscales showed high internal consistency with alpha value ranging from 0.90 to 0.96.

The Physical Education Enjoyment Processes Questionnaire (PEEPQ) [8] was used to measure the self-referent competency (SRC) (4 items; e.g. "My sport skills have improved from doing PE"), other-referent competency (ORC) (4 items; e.g. "Other students think I am good at PE"), activity-generated excitement (AGE) (5 items; e.g. "I am enthusiastic about PE activities"), teacher-generated excitement (TGE) (3 items; e.g. "My PE teachers make PE class interesting experience for me"), peer interaction (PI) (2 items; e.g. "PE gives me a chance to socialize with my classmates"), parental encouragement (PE) (2 items; e.g. "My parents are interested in the PE activities I do at school"), and PE enjoyment (2 items; e.g. "PE is fun"). Participants responded to all measures on a 5-point Likert scale ranging from 1 (Strongly disagree) to 5 (Strongly agree). All subscales and overall PEEPQ demonstrated high internal consistency with alpha reliability coefficients ranging from 0.62 to 0.95.

3 Results

Table 1 shows the demographic characteristics of the participants. A total of 260 participants were involved in this study, 133 (51.2%) males and 127 (48.8%) females. The age groups categories showed that 90 (34.6%) of the participants were 18–19 years old and 22–23 years old, and 80 (30.8%) were 20–21 years old. In terms of ethnicity, the Malay formed the largest percentage with 43.5%, followed by Native and Chinese who comprised of 33.5% and 23.1% of the population, respectively. Totally, 146 participants (56.2%) took up physical education while 114 (43.8%) of them took up sports science. Most of them came from UiTM (43.8%), followed by IPGK Ilmu Khas (23.1%), IPGK Batu Lintang (16.9%), and IPGK Tun Abdul Razak (16.2%), respectively. Based on the body mass index (BMI), the majority of the participants were in normal weight category with a total of 196 (75.4%), followed by underweight category with 36 (13.8%), overweight category with 26 (10.0%), and obese category with 2 (0.8%). Whereas, in terms of the frequency of exercise, a high percentage of the participants exercise 1 to 3 times (56.5%) a week averaging between 31 and 60 min (46.2%).

Table 2 shows the mean, standard deviation, and reliability of *AEQ-PE, Q-LES-Q,* and *PEEPQ*. For AEQ-PE, pride (4.24 $\pm$ 0.63) was the most essential factor that influenced the achievement emotions experienced by trainee teachers, followed by enjoyment (4.05 $\pm$ 0.63), anxiety (2.91 $\pm$ 0.80), boredom (2.23 $\pm$ 0.46), hopelessness (1.96 $\pm$ 0.69), and anger (1.94 $\pm$ 0.67). The overall AEQ-PE showed adequate internal consistency of 0.56 while subscales alpha coefficients for pride, enjoyment, anger, anxiety, hopelessness, and boredom demonstrated good reliability of 0.72,

Table 1 Demographic characteristics of the participants ($N = 260$)

Characteristics	Frequency (F)	Percentage (%)	M (SD)
Gender			1.49 (0.50)
Male	133	512	
Female	127	48.8	
Age			2.00 (0.83)
18–19	90	34.6	
20–21	80	30.8	
22–23	90	34.6	
Ethnicity			1.90 (0.87)
Malay	113	43.5	
Chinese	60	23.1	
Native	87	33.5	
Education			1.44 (0.50)
Physical education	146	56.2	
Sport science	114	43.8	
University			2.37 (1.16)
UiTM	114	43.8	
IPGK Ilmu Khas	60	23.1	
IPGK Batu Lintang	44	16.9	
IPGK Tun Abdul Razak	42	16.2	
Body mass index (BMI)			1.98 (0.52)
Normal (18.5–24.9)	196	75.4	
Underweight (≤ 18.49)	36	13.8	
Overweight (25.0–29.9)	26	10.0	
Obese (30.0–34.9)	2	0.8	
Frequency of exercise per week			1.43 (0.50)
1–3 times	147	56.5	
4–6 times	113	43.5	
Minutes exercise			2.25 (0.85)

(continued)

Table 1 (continued)

Characteristics	Frequency (*F*)	Percentage (%)	*M* (SD)
0–30 min	48	18.5	
31–60 min	120	46.2	
61–90 min	71	27.3	
91–120 min	21	8.1	

0.71, 0.63, 0.64, 0.66, and 0.66, respectively. For Q-LES-Q, leisure time activities (3.88 ± 0.66) were the most crucial construct that influenced the trainee teachers' degree of enjoyment and satisfaction, followed by feelings (3.87 ± 0.65), physical health (3.84 ± 0.69), and homework (3.62 ± 0.69). The overall Q-LES-Q revealed a good internal consistency of 0.96, while subscales alpha coefficients for physical health, feelings, homework, and leisure time activities demonstrated high reliability of 0.93, 0.92, 0.92, and 0.90, respectively. For PEEPQ, peer interaction (4.22 ± 0.75) is the most vital factor that influences the students' enjoyment towards physical activity, followed by PE enjoyment (4.20 ± 0.77), activity-generated excitement (4.09 ± 0.76), teacher-generated excitement (3.99 ± 0.79), parental encouragement (3.96 ± 0.87), self-referent competency (3.79 ± 0.82), and other-referent competency (3.26 ± 0.90). The overall PEEPQ and seven subscales demonstrated high reliability which ranged from 0.62 to 0.95.

Table 3 shows the independent sample *t*-test of *AEQ-PE, Q-LES-Q,* and *PEEPQ* based on gender. For AEQ-PE, the results revealed no significant difference for pride, $p = 0.22$, enjoyment, $p = 0.28$, anger, $p = 0.13$, hopelessness, $p = 0.53$, and boredom, $p = 63$. However, there was a statistically significant difference for anxiety, $p = 0.04$. The male (3.01 ± 0.76) rated 0.20 point higher than the female (2.81 ± 0.84). For Q-LES-Q, the results revealed there was no statistically significant for physical health, $p = 0.68$, feelings, $p = 0.12$, homework, $p = 0.65$, and leisure time activities, $p = 0.14$, respectively. For PEEPQ, the results also showed no significant difference between gender for self-referent competency, $p = 0.94$, other-referent competency, $p = 0.36$, activity-generated excitement, $p = 0.71$, teacher-generated excitement, $p = 0.07$, peer interaction, $p = 0.50$, and parental encouragement, $p = 0.76$, PE enjoyment, $p = 0.20$, respectively.

Table 4 shows the one-way ANOVA of *AEQ-PE, Q-LES-Q,* and *PEEPQ* based on age groups. For AEQ-PE, the results revealed that there was significant difference between the three age groups for pride, $p = 0.019$. Post-hoc Tukey adjusted comparisons for pride indicated that the mean score for age group of 18–19 (4.36 ± 0.63) was significantly higher that age group of 22–23 (4.09 ± 0.69), $\mu = 0.26$, $p = 0.015$. However, there was no significant difference between the three age groups for enjoyment, $p = 0.20$, anger, $p = 0.17$, anxiety, $p = 0.15$, and hopelessness, $p = 0.16$. Moreover, there was significant difference between the three age groups for boredom, $p = 0.029$, whereby the mean score for age group of 18–19 (2.14 ± 0.43) was significantly lower that age group of 22–23 (2.31 ± 0.47), $\mu = -0.18$, $p = 0.024$. For Q-LES-Q, the results revealed that there was significant difference between the three age groups for physical health, $p < 0.001$. Post-hoc Tukey adjusted

Table 2 Mean, standard deviation, and reliability for AEQ-PE, Q-LES-Q, and PEEPQ

Variables	M	SD	α
Overall AEQ-PE			0.56
Pride	4.24	0.63	0.72
Enjoyment	4.05	0.63	0.71
Anger	1.94	0.67	0.63
Anxiety	2.91	0.80	0.64
Hopelessness	1.96	0.69	0.66
Boredom	2.23	0.46	0.66
Overall Q-LES-Q			0.96
Physical health	3.84	0.69	0.93
Feelings	3.87	0.65	0.92
Homework	3.62	0.69	0.92
Leisure time activities	3.88	0.66	0.90
Overall PEEPQ			0.95
Self-referent competency	3.79	0.82	0.88
Other-referent competency	3.26	0.90	0.86
Activity-generated excitement	4.09	0.76	0.91
Teacher-generated excitement	3.99	0.79	0.82
Peer interaction	4.22	0.75	0.78
Parental encouragement	3.96	0.87	0.62
PE enjoyment	4.20	0.77	0.70

comparisons for physical health indicated that the mean score for age group of 18–19 (4.05 ± 0.62) was significantly higher that age group of 22–23 (3.28 ± 0.90), $\mu = 0.41$, $p < 0.001$. Besides, feelings were significant difference between the three age groups, $p = 0.001$, whereby the mean score for age groups of 18–19 (4.03 ± 0.59) was significantly higher that age group of 22–23 (3.66 ± 0.64), $\mu = 0.37$, $p < 0.001$. Furthermore, there was significant difference between the three age groups for homework, $p = 0.026$, whereby the mean score for age group of 18–19 (3.76 ± 0.72) was significantly higher that age group of 22–23 (3.48 ± 0.67), $\mu = 0.28$, $p = 0.019$. Lastly, leisure time activities were significant between the three age groups, $p = 0.014$, whereby the mean score for age group of 18–19 (4.03 ± 0.67) was significantly higher that age group of 22–23 (3.88 ± 0.66), $\mu = 0.28$, $p = 0.013$. For PEEPQ, the results revealed that there was significant difference between the three age groups for self-referent competency, $p = 0.002$, whereby the mean score for age group of 18–19 (3.87 ± 0.88) was significantly higher that age group of 22–23 (3.55 ± 0.78), $\mu = 0.32$, $p = 0.023$, and age group of 22–23 (3.55 ± 0.78) was significant lower than age group of 20–21 (3.97 ± 0.73), $\mu = 0.41$, $p = 0.003$. Besides, there was no significant difference between the three age groups for other-referent competency, $p = 0.57$, but significant difference for activity-generated excitement, $p = 0.002$, whereby the mean score for age group of 20–21 (4.32 ± 0.67) was significantly

Table 3 Independent t-test of AEQ-PE, Q-LES-Q, and PEEPQ based on gender

Variables	Male		Female		*t-test*	
	M	SD	M	SD	*t*-statistics (df)	*p*
AEQ-PE						
Pride	4.19	0.62	4.29	0.64	−1.22 (258)	0.22
Enjoyment	4.01	0.65	4.09	0.60	−1.09 (258)	0.28
Anger	2.00	0.68	1.88	0.66	1.52 (258)	0.13
Anxiety	3.01	0.76	2.81	0.84	2.07 (258)	0.04*
Hopelessness	1.99	0.65	1.93	0.72	0.63 (258)	0.53
Boredom	2.25	0.50	2.22	0.40	0.49 (250.77)	0.63
Q-LES-Q						
Physical health	3.83	0.71	3.86	0.67	−0.41 (258)	0.68
Feelings	3.80	0.70	3.93	0.59	−1.56 (254.51)	0.12
Homework	3.60	0.74	3.64	0.63	−0.45 (255.46)	0.65
Leisure time activities	3.93	0.71	3.81	0.61	1.47 (258)	0.14
PEEPQ						
Self-referent competency	3.79	0.90	3.79	0.73	0.08 (251.39)	0.94
Other-referent competency	3.21	1.02	3.31	0.75	−0.93 (242.92)	0.36
Activity-generated excitement	4.11	0.79	4.07	0.73	0.38 (258)	0.71
Teacher-generated excitement	4.08	0.79	3.90	0.79	1.83 (258)	0.07
Peer interaction	4.25	0.74	4.19	0.78	0.67 (258)	0.50
Parental encouragement	3.94	0.91	3.97	0.83	−0.30 (258)	0.76
PE enjoyment	4.26	0.78	4.14	0.76	1.27 (258)	0.20

* $p < 0.05$

higher that age group of 22–23 (3.91 ± 0.76), $\mu = 0.41$, $p = 0.001$. Furthermore, there was a significant between the three age groups for teacher-generated excitement, $p = 0.007$, whereby the mean score for age group of 20–21 (4.23 ± 0.73) was significantly higher that age group of 18–19 (3.89 ± 0.88), $\mu = 0.34$, $p = 0.014$, and age group of 22–23 (3.90 ± 0.73), $\mu = 0.33$, $p = 0.02$. However, there was no significant difference between the three age groups for peer interaction, $p = 0.10$, and parental encouragement, $p = 0.07$. In addition, PE enjoyment was significant difference between the three age groups, $p = 0.005$, whereby the mean score for age groups of 20–21 (4.43 ± 0.66) was significantly higher that age group of 18–19 (4.07 ± 0.82), $\mu = 0.36$, $p = 0.005$, and age group of 22–23 (4.14 ± 0.77), $\mu = 0.29$, $p = 0.033$.

Table 5 shows the one-way ANOVA of *AEQ-PE, Q-LES-Q,* and *PEEPQ* based on ethnicity. For AEQ-PE, results revealed that there was no statistically significant difference between the three ethnicities for pride, $p = 0.12$, enjoyment, $p = 0.18$, anger, $p = 0.89$, anxiety, $p = 0.32$, and hopelessness, $p = 0.54$. Conversely, there was a statistically significant difference between the three ethnicities for boredom, $p =$

Table 4 One-way ANOVA of AEQ-PE and Q-LES-Q based on age groups

Variables	Age groups			F (2, 257)	p
	18–19	20–21	22–23		
AEQ-PE					
Pride	4.36 (0.63)	4.27 (0.53)	4.09 (0.69)	4.05	0.019*
Enjoyment	4.14 (0.63)	4.03 (0.58)	3.97 (0.67)	1.64	0.20
Anger	1.84 (0.67)	2.03 (0.65)	1.97 (0.68)	1.79	0.17
Anxiety	2.81 (0.94)	3.04 (0.76)	2.90 (0.67)	1.90	0.15
Hopelessness	1.85 (0.69)	2.03 (0.69)	2.01 (0.68)	1.85	0.16
Boredom	2.14 (0.43)	2.25 (0.45)	2.31 (0.47)	3.58	0.029*
Q-LES-Q					
Physical health	4.05 (0.62)	3.83 (0.79)	3.65 (0.59)	8.26	< 0.001*
Feelings	4.03 (0.59)	3.91 (0.68)	3.66 (0.64)	7.76	0.001*
Homework	3.76 (0.72)	3.61 (0.65)	3.48 (0.67)	3.70	0.026*
Leisure time activities	4.03 (0.67)	3.83 (0.69)	3.88 (0.66)	4.33	0.014*
PEEPQ					
Self-referent competency	3.87 (0.88)	3.97 (0.73)	3.55 (0.78)	6.32	0.002*
Other-referent competency	3.32 (0.92)	3.28 (0.93)	3.18 (0.83)	0.57	0.57
Activity-generated excitement	4.07 (0.79)	4.32 (0.67)	3.91 (0.76)	6.58	0.002*
Teacher-generated excitement	3.89 (0.88)	4.23 (0.73)	3.90 (0.73)	5.00	0.007*
Peer interaction	4.28 (0.72)	4.31 (0.84)	4.08 (0.70)	2.36	0.10
Parental encouragement	3.90 (0.90)	4.14 (0.82)	3.85 (0.86)	2.63	0.07
PE enjoyment	4.07 (0.82)	4.43 (0.66)	4.14 (0.77)	5.43	0.005*

0.046. Post-hoc Tukey adjusted comparisons for boredom indicated that Malay (2.17 ± 0.43) was significantly lower than the Chinese (2.35 ± 0.49), $\mu = -0.18$, $p = 0.036$. For Q-LES-Q, there was significant difference between the three ethnicities for physical health, $p = 0.043$. Post-hoc Tukey adjusted comparisons for physical health indicated that Malay (3.95 ± 0.64) was significantly higher that the Native (3.70 ± 0.72), $\mu = 0.25$, $p = 0.033$. Results also revealed that there was no statistically significant difference between the three ethnicities for feelings, $p = 0.51$, homework, $p = 0.35$, and leisure time activities, $p = 0.58$. For PEEPQ, the results revealed that there was significant difference between the three ethnicities for self-referent competency, $p < 0.001$. Post-hoc Tukey adjusted comparisons for self-referent competency indicated that the mean score for Chinese (3.35 ± 0.79) was significantly lower that Malay (3.79 ± 0.79), $\mu = 0.45$, $p = 0.001$, and Native (4.09 ± 0.73), $\mu = 0.74$, $p < 0.001$. The mean score for Native (4.09 ± 0.73) was significantly higher than Malay, $\mu = 0.29$, $p = 0.022$. Besides, there was significant difference between the three ethnicities for other-referent competency, $p < 0.001$. Post-hoc Tukey adjusted comparisons for other-referent competency indicated that the mean score for Chinese (2.86 ± 0.83) was significantly lower that Malay (3.42 ± 0.87), $\mu = 0.58$, $p < 0.001$, and Native (3.32 ± 0.90), $\mu = 0.46$, $p = 0.005$. Furthermore, there was significant difference between the three ethnicities for activity-generated excitement, $p < 0.001$. Post-hoc Tukey adjusted comparisons for activity-generated excitement indicated that the mean score for Native (4.43 ± 0.63) was significantly higher that Malay (3.99 ± 0.77), $\mu = 0.44$, $p < 0.001$, and Chinese (3.78 ± 0.74), $\mu = 0.66$, $p < 0.001$. Moreover, there was a significant between the three ethnicities for teacher-generated excitement, $p < 0.001$. Post-hoc Tukey adjusted comparisons for teacher-generated excitement indicated that the mean score for Native (4.30 ± 0.67) was significantly higher that Malay (3.89 ± 0.85), $\mu = 0.41$, $p = 0.001$, and Chinese (3.77 ± 0.73), $\mu = 0.53$, $p < 0.001$. Similarly, there was significant difference between the three ethnicities for peer interaction, $p = 0.008$. Post-hoc Tukey adjusted comparisons for peer interaction indicated that the mean score for Native (4.40 ± 0.69) was significantly higher that Chinese (4.20 ± 0.79), $\mu = 0.39$, $p = 0.006$. Likewise, there was a significant between the three ethnicities for parental encouragement, $p < 0.001$. Post-hoc Tukey adjusted comparisons for parental encouragement indicated that the mean score for Native (4.25 ± 0.77) was significantly higher that Malay (3.91 ± 0.90), $\mu = 0.34$, $p = 0.013$, and Chinese (3.61 ± 0.82), $\mu = 0.64$, $p < 0.001$. Lastly, there was a significant between the three ethnicities for PE enjoyment, $p < 0.001$. Post-hoc Tukey adjusted comparisons for PE enjoyment indicated that the mean score for Native (4.48 ± 0.61) was significantly higher that Malay (4.07 ± 0.83), $\mu = 0.42$, $p < 0.001$, and Chinese (4.06 ± 0.75), $\mu = 0.42$, $p = 0.002$.

Table 6 shows the two-way ANOVA for *AEQ-PE, Q-LES-Q,* and *PEEPQ* based on gender and age groups. For AEQ-PE, the between-subjects ANOVA for pride revealed that there was no significant main effect of gender, $F(1, 254) = 1.76$, $p = 0.19$, $\eta_p{}^2 = 0.007$, but significant main effect of age groups, $F(2, 254) = 4.57$, $p = 0.011$, $\eta_p{}^2 = 0.035$. Besides, there was significant interaction between Gender × Age on pride, $F(2, 254) = 3.76$, $p = 0.025$, $\eta_p{}^2 = 0.029$. The between-subjects ANOVA for enjoyment revealed that there was no significant main effect of gender, $F(1, 254) = 1.28$, $p = 0.26$, $\eta_p{}^2 = 0.005$, and age groups, $F(2, 254) = 1.78$, $p = 0.17$, $\eta_p{}^2 = 0.014$. However, there was significant interaction between Gender × Age on enjoyment, $F(2, 254) = 3.66$, $p = 0.027$, $\eta_p{}^2 = 0.028$. The between-subjects ANOVA for anger revealed that there was no significant main effect of gender, $F(1, 254) = 1.67$, $p = 0.20$, $\eta_p{}^2 = 0.007$, and age groups, $F(2, 254) = 1.71$, $p = 0.18$, $\eta_p{}^2 = 0.013$. Moreover, there was no significant interaction between Gender × Age on anger, $F(2, 254) = 2.18$, $p = 0.12$, $\eta_p{}^2 = 0.017$. The between-subjects ANOVA for anxiety revealed that there was no significant main effect of gender, $F(1, 254) = 3.27$, $p = 0.07$, $\eta_p{}^2 = 0.013$, and age groups, $F(2, 254) = 1.45$, $p = 0.24$, $\eta_p{}^2 = 0.011$. Furthermore, there was no significant interaction between Gender × Age on anxiety, $F(2, 254) = 0.07$, $p = 0.94$, $\eta_p{}^2 = 0.001$. The between-subjects ANOVA for hopelessness revealed that there was no significant main effect of gender, $F(1, 254) = 0.18$, $p = 0.67$, $\eta_p{}^2 = 0.001$, and age groups, $F(2, 254) = 1.71$, $p = 0.18$, $\eta_p{}^2 = 0.013$. Besides, there was no significant interaction between Gender × Age on hopelessness, $F(2, 254) = 0.12$, $p = 0.89$, $\eta_p{}^2 = 0.001$. The between-subjects ANOVA for boredom revealed that there was no significant main effect of gender $F(1, 254) = 0.12$, $p = 0.74$, $\eta_p{}^2 < 0.001$, but significant main effect on age groups,

Table 5 One-way ANOVA of AEQ-PE, Q-LES-Q, and PEEPQ based on ethnicity

Variables	Ethnicity			F (2, 257)	p
	Malay	Chinese	Native		
AEQ-PE					
Pride	4.31 (0.58)	4.10 (0.68)	4.25 (0.65)	2.18	0.12
Enjoyment	4.11 (0.60)	3.93 (0.68)	4.05 (0.63)	1.76	0.18
Anger	1.95 (0.62)	1.97 (0.71)	1.92 (0.70)	0.12	0.89
Anxiety	2.98 (0.84)	2.79 (0.66)	2.91 (0.83)	1.13	0.32
Hopelessness	1.91 (0.69)	2.02 (0.73)	1.98 (0.66)	0.62	0.54
Boredom	2.17 (0.43)	2.35 (0.49)	2.22 (0.45)	3.11	0.046*
Q-LES-Q					
Physical health	3.95 (0.64)	3.85 (0.72)	3.70 (0.72)	3.18	0.043*
Feelings	3.91 (0.58)	3.79 (0.75)	3.86 (0.68)	0.67	0.51
Homework	3.66 (0.71)	3.51 (0.65)	3.64 (0.68)	1.07	0.35
Leisure time activities	3.92 (0.64)	3.87 (0.72)	3.82 (0.07)	0.55	0.58
PEEPQ					
Self-referent competency	3.79 (0.79)	3.35 (0.79)	4.09 (0.73)	16.45	< 0.001*
Other-referent competency	3.42 (0.87)	2.86 (0.83)	3.32 (0.90)	8.38	< 0.001*
Activity-generated excitement	3.99 (0.77)	3.78 (0.74)	4.43 (0.63)	16.77	< 0.001*
Teacher-generated excitement	3.89 (0.85)	3.77 (0.73)	4.30 (0.67)	10.48	< 0.001*
Peer interaction	4.20 (0.79)	4.01 (0.72)	4.40 (0.69)	4.93	0.008*
Parental encouragement	3.91 (0.90)	3.61 (0.82)	4.25 (0.77)	10.78	< 0.001*
PE enjoyment	4.07 (0.83)	4.06 (0.75)	4.48 (0.61)	9.14	< 0.001*

* $p < 0.05$

$F(2, 254) = 3.81$, $p = 0.023$, $\eta_p{}^2 = 0.029$. In addition, there was no significant interaction between Gender × Age on boredom, $F(2, 254) = 1.71$, $p = 0.18$, $\eta_p{}^2 = 0.013$. For Q-LES-Q, the between-subjects ANOVA for physical health revealed that there no was significant main effect of gender, $F(1, 254) = 0.05$, $p = 0.82$, $\eta_p{}^2 < 0.001$, whereas significant main effect of age groups, $F(2, 254) = 8.40$, $p < 0.001$, $\eta_p{}^2 = 0.062$. Besides, there was no significant interaction between Gender × Age on physical health, $F(2, 254) = 1.02$, $p = 0.36$, $\eta_p{}^2 = 0.008$. The between-subjects ANOVA for feelings revealed that there was no significant main effect of gender, $F(1, 254) = 2.32$, $p = 0.13$, $\eta_p{}^2 = 0.009$, but significant main effect of age groups, $F(2, 254) = 8.06$, $p < 0.001$, $\eta_p{}^2 = 0.060$. In addition, there was no significant interaction between Gender × Age on feelings, $F(2, 254) = 1.31$, $p = 0.27$, $\eta_p{}^2 = 0.010$. The between-subjects ANOVA for homework revealed that there was no significant main effect of gender, $F(1, 254) = 0.12$, $p = 0.73$, $\eta_p{}^2 < 0.001$, yet significant main effect of age groups, $F(2, 254) = 3.75$, $p = 0.025$, $\eta_p{}^2 = 0.029$. Furthermore, there was no significant interaction between Gender × Age on homework, $F(2, 254) = 0.68$, $p = 0.51$, $\eta_p{}^2 = 0.005$. The between-subjects ANOVA for leisure time activities revealed that there was no significant main effect of gender, $F(1, 254) = 3.23, p = 0.07, \eta_p{}^2 = 0.013$, whereas significant main effect of age groups, $F(2, 254) = 4.99, p = 0.007, \eta_p{}^2 = 0.038$. Moreover, there was no significant interaction between Gender × Age on leisure time activities, $F(2, 254) = 0.52$, $p = 0.60$, $\eta_p{}^2 = 0.004$. For PEEPQ, the between-subjects ANOVA for self-referent competency revealed that there was no significant main effect of gender, $F(1, 254) = 0.08$, $p = 0.77$, $\eta_p{}^2 < 0.001$, but significant main effect of age group, $F(2, 254) = 8.10$, $p < 0.001$, $\eta_p{}^2 = 0.060$. Besides, there was significant interaction between Gender × Age on self-referent competency, $F(2, 254) = 6.72$, $p = 0.001$, $\eta_p{}^2 = 0.050$. The between-subjects ANOVA for other-referent competency revealed that there was no significant main effect of gender, $F(1, 254) = 1.66, p = 0.20, \eta_p{}^2 = 0.006$, and age group, $F(2, 254) = 1.23$, $p = 0.30$, $\eta_p{}^2 = 0.010$. Conversely, there was significant interaction between Gender × Age on other-referent competency, $F(2, 254) = 13.07$, $p < 0.001$, $\eta_p{}^2 = 0.093$. The between-subjects ANOVA for activity-generated excitement revealed that there was no significant main effect of gender, $F(1, 254) = 0.04, p = 0.85, \eta_p{}^2 < 0.001$, whereas significant main effect of age group, $F(2, 254) = 8.89, p < 0.001, \eta_p{}^2 = 0.065$. Moreover, there was significant interaction between Gender × Age on activity-generated excitement, $F(2, 254) = 8.88$, $p < 0.001$, $\eta_p{}^2 = 0.065$. The between-subjects ANOVA for teacher-generated excitement revealed that there was no significant main effect of gender, $F(1, 254) = 1.54$, $p = 0.22$, $\eta_p{}^2 = 0.006$, but significant main effect of age group, $F(2, 254) = 6.25, p = 0.002, \eta_p{}^2 = 0.047$. There was significant interaction between Gender × Age on teacher-generated excitement, $F(2, 254) = 10.02, p < 0.001, \eta_p{}^2 = 0.073$. The between-subjects ANOVA for peer interaction revealed that there was no significant main effect of gender, $F(1, 254) = 0.14$, $p = 0.71$, $\eta_p{}^2 = 0.001$, yet significant main effect of age group, $F(2, 254) = 3.49$, $p = 0.032$, $\eta_p{}^2 = 0.027$. Besides, there was significant interaction between Gender × Age on peer interaction, $F(2, 254) = 7.89, p < 0.001, \eta_p{}^2 = 0.058$. The between-subjects ANOVA for parental encouragement revealed that there was no significant main effect of gender, $F(1, 254) = 0.71, p = 0.40, \eta_p{}^2 = 0.003$, whereas significant main effect of age group, $F(2, 254) = 4.13$, $p = 0.017$, $\eta_p{}^2 = 0.031$. In addition, there was significant interaction between Gender × Age on parental encouragement, $F(2, 254) = 8.78$, $p < 0.001$, $\eta_p{}^2 = 0.065$. The between-subjects ANOVA for PE enjoyment revealed that there was no significant main effect of gender, $F(1, 254) = 0.43$, $p = 0.51$, $\eta_p{}^2 = 0.002$, yet significant main effect of age group, $F(2, 254) = 8.24, p = 0.002, \eta_p{}^2 = 0.047$. Lastly, there was significant interaction between Gender × Age on PE enjoyment, $F(2, 254) = 7.71, p = 0.001, \eta_p{}^2 = 0.057$.

Table 7 shows the two-way ANOVA for *AEQ-PE, Q-LES-Q,* and *PEEPQ* based on gender and ethnicity. For AEQ-PE, the between-subjects ANOVA for pride revealed that there was no significant main effect of gender, $F(1, 254) = 0.68, p = 0.41, \eta_p{}^2 = 0.003$, and ethnicity, $F(2, 254) = 1.82, p = 0.17, \eta_p{}^2 = 0.014$. Besides, there was no significant interaction between Gender × Ethnicity on pride, $F(2, 254) = 0.21$, $p = 0.82$, $\eta_p{}^2 = 0.002$. The between-subjects ANOVA for enjoyment revealed that there was no significant main effect of gender, $F(1, 254) = 0.62, p = 0.43, \eta_p{}^2 = 0.002$, and ethnicity, $F(2, 254) = 1.71, p = 0.18, \eta_p{}^2 = 0.013$. Furthermore, there

Table 6 Main effect of gender versus age groups for AEQ-PE, Q-LES-Q, and PEEPQ

Variables	Gender			Age				Gender versus Age
	M	F	*p*	18 – 19	20 – 21	22 – 23	*p*	*p*
AEQ-PE								
Pride	4.20	4.31	0.19	4.36	4.31	4.09	0.011*	0.025*
Enjoyment	4.02	4.11	0.26	4.15	4.07	3.97	0.17	0.027*
Anger	1.99	1.88	0.20	1.83	2.01	1.97	0.18	0.12
Anxiety	3.00	2.82	0.07	2.82	3.03	2.90	0.24	0.94
Hopelessness	1.98	1.94	0.67	1.85	2.02	2.01	0.18	0.89
Boredom	0.04	2.22	0.74	2.13	2.24	2.31	0.023*	0.18
Q-LES-Q								
Physical health	3.84	3.86	0.82	4.06	3.84	3.65	< 0.001*	0.36
Feelings	3.82	3.94	0.13	4.03	3.94	3.66	< 0.001*	0.27
Homework	3.61	3.64	0.73	3.76	3.63	3.48	0.025*	0.51
Leisure time activities	3.95	3.80	0.07	4.05	3.81	3.76	0.007*	0.60
PEEPQ								
SRC	3.81	3.83	0.77	3.89	4.02	3.55	< 0.001*	0.001*
ORC	3.22	3.36	0.20	3.33	3.37	3.18	0.30	< 0.001*
AGE	4.12	4.13	0.85	4.09	4.38	3.90	< 0.001*	< 0.001*
TGE	4.09	3.97	0.22	3.92	4.27	3.90	0.002*	< 0.001*
PI	4.26	4.23	0.71	4.30	4.36	4.08	0.032*	< 0.001*
PE	3.95	4.04	0.40	3.92	4.21	3.85	0.017*	< 0.001*
PE enjoyment	4.27	4.21	0.51	4.10	4.47	4.14	0.002*	0.001*

* $p < 0.05$

was no significant interaction between Gender × Ethnicity on enjoyment, $F(2, 254) = 1.84$, $p = 0.16$, ${\eta_p}^2 = 0.014$. The between-subjects ANOVA for anger revealed that there was no significant main effect of gender, $F(1, 254) = 2.93$, $p = 0.09$, ${\eta_p}^2 = 0.011$, and ethnicity, $F(2, 254) = 0.08$, $p = 0.92$, ${\eta_p}^2 = 0.001$. Moreover, there was no significant interaction between Gender × Ethnicity on anger, $F(2, 254) = 0.53$, $p = 0.59$, ${\eta_p}^2 = 0.004$. The between-subjects ANOVA for anxiety revealed that there was significant main effect of gender, $F(1, 254) = 5.38$, $p = 0.021$, ${\eta_p}^2 = 0.021$, but no significant main effect of ethnicity, $F(2, 254) = 2.29$, $p = 0.10$, ${\eta_p}^2 = 0.018$. In addition, there was no significant interaction between Gender × Ethnicity on anxiety, $F(2, 254) = 2.14$, $p = 0.12$, ${\eta_p}^2 = 0.017$. The between-subjects ANOVA for hopelessness revealed that there was no significant main effect of gender, $F(1, 254) = 0.59$, $p = 0.44$, ${\eta_p}^2 = 0.002$, and ethnicity, $F(2, 254) = 0.38$, $p = 0.69$, ${\eta_p}^2 = 0.003$. At the same time, there was no significant interaction between Gender ×

Ethnicity on hopelessness, $F(2, 254) = 0.70, p = 0.50, \eta_p{}^2 = 0.005$. The between-subjects ANOVA for boredom revealed that there was no significant main effect of gender $F(1, 254) = 0.01, p = 0.93, \eta_p{}^2 < 0.001$, and ethnicity, $F(2, 254) = 2.82, p = 0.06, \eta_p{}^2 = 0.022$. Furthermore, there was no significant interaction between Gender × Ethnicity on boredom, $F(2, 254) = 0.07, p = 0.93, \eta_p{}^2 = 0.001$. For Q-LES-Q, the between-subjects ANOVA for physical health revealed that there was no significant main effect of gender, $F(1, 254) = 0.23, p = 0.63, \eta_p{}^2 = 0.001$, but significant main effect of ethnicity, $F(2, 254) = 3.15, p = 0.044, \eta_p{}^2 = 0.024$. Besides, there was no significant interaction between Gender × Ethnicity on physical health, $F(2, 254) = 2.48, p = 0.09, \eta_p{}^2 = 0.019$. The between-subjects ANOVA for feelings revealed that there was no significant main effect of gender, $F(1, 254) = 2.57, p = 0.11, \eta_p{}^2 = 0.010$, and ethnicity, $F(2, 254) = 0.19, p = 0.83, \eta_p{}^2 = 0.002$. Moreover, there was no significant interaction between Gender × Ethnicity on feelings, $F(2, 254) = 1.17, p = 0.31, \eta_p{}^2 = 0.009$. The between-subjects ANOVA for homework revealed that there was no significant main effect of gender, $F(1, 254) = 0.30, p = 0.59, \eta_p{}^2 = 0.001$, and ethnicity, $F(2, 254) = 0.56, p = 0.57, \eta_p{}^2 = 0.004$. Furthermore, there was no significant interaction between Gender × Ethnicity on homework, $F(2, 254) = 0.81, p = 0.45, \eta_p{}^2 = 0.006$. The between-subjects ANOVA for leisure time activities revealed that there was significant main effect of gender, $F(1, 254) = 1.28, p = 0.26, \eta_p{}^2 = 0.005$, and ethnicity, $F(2, 254) = 0.68, p = 0.51, \eta_p{}^2 = 0.005$. Result also revealed that there was no significant interaction between Gender × Ethnicity on leisure time activities, $F(2, 254) = 1.25, p = 0.29, \eta_p{}^2 = 0.010$. For PEEPQ, the between-subjects ANOVA for self-referent competency revealed that there was no significant main effect of gender, $F(1, 254) = 0.11, p = 0.75, \eta_p{}^2 < 0.001$, but significant main effect of ethnicity, $F(2, 254) = 13.44, p < 0.001, \eta_p{}^2 = 0.096$. Besides, there was significant interaction between Gender × Ethnicity on self-referent competency, $F(2, 254) = 4.33, p = 0.014, \eta_p{}^2 = 0.033$. The between-subjects ANOVA for other-referent competency revealed that there was no significant main effect of gender, $F(1, 254) = 0.54, p = 0.46, \eta_p{}^2 = 0.002$, yet significant main effect of ethnicity, $F(2, 254) = 5.83, p = 0.003, \eta_p{}^2 = 0.044$. Furthermore, there was no significant interaction between Gender × Ethnicity on other-referent competency, $F(2, 254) = 1.88, p = 0.15, \eta_p{}^2 = 0.015$. The between-subjects ANOVA for activity-generated excitement revealed that there was no significant main effect of gender, $F(1, 254) = 0.24, p = 0.63, \eta_p{}^2 = 0.001$, whereas significant main effect of ethnicity, $F(2, 254) = 15.13, p < 0.001, \eta_p{}^2 = 0.106$. Moreover, there was no significant interaction between Gender × Ethnicity on activity-generated excitement, $F(2, 254) = 1.83, p = 0.16, \eta_p{}^2 = 0.014$. The between-subjects ANOVA for teacher-generated excitement revealed that there was no significant main effect of gender, $F(1, 254) = 3.80, p = 0.05, \eta_p{}^2 = 0.001$, but significant main effect of ethnicity, $F(2, 254) = 10.30, p < 0.001, \eta_p{}^2 = 0.075$. Additionally, there was no significant interaction between Gender × Ethnicity on teacher-generated excitement, $F(2, 254) = 0.42, p = 0.66, \eta_p{}^2 = 0.003$. The between-subjects ANOVA for peer interaction revealed that there was no significant main effect of gender, $F(1, 254) = 0.62, p = 0.43, \eta_p{}^2 = 0.002$, yet significant main effect of ethnicity, $F(2, 254) = 4.36, p = 0.014, \eta_p{}^2 = 0.033$. Besides, there was no significant interaction between Gender × Ethnicity on peer interaction, $F(2, 254) = 0.69, p = 0.50, \eta_p{}^2 = 0.005$. The between-subjects ANOVA for parental encouragement revealed that there was no significant main effect of gender, $F(1, 254) = 0.04$, $p = 0.83$, $\eta_p{}^2 < 0.001$, whereas significant main effect of ethnicity, $F(2, 254) = 9.01, p < 0.001, \eta_p{}^2 = 0.066$. Next, there was no significant interaction between Gender × Ethnicity on parental encouragement, $F(2, 254) = 0.90$, $p = 0.41$, $\eta_p{}^2 = 0.007$. The between-subjects ANOVA for PE enjoyment revealed that there was no significant main effect of gender, $F(1, 254) = 0.73$, $p = 0.39$, $\eta_p{}^2 = 0.003$, yet significant main effect of ethnicity, $F(2, 254) = 8.24$, $p < 0.001$, $\eta_p{}^2 = 0.061$. Furthermore, there was no significant interaction between Gender × Ethnicity on PE enjoyment, $F(2, 254) = 2.08, p = 0.13, \eta_p{}^2 = 0.016$.

Table 7 Main effect of gender versus ethnicity for AEQ-PE, Q-LES-Q, and PEEPQ

Variables	Gender (G)			Ethnicity (E)				G versus E
	M	F	*p*	Malay	Chinese	Native	*p*	*p*
AEQ-PE								
Pride	4.19	4.25	0.41	4.30	4.10	4.25	0.17	0.82
Enjoyment	4.00	4.06	0.43	4.12	3.93	4.05	0.18	0.16
Anger	2.01	1.85	0.09	1.95	1.92	1.92	0.92	0.59
Anxiety	3.00	2.76	0.021*	3.01	2.73	2.91	0.10	0.12
Hopelessness	1.99	1.92	0.44	1.91	1.98	1.98	0.69	0.50
Boredom	2.25	2.25	0.93	2.17	2.35	2.22	0.06	0.93
Q-LES-Q								
Physical health	3.82	3.87	0.63	3.94	3.89	3.70	0.044*	0.09
Feelings	3.80	3.94	0.11	3.90	3.85	3.86	0.83	0.31
Homework	3.59	3.64	0.59	3.67	3.55	3.64	0.57	0.45
Leisure time activities	3.93	3.83	0.26	3.93	3.89	3.82	0.51	0.29
PEEPQ								
SRC	3.78	3.74	0.75	3.79	3.40	4.09	< 0.001*	0.014*
ORC	3.19	3.27	0.46	3.42	2.93	3.32	0.003*	0.15
AGE	4.10	4.05	0.63	3.99	3.81	4.43	< 0.001*	0.16
TGE	4.08	3.88	0.05	3.90	3.75	4.29	< 0.001*	0.66
PI	4.25	4.17	0.43	4.20	4.02	4.40	0.014*	0.50
PE	3.93	3.95	0.83	3.92	3.65	4.25	< 0.001*	0.41
PE enjoyment	4.26	4.18	0.39	4.08	4.09	4.48	< 0.001*	0.13

* $p < 0.05$

4 Discussion

The study showed that anxiety remains as the predominant variables which showed males had higher anxiety-related emotions unlike the females would tend to perceive PE as progressively challenging have higher confidence in their self-abilities and mastery-oriented. The younger age group perceived to high quality of life and satisfaction implied that social and emotional support were more positive in terms of invaluable social support that can enhance better participation and well-being towards PE. Thus, the younger age groups were in their first years before transitioned to their second or third year in which they would need to make the necessary adjustment and well-being during the duration as students. Adversely, the older age groups were more self-referent in their competence as they possessed more experience and knowledge in acquiring more challenging tasks and skills as they transit from first year to second and third year during their study duration period. The intrinsic achievement derived from their self-referent were related with enjoyment in PE according to their level of competence and ability in which the activities were systemically structure excitement, stimulation, and involvement. The natives were found to have higher level of enjoyment and satisfaction in which competence provide the aim in providing enjoyment in PE. The enjoyment and social relatedness enable some forms of satisfaction and autonomy, whereby the natives are generally more self-reliant and active in PA which have been embedded in PE. Thus, the practices of teaching PE needs to provide equality without discrimination as well as plan according to the creativity and inclusivity that would lead to better development of PE among the students. The limitations of the study were the utilization of convenience sampling who do not represent the total distribution of diversities of the races. It is recommended that longitudinal study incorporating more schools from different regions in Malaysia would serve a better understanding of the students' enjoyment and satisfaction in PE.

5 Conclusion

In order to cater for the multicultural Malaysian societies, the PE teachers would need to stimulate better interactive and creative approaches in the development of PE. In addition, there should be distancing from stereotyping and discrimination in order to communicate the values of practices, activities, and exercises in PE in order to develop awareness of the importance of PE.

References

1. Kahu ER, Nelson K (2018) Student engagement in the educational interface: understanding the mechanisms of student success. High Educ Res Dev 37(1):58–71. https://doi.org/10.1080/07294360.2017.1344197
2. Diener E (1994) Assessing subjective well-being: progress and opportunities. In: Social indicators research, vol 31. Springer, pp 103–157
3. Pekrun R (2006) The control-value theory of achievement emotions: assumptions, corollaries, and implications for educational research and practice. Educ Psychol Rev 18:315–341
4. Sotardi VA, Bosch J, Brogt E (2020) Multidimensional influences of anxiety and assessment type on task performance. Soc Psychol Educ 23:499–522
5. Chen P, Mao L, Nassis GP, Harmer P, Ainsworth BE, Li F (2020) Coronavirus disease (COVID-19): the need to maintain regular physical activity while taking precautions. J Sport Health Sci 9(2):103–104. https://doi.org/10.1016/j.jshs.2020.02.001
6. Törmänen T, Järvenoja H, Mänty K (2021) All for one and one for all—how are students' affective states and group-level emotion regulation interconnected in collaborative learning? Int J Educ Res 69(5) https://doi.org/10.1016/j.ijer.2021.101861
7. Scanlan TK, Simons JP (1992) The construct of sport enjoyment. In: Roberts GC (ed) Motivation in sport and exercise. Champaign, Human Kinetics, pp 202–203
8. Hashim HA, Grove B, Whipp P (2008) Validating the youth sport enjoyment construct in high school physical education. Res Q Exerc Sport Sci 79(2):183–194
9. Ibrahim MF, Kuan G, Hashim HA, Hamzah NA, Kueh YC (2021) Measuring achievement emotions questionnaire for physical education (AEQ-PE): a confirmatory study in Malay language. BMC Public Health 21(1):1–8. https://doi.org/10.1186/s12889-021-11869-4
10. Ibrahim MF, Kuan G, Hashim HA, Hamzah NA, Kueh YC (2021) Psychometric properties of the Malay-language quality of life enjoyment and satisfaction scale: a confirmatory study on Malaysian children. Int J Environ Res Public Health 18:1–10

Cross-Cultural Validation and Psychometric' Evaluation of the Athletic Mental Energy Scale (AMES)

Chatkamon Singnoy, Naruepon Vongjaturapat, Tanida Julvanichpong, Li-Chin Yeh, Cen-Yi Lin, Tsung-Kuo Tien-Liu, and Frank J. H. Lu

Abstract Athletic mental energy influences athletes' cognition, emotion, and performance. It is essential to have a valid and reliable psychological measure to start the research. Thus, this study aimed to validate Athletic Mental Energy Scale (AMES) into Thai and examine its psychometric properties. We examined content validity by inviting 15 participants (sport/language experts = 5; athletes = 10) to examine the appropriateness of the content of a translated AMES-Thai. Moreover, we examined the factorial structure and reliability of the AMES-Thai by item analysis (IA), confirmatory factor analysis (CFA), composite reliability (CR) analysis, and average variance extracted (AVE). We suggest Thai scholars and sports professionals may use AMES-Thai for further research and practice in future.

Keywords Optimal state of mind · Psychology of sport excellence · Cross-cultural validation · Peak performance

1 Introduction

In order to explore athletes' optimal psychological states in sports, Lu and his colleagues [1] developed a new concept termed "athletic mental energy (AME)"

C. Singnoy (✉) · N. Vongjaturapat · T. Julvanichpong
Burapha University, Saen Suk, Thailand
e-mail: chatkamon@gmail.com

L.-C. Yeh
Chungyu University of Film and Arts, Keelung, Taiwan

C.-Y. Lin
National Taiwan Sport University, Taoyuan, Taiwan

T.-K. Tien-Liu
Fu Jen Catholic University, New Taipei, Taiwan

F. J. H. Lu
Chinese Culture University, New Taipei, Taiwan

G. Kuan et al. (eds.), *Advancing Sports and Exercise via Innovation*, Lecture Notes in Bioengineering, https://doi.org/10.1007/978-981-19-8159-3_41

and defined AME as "an athlete's perceived existing state of energy which is characterized by its intensity in motivation, confidence, concentration, and mood [1]." Under this initial work, Lu and colleagues [1] contended that athletic mental energy comprises both cognitive and affective components including vigor, calmness, tiredness, concentration, confidence, and motivation. Lu and colleagues [1] used six studies and developed a six-factor, 18-item Athletic Mental Energy Scale (AMES) which had adequate and reliability.

Athletic mental energy is an important psychological construct because it influences athletes' emotions, cognition, and behavior. Past research in mental energy is dominated by nutrition scientists [2, 3]. The reasons that nutrition scientists and other researchers dedicated themselves to studying mental energy are because mental energy is "an individual's ability to continue long hours of thinking, concentrating attention, and blocking distractions to achieve a given task [4]." According to [4] many astonishing scientific works and advanced technology can be done by scientists are because they had strong mental energy. Further, [4] argued that if anyone wants to be successful in his/her works, he or she needs extraordinary mental energy.

In the athletic domain, many sport psychologists are also interested in this athletic mental energy. For example, [5] developed a special psychological skill termed "centering" and claimed that when athletes apply this skill in competition/training it can make him/her in a state of being calm, energetic, and confident. Similarly, a well-known psychologist, [6] also developed a psychological skill termed "Visual Motor Behaviour Rehearsal (VMBR)" to teach Olympic athletes. Suinn [6] contended that after athletes master this skill they can become concentrated and confident when entering competitions. Furthermore, another sport psychologist, Loehr [7], proposed a pyramid model of energy to depict how different types of energy contributing to athletes' peak performance. In this model, the base is physical energy. Then, emotional energy is at the second level. At the third level, it is athletes' mental energy. Finally, and at the top is spiritual energy. Among all types of energy, [7] considered mental energy is very important because it relates to athletes' higher-order functioning such as abstract thinking, awareness, and self- regulation.

In line with the research into elite athletes and their optimal state of mind, Lu and colleagues [1] proposed that athletic mental energy comprised of two major emotional and cognitive components which are frequently mentioned in sport psychology literature. In terms of the emotional component, Lu and colleagues' [1] AMES includes tirelessness, vigor, and calmness as major emotional components of athletic mental energy. These components echo past research on athletes' optimal psychological states in peak performance. For example, Morgan's [8, 9] iceberg profile model suggested that successful US Olympic candidates had high scores on vigor but low scores on fatigue, anger, confusion, and anxiety compared to less successful athletes. Thus, the vigor and tirelessness in AMES echoed Morgan's [8, 9] research. In addition to vigor and tirelessness, calm is also very important in elite sport. For example, [7, 10] reported that when performing at their best, elite athletes reported they felt low anxiety, relaxed, and experienced no fear of loss. Recently, in their investigation into the resilience of Olympic gold medalists, Anderson and colleagues [11] reported that Australian Olympic champions were more emotionally stable, less anxious, and optimistic compared to less successful athletes. Thus, the emotional factors of vigor, tirelessness, and calmness in AMES seem very meaningful for elite athletes.

Since the AMES has been developed, researchers use it to examine its' prediction on athletic success. For example, in Lu and colleagues' [1] study 5, the researchers administered the AMES to 78 Malaysian martial artists one day before a national tournament. Then, they collected the martial artists' wins and losses from the tournament. Using a logistic regression statistical analysis, Lu and colleagues' [1] found the subscales of AMES (i.e., confidence, motivation, tirelessness, and calmness) predicted competition outcomes.

Recently, Chiu and colleagues [12] sampled two groups of college student-athletes (Study 1 = 230; Study 2 = 159) to examine the moderating effects of athletic mental energy on the stress–burnout relationship. Participants were measured by life stress, athletic mental energy (i.e., AMES), and burnout. Both studies found that the emotional and cognitive components of athletic mental energy moderated the athletes' life stress–burnout relationship. Chiu and colleagues [12] concluded that athletic mental energy is an important psychological variable that can reduce athletes' stress–burnout relationship.

Thus, we intended to translate Lu and colleagues' [1] AMES into Thai and examine its psychometric properties. We followed Cohen and Swerdlik's [13] suggestions in validating a psychological test. We used back translation to make an AMES draft and examine its' content validity. Moreover, we examined the factor structure and basic indices of the reliability.

2 Methods and Results

2.1 Phase 1

The objective of Phase 1 was to translate Lu and colleagues' [1] AMES into Thai and examine its content validity.

Participants. Participants were 4 experts in Thai, English, and sports psychology; and 1 native English speaker. Also, we invited 10 student-athletes to read the final version of the translated AMES.

Procedures. We adopted Birslin's [14] suggestions to translate all 18 items of AMES into Thai by following steps. In step 1, two bilingual translators who are experts in Thai, English, and sport psychology were invited to translate AMES into Thai. Then, two translators compared each translated version and discussed the differences. Next, they revised several parts until they agreed upon one tentative version called "AMES-Thai." In step 2, two experts who are proficient in Thai, English, and sport psychology were invited to translate AMES-Thai into English. In step 3, one native English speaker compared the original AMES and translated English AMES-Thai

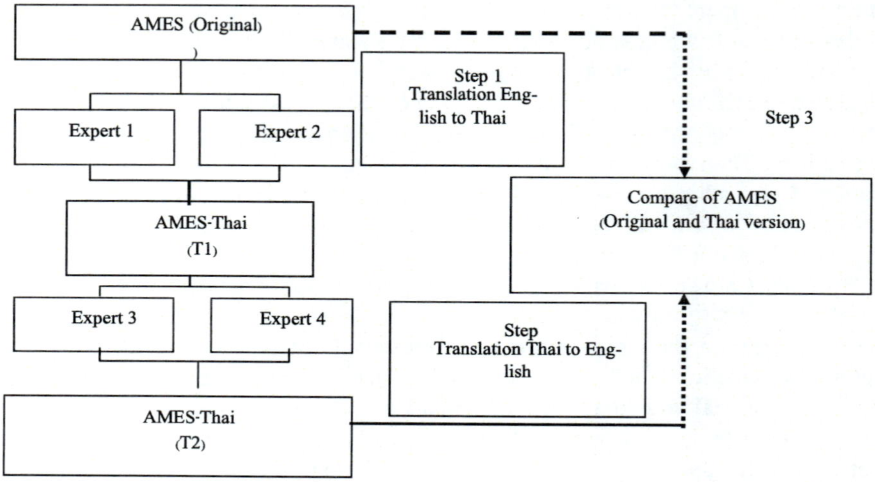

Fig. 1 Back translation procedures

to see whether there was any difference between the two. In addition, two Thai co-authors and this native English speaker discussed the semantic meanings of translated English AMES-Thai and AMES-Thai, and then they made a final version of AMES-Thai. After these three steps, we invited 10 Thai student-athletes to read the contents of AMES-Thai, and asked them to comment about the contents of the AMES-Thai. Finally, we made the last minor revision after 10 student-athletes' feedback. All back translation procedures are illustrated in Figure 1.

Results. After three phases of translation, the final version of the AMES-Thai reflected the original meaning of the AMES. Especially, in the last step of back translation, the native English speaker gave us much valuable feedback about the original AMES and translated English AMES-Thai. We accepted his suggestions by revising several places to make it clear and understandable. Besides, we adopted [15] suggestions to examine the cultural equivalence, functional equivalence, item equivalence, and scalar equivalence of the AMES-Thai. Furthermore, the 10 student-athletes reported that they had no difficulty reading and understanding the content of the final version of AMES-Thai in terms of fluency, clarity, and understandability. Thus, we prepared AMES-Thai for the next field testing.

Conclusion. The objective of Study 1 was to translate an English version of AMES into Thai and examine its content validity. By using a back translation, we made an AMES-Thai draft with appropriate content validity. Thus, we gained a basic tool for assessing athletic mental energy scale for field testing in Thailand. However, we still need other reliability and validity indices to confirm the psychometric properties of the AMES-Thai. Therefore, we conducted Study 2.

2.2 Phase 2

Objective. The objective of Study 2 was to examine the factor validity and reliability of the AMES-Thai. We used item analysis (IA), confirmatory factor analysis (CFA), average variance extracted (AVE), and composite reliability (CR) to achieve our objective.

2.3 Methods

Participants. We sampled 572 Thai student-athletes (males = 359, females = 213) from 12 Universities with a mean age of 20.12 ± 1.58 years and 6 years' average experience in sports. They participated in different team sports (including football, basketball, volleyball, futsal, hockey, beach volleyball, handball, sepak takraw, rugby, cricket, and softball) and individual sports (including taekwondo, track and field, wrestling, boxing, Thai boxing, shooting, tennis, swimming, petanque, badminton, pencak silat, and e-sport). The number of days spent training was 4.8 days per week with 3.2 h of training per day. We adopted [16] suggestions to recruit enough participants. According to [16], a sample size of 500 cases is excellent for the CFA.

Measurements and Procedures. Before collecting data, we gained ethical approval from the Burapha University Human Research Ethics Committee (HU 138/2563). We contacted targeted teams' head coaches through emails and phone calls. We informed them that this is research to examine athletes' responses about an English psychological test. After their permission, we visited the team at an appointed date to collect data. Before field testing, we informed participants of the general purpose of the study, the procedure to complete the survey packet, the right of the participation, the confidentiality, and the anonymity of all data. If participants interested in this research and agreed to participate, they signed informed consent forms, then a demographic questionnaire, and AMES-Thai as follows.

Demographic questionnaire. The demographic questionnaire included participants' general information such as gender, age, sports, training load (i.e., frequency, duration per day/week), and years of sports experience.

Athletic Mental Energy Scale-Thai (AMES-Thai): The AMES-Thai assess participants' perception of their existing energy state. The AMES-Thai includes six factors motivation, confidence, concentration, vigor, tirelessness, and calmness. Each factor included three items. The sample question for each factor as follows: (a) vigor (e.g., I feel there is an endless energy coming from my body), (b) confidence (e.g., I can control all sports movements and skills), (c) motivation (e.g., I feel excited in future competitions), (d) tirelessness (e.g., Even the training is over I still feel I have endless energy to use), (e) concentration (e.g., I am free of distraction during competition and training), and (f) calmness (e.g., Even facing to a tough opponent I don't feel anxious). Participants read each item and identify their perceptions on a 6-point Likert scale that ranges from 1 (not at all) to 6 (completely so).

Statistical Analyses. Before formal statistical analyses, we examined descriptive information about data including means, standard deviations, skewness, kurtosis, and outliers to check whether there is an abnormal data. Next, a Pearson product-moment correlation analysis was used to examine the relationships of all items. Further, an item analysis was used to examine participants' responses on all items between high and low scores. Moreover, we used AMOS version 22 statistical software to perform a CFA analysis by the following criteria suggested by [17]: (1) the χ^2/DF ratio (between 1 and 3); (2) the Goodness of Fit Index (GFI, greater than 0.90); (3) the Comparative Fit Index (CFI, greater than 0.90); (4) the root mean square error of approximation (RMSEA, less than 0.08); (5) the standardized root mean square residual (SRMR, less than 0.05).

Results. It was found no outliers and the mean of all items was between 4.05 and 4.89 (SD = 0.93–1.24); the skewness was around − 1.12–0.25 (kurtosis = 0.11–1.09). Thus, the preliminary analyses indicated that the raw data fit the statistical assumptions. Further, we used Mardia's normalized estimate to examine multivariate normality. Results found the estimate was less than 3 which met the assumption of multivariate normality [18]. In addition, the results form Pearson product-moment correlation analysis found all items were correlated ($r = 0.31$–0.79) which exceeded the minimum requirement of 0.30 [19].

Moreover, if was found that all items had significant differences between those scored higher than 75% and lower than 25% as Table 1 illustrated. Thus, it demonstrated that all of AMES-Thai's items can differentiate those who scored high and low [20]. Moreover, as Figure 1 illustrated, the measurement model of AMES-Thai indicated that a six-factor, 18-item was satisfactory (χ^2/DF = 149.76, CFI = 0.99, GFI = 0.97, RMSEA= 0.004, SRMR = 0.003). The composite reliability for each subscale was calculated as follows: vigor (0.71), confidence (0.80), motivation (0.63), concentration (0.57), tirelessness (78), and calmness (0.79). The average variance extracted was also calculated as follows: vigor (0.58), confidence (0.66), motivation (0.52), concentration (48), tirelessness (0.64), and calmness (0.65), which were all above the acceptable standard (0.25) as [21] suggested.

In phase 2, we examined the factor structure of the AMES-Thai and its related reliabilities. We sampled 572 University athletes and administering them with AMES-Thai. The preliminary data examination showed that the raw data fits the statistical assumptions. Also, item analysis found that there is a significant difference between high vs. low scores among participants. The item analysis showed that the overall quality of the items of the AMES-Thai was appropriate. Moreover, CFA results illustrated that AMES-Thai matches the factor structure of the original AMES. Also, the indices of composite reliability and average variance extracted illustrated that AMES-Thai has appropriate reliability. According to Netemeyer and colleagues [22], composite reliability (CR) is referred to construct reliability. Like Cronbach's alpha, it represents a measure's internal consistency. On the other hand, average variance extracted (AVE) is a measure of the amount of variance due to measurement error

Table 1 Item analysis of the 18-item AMES-Thai

	High versus low	*N*	*M*	SD	*t*-value
AMES1	1	167	3.23	1.288	– 14.0[illegible]*
	2	155	4.94	0.865	
AMES2	1	167	3.71	0.958	– 16.28*
	2	155	5.18	0.639	
AMES3	1	167	3.41	1.048	– 18.7[illegible]*
	2	155	5.17	0.583	
AMES4	1	167	3.86	1.199	– 12.38*
	2	155	5.25	0.784	
AMES5	1	167	3.72	0.931	– 17.57*
	2	155	5.28	0.652	
AMES6	1	167	4.19	1.283	– 6.53*
	2	155	5.07	1.111	
AMES7	1	167	4.07	1.309	– 11.59*
	2	155	5.43	0.738	
AMES8	1	167	3.31	1.040	– 17.27*
	2	155	5.05	0.763	
AMES9	1	167	3.27	0.915	– 20.63*
	2	155	5.10	0.662	
AMES10	1	167	4.03	1.337	– 12.93*
	2	155	5.51	0.607	
AMES11	1	167	3.29	0.919	– 18.91*
	2	155	5.01	0.712	
AMES12	1	167	3.01	0.997	– 21.09*
	2	155	5.01	0.693	
AMES13	1	167	3.53	0.870	– 20.00*
	2	155	5.15	0.556	
AMES14	1	167	3.41	1.082	– 16.70*
	2	155	5.09	0.658	
AMES15	1	167	3.71	0.872	– 19.06*
	2	155	5.26	0.556	
AMES16	1	167	3.81	1.225	– 13.32*
	2	155	5.25	0.641	
AMES17	1	167	3.46	1.016	– 15.13*
	2	155	4.99	0.794	
AMES18	1	167	3.60	1.086	– 12.29*
	2	155	4.97	0.90	

Note #1. $*p < 0.01$
#2. 1 = low score group; 2 = high score group

of a psychological construct. Thus, the illustration of CR and AVE indicated that AMES-Thai had appropriate reliability.

Despite this initial psychometric evidence, we still need the nomological validity of the six-factor, 18-item AMES-Thai. By doing so, we hope to provide more psychometric evidence of the six-factor, 18-item AMES-Thai.

3 Discussion

Theoretical contributions/implications

The objective of the present study was to introduce a reliable and valid measure of athletic mental energy into Thailand. By conducting three studies, we found a six-factor 18-item AMES-Thai had appropriate factor structure, composite reliability, internal consistency, convergent validity, and discriminant validity. We believe our work offers Thai researchers a useful tool for the study of athletic mental energy. Future researchers may use this six-factor, 18-item AMES-Thai to examine its' predictivity in athletic performance. Also, practitioners may use it for counseling or diagnosing Thai athletes' mental energy.

Generally, the six-factor, 18-item AMES-Thai maintained the same six-factor structure of the original AMES [1]. In particular, the results from convergent validity and discriminant validity echo the theoretical explanation and empirical studies in sport psychology [1, 12]. The results of the convergent validity analysis found athletic mental energy positively correlated with student-athletes' positive state of mind; this is a meaningful in theory. According to [23], a positive state of mind is having a positive mood, productive, creative, and free of worry, depression, and hostility. Thus, the convergent validity provides a significant finding that athletes who experience high mental energy also have a high positive state of mind.

The results of the discriminant validity analysis also supported past research that athletes high in athletic mental energy would be low in life stress [1] and burnout [12]. Specifically, Chiu and colleagues' [12] found that athletes high in athletic mental energy decreased the relationship between athletes' stress–burnout. Thus, sports professionals need to boost athletes' mental energy, not only to enhance their performance but also to promote their overall well-being.

The other implication for this study is the cross-cultural validation. Sun and colleagues [24] contended that culture influences one's values and attitudes, which in turn, affect one's emotional responses and behavior. Markus and Kitayama [25] also contended that culture has a strong influence in perceptions about self and others. Thus, it needs careful procedures to introduce an existing psychological measure to another culture. In addition, research suggested [26, 27], when transferring an existing concept and measure to another culture, much adaptation and validation are needed. This includes appropriate translation and field testing. Our study adopted [14] suggestion to translate all 18 items of AMES into Thai to establish content validity. Also, we performed two studies to gain solid indices of reliability and validity. All of these efforts are to make sure that we gain a valid and reliable psychological measure to use in Thailand. We believe that the validated AMES-Thai is suitable for Thai researchers and practitioners to use in future.

4 Limitations and Future Suggestions

Several limitations must be addressed. First, although the AMES-Thai gains its initial psychometric indices, it still needs more work in future. It is suggested that any development of a psychological measure is an ever-lasting procedure [13]. A psychological measure always needs more work to confirm its validity and reliability. For example, because we did not examine the AIMS-Thai's measurement invariance, we suggest that future study may examine measurement invariance of the AMES-Thai across gender or sports. Further, our study sampled college student-athletes as participants. Thus, whether our results can be generalized to adolescent athletes or professional athletes need to be further examined. Moreover, because the athletic mental energy is a "state-like" experience, it may change from day to day under certain physical conditions (e.g., fatigue). Thus, we did not examine test-retest reliability. We suggest future studies may examine other reliability indices such as inter-method reliability and half-split reliability [28].

For the future investigation, we suggest Thai researchers may use the six-factor, 18-item AMES-Thai to examine whether there is an association between athletic mental energy and other psychological variables such as psychological skills [29], mental toughness [30], injury [31], and burnout [32, 33]. In addition, we suggest Thai researchers may use AMES-Thai to examine its' prediction on athletic performance in different sports.

In terms of application, we suggest Thai coaches, sports administrators, or sport psychology consultants may use AMES-Thai to assess Thai athletes' mental energy states before a competition. Also, they can use AMES-Thai to diagnose Thai athletes' mental energy after heavy training. By doing so, they can offer appropriate suggestions for coaches. To enhance athletic mental energy, Lu and colleagues [1] suggested that adequate life management, proper sports training, and psychological skills training can enhance athletes' athletic mental energy.

5 Conclusion

The objective of the present study was to obtain a reliable and valid measure of athletic mental energy for Thai researchers. To achieve this objective, we conducted three studies to validate AMES-Thai. We believe that this six-factor, 16-item AMES-Thai is just the beginning of the athletic mental energy research in Thailand. We hope more Thai researchers engage in athletic mental energy research, not only to enhance Thai athletes' performance but also to promote Thai athletes' psychological well-being.

References

1. Lu F, Gill DL, Yang C, Lee PF, Chiu YH, Hsu YW, Kuan G (2018) Measuring athletic mental energy (AME): instrument development and validation. Front Psychol 9:1–15. https://doi.org/10.3389/fpsyg.2018.02363
2. Dammann KW, Bell M, Kanter M, Berger A (2013) Effects of consumption of sucromalt, a slowly digestible carbohydrate, on mental and physical energy questionnaire responses. Nutrition Neurosci 16(2):83–95. https://doi.org/10.1179/1476830512Y.0000000034
3. Mohajeri MH, Wittwer J, Vargas K, Hogan E, Holmes A, Rogers PJ, Goralczyk R, Gibson EL (2015) Chronic treatment with a tryptophan-rich protein hydrolysate improves emotional processing, mental energy levels and reaction time in middle-aged women. Br J Nutr 113(2):350–365. https://doi.org/10.1017/S0007114514003754
4. Lykenn TD (2015) Mental energy. Intelligence 33:331–335
5. NidefferR (1985) Athletes'guide to mental training. Human Kinetics, Champaign, IL
6. Suinn R (1986) Seven steps to peak performance. Hans Huber, Toronto
7. Loehr JE (2005) Leadership: full engagement for success. In: Murphy S (ed) The sport psychology handbook. Human Kinetics, Champaign IL, pp 157–158
8. Morgan WP (1979) Prediction of performance in athletics. In: Klavora P, Daniel JV (eds) Coach, athlete, and the sport psychologist. Human Kinetics, Champaign IL, pp 173–186
9. Morgan WP (1980) The trait psychology controversy. Res Q Exerc Sport 51:50–76. https://doi.org/10.1080/02701367.1980.10609275
10. Ravizz K (1977) Peak experience in sport. J Humanist Psychol 17:35–40
11. Anderson R, Hanrahan SL, Mallett CJ (2014) Investigating the optimal psychological state for peak performance in Australian elite athletes. J Appl Sport Psychol 26:318–333. https://doi.org/10.1080/10413200.2014.885915
12. Chiu SS, Hsu YW, Chiu YH, Chou CC, Gill DL, Lu FJH (2020) Seeking positive strengths in buffering athletes' life stress-burnout relationship: the moderating roles of athletic mental energy. Front Psychol 10:1–12. https://doi.org/10.3389/fpsyg.2019.03007
13. Cohen RJ, Swerdlik ME (2018) Psychological testing and assessment: introduction to tests and measurement, 9th edn. McGraw Hill Education, New York
14. Birslin RW (1970) Back-translation for cross-cultural research. J Cross-Cultural Psychol 1(3):185–216. https://doi.org/10.1177/135910457000100301
15. Jones PS, Lee JW, Phillips LR, Zhang XE, Jaceldo KB (2001) An adaptation of Brislin's translation model for cross-cultural research. Nurs Res 50(5):300–304
16. Comrey AL, Lee HB (1992) A first course in factor analysis, 2nd edn. Lawrence Erlbaum Associates, Inc., Hillsdale, NJ, US
17. Cheung GW, Rensvold RB (2002) Evaluating goodness-of-fit indexes for testing measurement invariance. Struct Equation Model 9(2):233–255. https://doi.org/10.1207/S15328007SEM0902_5
18. Bentler PM, Wu E (2002) EQS 6 for Windows user's guide. Multivariate Software, Encino, CA
19. Tabachnick BG, Fidell LS (2007) Using multivariate statistics, 4th edn. Allyn & Bacon, Boston, MA
20. Kline RB (1998) Principles and practice of structural equation modeling. Guilford Press, New York, NY
21. Fornell C, Larcker DF (1981) Evaluating structural equation models with unobservable variables and measurement error. J Mark Res 18(1):39–50. https://doi.org/10.2307/3151312
22. Netemeyer RG, Bearden WO, Sharma SO (2003) Scaling procedures: issues and applications. Sage Publications
23. Horowitz M, Alder N, Kegeles S (1988) A scale for measuring the occurrence of positive state of mind: a preliminary report. Psychosom Med 50:477–483. https://doi.org/10.1097/00006842-198809000-00004
24. Sun T, Horn M, Merritt D (2004) Values and life styles of individualists and collectivists: a study on Chinese, Japanese, British, and US consumers. J Consum Mark 21(5):318–331

25. Markus HR, Kitayama S (1991) Culture and the self: implications for cognition, emotion, and motivation. Psychol Rev 98(2):224–253
26. Sirisawasd P, Chaiear N, Johns N, Khiewyoo J (2014) Validation of the Thai version of a work-related quality of life scale in the nursing profession. Saf Health Work 5:80–85. https://doi.org/10.1016/j.shaw.2014.02.002
27. Tongprasert S, Rapipong J, Buntragulpoontawee M (2014) The cross-cultural adaptation of the DASH questionnaire in Thai (DASH-TH). J Hand Ther 27:49–54. https://doi.org/10.1016/j.jht.2013.08.020
28. Murphy KR, Davidshofer CO (2009) Psychological testing: principles and applications, 6th edn. Pearson/Prentice Hall, Upper Saddle River, N.J.
29. Durand-Bush N, Salmela JH, Green-Demers I (2001) The Ottawa mental skills assessment tool (OMSAT-3*). Sport Psychol 15(1):1–19. https://doi.org/10.1123/tsp.15.1.1
30. Sheard M, Golby J, Wersch A (2009) Progress toward construct validation of the sports mental toughness questionnaire (SMTQ). Euro J Psychol Measure 25(3):186–193
31. Maddison R, Prapavessis H (2005) A psychological approach to the prediction and prevention of athletic injury. J Sport Exerc Psychol 27(3):289–310
32. Gustafsson H, Skoog T (2012) The mediational role of perceived stress in the relation between optimism and burnout in competitive athletes. Anxiety Stress Coping 25(2):183–199
33. Lu FJH, Lee WP, Chang YK, Chou CC, Hsu YW, Lin JH, Gill DL (2016) Interaction of athletes' resilience and coaches' social support on the stress-burnout relationship: a conjunctive moderation perspective. Psychol Sport Exerc 22:202–209

The Effect of a Marathon Course on Motivation and Running Behaviour

Zhou Yali and Garry Kuan

Abstract In China, there are few colleges and universities offering Marathon course as an elective. Therefore, it is imperative to conduct research on this course from various aspects, including timetable scheduling, teaching effectiveness and student feedback. The purpose of this study is to examine the effectiveness of Marathon course on running behaviours and to provide theoretical basis for the promotion and development of the course. A total of 511 students (boys = 319, girls = 192), who had completed this course volunteered to answer the questionnaires. The questionnaires included students' motivation, self-assessment of participation, and fitness levels. The results revealed that 70.6% of the students continued to run after participating in the course, of which 56.4% started running after enrolling in the course, 14.2% of the students were running prior to taking the course. It is found that time spent exercising is disproportionate to fun effect ($p = -0.23$), and exercise persistence is disproportionate to ability effect ($p = -0.17$). Significant differences present between individuals with different fatness with regard to appearance motivation ($p = 0.0001$) and institutional motivation ($p = 0.004$), while it has a significant disproportionate relationship with institutional motivation ($r = -0.114$*) and a significant proportional relationship with appearance motivation ($r = 0.189$**). Many students were concerned about their fitness and appearance, and it has an impact on their motivation to enrol in this course. Therefore, considerable emphasis can be placed on the course's content, design and promotion, to increase enrolment in the marathon course. From the results, it was concluded that introducing marathon as an elective course could increase the number of students to start running and engage in physical activity.

Keywords Marathon · Elective course · Motivation · Fitness · Exercise behaviours

---

Z. Yali (✉) · G. Kuan
Exercise and Sports Science Programme, School of Health Sciences, Universiti Sains Malaysia, Kubang Kerian, Kelantan, Malaysia
e-mail: zhouyali@student.usm.my

Z. Yali
School of Health Sciences, Universiti Sains Malaysia, 16150 Kubang Kerian, Kelantan, Malaysia

G. Kuan et al. (eds.), *Advancing Sports and Exercise via Innovation*, Lecture Notes in Bioengineering, https://doi.org/10.1007/978-981-19-8159-3_42

1 Introduction

In order to actively promote the deepening of educational reform and strengthen the spirit of physical education and extracurricular exercise proposed in *Opinions on Strengthening Physical Education in Schools to Promote All-round Development of Students' Physical and Mental Health* [1], Chongqing University added marathon course as an elective. Based on the perspective of event-group theory, Marathon is an endurance sport dominated by physical capacity, and endurance quality has an important impact on people's living ability and their sports ability [2]. A number of studies have shown that aerobic exercise has a significant effect on promoting people's physical and mental health, comprehensively enhancing physical fitness, improving body shape, controlling cardiovascular diseases and enhancing cardiopulmonary function [3–6]. What is worth mentioning, cardiopulmonary endurance is an objective physiological index of physical activity levels and is highly correlated with the all-cause mortality rates for all groups, and it is the core element of each component of physical health [7]. Therefore, the promotion of marathon in colleges and universities will help to strengthen the physique of students and to provide strong guarantee for their physical health. Because there are few colleges and universities offering marathon as an elective course in China, it is imperative to conduct research on this course from various aspects, including timetable scheduling, teaching effectiveness and student feedback. Through correlation analysis and variance analysis on students' participation motivation, self-assessment of participation, exercise behaviour and fitness levels, this paper preliminarily discusses the relationship among these factors and provides a theoretical basis for the further promotion and development of marathon course.

2 Research Object and Method

2.1 Research Object

The research object was undergraduate students who have chosen marathon as an elective course in Chongqing University. The total number of the samples was 511, including 319 boys (62.4%) and 192 girls (37.6%), and their ages ranged from 17 to 24 (19.49 ± 1.132), and their BMI was between 16.2 and 31.79 (20.75 ± 2.54).

2.2 Measurement Instrument

Students' participation motivation for marathon course was investigated with the Motivation for Physical Activities Measure (reversed edition) (MPAM-R) [8] and this was internal motivation, including ability motivation, fun motivation, healthy motivation, social motivation and appearance motivation, with 3 questions each. External motivation was investigated with the External Motivation for Physical Activities Measure by Chen Shanping [9], and this included institutional motivation of students to pursue academic achievement and their obedience motivation to comply with the expectations of other important people, with 3 questions each. Likert 5-point scale has been used in all these questions, with motivational intensity ranging from "very strong motivation" to "no motivation". The Cronbach α reliability coefficient of each sub-scale was from 0.756 to 0.841, and the overall Cronbach α coefficients were 0.923 (internal motivation) and 0.798 (external motivation), respectively, indicating that the scale had good internal consistency.

Self-assessment of participation adopted the Exercise Effect Inventory (EEI) [10], which measured the positive effects of physical exercise on students' ability, fun, health, social interaction and appearance, with 2 questions each. Likert 5-point scale has been used in all these questions, with effect ranging from "very obvious effect" to "no effect". The Cronbach α reliability coefficient of each sub-scale was from 0.773 to 0.909, and the overall Cronbach α coefficient was 0.924, indicating that the scale had good internal consistency.

To test the relationship between different participation motivation and exercise behaviour and the relationship between self-assessment of participation and exercise behaviour, the scale used 3 indicators (exercise time, exercise frequency and exercise intensity) for measuring the sports population [11]. And it used indicators for measuring exercise adherence, which was complied according to the weeks of having marathon course and the characteristics of the project at the moment of issuing the questionnaires. The question was based on the definition of regular running. Regular running referred to physical activity that you continued to run 3 or more than 3 times a week and run for at least 20 min in a time and after that you would sweat. Based on this definition, the following six answers were listed. First, no, I am not going to start running. Second, no, I have started running but I run irregularly. Third, no, I have started running and I plan to run regularly in a month. Fourth, yes, I have started running regularly but it lasts for less than 2 months. Fifth, yes, I have started running regularly but it lasts for less than 6 months. Sixth, yes, I have been running regularly for more than 6 months.

In order to figure out the relationship between students with different fatness with regard to participation motivation and self-assessment of participation, students' fitness levels were based on Body Mass Index (BMI) which was used internationally as a standard of measuring body shape. According to Chinese standards, BMI less than 18.5 meant slightly thin; BMI greater than or equal to 18.5 and less than 24.0 meant normal; BMI greater than or equal to 24.0 and less than 28.0 meant slightly fat; and BMI greater than or equal to 28.0 meant fat [12].

2.3 Research Process

The research data were collected by sending questionnaires to those classes with marathon course. A total of 600 questionnaires were sent out before class and the questionnaires were filled out and collected on the spot. 600 pieces were received with the recovery rate of 100%. 89 blank and invalid questionnaires were removed, and 511 questionnaires were used for statistical analysis.

2.4 Data Analysis

Excel software was used to summarize the data, and SPSS27.0 was used for descriptive statistical analysis, correlation analysis and one-way analysis of variance [13].

3 Result and Analysis

3.1 Exercise Adherence Descriptive Statistics

Exercise adherence refers to the duration of an individual's long-term persistence to running, that is, the duration of an individual's continuous running for 3 or more than 3 times per week. Table 1 showed students' exercise adherence after participating in marathon course. Classification 2 indicated that 14.9% of students have started running but they didn't continue it. Classification 3 indicated that 14.5% of students have started running and they planned to keep running in the next month. Classification 4 indicated that 56.4% of students have started running regularly but it lasted for less than 2 months. Classification 5 indicated that 11.7% of students have started running regularly but it lasted for less than 6 months. Classification 6 indicated that 2.5% of students have started running regularly and it lasted for more than 6 months.

Table 1 Exercise adherence descriptive statistics

Classification	2	3	4	5	6	Total
N	76	74	288	60	13	511
%	14.9	14.5	56.4	11.7	2.5	100%

Table 2 Correlation analysis of participation motivation and exercise behaviour. (Pearson correlation)

	Healthy motivation	Appearance motivation	Fun motivation	Ability motivation	Social motivation	Institutional motivation	Obedience motivation
Exercise time	0.023	0.156**	0.046	– 0.001	– 0.001	– 0.042	0.043
Exercise frequency	0.091*	0.117**	0.143**	0.109*	0.066	0.006	– 0.033
Exercise intensity	– 0.056	0.052	– 0.046	– 0.040	– 0.040	0.046	0.020
Exercise persistence	0.199**	0.175**	0.203**	0.209**	0.103*	0.019	– 0.028

**$p < 0.01$, *$p < 0.05$

3.2 Correlation Analysis of Participation Motivation and Exercise Behaviour

Participation motivation is the psychological motivation to participate in exercise, so participation motivation should have a proportional relationship with exercise behaviour [9]. Table 2 showed the relationship between participation motivation and exercise behaviour. Internal motivation almost had a significant proportional relationship with exercise behaviour. But exercise time had a disproportionate relationship with ability motivation and social motivation. Besides, exercise intensity had a disproportionate relationship with healthy motivation, fun motivation, ability motivation and social motivation. External motivation was the motivation produced by the action of external forces. Based on the characteristics of college students, their external motivation included institutional motivation to pursue academic achievement and obedience motivation to comply with the expectations of other important people [9]. According to Table 2, there was no significant correlation between external motivation and exercise behaviour. Exercise time had a disproportionate relationship with institutional motivation. Obedience motivation had a disproportionate relationship with both exercise frequency and exercise persistence.

3.3 Correlation Analysis of Self-Assessment of Participation and Exercise Behaviour

Self-assessment of participation refers to the positive effects that individuals experience or recognize through physical exercise [10]. Self-assessment of participation should have a proportional relationship with exercise behaviour. According to Table 3, exercise frequency had a significant proportional relationship with healthy effect. Exercise intensity had a significant proportional relationship with appearance effect.

Table 3 Correlation analysis of self-assessment of participation and exercise behaviour. (Pearson Correlation)

	Healthy effect	Appearance effect	Fun effect	Ability effect	Social effect
Exercise time	0.051	0.029	− 0.023	0.022	0.032
Exercise frequency	0.094*	0.018	0.074	0.019	0.001
Exercise intensity	0.086	0.092*	0.005	0.039	0.045
Exercise persistence	0.008	0.002	0.048	− 0.017	0.038

*$p < 0.05$

Exercise time had a disproportionate relationship with fun effect. Exercise persistence had a disproportionate relationship with ability effect.

3.4 *Comparison and Correlation Analysis of Participation Motivation Between Different Fatness*

According to Table 4, correlation analysis and one-way analysis of variance showed that among five internal motivation, the intensity of appearance motivation and the intensity of institutional motivation had critical differences in fitness levels. Appearance motivation of slightly fat students was stronger than that of slightly thin students ($P = 0.000$) and normal students ($P = 0.024$). Appearance motivation of slightly thin students was weaker than that of normal students ($P = 0.000$) and slightly fat students ($P = 0.000$). Institutional motivation of slightly thin students was stronger than that of normal students ($P = 0.000$). Obedience motivation of fat students was slightly stronger than that of slightly thin students ($P = 0.029$) and normal students ($P = 0.037$). Besides, appearance motivation had a proportional relationship with fitness levels, while institutional motivation had a disproportionate relationship with fitness levels.

3.5 *Comparison and Correlation Analysis of Participation Effect Between Different Fatness*

According to Table 5, self-assessment of participation effect between students of different fatness had critical differences in appearance effect and fun effect. As for appearance effect, slightly thin students were more obvious than that of slightly fat students ($p = 0.024$) and fat students ($p = 0.029$). And normal students were more

Table 4 Comparison of participation motivation between different fatness

BMI	N	Healthy motivation	Appearance motivation	Fun motivation	Ability motivation	Social motivation	Institutional motivation	Obedience motivation
BMI < 18.5	84	3.77 ± 0.89	2.98 ± 1.05	3.39 ± 1.03	3.08 ± 1.10	2.85 ± 1.05	3.71 ± 0.93	1.96 ± 0.92
18.5 ≤ BMI < 24.0	378	3.84 ± 0.75	3.44 ± 1.01	3.52 ± 0.88	3.18 ± 0.89	2.94 ± 0.94	3.29 ± 0.97	2.02 ± 0.99
24.0 ≤ BMI < 28.0	41	3.76 ± 0.85	3.81 ± 0.88	3.34 ± 0.76	2.90 ± 0.94	2.86 ± 1.05	3.40 ± 1.00	2.12 ± 0.96
BMI ≥ 28	8	3.42 ± 0.71	3.54 ± 1.02	2.92 ± 0.66	2.83 ± 0.89	3.04 ± 0.88	3.17 ± 0.78	2.75 ± 0.97
Total	511	3.82 ± 0.79	3.39 ± 1.03	3.48 ± 0.90	3.14 ± 0.93	2.92 ± 0.96	3.37 ± 0.97	2.03 ± 0.98
F		1.03	7.44	1.91	1.59	0.31	4.58	1.74
Sig.		0.379	0.000	0.128	0.191	0.815	0.004	0.158
R		− 0.019	0.189**	− 0.03	− 0.035	0.021	− 0.114*	0.077

$^{**}p < 0.01$, $^{*}p < 0.05$

Table 5 Comparison of participation effect between different fatness

	N	Healthy effect	Appearance effect	Fun effect	Ability effect	Social effect
BMI < 18.5	84	3.68 ± 0.83	3.10 ± 0.98	3.55 ± 0.86	3.34 ± 0.94	2.83 ± 1.07
18.5 ≤ BMI < 24.0	378	3.67 ± 0.82	3.02 ± 1.04	3.57 ± 0.96	3.26 ± 0.97	2.77 ± 1.10
24.0 ≤ BMI < 28.0	41	3.49 ± 0.85	2.65 ± 1.15	3.26 ± 1.14	3.09 ± 0.99	2.59 ± 1.15
BMI ≥ 28	8	3.13 ± 0.64	2.25 ± 1.00	2.75 ± 0.71	2.75 ± 0.85	2.19 ± 0.92
总计	511	3.65 ± 0.82	2.99 ± 1.05	3.53 ± 0.96	3.25 ± 0.97	2.76 ± 1.10
F		1.73	3.25	3.08	1.38	1.21
Sig.		0.159	0.022	0.027	0.249	0.307
R		− 0.076	− 0.117**	− 0.093*	− 0.081	− 0.073

obvious than that of slightly fat students ($p = 0.028$) and fat students ($p = 0.038$). As for fun effect, slightly thin students were more obvious than that of fat students ($p = 0.025$). And normal students were more obvious than that of slightly fat students ($p = 0.049$) and fat students ($p = 0.017$). What's more, different fatness had a significant disproportionate relationship with both appearance effect and fun effect.

4 Discussion

4.1 Characteristics of Students' Exercise Adherence in the Marathon Course

Research results showed that all students would run after participating in the marathon course, but their exercise adherence was different. 29.4% of students didn't continue to run, while 70.6% of students continued to run after participating in the course. Among 70.6% of students, 11.7% of students continued to run for less than 6 months and 2.5% of students continued to run for more than 6 months. The questionnaires were handed out to students after starting the marathon course for 1.5 months. 56.4% of students continued to run for less than 2 months, which revealed that these students started running after enrolling in the course, while 14.2% of students were running prior to taking the course. The reason why 70.6% of students continued to run after participating in the course may be related to the content of the course. Chongqing University required that students should run 100 kms in 3 months, including running 3 kms for 25 times and running 5 kms for 5 times, so that they can take an examination. In addition, 14.2% of students may regard running as their personal hobbies or exercise habits.

4.2 The Relationship Between Participation Motivation and Exercise Behaviour

In this research, more than 50% of internal motivation had a significant proportional relationship with exercise behaviour, which was consistent with the research results of Chen [14] on the behaviour and exercise motivation of fat people. But part of internal motivation had a disproportionate relationship with both exercise time and exercise intensity, which may be related to running time and running volume. In running, longer distance per unit time was considered as a higher level of exercise, so ability motivation had a disproportionate relationship with exercise time. Exercise intensity was to assess the physical change of students after they ran every time according to their own conditions. And the evaluation standard was their degree of sweating. 64% of students were above moderate degree of sweating. The stronger healthy motivation, appearance motivation, fun motivation and social motivation, the more likely it may be to control exercise intensity. Besides, excessive exercise intensity may also inhibit exercise motivation.

4.3 The Relationship Between Self-Assessment of Participation and Exercise Behaviour

Marathon, as an endurance sport dominated by physical capacity [2], can greatly cultivate students' will and quality and it requires students to have higher endurance and willpower. Even though it can improve students' health, it will not achieve the significant results if exercise adherence is not enough. Those students, who had long-term exercise adherence, had certain knowledge for physical exercise, so their ability effect may not be improved. Because running was less fun than other sports, the longer exercise time, the less obvious fun effect may be.

4.4 The Relationship Between Students with Different Fatness and Participation Motivation

Appearance motivation means that people participate in physical exercise in order to improve their body shape, control their weight and make their body more attractive [15]. Slightly fat students obviously had stronger appearance motivation than others. However, the need for slightly thin students to control weight was not as strong as that of other students, so their appearance motivation was obviously weaker than others. Obedience motivation of fat students was slightly stronger than other students. The statistical results were mainly affected by the expectations of other important people to students, so the analysis was that students paid more attention to others' views on themselves.

4.5 The Relationship Between Students with Different Fatness and Participation Effect

The marathon course had certain requirements on running duration and running distance and it may be more difficult for slightly fat students and fat students to complete the course than students of other fitness levels, so their fun effect was significantly lower than other students. Long-term exercise adherence can help people get obvious and satisfactory effect in weight loss and improvement of body shape. Appearance effect of slightly fat students and fat students was also lower than other students, perhaps because they did not continue to run for a long time.

5 Conclusion

70.6% of the students continued to run after participating in marathon course, of which 56.4% started running after enrolling in the course. It showed that introducing marathon as an elective course could increase the number of students to start running and could have a positive influence on universities' physical education to achieve the goal of strengthening students' physique.

The relationship between participation motivation and exercise behaviour and the relationship between self-assessment of participation and exercise behaviour were different from other scholars' research results. Part of participation motivation had a disproportionate relationship with exercise time and exercise intensity. Exercise time had a disproportionate relationship with fun effect. Exercise persistence had a disproportionate relationship with ability effect. Students who continued to run for a long time had certain knowledge for physical exercise, so their ability effect may not be improved. Since running was less fun than other sports, fun effect would be less obvious if students exercised longer and longer.

Students of different fatness had significant differences in appearance motivation and institutional motivation. Fitness levels had a critical disproportionate relationship with institutional motivation, while it had a significant proportional relationship with appearance motivation. Self-assessment of participation effect between students of different fatness had critical differences in appearance effect and fun effect. Different fatness had a significant disproportionate relationship with both appearance effect and fun effect. As for appearance motivation, slightly fat students were more motivated than others to control their weight and improve their body shape. Therefore, considerable emphasis can be placed on the course's content, design and promotion to increase students' enrolment and enthusiasm in the marathon course.

References

1. General Office of the State Council of the People's Republic of China (2016) Opinions on strengthening physical education in schools to promote all-round development of students' physical and mental health
2. Tian MJ (2006) Sports training science. Higher Education Press, Beijing, pp 188–194
3. Li LS, Wu BL, Ji L, Wang XZ (2010) An empirical research on the impact of aerobic exercise on the health condition of young & mid-age woman. J Beijing Sport Univ 4:66–68+72
4. Tian HY (2007) Brief analysis of aerobic metabolism exercise and state of health. J Beijing Sport Univ s1:219+226
5. Deng JP, Liu XX (2014) Effects of regular aerobic exercise on physical fitness and quality of life of older people in community. Chin J Gerontol 14:3994–3996
6. Zhang BN, Guo YJ, Liu XJ (2009) An experimental research on the influence of aerobic exercise on female university students' physical condition. J Beijing Sport Univ 4:72–74
7. Xie MH, Li HJ, Wang ZZ, Wang J, Sui XM, Blair Steven N (2011) Cardiorespiratory fitness: a core component of health-related physical fitness—introduction to the aerobics center longitudinal study. J Beijing Sport Univ 2:1–7
8. Chen SP, Wang YB, Rong JZ, Pan XG, Bao J (2008) The simplified version of the MPAM-R: reliability and validity. J Beijing Sport Univ 2:66–70+78
9. Chen SP, Li XS, Rong JZ (2008) The internal and external motivation of college student's physical exercise. China Sport Sci Technol 4:135–138+143
10. Chen SP, Pan XG, Zhang P, Zhang ZJ (2008) On the designation reliability and validity of exercise effects inventory (EEI) for Chinese college student. J Beijing Sport Univ 10:1404–1406
11. Chou J (2002) Theoretical exploration and empirical research on China's sports population. Beijing Sport University Press, Beijing, pp 36
12. National Health and Family Planning Commission of the People's Republic of China (2013) WS/T428-2013 Criteria of weight for adults. People's Medical Publishing House, Beijing
13. Li X, Zhang MM (2015) From introduction to mastery of SPSS 22.0 statistical analysis. Electronic Industry Press, Beijing
14. Zhang P, Chen SP (2010) Comparative study on exercise behavior and motivation of obese people—taking the intellectuals of Xi'an city as the example. J Capital Univ Phys Educ Sports 22(5):89–92
15. Chen SP (2010) Measurement tools and application in the study of physical exercise. Xi'an Jiaotong University Press, Xi'an

Examining Body Mass Index and Strength Endurance Abilities of Adolescent Students

Victor P. L. Salanggang, Chin Ngien Siong, and Teo Boon Sian

Abstract The rate of obesity among adolescents has been of public health concern which is linked to the increasing prevalence of health-related non-communicable diseases. The health of adolescents are often overlooked which has implication to their health and mental well-being during adulthood. The study examined the prevalence of weight status and strength endurance among adolescent students aged 18 to 20 years old (1.36 ± 0.60). The Body Mass Index (BMI) and Burpee Test Battery were utilized to measure the body composition and strength endurance of the adolescent students. The BMI test revealed that there were more females as compared to males in the normal weight, underweight, overweight, obese, and extreme obese BMI. In addition, the burpee test revealed that females performed better level of strength endurance than males. Compulsory school-based physical activity (PA) interventions need to be prioritized for the prevention of obesity and promotion of PA in schools in order to instill a better and healthier educational environment.

Keywords Body mass index · Strength endurance · Adolescents

1 Introduction

Adolescence is a critical period in one's life whereby everyone acquires the skills needed for a productive, healthy, and satisfying life throughout this time [1]. Adolescents require access to physical and health education in order to make a healthy transition into adulthood. Every adolescent would need to adapt to the new environment and situation as they transit from one institution of learning to another such as high school to universities that represent a substantial life adjustment [2]. Many students find the move to any institution of higher learning to be stressful and demanding due

V. P. L. Salanggang
Institute of Teacher Education Keningau, Keningau, Sabah, Malaysia

C. N. Siong (✉) · T. B. Sian
Institute of Teacher Education Batu Lintang Campus, Kuching, Sarawak, Malaysia
e-mail: ngiensiong@gmail.com

G. Kuan et al. (eds.), *Advancing Sports and Exercise via Innovation*, Lecture Notes in Bioengineering, https://doi.org/10.1007/978-981-19-8159-3_43

to academic performance, workload, and tests [3]. The impact of this transition on university students has a significant impact on their level of physical activity (PA) during their time at university. According to a study conducted by Bray and Born [2], one-third of high school active students turned inactive after entering university. Furthermore, a meta-analysis of previous studies conducted by Keating et al. [4] revealed that university students are sedentary, resulting in a decrease in PA. Additional research in Romania [5] and Saudi Arabia [3] indicated that 60% of Romanian university students were not involved in sports and that PA activity decreased after entering college or university. Problems such as poor health outcomes experienced by the secondary school students since the bulk of them are physically sedentary while being close to PA facilities [4] were linked to low grades, concentration, anxiety, and school failure [4, 6].

According to Zi and Cardinal [7], Asian students, who make up the largest international student population in American institutions and colleges, participated in PA at a low level, which negatively impacted their physical and mental wellness. Furthermore, 28.1% of Asian female foreign university students and 11.7% of Asian male international university students were physically inactive, according to the study. Furthermore, Mincey [8] discovered in an urban state university study that Black male college students were at a higher risk than their peers due to their unhealthy lives. Asian female foreign students are the least active group in PA, averaging 1.3 h of PA weekly [9] and less than 2.5 h of moderate-to-vigorous PA, meaning that they are the least active group in PA [10]. In terms of gender, Asian female university students were the least active in PA, averaging only 1.3 h per week [9]. Furthermore, Asian female students have been shown to be physically inactive, averaging 1.3 h of PA per week and less than the recommended 2.5 h of moderate-to-vigorous exercise per week.

Malaysia has been a trailblazer in terms of high adult obesity rates, with an increase from 4.4 to 17.7% between 1996 and 2015 [11]. According to the 2019 National Health and Morbidity Survey (NHMS), the current obesity rate among Malaysian adults is 19.7% [11]. As a result, these findings do not bode well with the existing participation of PA among Malaysian students as they progress from secondary school to university. Malaysia is one of the top ten most physically inactive countries, with an average of 61.4% of people aged 15 and up being physically inactive [12]. In general, PA is reduced to compensate for high workloads and a lack of spare time among Malaysian university students, with males being marginally more engaged in PA than females [13, 14]. Obesity has always been a serious health concern in Malaysia, and it is especially prevalent among Malaysian university students, having a negative impact on their psychological and physical health. Obesity has a negative impact on students' social, emotional, and mental health by lowering their overall quality of life [3, 15]. Adolescent and juvenile obesity may be a precursor to adult obesity, affecting morbidity and death [16]. As a result, it is critical to manage obesity early in childhood in order to reduce the impact of major health consequences later in life. Due to limited studies which show the effects of COVID-19 pandemic on BMI and fitness level among adolescent students, the purpose of this study is to determine the BMI and strength endurance of adolescent students in Sarawak.

2 Materials and Methods

2.1 Participants

The participants comprised of 446 adolescent students (106 males, 340 females) aged 18–20 years old. The selection of the participants was based on a convenience sampling method. The study approval was obtained by the Institution Ethics Board and informed consent was also obtained from the participants and relevant authorities as well as assured of their confidentiality prior to the study.

2.2 Measures

The self-reported BMI was utilized to measure the anthropometric measurements which is the body composition in terms of the body height and weight. The values of the weight and height were recorded to the nearest kilograms and centimeter. The measurement for BMI test is weight (kg) divided by height (m^2). The BMI test was abide by the WHO guidelines according to the WHO norms [17]. BMI $\leq$ 18.49 is defined as underweight, 18.50 $\leq$ BMI $\leq$ 24.99 is classified as normal weight, 25.00 $\leq$ BMI $\leq$ 29.99 as overweight, 30.00 $\leq$ BMI $\leq$ 34.99 as obese and BMI $\geq$ 35.00 is considered as extremely obese. Whereas, the 1-min burpee test was utilized to test strength endurance of the participants. Procedures on performing the burpee test were provided to ensure that it was perform accurately. It is based on the number of burpee cycles that the participants can perform in 1 min [18, 19]. The norms of the burpee test was based on the Manual Pelaksanaan & Penilaian Video Kecergasan Fizikal [20].

2.3 Statistical Analyses

Data collection was performed and computed through Statistical Package for Social Science (SPSS) version 28. Descriptive statistical analysis was conducted to analyze frequency, percentage, mean, and standard deviation. Statistical significance was set at $p < 0.01$.

3 Results

Table 1 shows the demographic characteristics of the participants. A total of 446 participants involved in this study, comprising of 106 (23.8%) males and 340 (76.2%) females. The majority of the participants were from Sabah with a total of 254 (57.0%), followed by Sarawak at 192 (43.0%) participants. Based on the Body Mass Index (BMI), 265 (59.4%) were in the normal BMI, 102 (22.9%) were in the underweight BMI, 53 (11.9%) were in the overweight BMI, 19 (4.3%) were in the obese BMI, and only 7 (1.6%) were in the extreme obese BMI.

Table 2 shows the BMI test based on gender. Results revealed that there were more females (43.5%) than males (15.9%) in the normal BMI level. Besides, more females (20.9%) were in the underweight BMI level as compared to males (2.0%). Furthermore, there were slightly more females (7.4%) than males (4.5%) in the overweight BMI level. Moreover, slightly more females (3.4%) were in the obese BMI level as compared to males (0.9%). Lastly, only 5 (1.1%) females and 2 (0.5%) males were in the extreme obese BMI level.

Table 3 shows the burpee test based on gender. Results revealed that there were slightly more females (1.8%) than males (0.9%) in the excellent level of strength endurance in the burpee test. Besides, there were more females (15.7) than males (6.1%) in the good level of strength endurance in the burpee test. Furthermore, more females (36.5%) were in the average level of strength endurance as compared to males (9.4%) in the burpee test. Moreover, there were slightly more females (14.3%) than males (5.8%) were in the poor level of strength endurance in the burpee test. Lastly, a total of 35 (7.8%) females and 7 (1.6%) males were in the very poor level of strength endurance in the burpee test.

4 Discussion

The study examined the Body Mass Index and strength endurance abilities of adolescent students based on gender. In this study, the overall BMI test revealed that 59.4% of the adolescent students were in the normal weight BMI level. Both intrinsic and extrinsic motivation may contribute to the predominance of normal weight BMI. For instance, females were driven by external factors such as ill-health avoidance, weight control, and attractiveness, whereas males were driven by internal factors such as social affiliation, competition with others, and challenges [21]. The overall 1-min burpee test showed that slightly over 70% of the students were under excellent, good and average level, whereas the remaining percentage of the students were under the poor and very poor level of strength endurance. Results revealed that females had greater levels of physical fitness scores in burpee test as compared to males in terms of strength endurance. This could be due to females are more reliant on interpersonal reasons from community or social support such as family support and friend support as compared to males in engaging in PA.

On the other hand, more females than males were in the underweight category. One of the reasons why adolescent females were underweight was a diet deficient in calories. This is corroborated by the findings of Hakim et al. [22], who discovered that low calorie consumption may lead to insufficient energy intake which would result in poor intake of vital nutrients such as fat, calcium, iron, and protein. According to Hakim et al. [22], there were more females (90.5%) as compared to the males

Table 1 Demographic characteristics of the participants ($N = 446$)

Characteristics	Frequency (F)	Percentage (%)	Mean (SD)
Gender			1.76 (0.43)
Male	106	23.8	
Female	340	76.2	
State			1.57 (0.50)
Sarawak	192	43.0	
Sabah	254	57.0	
BMI test			1.66 (0.95)
Normal (18.50–24.99)	265	59.4	
Underweight (≤ 18.49)	102	22.9	
Overweight (25.00–29.99)	53	11.9	
Obese (30.00–34.99)	19	4.3	
Extreme obese (≥ 35.0)	7	1.6	
Burpee test (1 min)			
Male			3.05 (0.96)
Excellent (≥ 27)	4	3.8	
Good (21–26)	27	25.5	
Average (14–20)	42	39.6	
Poor (8–13)	26	24.5	
Very poor (≤ 7)	7	6.6	
Female			3.14 (0.94)
Excellent (≥ 22)	8	2.4	
Good (17–21)	70	20.6	
Average (11–16)	163	47.9	
Poor (6–10)	64	18.8	
Very poor (≤ 5)	35	10.3	

Table 2 BMI test based on gender

Gender	Level										∑n
	Normal		Underweight		Overweight		Obese		Extreme Obese		
	n	%	*n*	%	*n*	%	*n*	%	*n*	%	
Male	71	15.9	9	2.0	20	4.5	4	0.9	2	0.5	106
Female	194	43.5	93	20.9	33	7.4	15	3.4	5	1.1	340
Total	265	59.4	102	22.9	53	11.9	19	4.3	7	1.6	446

Table 3 Burpee test based on gender

Gender	Level										∑n
	excellent		Good		Average		Poor		Very Poor		
	n	%	*n*	%	*n*	%	*n*	%	*n*	%	
Male	4	0.9	27	6.1	42	9.4	26	5.8	7	1.6	106
Female	8	1.8	70	15.7	163	36.5	64	14.3	35	7.8	340
Total	12	2.7	97	21.8	205	45.9	90	20.1	42	9.4	446

(30.3%) among university students failed to fulfill RNI recommendations for iron consumption. Moreover, breakfast is an essential meal that aids in refueling the body and maintaining a healthy weight, as well as boosting concentration and mood [23]. Several studies found that the habit of skipping breakfast regularly will result in a lowered intake of nutrients and calories [22]. This is supported by Liyanage et al. [24] and Vishnukumar et al. [25] revealed that the most prevalent reason why the students were more likely to skip breakfast in the morning was primarily due to lack of time because they typically stayed up late into the night to owe their assignments and studying. Similarly, Ahmad et al. [26] also stated that the university students' lack of time to have breakfast was most likely attributable to the rigorous academic curriculum load. As a result, the proportion of students who consume breakfast rapidly declines as they have demanding class and work schedules, homesickness, and peer group pressure which can lead to an irregular eating pattern [25]. According to Taha and Rashed [27], the most common reasons why majority of the students who skipped breakfast was due to insufficient time (37%), followed by feeling less appetite (33%), not having breakfast cooked at home (15%), preventing excessive weight gain (7%), and disliking for the food cooked at home (5%).

Apart from that, there were slightly more females as compared to males in the overweight, obese, and extreme obese BMI. The prevalence of overweight, obese, and extreme obese BMI showed that they were physical inactive as most of them embraced sedentary behavior as they spent most of the time lying down which accounted for 33% and sitting accounted for 41% during their daily routine, which may ultimately result in a sedentary lifestyle [28]. These findings were supported by the previous research by Jia et al. [29] which indicated that physical inactivity was significantly related to being overweight. For example, the adolescent students maximized their free time by watching television, playing online games, and surfing the web in addition to their typical daily study schedules [30]. In addition, adolescent students were physically inactive both during and after their time in schools due to time constraints, which prohibited them from engaging in PA that was not prioritized in comparison with their academic work [31]. Previous research among university students in Kuwait and Muscat revealed that not having sufficient time is one of the major barriers to not participating in PA [32, 33]. Similarly, 35.5% of the adolescent students did not take part in PA due to time constraints [34]. Consequently, they are susceptible to significant non-communicable diseases such as high blood pressure, hypertension, coronary heart disease, and type-2 diabetes [35].

5 Conclusion

Without more compulsory and adaptive school-based obesity prevention programs, existing trends in adolescents being overweight, and obesity are unlikely to be reversed in Malaysia. It is due to the substantial amount of time students spend in school, research suggests focusing obesity prevention efforts on modifiable elements of school environments that put students at risk. The current study's findings imply that it may be prudent to direct obesity prevention efforts toward schools that do not provide students with access to leisure facilities during school hours or that do not encourage active participation for adolescents.

It is critical to understand the school features that may influence behavior because if a school program or policy can be altered to have even a minor impact on changing or normalizing the distribution of a risk factor across all schools, the effect on all students could be significant.

References

1. Laski L, Expert Consultative Group for Every Woman Every Child on Adolescent Health (2015) Realising the health and wellbeing of adolescents. BMJ (Clin Res Ed) 351:h4119. https://doi.org/10.1136/bmj.h4119
2. Bray SR, Born HA (2004) Transition to university and vigorous physical activity: implications for health and psychological well-being. J Am Coll Health 52(4):181–188
3. Sarah AA, Najwa FA, Ghadeer NL, Shahad ZK, Lujain KK, Murooj YS, Osama YS (2019) Pattern of physical exercise practice among university students in the Kingdom of Saudi Arabia (before beginning and during college): a cross-sectional study. BMC Publ Health 19:1716. https://doi.org/10.1186/s12889-019-8093-2
4. Keating XD, Guan J, Piñero JC, Bridges DM (2005) A meta-analysis of college students' physical activity behaviors. J Am Coll Health 54(2):116–125
5. Massiera B, Petracovschi S, Jessica C (2013) Ideological challenges to developing leisure sport in Romania: a cultural and historical analysis of the impact of elite sport on popular sport practice. Soc Leis 36(1):111–126. https://doi.org/10.1080/07053436.2013.805578

6. Matingwina T (2018) Health, academic achievement and school-based interventions. In: Health and Academic Achievement. Intech Open. https://doi.org/10.5772/intechopen.76431
7. Zi Y, Cardial BJ (2013) Increasing Asian international college students' physical activity behavior: a review of the youth physical activity promotion model. Spring 45(1):35–45
8. Mincey K, Turner BL, Brown A, Maurice S (2010) Understanding barriers to health behaviors in college black men. J Am Coll Health 65(8):567–574
9. Yoh T, Yang H, Gordon B (2008) Status of participation in physical activity among international students attending colleges and universities in the United States. Coll Stud J 42:1110–1117. http://www.questia.com/library/jp-college-student-journal.jsp?
10. Haskell WL, Lee I, Pate RR, Powell KE, Blair SN, Franklin BA, Bauman A (2007) Physical activity and public health. Updated recommendation for adults from the American college of sports medicine and the American Heart Association. Med Sci Sports Exerc 39:1423–1434
11. Mohd SS, Lekhraj R, Foo CN (2021) Prevalence, associated factors and psychological determinants of obesity among adults in Selangor, Malaysia. Int J Environ Res Publ Health 18:868. https://doi.org/10.3390/ijerph1803086
12. Keating XD, Castro-Piñero J, Centeio E, Harrison JL, Ramirez T, Chen L (2010) Health related fitness knowledge and its relation to student physical activity patterns at a large US southern state university. ICHPER-SD J Res Health Phys Educ 5(2):3–9
13. Norlena S, Mohd TH (2013) Physical activity index among Malaysian youth. Asian Soc Sci 9(12)
14. Zainudin AB, Halijah I, Zakiah MA (2019) Physical activity among Malaysian university students: the motives. Indian J Publ Health Res Dev 10(6):1305–1308
15. Black DR, Coster DC, Paige SR (2017) Physiological health parameters among college students to promote chronic disease prevention and health promotion. Prev Med Rep 7:64–73
16. Dietz WH (2004) Overweight in childhood and adolescence. New England J Med 350(9):855–857. https://doi.org/10.1056/NEJMp048008
17. World Health Organization (2004) Expert consultation: appropriate body mass index for Asian populations and its implication for policy and intervention strategies. Lancet 363(9403):157–163.https://doi.org/10.1016/S0140-6736(03)15268-3; Podstawski R, Markowski P, Clark CCT, Choszcz D, Ihász F, Stojiljković S, Gronek P (2019) International standards for the 3-minute burpee test: high-intensity motor performance. J Hum Kinet 69(1):137–147
18. Podstawski R, Żurek P, Clark CC, Laukkanen JA, Markowski P, Gronek P (2019) A multifactorial assessment of the 3-minute burpee test. J Phys Educ Sport 19(2):1083–1109
19. Manual Pelaksanaan & Penilaian Video Kecergasan Fizikal (2022) Institut Pendidikan Guru Malaysia
20. Egli T, Bland HW, Melton BF, Czech DR (2011) Influence of age, sex, and race on college students' exercise motivation of physical activity. J Am Coll Health 59(5):399–406. https://doi.org/10.1080/07448481.2010.513074
21. Hakim NA, Muniandy ND, Danish A (2012) Nutritional status and eating practices among university students in selected universities in Selangor, Malaysia. Asian J Clin Nutr 4(3):77–87. https://doi.org/10.3923/ajcn.2012.77.87
22. Sofar SM, El HNA (2019) Relations between skipping breakfast, academic performance and body index mass undergraduate university nursing students. IOSR J Nurs Health Sci 8(4):32–42
23. Liyanage G, Siriwardana HD, Wettasinghe CA, Kumarasiri MKHN, Niwanthika TKI (2017) Breakfast habits and it's relationship to body mass index, cognitive function among final year medical students. J GP 5(3):1–4. https://doi.org/10.4172/2329-9126.1000311
24. Vishnukumar S, Sujirtha N, Ramesh R (2017) The effect of breakfast on academic performance and behaviour in schoolchildren from Batticaloa district. J Nutr 110:159–165
25. Ahmad S, Damit NA, Rahman HA (2019) Breakfast intake habits among Universiti Brunei Darussalam students. Pakistan J Nutr 18(9):817–823. https://doi.org/10.3923/pjn.2019.817.823
26. Ahmad S, Damit NA, Rahman HA (2019) Breakfast intake habits among Universiti Brunei Darussalam students. Pakistan J Nutri 18(9):817–823. https://doi.org/10.3923/pjn.2019.817.823

27. Emerald SH, Thomas M, Raju A, Tomy T, Frank R, Leena KC (2019) Body mass index and eating habits among the university students. Int J Health Sci Res 9(3):164–167
28. Taha Z, Rashed AS (2017) The effect of breakfast on academic performance among high school students in Abu Dhabi. Arab J Nutr Exerc (AJNE) 2(1):40–49. https://doi.org/10.18502/ajne.v2i1.1243
29. Poh BK, Safiah MY, Tahir A, Siti Haslinda MD, Siti Norazlin N, Norimah AK, Wan Manan WM, Mirnalini K, Zalilah MS, Azmi MY, Fatimah S (2010) Physical activity pattern and energy expenditure of Malaysian adults: findings from the Malaysian Adult Nutrition Survey (MANS). Malaysian J Nutr 16(1):13–37
30. Jia H, Zack MM, Gottesman II, Thompson WW (2018) Associations of smoking, physical inactivity, heavy drinking, and obesity with quality-adjusted life expectancy among US adults with depression. Value Health 21(3):364–371. https://doi.org/10.1016/j.jval.2017.08.002
31. Moulin MS, Irwin JD (2017) An assessment of sedentary time among undergraduate students at a Canadian University. Int J Exerc Sci 10(8):1116–1129
32. Alkhateeb SA, Alkhameesi NF, Lamfon GN, Khawandang SZ, Kurdi LK, Faran MY, Khoja AA, Bukhari LM, Aljahdali HR, Ashour NA, Bagasi HT, Delli LM, Khoja OA, Safdar OY (2019) Pattern of physical exercise practice among university students in the Kingdom of Saudi Arabia (before beginning and during college): a cross-sectional study. BMC Publ Health 19(1):1–7. https://doi.org/10.1186/s12889-019-8093-2
33. Musaiger AO, Al-Kandari FI, Al-Mannai M, Al-Faraj AM, Bouriki FA, Shehab FS, Al-Dabous LA, Al-Qalaf WB (2014) Perceived barriers to weight maintenance among university students in Kuwait: the role of gender and obesity. Environ Health Pre Med 19(3):207–214
34. Youssef RM, Al Shafie K, Al-Mukhaini M, Al-Balushi H (2013) Physical activity and perceived barriers among high-school students in Muscat, Oman. East Mediterr Health J 19(9):759–768. https://doi.org/10.26719/2013.19.9.759
35. El-Gilany AH, Badawi K, El-Khawaga G, Awadalla N (2011) Physical activity profile of students in Mansoura University, Egypt. East Mediterr Health J 17(8):694–702

Attributes Contributing to the Travel Behaviour of Trail Runners

Mohamad Nizam Nazarudin, Sabariah Bte Sharif, Mohd Khairuddin Jerry Abdullah, Soon Singh Bikar Singh, Tini Maizura Mohtar, and Bamini Kpd Balakrishnan

Abstract Sabah was a renowned trail running destination before the COVID-19 outbreak halted all tourism activity. The primary goal of this research was to determine which trail areas and trail qualities influence trail runners' travel behaviour following the COVID-19 pandemic. The research was exploratory in character, with a quantitative methodology. Surveys were conducted in twelve distinct trail running locations in Kundasang, Sabah, where 232 questionnaires were collected from trail runners and analysed. The top five event, location, and trail qualities that people look for when selecting an event are a trail that makes them feel good, a well-planned event, a path that goes through a variety of landscapes and natural features, a trail that gives them something to remember, and a well-marked trail, according to the findings. The most essential trail aspects that runners look for when selecting a trail running event are, in order of priority, how the trail is managed, how much the event costs, how simple it is to get to, what the route is like, what the destination is like, and how the event is conducted. The findings of this study aided both trail running events and host destination marketers by highlighting the features that should be used in their events and host destination marketing to meet the needs of trail runners and encourage the growth of sustainable trail running events in Kundasang, Sabah. This study also discovered that trail qualities are distinct features that trail runners want. This means that the organisers and marketers of trail running events, as well as the organisers and marketers of the host destination, should incorporate these features into their marketing and event planning.

Keywords Sport tourism · Endurance sport · Trail running · Travel behaviour · Trail attributes

M. N. Nazarudin (✉)
Universiti Kebangsaan Malaysia, Bangi, Selangor, Malaysia
e-mail: mohdnizam@ukm.edu.my

S. B. Sharif · M. K. Jerry Abdullah · S. S. Bikar Singh · T. M. Mohtar · B. K. Balakrishnan
Universiti Malaysia Sabah, Kota Kinabalu, Sabah, Malaysia

G. Kuan et al. (eds.), *Advancing Sports and Exercise via Innovation*, Lecture Notes in Bioengineering, https://doi.org/10.1007/978-981-19-8159-3_44

1 Introduction

Trail running is an outdoor activity characterised by the authenticity of the terrain and the simplicity of appreciating nature while also testing one's mind and body [1]. Trail running is a pedestrian race in a natural setting with as little concrete or asphalt road as feasible, and the route must be well-marked. Track running, for the sake of this study, refers to routes in natural areas that may consist of a muddy path or a trail of loose rocks and gravel. Furthermore, trail running takes place in a range of settings, including along the beach, desert, and savannah. As the number of trail running events in Sabah grows each year, it is critical to understand the features trail runners want when choosing an event. An attribute is a quality of anything that is thought to be appealing to visitors and might influence their decision-making in either a positive or negative way [2]. People perceive their journey as a decision-making process influenced by psychological elements such as stress [3]. People travel because they want to address unmet needs. Travel behaviour refers to how visitors behave around a specific product or service and how they react when using that product or service. Travel behaviour is the study of why travellers buy specific items and how they arrive at those decisions [3]. Based on the foregoing and how it relates to the current study, travel behaviour can be defined as a tourist's movement and actions outside of their permanent location, such as the distance travelled, the number of times they've been there before, the length of their stay, the size of their group, the number of years they've been going, the number of events they attend each year, and the total amount they spend, all of which are attempts to satisfy various motivations and desires.

Trail running events have various characteristics that distinguish them and attract certain racers. There are, for example several beginning sites and distances for everyone from beginners to advanced runners. There are additional unique aspects to the destination and the trail itself [4]. Trail running events differ from other endurance sports in that they are more dynamic and involve more obstacles, such as running up and down hills [4]. Another advantage of trail running events is that they are attended by smaller numbers of people at a time. According to what has been discussed thus far, trail running competitions are distinct from other endurance sports. This makes them the ideal sport endurance group to research. It is critical to understand what motivates a participant to attend an event since these factors can be used to influence a participant's travel behaviour [5]. Travel behaviour refers to how visitors behave prior to, during, and after a journey based on their attitudes [5]. Travel behaviour in the context of this study can be defined as a tourist's movement and acts outside of their permanent place. They do this in an attempt to address a wide range of wants and motives, any of which may be the most essential at any particular time.

The purpose of this study is to contribute to the literature on endurance sport tourism by learning more about the factors that influence how trail runners travel, specifically which event, location, and trail factors have the greatest impact. Trail runners, their travel habits, and the things that matter to them are all ideal research topics. If event and destination marketers collaborate to learn more about the people who participate in trail running events, specialised marketing tactics can be developed and implemented [6, 7]. The information in this study can be used by marketing and management teams for trail running events and host destinations to create effective marketing plans and management strategies that meet the needs of trail runners while also promoting and running the event and host destination as a whole [8]. Furthermore, trail running events can provide runners with memorable experiences, causing them to become loyal to the host place and return to participate in the event again [9]. Finally, travel behaviour can be beneficial to the event host destination because, as previously said, if the event has a positive impact on how people travel, they may stay longer and spend more money in order to train and prepare for the trail running event at the event host destination.

1.1 Models and Theories of Vacationing Behaviour

A person's behaviour is the result of several factors, any one of which may be the most significant at any one time [10]. Attendees' actions are motivated by the pursuit of a satisfying, memorable experience that allows them to get personal value from the sporting or event [11]. Prior research has led to the development of theories and models based on these categorisations of visitors and their journeys. Relevant theories and models include the Theory of Planned Behaviour (TPB), the Schmoll model, the five-stage model of travel purchase behaviour developed by Mathieson and Wall, and the multi-level model of outbound trip behaviour. It has been theorised and modelled that people's method of transportation selection is a series of choices. Theories and models of travel behaviour have revealed that factors such as the nature of a given location, one's level of comfort with spending, and the state of one's demographics can all influence a person's propensity to travel [12, 13].

1.1.1 The Theory of Planned Behaviour (TPB)

TPB estimates whether a person will do an action at a given time and location [14]. The TPB examines three aspects of tourists' behaviour when they travel: their psychological and social characteristics, as well as their perceptions of their own capacities to carry out the behaviour [15].

1.1.2 Schmoll Model

This theory holds that travellers' individual and collective backgrounds influence their decisions and actions while away [16]. The model aids in understanding the decision-making and information-processing processes of humans. Factors both internal and external to the potential traveller, as well as the qualities of the potential vacation destination, all play a role in the ultimate decision-making process [16].

1.1.3 Mathieson and Wall's 5-Phase Model of Travel Buying Behaviour

This model outlines the five considerations travellers make before making any plans for their vacation. There are five distinct phases of a tourist's experience, beginning with (1) the realisation that they want to travel to a particular place, (2) gathering information from official and unofficial sources, (3) settling on a course of action, (4) preparing for the trip, (5) evaluating the quality of their trip. Four factors—the tourist's profile, the measures taken during the trip, the destination's characteristics and resources, and the nature of the trip itself—influence these five stages [17]. Multi-Level Model of Outbound Travel Behaviour is the last model. Its goal was to learn whether and how the economy has influenced the vacation habits of travellers.

The model predicted that vacationers' actions would be influenced by both local and global economic conditions. Changes in the economy were also predicted to have an impact on tourists' preferences in where they choose to go and how they spend their money. It is obvious from the many ideas and models that a variety of factors can influence the choices and behaviours of tourists. Therefore, it is essential to investigate the elements that influence people's travel habits. These five steps are influenced by four factors: the nature of the tourists, the features of the place, and the available tools [17].

1.1.4 Multi-Level Outbound Travel Behaviour Model

This model was developed to investigate the interplay between demographic and economic shifts and their impact on vacationers. The model predicted that vacationers' spending habits would be influenced by both personal and national economic conditions. The model also predicted that the economy would have an impact on vacationers' inclinations and spending patterns. It is obvious from the many ideas and models that a variety of factors can influence the decisions and behaviours of tourists. This highlights the significance of researching the elements that influence people's travel choices.

1.2 Factors that Might Affect Sport Participants Travel Choices

Attributes are typically either helpful or harmful in influencing a choice [2]. Location, climate, demographics, circumstances, lifestyle, culture, motivations, and advertising are also significant [18]. Participants' and viewers' decision-making processes can benefit from these factors, leading them to wish to travel [13].

2 Methods

The empirical data was collected in Kundasang, Sabah, at the 12 trails area. Kundasang is a town in Sabah, Malaysia. It is in the Ranau district and is near the Kundasang Valley. Kinabalu National Park is about 6 km away, and Ranau is about 12 km away. It is the closest settlement to Mount Kinabalu and has the best view of it. The hikers were given a structured questionnaire to fill out. About 250 questionnaires were sent out, and 232 valid questionnaires were sent back. This means that the response rate was about 93%. The questionnaires [19] have 35 questions about the event, its location, and the features of the trail that make trail runners decide whether to go and how they will get there. On a five-point Likert scale of importance, 1 meant the item wasn't important at all, 2 meant it was a little important, 3 meant it wasn't important either way, 4 meant it was important, and 5 meant it was very important. The Statistical Process for Social Sciences (SPSS) version 28 was used to look at the data. Initial data analyses (descriptive statistics) were used to look at the characteristics of the event, the destination, and the trail. Two exploratory factor analyses were used to look at the event, destination, and trail attributes that trail runners think are important when choosing an event, as well as how trail runners' travel.

3 Result

3.1 The Qualities of the Event, the Setting, and the Route that Participants Value Most

Table 1 shows what trail runners think are the most important things about an event, its location, and the trail itself when choosing to take part in one.

3.2 The Effect of Event Participation on Trail Running Participants' Vacation Behaviour

This section presents the results of the initial exploratory factor analysis conducted to determine the elements (event, destination, and trail qualities) that participants consider when selecting a trail running event. The pattern matrix of the principal component factor analysis utilising a Varimax rotation with Kaiser Normalisation revealed five attribute components labelled in accordance with similar qualities (Table 2). About 70% of the variance was accounted for by these factors. All the attribute components have relatively high reliability coefficients (Cronbach's Alpha), ranging from 0.78 (the lowest) to 0.95 (the highest). The attribute component inter-item correlation coefficients ranged between 0.49 and 0.65. All the reliability factors

Table 1 Event, destination, and trail attributes participants seek

	Factors that participants in an event look for include the event's popularity, the location of the event, and the quality of the trail. Five-point Likert scale of importance (1 = not at all important; 2 = slightly important; 3 = neutral; 4 = important and 5 = extremely important)	Mean value	Std. deviation
1	A path that makes you feel good	4.44	0.89
2	An event that is well-planned	4.41	0.88
3	A path that goes through a variety of landscapes and natural features	4.38	0.82
4	A trail that gives you something to remember (beautiful views or wildlife)	4.21	0.79
5	A trail with clear signs	4.20	0.81
6	A trail with a lot of security and safety	3.99	0.92
7	The length of the trail's route (km)	3.94	0.90
8	A trail that is well-kept	3.90	0.89
9	How cheap it is to sign up for an event	3.90	0.92
10	A hard path to walk	3.90	0.85
11	How well the registration process works	3.89	0.95
12	An event that is well-organised	3.89	1.03
13	A place for an event that doesn't cost a lot	3.89	0.94
14	The planning of the event (the date on which the event takes place)	3.87	0.99
15	A place for an event that gives me the chance to go to a new place	3.84	1.00
16	Costs of getting to the event	3.82	0.92
17	A path with sections that go uphill and downhill	3.81	0.96
18	A place for events with friendly locals and great service	3.80	0.99
19	A successful event	3.78	1.08
20	The fact that the event has a good name (most difficult trail in the state)	3.75	1.00
21	A place for an event that has enough places to stay	3.75	1.03
22	A place for an event with lots to do and see	3.70	1.03
23	A trail with rocks, roots, and other obstacles at different points	3.67	1.00
24	A place for an event that is perfect in terms of weather and height	3.60	0.98
25	A place for an event that has a city government that supports the hosting of sports events	3.59	1.16
26	A trail with a complicated layout	3.55	1.08
27	A trail with a lot of single-track parts	3.53	1.03
28	A place for an event that is easy to get to and from	3.53	1.10

(continued)

Table 1 (continued)

	Factors that participants in an event look for include the event's popularity, the location of the event, and the quality of the trail. Five-point Likert scale of importance (1 = not at all important; 2 = slightly important; 3 = neutral; 4 = important and 5 = extremely important)	Mean value	Std. deviation
29	An event that can be put into different groups I can participate in (professional, amateur, charity)	3.47	1.18
30	Because the event was close to where I live	3.43	1.13
31	A place for events that has many high-quality entertainment options	3.42	1.16
32	A place for events with several tourist attractions	3.32	1.15
33	An event that helps with making travel plans	3.30	1.28
34	A competition that helps me get into other competitions	3.28	1.22
35	A place for events that has good sports and training facilities	3.20	1.30

fell within the recommended range. In consideration of the factorability of the data, the Kaiser–Meyer–Olkin measure of sampling adequacy evaluated 0.93 for the attribute factors, exceeding the suggested threshold of 0.60 [20]. Bartlett's test of sphericity likewise obtained statistical significance ($p < 0.001$) with 187, confirming the factorability of the correlation matrix [21]. The subsequent section describes each of the five characteristic factors independently, following the presentation of Table 2.

4 Discussion and Conclusion

First, with a mean score of 4.44, participants rated the event's trail's ability to make them feel good as one of the most essential aspects to consider when selecting an event to participate in. A desire to have a good time is a major factor in selecting an endurance sport event to participate in, according to several studies [6, 7]. Whether a participant returns to a certain activity depends on factors like how entertaining the trail itself is. Thus, it can influence whether a trail runner returns to an annual race they have previously enjoyed. Second, the most significant quality of a successful event is meticulous preparation (4.41). For trail runners, a well-organised event has to do with the way the trail is set up, which is consistent with the findings of several research on endurance sport events [8, 22] and a trail running study [7]. According to prior studies on trail running [7, 23], thirdly, a route that passes through a diversity of landscapes and natural elements is valued (4.38).

The event's trail should also provide trail runners with an experience they won't forget, which brings us to our next point (4.21). Previous studies have shown that good memories and/or past events can lead to enjoyable trails and events in future [22, 24]. Researchers found that an endurance sport event or trail running event that

Table 2 Results of the exploratory factor analysis: event, destination, and trail characteristics that participants deem crucial while selecting a trail running event

Important event, destination and trail attributes endurance trail running participants seek when choosing an event	Factor 1: destination attributes	Factor 2: trail management	Factor 3: trail attributes	Factor 4: event expenses and accessibility	Factor 5: event management
A place for an event with lots to do and see options	0.824				
A place for events with several tourist attractions	0.816				
A place for events that has good sports and training facilities	0.781				
A place for an event that has enough places to stay	0.762				
A place for events that has many high-quality entertainment	0.755				
A place for an event that is easy to get to and from	0.753				
A place for events with friendly locals and great service	0.731				
A place for an event that gives me the chance to go to a new place	0.712				

(continued)

Table 2 (continued)

Important event, destination and trail attributes endurance trail running participants seek when choosing an event	Factor 1: destination attributes	Factor 2: trail management	Factor 3: trail attributes	Factor 4: event expenses and accessibility	Factor 5: event management
A place for an event that has a city government that supports the hosting of sports events	0.659				
A place for an event that is perfect in terms of weather and height	0.647				
A place for an event that doesn't cost a lot	0.588				
A trail that gives you something to remember (beautiful views or wildlife)		0.763			
A trail with clear signs		0.734			
A path that makes you feel good		0.656			
A trail with a lot of security and safety		0.601			
A path that goes through a variety of landscapes and natural features		0.590			

(continued)

Table 2 (continued)

Important event, destination and trail attributes endurance trail running participants seek when choosing an event	Factor 1: destination attributes	Factor 2: trail management	Factor 3: trail attributes	Factor 4: event expenses and accessibility	Factor 5: event management
An event that is well-planned		0.570			
A trail that is well-kept		0.521			
A trail with rocks, roots, and other obstacles at different points			0.801		
A trail that is very difficult to follow			0.798		
A trail that has a lot of single-track parts			0.743		
A hard path to walk			0.712		
A path with sections that go uphill and downhill			0.689		
The length of the trail's route (km)			0.597		
The affordability of event registration				0.832	
The travel costs associated with participating in the event				0.799	

(continued)

Table 2 (continued)

Important event, destination and trail attributes endurance trail running participants seek when choosing an event	Factor 1: destination attributes	Factor 2: trail management	Factor 3: trail attributes	Factor 4: event expenses and accessibility	Factor 5: event management
The proximity of the event to my home city or town				0.604	
The effectiveness of the registration process				0.598	
An event with excellent logistics					0.750
A well-marketed event					0.689
An event that assists with travel arrangements					0.566
An event that helps me qualify for other events					0.552
Total variance explained (Eigen value > 1	70%				
Average inter-item correlation	0.65	0.53	0.51	0.52	0.49
Reliability coefficient (Cronbach's alpha)	0.95	0.85	0.85	0.78	0.73
Mean value	3.70	4.31	0.80	3.83	3.63

offers a memorable experience (such as beautiful views or wildlife) was seen as an important factor. Thus, a positive recollection of a positive experience or a positive experience in the past can aid a participant in deciding about an event in future [18]. Finally, having a clearly designated trail is crucial, with an average value of (4.20). There haven't been many investigations done along a well-defined path. But similar research by found that hikers value having a well-marked trail [25].

Five qualities of events, venues, and courses that trail runners consider unimportant or neutral while deciding whether to participate are also listed in Table 1. The average score for an event venue that provides multiple, high-quality entertainment options for trail runners was 3.42. Mountain bikers and trail runners were found to place moderate importance on the availability of a wide variety of high-quality entertainment options at the event venue [26]. The presence of multiple tourist attractions at the event venue is the second feature that is not given much weight (3.32). In contrast to the findings of this study, trail runners only gave moderate weight to a host destination's variety of tourist attractions. There is a third occasion that facilitates my trip planning. The ability to make travel plans easier (3.30) is also considered to be of low importance. Both trail runners and mountain bikers found an event that helps with journey planning to be slightly significant [26].

A mountain biking event has a lower mean value of 3.31 because "it helps me prepare for more big events," as was found in a study by [26]. Lastly, participants rated first-rate sporting and training facilities as the least crucial feature of an event's location (mean score: 3.20). Although participants in trail running rated the quality of sport and training facilities as the least important factor in this study, it is a very important predictor of the intention to return to the endurance sport event [18]. High-quality sporting and training facilities may not have been as crucial to trail runners as they were to participants in other endurance sports because trail runners train on the trails themselves [27].

Attributes of the destination scored the second-lowest average. In contrast to the results of this study, very little prior research has demonstrated that exploration of a potential destination serves as a significant predictor of the intention to return to an endurance activity [28]. Similarly, to the current study, previous research [26] on mountain biking and trail running indicated that these activities are so reliant on hard terrain that the idea of destination qualities is only minimally important, and the route aspects have higher relevance. Participants in endurance sports care more about the difficulty and prestige of the event than they do about the weather and setting [6]. Participants' top priority when picking a trail running race was the quality of the trail management. Trail runners place a high value on features like well-marked paths, opportunities to see interesting things along the way (such scenery or wildlife), and a good time overall. Trail quality [including trail design and maintenance, the trail's naturalness, and the trail's memorable scenery [including lakes and birdlife], amongst other factors, has been shown by previous researchers to be an extremely important motivator that can influence a participant's choice of sport event [1].

The mean of trail characteristics was third highest. Researchers have found only a small number of studies that have attempted to quantitatively characterise trail features [23, 29]. However, prior research supports this notion, showing that "trail-related attributes/motivations" (a tough path) are crucial for endurance sport participants [22]. The trail itself is one of the attribute variables that drive trail runners to participate in the event; hence, trail-related features are crucial, according to studies focussed on trail running [23]. The median value for event costs and convenience was second highest. This is consistent with findings from a study [30] showing event prices are a key element to consider when picking an endurance cycle sport event. Accessibility and cost have been determined to be particularly important attribute factors that influence the choices of endurance sport participants to take part in a certain endurance sport event [22, 28]. Further, contrary to the findings of this research, participants in trail running and mountain biking cite ease of access to information about the event (event attributes) as a more crucial attribute element [26].

Due to the recent reopening of trail operators, this study suffers from a low sample size. To improve these results of this study, an interviewer session should be organised. This study's results helped event organisers and destination promoters in Kundasang, Sabah, by illuminating the elements that should be incorporated into trail running events and host destination marketing to cater to the needs of trail runners and foster the long-term success of trail running events in the area. It was also found in this research that path attributes are an important factor for trail runners. Trail running event organisers and promoters, as well as those responsible for promoting the host location, would do well to keep these considerations in mind when designing and promoting their respective events.

References

1. Perić M, Slavic N (2019) Event sport tourism business models: the case of trail running. Sport Bus Manag Int J 9(2):164–184
2. Qiu H, Hsu C, Li M, Shu B (2018) Self-drive tourism attributes: influences on satisfaction and behavioural intention. Asia Pac J Tour Res 23(4):395–407
3. Hudson S (1999) Consumer behavior related to tourism. Con Behav Travel Tour:7–32
4. Robillard J (2014) The ultimate guide to trail running and ultramarathons: expert advice, and some humor, on training, competing, gummy bears, snot rockets, and more. Simon and Schuster
5. Van Vuuren C, Slabbert E (2011) Travel behaviour of tourists to a South African holiday resort. Afr J Phys Health Educ Recreat Dance 17(si-1):694–707
6. Myburgh E (2017) A framework fostering commitment for endurance sport tourism in South Africa. North-West University, Potchefstroom (Thesis—Ph.D.)
7. Perić M, Slavić N (2019) Event sport tourism business models: the case of trail running. Int J Sport Bus Manag 9(2):164–184
8. Dixon AW, Backman S, Backman K, Norman W (2012) Expenditure-based segmentation of sport tourists. J Sports Tour 17(1):5–21
9. Kruger M, Saayman M (2012) Creating a memorable spectator experience at the two oceans. J Sports Tour 17(1):63–77
10. Heyns CM (2010) Student travel behaviour: North-West University. North-West University, Potchefstroom (Thesis—MA)
11. Funk D, Alexandris K, McDonald H (2008) Consumer behaviour in sport and events: marketing action. Routledge, New York, NY

12. Wong IA, Fong LHN, Law R (2016) A longitudinal multilevel model of tourist outbound travel behavior and the dual-cycle model. J Travel Res 55(7):957–970
13. Luenendonk M (2019) Theory of planned behavior: definition, explained, examples. https://www.cleverism.com/theory-of-planned-behavior/. Last accessed 2022/6/21
14. LaMorte WW (2022) The theory of planned behavior. http://sphweb.bumc.bu.edu/otlt/MPHModules/SB/BehavioralChangeTheories/BehavioralChangeTheories3.htm
15. Moghtaderi F, Burke M, Troelsen J (2015) Application of a theoretical framework as a context for a travel behavior change policy intervention. Int J Archit Environ Eng 9(2):607–614
16. Minciu R, Moisă CO (2009) Fundamental aspects regarding youth and their decision to practice tourism. Annales Universitatis Apulensis Series Oeconomica 11(2):1009–1018
17. Juvan E, Omerzel DG, Maravić MU (2017) Tourist behaviour: an overview of models to date. In: Management international conference, Italy
18. Ferreira M (2015) A critical assessment of sport consumption at endurance events in South Africa. North-West University, Potchefstroom (Thesis—Ph.D.)
19. Jordaan T (2021) Attributes contributing to the travel behaviour of trail runners. Doctoral dissertation, North-West University (South Africa)
20. Kaiser HF (1974) An index of factorial simplicity. Psychometrika 39(1):31–36
21. Pallant J (2016) SPSS survival manual: a step-by-step guide to data analysis using IBM SPSS, 6th edn. McGraw Hill Education, Maidenhead, UK
22. Myburgh E, Kruger M, Saayman M (2018) Aspects influencing the commitment of endurance athletes: a tourism perspective. J Sport Tour 22(4):275–301
23. Hodeck A, Kühnast J, Wohlfart O (2018) Comparing two types of nature sport (event) tourists in Germany based on travel motivation and behaviour—the case of ski tourers vs. trail runners. Paper delivered at the 26th European sport management conference managing sport in a changing Europe, Malmo University, Sweden
24. Ferreira M (2019) Role of destination management and destination marketing in tourism. https://pragueeventery.com/role-of-destinationmanagement/
25. Kołodziejczyk K (2019) Networks of hiking tourist trails in the Krkonoše (Czech Republic) and Peneda-Gerês (Portugal) national parks—comparative analysis. J Mt Sci 16(4):725–743
26. Getz D, McConnell A (2014) Comparing trail runners and mountain bikers: motivation, involvement, portfolios, and event-tourist careers. J Conv Event Tour 15(1):69–100
27. Hadfield J (2021) How to train for your first trail race. https://www.jennyhadfield.com/how-to-train-for-your-first-trail-race/(2021). Last accessed 2022/6/16
28. Aicher TJ, Newland BL (2018) Exploring sport participants' event and destination choices. J Sport Tour 22(2):131–149
29. Atlas R, Nuraini Putit PA, Puem LBG, Enggong TS (2018) Sports tourism: factors influencing runners joining marathon events. Int J Acad Res Bus Soc Sci 8(16):218–230
30. Fotiadis A, Vassiliadis C, Yeh SP (2016) Participant's preferences for small-scale sporting events: a comparative analysis of a Greek and a Taiwanese cycling event. Euro Med J Bus 11(2):232–247
31. Boit J, Doh M (2014) The role of destination attributes and visitor satisfaction on tourist repeat visit intentions: the case of Lake Nakuru National Park, Kenya. 26th annual North-Eastern recreation research symposium of New York, Cooperstown

The Influence of Physical Activity Motivations Upon Different Stages of Change Among Malaysian Type 2 Diabetes Patients

Aizuddin Hidrus, Yee Cheng Kueh, Ke Zhou, Garry Kuan, and Syed Sharizman Syed Abdul Rahim

Abstract Motivation has been reported to play a significant effect in physical activity (PA) or exercise engagement on numerous occasions. Different stages of behaviour change (SOC) (i.e., pre-contemplation, contemplation, preparation, action, maintenance, and relapse) may attribute to different levels of motivation among type-2 diabetes mellitus (T2DM) patients. Thus, this study aimed to explore the different motivations toward PA across each SOC among Malaysian T2DM patients. A combination of cross-sectional study design and purposive sampling method was adopted in recruiting the participants from Hospital Universiti Sains Malaysia (HUMS). A total of 331 participants (172 males and 159 females) with a mean (SD) age of 62.64 (10.29) completed the study questionnaires, which comprised the 40-item Physical Activity and Leisure Motivation Scale (PALMS) and the 6-item SOC Scale. Multivariate analysis of variance (MANOVA) was used to analyse the data. The results showed a significant difference in the mean score of eight motives between the six stages of change [$F(\text{df}) = 3.805\ (40, 1610), p < 0.001$]. Mean scores for the competition (15.78), affiliation (17.11), mastery (17.89), and enjoyment were found highest at the maintenance stage. However others' expectation (16.50), physical condition (22.50), and psychological conditions (21.00) were found highest at the stage of action. The motive of appearance (18.43) alone was found highest at the relapse

A. Hidrus (✉) · S. S. Syed Abdul Rahim
Department of Public Health Medicine, Faculty of Medicine and Health Sciences, Universiti Malaysia Sabah, Kota Kinabalu, Sabah, Malaysia
e-mail: aizuddin88@ums.edu.my

Y. C. Kueh
Unit of Biostatistics and Research Methodology, School of Medical Sciences, Universiti Sains Malaysia, Kubang Kerian, Kelantan, Malaysia

K. Zhou
School of Physical Education, Sports Reform and Development Research Center of Henan University, Henan University, Kaifeng, China

G. Kuan
Exercise and Sports Science, School of Health Sciences, Universiti Sains Malaysia, Kubang Kerian, Kelantan, Malaysia

Department of Life Sciences, Brunel University, London UB8 3PH, UK

G. Kuan et al. (eds.), *Advancing Sports and Exercise via Innovation*, Lecture Notes in Bioengineering, https://doi.org/10.1007/978-981-19-8159-3_45

stage. Hence, we conclude that different stages of change in exercise would have different motives for physical activity for T2DM patients. The motives of PA may influence an individual with T2DM to either maintain their new active lifestyle or revert to their previous sedentary behaviours.

Keywords Motivation · Stages of change · Physical activity · Type-2 diabetes · MANOVA

1 Introduction

The ability of physical activity (PA) and exercise to improve the health status of patients with any disease, including diabetes, has been empirically recognised by physicians and academics for many years. PA can achieve the following objectives for diabetic patients: improved glycaemic control, decreased insulin resistance, improved lipid profiles, decreased blood pressure (BP), and maintenance of weight loss. [1, 2]. Moreover, significant decreased morbidity and death are linked to moderate to vigorous levels of PA and cardiorespiratory fitness in diabetics [3]. Through a systematic review and meta-analysis of the therapeutic effect of exercise on glycaemic control, Shah et al. [4] concluded that exercise is crucial for T2DM patients' glycaemic control optimization as well as for enhancing their quality of life (QoL), BMI, and waist circumference [4]. In addition, PA intervention was found to influence other psychological constructs such as exercise self-efficacy and decisional balance towards a healthy active lifestyle among T2DM patients [5, 6].

Motivation has frequently been cited as playing a key influence in physical/sports involvement [7]. It is a crucial component of the psychological process by which individuals decide whether or not to participate in physical activities. The same is true for exercise and physical activity, where motivation plays a significant role in both increasing and maintaining participation [8, 9]. It has been demonstrated that different sorts of motivation influence people's efforts during exercise sessions and their intentions to continue exercising [10]. The correlation between PA and motivation was also proven as Hidrus et al. [11] found that fun and flexible PA intervention could improve T2DM patients' motivation towards active habits [11].

The Physical Activity and Leisure Motivations Scale (PALMS) was developed by Rogers and Morris [12] to measure an individual's motivation for PA [12]. Originally comprised of 73 questions and known as the Recreational Exercise Motivation Measure (REMM), it was reduced to 40 elements and renamed the PALMS. In the version designed for adults, these 40 things are divided into eight distinct aspects or motivations: enjoyment, mastery, competition/ego, attractiveness, affiliation, others' expectations, psychological condition, and physical condition. The adult version of the PALMS was translated into Malay and validated by confirmatory factor analysis (CFA), resulting in the creation of the PALMS-M [13].

Stages of change (SOC) is an integrated model of cognitive, affective, and behavioural mechanisms articulated in other well-established theories of deliberate change [14]. It is also the organising structure of the transtheoretical model (TTM) and the theory of cyclical development and recession [15]. From a different perspective, it is an accepted and implemented component of the TTM that provides a framework for the relationships between specific cognitive and motivational mechanisms and health behaviours, including smoking cessation [16]. Marcus et al. [17] created the initial version of the SOC scale, which consists of five stages: pre-contemplation, contemplation, preparation, action, and maintenance. Dishman et al. [18] are one of the study groups that have applied the SOC scale.

In addition, another version of the SOC scale is reconstructed with six distinct stages, with the sixth stage being either relapse [15] or termination [19, 20]. Termination is the stage at which individuals with 100% self-efficacy have no desire to return to an inactive lifestyle [21] and have been routinely active for more than five years [20]. While relapse is the stage in which individuals may revert to their previous sedentary lifestyle [15]. According to Prochaska and Prochaska [22], the termination stage is deemed unattainable for the majority of persons. Consequently, Middelkamp [15] identified the sixth stage of SOC as relapse with the description that individuals may retain their active behaviour but may also revert to their former behaviour and return to the earlier stage of SOC [15].

Changing an unhealthy lifestyle into a healthy lifestyle is not an easy task for patients [23]. Only motivated patients may benefit from behaviour therapy, and sex and age differences in desire for diets and exercise may affect treatment outcomes as the key factor in behaviour change is the personal motivation for change [24]. Motivation and SOC are correlated with each other especially when it comes to a healthier lifestyle. To support the change in all domains of the lifestyle intervention, motivation should be maintained throughout all lifestyle interventions [25]. Recent research on type 2 diabetes patients found that the phases of change were required to modify behaviours toward a balanced diet and regular physical activity, and that patients had difficulty identifying the need to enhance physical activity rather than eating patterns. For a high proportion of individuals, motivation to change remained a challenge [25, 26]. Furthermore, the substantial variability of response to lifestyle intervention emphasises the need of investigating patterns of change in PA and healthy nutrition [27]. Hence, this study aimed to explore the different motivations towards PA across each SOC among Malaysian T2DM patients. Output yielded from this study could be beneficial for future intervention-based studies as a guide to developing a new intervention programme that is specific for different SOC with different motivations.

2 Methods

2.1 Study Design and Recruitment

Through the dissemination of a questionnaire to type 2 diabetes mellitus (T2DM) patients at Hospital Universiti Sains Malaysia, a combination of cross-sectional study design and purposive sampling was implemented (HUSM). Patients that were approached were screened for eligibility. Those who were eligible were invited to voluntarily participate in the study. The study process was explained to the participants, and written consent was obtained. Outpatients' clinics and inpatients' wards were selected in order to collect a study sample of T2DM patients from various age groups, genders, jobs, other sociodemographic characteristics, and disease severity levels.

2.2 Eligible Participants

This study comprised Malaysians (18 years and older) who were clinically diagnosed with T2DM, could read and comprehend Bahasa Malaysia, and could comprehend the material presented by the researcher. However, those who were found diagnosed with any mental disorders were excluded. At the end of recruitment, a total of 331 T2DM patients voluntarily participated in this study.

2.3 Data Collection and Analysis

Data collection was carried out by a self-administered questionnaire answered by the recruited participants. The questionnaire of this study consisted of three sections: Section A: Socio-demographic with twelve questions, Section B: SOC with one question (six stages), and Section C: PALM scale with forty questions. The questionnaire was all in the Malay language.

There were two strategies to approach the T2DM patients. The first was to immediately approach the individuals who were waiting for their health review number to be called by the doctor. They were briefed on the current study and asked if they were interested in participating. Posters affixed to the signboards of T2DM clinics and wards constituted the second method. On the poster was a brief explanation of the study and the contact information for the researcher. Those who were interested, willing, and eligible to participate were then briefed on the study and needed to sign an informed consent letter before completing the questionnaire. Ten to fifteen minutes are required to complete the questions.

Collected data were entered into Microsoft Excel and then transferred into IBM SPSS version 26.0. At first, data cleaning and preparation were performed to check for any missing data before preliminary analysis was done. Descriptive analyses were then carried out for the data normality test and to obtain data descriptive output. Multivariate analysis of variance (MANOVA) was chosen to be the inferential statistical analysis for this study and performed to determine any differences in motivations for PA in each stage of behaviour change among T2DM patients.

2.4 *Measurement Tools*

2.4.1 Stages of Change (SOC) Scale

According to Marcus et al. [28], SOC is a questionnaire with a single question and six levels of response. On the basis of the most recent stage of exercise behaviour, a dichotomous scale (yes/no) was used to evaluate it. Individuals are grouped according to the ordinal level of SOC behaviour change (pre-contemplation, contemplation, preparation, action, maintenance, and relapse). The SOC has been translated into Malay, and its content-based validity is satisfactory [28].

2.4.2 Physical Activity and Leisure Motivation Scale (PALMS)

The PALMS measures motives for participating in PA and leisure [12]. The PALMS questionnaire used in this study consisted of 40 items measuring the aspects of competition, appearance, others' expectations, affiliation, physical condition, psychological condition, mastery, and enjoyment. Each factor consisted of 5 items, all scored on a 5-point Likert scale rated from 1 (*strongly disagree*) to 5 (*strongly agree*). The PALMS has shown good reliability with 4-week test–retest correlations ranging from 0.78 to 0.94 [29]. Internal consistency measured by Cronbach's alpha for all eight factors was good, ranging from 0.78 to 0.82. Molanorouzi et al. [29] examined the validity of PALMS using CFA and reported that the eight factors in the 40-item scale had acceptable goodness of fit [29]. The Malay version of PALMS (PALMS-M) was used in this study.

The validity and reliability of the PALMS-M were found satisfactory when tested among 634 Malaysian undergraduate students [13]. The validation was based on CFA and the majority of the fit indices achieved the threshold fit values: RMSEA = 0.041 (90% CI: 0.038, 0.044), CI fit RMSEA = > 0.950, CFI = 0.911, TLI = 0.901, SRMR = 0.052). The reliability based on composite reliability ranged from 0.648 to 0.846.

2.5 *Ethical Consideration*

Prior to the commencement of this investigation, the Human Research Ethics Committee USM (HRECU) was applied for approval. The HRECU got ethical approval on July 5, 2018 (USM/JEPeM/18040201) and conducted the research in accordance with the Declaration of Helsinki. After receiving ethical approval, we requested authorization from the Director of HUSM to gather data. After receiving approval from HRECU and the Director of HUSM, we commenced data gathering.

3 Results

A total of 331 participants were involved in this study and most of them were male ($n = 172$: 52%) with a mean (SD) age of 62.64 (10.29). Most of the participants were Malay ($n = 296$: 89.4%), with secondary school education background ($n = 158$: 47.7%) being the highest frequency and 139 (42%) were pensioners among them. Details of participants' demographic data are displayed in Table 1.

For the inferential analysis, MANOVA was performed to determine the differences in PA motives for each stage of behaviour change. From the analysis output, the overall MANOVA shows a significant difference in means of motives for PA (competition, appearance, others' expectations, affiliation, physical condition, psychological condition, mastery, and enjoyment), Pillai's Trace = 0.432, F(df) = 3.805 (40, 1610), $p < 0.001$.

The univariate ANOVA shows that there is a significant difference in the mean for all motives for PA (competition, appearance, others' expectations, affiliation, physical condition, psychological condition, mastery, and enjoyment) between the six SOC. More details of univariate ANOVA results and pairwise comparisons between the six SOC are displayed in Tables 2 and 3.

4 Discussion

Based on the results, it was noticed that both multivariate and univariate analyses presented significant differences in motives for PA across different SOC. At the maintenance stage, the mean score for competition (15.78), affiliation (17.11), mastery (17.89), and enjoyment were found highest. However others' expectation (16.50), physical condition (22.50), and psychological conditions (21.00) were found highest at the stage of action. The motive of appearance (18.43) alone was found highest at the relapse stage. The results indicate that each level of SOC has different motives for PA depending on the condition and state of the T2DM patients.

In this study, the majority of T2DM patients were in the planning stage of SOC, with 39.9% wanting to become more physically active. This result was nearly equivalent to the findings of Holmen et al. [30], who examined the level of SOC about PA performance and dietary intake habits among T2DM patients in Norway. Holmen et al. [30] discovered that the majority of study participants (58%) were in the pre-action phase of conducting PA. In addition, the results of a study conducted on a Malaysian population with health issues revealed that 82.9% of participants were in the pre-action phase for weight loss [31]. Thus, the results of this study on SOC among Malaysians with T2DM are consistent with prior research focusing on PA performance and health issues.

Comparing this study's results with a study done by Centis et al. [24], there is a slight difference. Centis et al. [24] aimed to determine the stage of change and associated psychological factors as a prerequisite to refining strategies to implement

Table 1 Demographic characteristics of T2DM patients in HUSM ($n = 331$)

Characteristics	Frequencies	Percentage (%)	Mean (SD)
Gender			
Male	172	52.0	
Female	159	48.0	
Age			62.64 (10.29)
Age group			
21–40 years old	9	2.7	
41–60 years old	124	37.5	
61–80 years old	189	57.1	
81–100 years old	9	2.7	
Ethnicity			
Malay	296	89.4	
Chinese	25	7.6	
Indian	6	1.8	
Others	4	1.2	
Education background			
Primary	87	26.3	
Secondary	158	47.7	
Diploma	60	18.1	
Bachelor's degree	26	7.9	
Occupation			
Working/business	92	27.9	
Pensioners	139	42.0	
Not working/housewife	100	30.1	
Diabetic period			
Less than 5 years	38	11.5	
$\geq$ 5 years	32	9.7	
$\geq$ 10 years	83	25.1	
$\geq$ 20 years	178	53.8	
BMI			27.48 (4.33)
BMI group			
Underweight	8	2.4	
Healthy weight	88	26.6	
Overweight	143	43.2	
Obese	92	27.8	
Stages of change			
Precontemplation	100	30.2	

(continued)

Table 1 (continued)

Characteristics	Frequencies	Percentage (%)	Mean (SD)
Contemplation	72	21.8	
Preparation	132	39.9	
Action	2	0.6	
Maintenance	18	5.4	
Relapse	7	2.1	

behaviour changes among non-alcoholic fatty liver patients. In contrast with this study, Centri et al. [24] found that most of the participants were in between the pre-contemplation and contemplation stages (over 50%) for habitual physical activity. Moreover, the mean score for habitual PA of the participants was highest between the contemplation and preparation stages, which is in contrast to this study where the mean score for most of the motivation for PA was found highest in the action and maintenance stages.

We anticipated that with the right and well-organised intervention, the highest percentage will be changed from preparation to action or maintenance among those participants as has been proven by Buratta et al. [25]. The study done by Buratta et al. [25] was among the participants of 100 overweight/obese outpatients who were found to show changes in their SOC before and after the intervention. Following the intervention, patients remained committed but intensified their action towards change, demonstrating a better degree of maintenance and habit acquisition, particularly in the physical domain of the new lifestyle [25].

Results also showed that there is no significant difference between action compared to maintenance and relapse stages for all eight motives for PA. Similar goes to the comparison of maintenance and relapse stages. This could be due to the possibility of where participants in the action and maintenance stages are having the same motivation level as in both stages they are performing PA regularly. While for relapse, this could be that participants in this stage are still motivated to perform PA regularly, but some constraints and hindrances such as time issues, logistics problems, and other morbidity complications change them into the relapse stage.

As for the MANOVA results of this study, they display an almost identical multivariate result to the study done by Wininger [32]. One of Wininger's [32] study objectives were to assess the differentiation among groups by employing the Stages of Change for engaging in regular exercise and using the subscales from the exercise motivation scale (EMS). The multivariate test result of the study showed there is a significant difference [$F(32, 484.70) = 3.31, p < 0.001$] in motivations across SOC. Upon ANOVA test checking, similar to this study, most of the motivations from EMF showed significant differences ($p < 0.001$) between each SOC except for the external subscale ($p = 0.274$).

Another study was done in Mexico displayed the same MANOVA result as this study. Zamarripa et al. [33] were to analyse variations in behavioural/motivational regulations using the SOC model with the adoption of Behavioural Regulation in

Table 2 Univariate ANOVA of motives for PA between the six SOC

Variables	Adjusted mean (95% CI)	*F* statistics (df)	*p*-value*
Competition		17.016 (5, 325)	< 0.001*
Pre-contemplation	10.510 (9.904, 11.116)		
Contemplation	12.431 (11.717, 13.144)		
Preparation	13.606 (13.079, 14.133)		
Action	14.500 (10.217, 18.783)		
Maintenance	15.778 (14.350, 17.205)		
Relapse	15.143 (12.854, 17.432)		
Appearance		14.253 (5, 325)	< 0.001*
Pre-contemplation	11.590 (10.835, 12.345)		
Contemplation	14.208 (13.318, 15.098)		
Preparation	14.803 (14.146, 15.460)		
Action	17.500 (12.160, 22.840)		
Maintenance	17.500 (15.720, 19.280)		
Relapse	18.429 (15.574, 21.283)		
Others' expectation		7.411 (5, 325)	< 0.001*
Pre-contemplation	12.040 (11.610, 12.470)		
Contemplation	12.361 (11.854, 12.868)		
Preparation	12.705 (12.330, 13.079)		
Action	16.500 (13.460, 19.540)		
Maintenance	14.833 (13.820, 15.847)		
Relapse	14.286 (12.661, 15.911)		
Affiliation		10.766 (5, 325)	< 0.001*
Pre-contemplation	11.850 (11.160, 12.540)		
Contemplation	14.208 (13.396, 15.021)		
Preparation	14.371 (13.771, 14.972)		
Action	13.500 (8.623, 18.377)		
Maintenance	17.111 (15.485, 18.737)		
Relapse	15.857 (13.250, 18.464)		
Physical condition		11.538 (5, 325)	< 0.001*
Pre-contemplation	15.540 (14.843, 16.237)		
Contemplation	16.597 (15.775, 17.419)		
Preparation	17.621 (17.014, 18.228)		
Action	22.500 (17.568, 27.432)		
Maintenance	21.333 (19.689, 22.977)		
Relapse	19.714 (17.078, 22.351)		
Psychological condition		15.770 (5, 325)	< 0.001*

(continued)

Table 2 (continued)

Variables	Adjusted mean (95% CI)	*F* statistics (df)	*p*-value*
Pre-contemplation	12.920 (12.217, 13.623)		
Contemplation	14.153 (13.324, 14.981)		
Preparation	14.909 (14.297, 15.521)		
Action	21.000 (16.030, 25.970)		
Maintenance	19.722 (18.066, 21.379)		
Relapse	19.143 (16.486, 21.799)		
Mastery		18.377 (5, 325)	< 0.001*
Pre-contemplation	11.670 (11.017, 12.323)		
Contemplation	14.236 (13.467, 15.005)		
Preparation	14.909 (14.341, 15.477)		
Action	14.500 (9.886, 19.114)		
Maintenance	17.889 (16.351, 19.427)		
Relapse	17.429 (14.962, 19.895)		
Enjoyment		17.770 (5, 325)	< 0.001*
Pre-contemplation	12.400 (11.723, 13.077)		
Contemplation	13.931 (13.133, 14.728)		
Preparation	15.015 (14.426, 15.604)		
Action	18.000 (13.215, 22.785)		
Maintenance	19.556 (17.960, 21.151)		
Relapse	17.857 (15.299, 20.415)		

One-way MANOVA: Pillai's Trace = 0.432, $F(\text{df}) = 3.805$ (40, 1610), $p < 0.001$
*One-way ANOVA was performed indicating there is at least one significant mean difference between SOC

Exercise Questionnaire-3 (BREQ-3) after the validation of it was carried out. The MANOVA multivariate test result indicated that there is a significant difference ($F = 13.47$, $p < 0.001$) in behavioural/motivational regulations across SOC [33]. In addition, most of the behavioural/motivational regulations of BREQ-3 found significant differences between each SOC except for the introjected regulation after the ANOVA test checking was performed.

5 Conclusion

Descriptive analysis and MANOVA results yielded from this study present a significant *p*-values as an indication of the existence of differences in motives for PA for each level of SOC. In comparison with the previous studies, there are slight differences in frequencies and percentages of descriptive results. Yet, there are previous

Table 3 Summary of comparison of SOC for the eight motives of PA in PALMS (post Hoc test)

Comparison of SOC	Competition	Appearance	Others' expectation	Affiliation	Physical condition	Psychological condition	Mastery	Enjoyment
	p-value	*p*-value	*p*-value	*p*-value	*p*-value	*p*-value	*p*-value	*p*-value
First–second	0.001*	< 0.001*	1.000	< 0.001*	0.818	0.394	< 0.001*	0.064
First–third	< 0.001*	< 0.001*	0.337	< 0.001*	< 0.001*	0.001*	< 0.001*	< 0.001*
First–fourth	1.000	0.478	0.068	1.000	0.095	0.025*	1.000	0.349
First–fifth	< 0.001*	< 0.001*	< 0.001*	< 0.001*	< 0.001*	< 0.001*	< 0.001*	< 0.001*
First–sixth	0.002*	< 0.001*	0.135	0.056	0.042*	< 0.001*	< 0.001*	0.001*
Second–third	0.144	1.000	1.000	1.000	0.743	1.000	1.000	0.482
Second–fourth	1.000	1.000	0.130	1.000	0.312	0.118	1.000	1.000
Second–fifth	0.001*	0.019*	< 0.001*	0.027*	< 0.001*	< 0.001*	0.001*	< 0.001*
Second–sixth	0.401	0.087	0.402	1.000	0.406	0.007*	0.234	0.063
Third–fourth	1.000	1.000	0.230	1.000	0.814	0.259	1.000	1.000
Third–fifth	0.079	0.082	0.002*	0.031*	0.001*	< 0.001*	0.006*	< 0.001*
Third–sixth	1.000	0.231	0.946	1.000	1.000	0.036*	0.766	0.509
Fourth fifth	1.000	1.000	1.000	1.000	1.000	1.000	1.000	1.000
Fourth–sixth	1.000	1.000	1.000	1.000	1.000	1.000	1.000	1.000
Fifth–sixth	1.000	1.000	1.000	1.000	1.000	1.000	1.000	1.000

First = Precontemplation, Second = Contemplation, Third = Preparation, Fourth = Action, Fifth = Maintenance, and Sixth = Relapse. *Significant p-value indicating there is a significant mean difference between the two SOC studies that showed identical results of MANOVA for inferential analysis. Hence, we conclude that different stages of change in exercise would have different motives for physical activity for T2DM patients. The motives of PA may influence an individual with T2DM to either maintain their new active lifestyle or revert to their previous sedentary behaviours.

Acknowledgements We want to thank all participants with T2DM in HUSM who volunteered and participated in the present study. We also want to convey our sincere gratitude to the staff in HUSM for their support and co-operation during the data collection.

The authors would like to acknowledge Ming-Kai Chin for his guidance and for providing the Brain breaks platform, under the GCH and HOPSports® Inc. This research was supported by the Ministry of Higher Education Malaysia for Fundamental Research Grant Scheme (FRGS) with Project Code: FRGS/1/2020/SKK06/USM/03/13.

Conflicts of Interest The authors declare no conflict of interest.

Funding Exercise Medicine Research Grant (EMRG/2019/19004).

References

1. Colberg SR, Sigal RJ, Yardley JE, Riddell MC, Dunstan DW, Dempsey PC, Horton ES, Castorino K, Tate DF (2016) Physical activity/exercise and diabetes: a position statement of the American Diabetes Association. Diabetes Care 39:2065–2079
2. Chudyk A, Petrella RJ (2011) Effects of exercise on cardiovascular risk factors in type 2 diabetes: a meta-analysis. Diabetes Care 34:1228–1237
3. Sigal RJ, Armstrong MJ, Bacon SL, Boulé NG, Dasgupta K, Kenny GP, Riddell MC (2018) Physical activity and diabetes. Can J Diabetes 42:S54–S63. https://doi.org/10.1016/j.jcjd.2017.10.008
4. Shah SZA, Karam JA, Zeb A, Ullah R, Shah A, Haq IU, Ali I, Darain H, Chen H (2021) Movement is improvement: the therapeutic effects of exercise and general physical activity on glycemic control in patients with type 2 diabetes mellitus: a systematic review and meta-analysis of randomized controlled trials. Diabetes Therapy 12:707–732. https://doi.org/10.1007/s13300-021-01005-1
5. Hidrus A, Kueh YC, Norsa'adah B, Chang Y-K, Kuan G (2022) Effects of technology-supported brain breaks videos on exercise self-efficacy among type 2 diabetes mellitus Malaysians. Sci Rep 12:11651. https://doi.org/10.1038/s41598-022-15142-5
6. Hidrus A, Kueh YC, Norsa'adah B, Chang YK, Kuan G (2021) Effects of brain breaks video intervention of decisional balance among Malaysians with type 2 diabetes mellitus: a randomised controlled trial. Int J Environ Res Publ Health 18.https://doi.org/10.3390/ijerph18178972
7. Gould D, Dieffenbach K, Moffett A (2002) Psychological characteristics and their development in olympic champions. J Appl Sport Psychol 14:172–204. https://doi.org/10.1080/10413200290103482
8. Aaltonen S, Leskinen T, Morris T, Alen M, Kaprio J, Liukkonen J, Kujala U (2012) Motives for and barriers to physical activity in twin pairs discordant for leisure time physical activity for 30 years. Int J Sports Med 33:157–163. https://doi.org/10.1055/s-0031-1287848
9. André N, Dishman RK (2012) Evidence for the construct validity of self-motivation as a correlate of exercise adherence in French older adults. J Aging Phys Act 20:231–245. https://doi.org/10.1123/japa.20.2.231

10. Wilson PM, Rodgers WM, Fraser SN, Murray TC (2004) Relationships between exercise regulations and motivational consequences in university students. Res Q Exerc Sport 75:81–91. https://doi.org/10.1080/02701367.2004.10609136
11. Hidrus A, Kueh YC, Norsaádah B, Chang YK, Hung TM, Naing NN, Kuan G (2020) Effects of brain breaks videos on the motives for the physical activity of Malaysians with type-2 diabetes mellitus. Int J Environ Res Publ Health 17.https://doi.org/10.3390/ijerph17072507
12. Rogers H, Morris T (2003) New approaches to exercise and sport psychology: theories, methods and applications. In: An overview of the development and validation of the recreational exercise motivation measure (REMM), XIth European congress of sport psychology, Copenhagen, Denmark, pp 22–27
13. Kueh YC, Kuan G, Morris T (2019) The physical activity and leisure motivation scale: a confirmatory study of the Malay language version. Int J Sport Exerc Psychol 17:250–265. https://doi.org/10.1080/1612197X.2017.1321029
14. Floyd J, Zebrowski PM, Flamme GA (2007) Stages of change and stuttering: a preliminary view. J Fluency Disord 32:95–120. https://doi.org/10.1016/j.jfludis.2007.03.001
15. Middelkamp J, van Rooijen M, Wolfhagen P, Steenbergen B (2017) Effects of a self-efficacy intervention on exercise behavior of fitness club members in 52 weeks and long-term relationships of transtheoretical model constructs
16. Keller CS, Mcgowan N (2001) Examination of the processes of change, decisional balance, self-efficacy for smoking and the stages of change in Mexican American women
17. Marcus BH, Rossi JS, Selby VC, Niaura RS, Abrams DB (1992) The stages and processes of exercise adoption and maintenance in a worksite sample. Health Psychol 11:386
18. Dishman RK, Vandenberg RJ, Motl RW, Nigg CR (2010) Using constructs of the transtheoretical model to predict classes of change in regular physical activity: a multi-ethnic longitudinal cohort study. Ann Behav Med 40:150–163. https://doi.org/10.1007/s12160-010-9196-2
19. Fallon EA, Hausenblas HA, Nigg CR (2005) The transtheoretical model and exercise adherence: examining construct associations in later stages of change. Psychol Sport Exerc 6:629–641. https://doi.org/10.1016/j.psychsport.2005.01.003
20. Horiuchi S, Tsuda A, Watanabe Y, Fukamachi S, Samejima S (2013) Validity of the six stages of change for exercise. J Health Psychol 18:518–527
21. Prochaska JO, Velicer WF (1997) The transtheoretical model of health behavior change. Am J Health Promotion 12
22. Prochaska JO, Prochaska JM (2019) Transtheoretical model. In: Lifestyle medicine. CRC Press, pp 219–228
23. Bellentani S, Dalle Grave R, Suppini A, Marchesini G, Bedogni G, Bugianesi E, Crocè LS, Gastaldelli A, Marra F, Perseghin G, Svegliati-Baroni G, Tell G, Tiribelli C (2008) Behavior therapy for nonalcoholic fatty liver disease: the need for a multidisciplinary approach. Hepatology 47:746–754
24. Centis E, Moscatiello S, Bugianesi E, Bellentani S, Fracanzani AL, Calugi S, Petta S, Dalle Grave R, Marchesini G (2013) Stage of change and motivation to healthier lifestyle in non-alcoholic fatty liver disease. J Hepatol 58:771–777
25. Buratta L, Reginato E, Ranucci C, Pippi R, Aiello C, Sbroma Tomaro E, Perrone C, Tirimagni A, Russo A, de Feo P, Mazzeschi C (2016) Stage of change and motivation to a healthier lifestyle before and after an intensive lifestyle intervention. J Obes 2016.https://doi.org/10.1155/2016/6421265
26. Centis E, Trento M, Dei Cas A, Pontiroli AE, de Feo P, Bruno A, Sasdelli AS, Arturi F, Strollo F, Vigili de' Kreutzenberg S, Invitti C, di Bonito P, di Mauro M, Pugliese G, Molteni A, Marchesini G (2014) Stage of change and motivation to healthy diet and habitual physical activity in type 2 diabetes. Acta Diabetol 51:559–566.https://doi.org/10.1007/s00592-013-0551-1
27. Yank V, Xiao L, Wilson SR, Stafford RS, Rosas LG, Ma J (2014) Short-term weight loss patterns, baseline predictors, and longer-term follow-up within a randomized controlled trial. Obesity 22:45–51. https://doi.org/10.1002/oby.20510
28. Kuan G (2019) Test of relationship between the transtheoretical model constructs and physical activity in Malaysian college students. Int J Human Move Sci 13:25–36

29. Molanorouzi K, Khoo S, Morris T (2014) Validating the physical activity and leisure motivation scale (PALMS)
30. Holmen H, Wahl A, Torbjørnsen A, Jenum AK, Småstuen MC, Ribu L (2016) Stages of change for physical activity and dietary habits in persons with type 2 diabetes included in a mobile health intervention: the Norwegian study in RENEWING HEALTH. BMJ Open Diabetes Res Care 4:193. https://doi.org/10.1136/bmjdrc-2016
31. Chang CT (2007) Applicability of the stages of change and weight efficacy lifestyle questionnaire with natives of Sarawak, Malaysia. Rural Remote Health 7:1–12
32. Wininger SR (2007) Self-determination theory and exercise behavior: an examination of the psychometric properties of the exercise motivation scale. J Appl Sport Psychol 19:471–486. https://doi.org/10.1080/10413200701601466
33. Zamarripa J, Castillo I, Baños R, Delgado M, Álvarez O (2018) Motivational regulations across the stages of change for exercise in the general population of Monterrey (Mexico). Front Psychol 9. https://doi.org/10.3389/fpsyg.2018.02368

Tennis Player's Coping Strategies at Duta International Tennis Academy During Their Different Career Phases: A Narrative Review

Yidan Wang, Le Ting Tan, and Garry Kuan

Abstract Sport athletes are confronted with demanding situations that necessitate specific coping skills to ensure success and high performance in their careers. This study aimed to identify the coping strategies utilised by tennis players from Duta International Tennis Academy during their different career phases. Literature was conducted using databases from Scopus, PubMed, ResearchGate, Science Direct, and ITF. Google Scholar was used as the primary search engine to gain a deeper understanding of the topic. The published articles used were from January, 1984 to April, 2022. A total of twenty-four studies were identified for this study. They revealed that tennis players employed physical, psychological, emotional, cognitive, social, and planning as their coping strategies, in different phases of their careers. Future studies should investigate additional studies on tennis especially amongst the Malaysian players and their coping strategies. In addition, different strategies such as how the tennis players cope with injury, health issues, and their mental toughness in sports should be considered.

Keywords Tennis players · Coping strategies · Athlete · Sport · Malaysia

1 Introduction

Research has shown that athletes are becoming conscious of their performances, so there is an increased need to administer coping strategies that will help to lower their stress and arousal levels before and during the game. Coping is a critical process that entails self-regulation and the volition of thoughts and actions to manage psychologically and physically demanding situations. It involves successful adaptation to the sports setting through emotional, cognitive, and behavioural skills [1]. This study

Y. Wang (✉)
Brunel Business School, Brunel University, London, UK
e-mail: ydanwang@icloud.com

Y. Wang · L. T. Tan · G. Kuan
Exercise and Sports Science Programme, School of Health Sciences, Universiti Sains Malaysia, Kubang Kerian, Kelantan, Malaysia

G. Kuan et al. (eds.), *Advancing Sports and Exercise via Innovation*, Lecture Notes in Bioengineering, https://doi.org/10.1007/978-981-19-8159-3_46

explores documented literature from various sources that helps conceptualise coping in sports and the tennis game. The study uses quantitative, qualitative, and combined literature from past studies, which will be reviewed as per the PRISMA guideline. The literature will take a thematic direction on players' identity, their aggression, and how they cope in a demanding sport like tennis. The literature will examine tennis players' psychological, physical, emotional, and social aspects and the corresponding coping strategies. According to Crocker et al. [1], there are many coping strategies in sport, including avoidance, social support, increased effort, changing tactics, confrontation, arousal control, relaxation, planning, and wishful thinking. This study will discuss the findings and provide recommendations and suggestions for future studies.

1.1 Tennis as a Sport

Tennis is a game that involves two or four players as a single or double opponent. The players use a tennis racket to strike the hollow rubber ball over a stretched net across the court. When a player strikes the ball over the net into the opponent's and the opponent fails to return the ball, a point is won. The scores are classified as 15, 30, and 40, and the winner must win four points with a margin of two points. The game was invented many years ago, as early as 1873, and the English people first played it as a crude parlour pastime game. Previously called lawn tennis, the game was invented by Major Walter Wingfield [2]. Later, the Englishman invented a celluloid ball and pebbled rubber cloth, which surfaced the racket blade. The game became popular after World War I attracting players globally, which led to the formation of an international governing body, Table Tennis Federation (TTF), to oversee the worldwide competitions. TTF adopted uniform rules and developed specifications for standardised equipment, and tennis became more popular [3]. Ever since, many countries have adopted tennis. Millions play the game, it is broadcasted on television and other media, and it has gone through tremendous transformations. It is currently governed by the International Tennis Federation (ITF), which oversees tennis administration, development, structure, regulations, competitions, and promotions. The rules of the game require the players not to touch the post and nets, not to cross to the opponent's side, not to hit the ball twice, not to carry the ball with the ratchet or touch it, and to wait until the ball passes the net before they return it [4]. Tennis is the only trailing soccer, and it has become the second most popular sport after football, attracting eighty-seven million fans from 19 countries [5].

2 Tennis Players in Malaysia

The Lawn Tennis Association of Malaysia (LTAM) commonly referred to as tennis Malaysia governs tennis in Malaysia. It is a member of ITF, which Mirzan Bin

Mahathir leads. The association develops and promotes nationwide tennis, and it works closely with the Ministry of Education's State Sports Officer. Its mission is to maintain a world-class tennis platform through the comprehensive development of youth and professional athletes. LTAM regulates fourteen affiliated state tennis associations. It functions as a whole sports association that ensures that necessary activities and programs for high performance and talent development are maintained. LTAM adopts strategies for talent identification, financial support, training and requirements, and competitive structure [6]. Malaysian tennis players are categorised into male and female players. Malaysia has several tennis academies in, such as DUTA, JJ academy in Kuala Lampur, Ace in Petaling Jaya, NSTE, PJ Union, and Tennis Town Academy. Tennis classes are also offered in PG sports club and star elite table tennis centre at an average duration of one month. Mohd Assri Merzuki is the best Malaysian tennis player, with 1333 single rankings and 795 double rankings from the Association of Tennis Professionals (ATP).

2.1 *Duta International Tennis Academy*

DUTA was established in 1999, the academy has proudly inspired over 1586 players, and it has had 12,848 proud coaching moments. It provides kids trial and adult classes, kids' tennis, junior tennis, elite's tennis, and high-performance programs. DUTA aims at building a robust and exciting tennis community that is open to all. The management team of DUTA includes the director of coaching, head coach, two assistant coaches, and the manager. The current Director is V. Selvam, a highly-ranked Malaysian tennis player with over thirty years of experience. Selvam has guided many junior players in his career to achieve their highest game potential. The head coach and his assistants are qualified coaches on the ITF circuit. They all bring their coaching skills in preparing the players to perform highly and cultivate new and high potentials of the junior players. The manager oversees the organisation's administration, operations, and customer relations. DUTA recognises tennis as a fun sport that brings discipline, teamwork, commitment, learning, and develop healthy lifestyles. The team is instilled life values that enable them to respect and belief in themselves with positive thinking in their daily lives. DUTA believes in playing smarter, and they have adopted Nick Bollettieri's mind-set of "Train to win" and aligned themselves with the Bounex system to develop their players. DUTA has a training system that provides verified coach ratings to optimise players' development through level-based play. Through ladders, tournaments, match play, and structured leagues, a player is matched against others at the same levels, which helps the player aim to become better. The Bounx system enables online booking through a mobile app where players can register for classes, book a coach, or other players to compete with [7].

3 Method

3.1 Search Strategy

The literature will be sourced from different computerised databases. The study will use literature from relevant websites such as Scopus, PubMed, ResearchGate, ScienceDirect, and ITF. Peer-reviewed scholarly articles on tennis sport and coping strategies will form the basis of this review. Online articles that bring a more profound understanding of the topic will also be used. Google Scholar will be used as the primary search engine for scholarly articles and journals. Keywords to be used in the study are tennis, Malaysia, DUTA, sport, athlete, players, strategies, and coping. The literature will take a thematic direction on tennis players' identity, aggression, and how they cope in a demanding sport, their psychological, physical, emotional, and social aspects. The study will perform manual search of referenced studies from the databases.

3.2 Filtering Criteria (Inclusion and Exclusion)

The study's objective is to identify coping strategies used by tennis players. The study will first analyse studies on coping strategies in sport and then evaluate whether the strategies correlate with tennis sports performance according to the moderate variables. The study will use PRISMA guidelines to search, select, and adopt the relevant studies. A total of 35 studies were reviewed; six of them did not meet the study criteria, so they were removed. Literatures from five sources had been included to help with understanding of tennis as a sport and its operational model in Malaysia in general but they were found not specific to the topic, so they were also removed. The remaining 24 literature materials cited in the reference list were found relevant to identify coping strategies undertaken by players and athletes in tennis and so they were used in this study.

4 Literature Review

4.1 Coping Strategies Conceptualization

The coping concept was introduced as early as 1984 by Lazarus and Folkman [8] in the transactional theories of stress and coping. However, by 1926 Freud had introduced the defense mechanism as a state-oriented method of coping in the earliest macro-analytic. The earliest studies on coping originated from physics and not psychology and physiology in the nineteenth century's beginnings of the psychoanalytic movement. Other theorists who introduced the coping concept are Pearlin and Schooler in 1978, Kobas in 1979, and Billings and Moss in 1984. They developed different approaches and identified ways that showed that coping was related to given outcomes [9]. The first studies focused chiefly on stress as the theme, the famous study being *Psychological Stress and the Coping Process* by Lazarus in 1966. In 1984, coping was defined as a constant changing behavioural and cognitive efforts to manage specific demands appraised to be exceeding a person's resources. They introduced avoidance, tolerance, acceptance, and stressor minimisation as coping strategies. Their coping conceptualisation had it that coping was a transaction between the person and the environment, a dynamic process that changes over time due to objective and subjective demands. That appraisal was a feature in the coping process [8]. According to Frydenberg [9], stress and coping are the most highly researched areas in psychology and between 1980s and 2014, the study has yielded 881,436 peer-reviewed journal articles today.

4.2 Studies that Used Quantitative Methods

Physical and Psychological Coping Strategies. Physical activity (PA) positively impacts sustained attention, especially in aerobic exercises with moderate to high intensity. Hajar et al. [10] conducted a systematic review of thirteen studies on the effects of physical activity on sustained attention using documented literature from Scopus, PubMed, Research Gate, ScienceDirect, and Google Scholar (Fig. 1). The results indicated that physical activity increases physical fitness by decreasing reaction time and impulsivity and increasing surveillance and cognitive functioning (Fig. 2). Sustained attention is significant in sport because it sustains performance over a prolonged or extended period. According to Sarter et al. [11], sustained attention enables people to maintain persistent response, vigilance, selective and focused attention, and continuous effort. It is a cognitive function that is crucial in gathering information and instructions.

Tennis is a game with high attention demand, and physical activity becomes a coping strategy that players use to ensure they can respond to their targets accordingly. Usually, humans cannot maintain optimal attention for long without becoming fatigued. Mental fatigue decreases the ability to suppress irrelevant information and irrelevant stimuli, and as a result, it increases reaction time and the number of incorrect responses [12]. Mental fatigue can also bring disorganisation, fidgetiness, or deficit hyperactivity disorder. Tennis players should therefore use physical activity to overcome mental fatigue and improve sustained attention [13]. PA also promotes cardiovascular and musculoskeletal system development, and it enhances learning and memory, improving people mentally, improving individuals' quality of life, and improving academic achievement.

A quantitative study involving 36 athletes from volleyball and handball was conducted by Cosma et al. [14] to ascertain the impact of coping strategies on sports

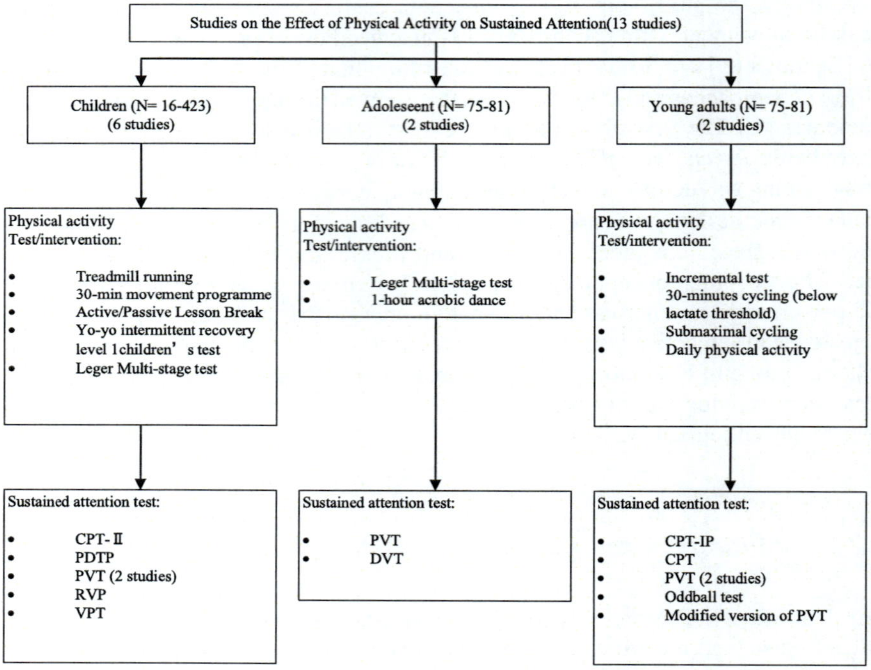

Fig. 1 Summary of study data extracted from previous sustained attention studies

performance. The study revealed that freedom from worry and coachability was the strongly represented coaching strategies while peaking under pressure and coping with adversity were presented least (Table 1). The study also identified males with more ability to accomplish tasks, higher confidence levels, higher motivation of dealing with stressful situations, and more mentally prepared for difficulties than their female counterparts (Table 2).

4.3 Studies that Used Qualitative Methods

Physical Coping Strategies. Tennis is an athletic sport that requires players to speed up the racket's movement, so a tennis player should be robust and fast. Tennis players run around 1000 m in every match, so their body morphology is significant. The players should also be people with resistance which is achieved through training by weight lifts, throwing and catching the ball, and applying other specific movements in the court. Physical exercises that can help to improve composite flexibility, such as static positions that lengthen muscles, are crucial for a tennis player. According to

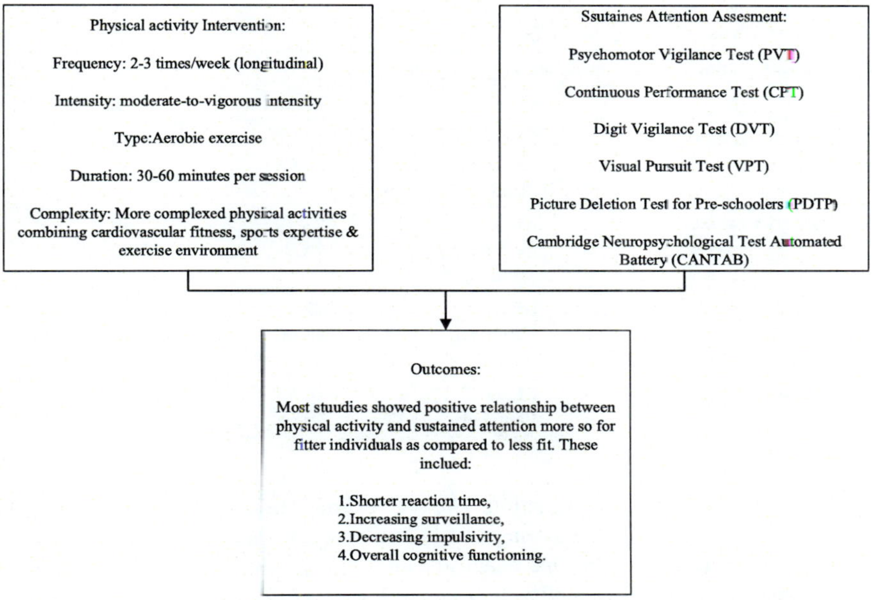

Fig. 2 Positive outcomes from physical activity and sustained attention relationship

Table 1 Impact of coping strategies on sport performance

Means, standard deviations, skewness, and kurtosis for ACSI

	Min	Max	*M*	SD	Skewness	Kurtosis
Coping with adversity	3	11	7.61	2.33	− 0.599	− 0.279
Peaking under pressure	0	11	6.28	2.43	− 0.494	0.107
Goal setting, mental preparation	2	12	8.25	2.73	− 0.707	0.064
Concentration	4	11	7.81	2.23	− 0.086	− 1.278
Freedom from worry	5	16	11.17	2.81	− 0.425	− 0.243
Confidence and achievement motivation	4	12	8.64	2.03	− 0.148	− 0.378
Coachability	5	14	11.06	2.47	− 0.716	0.170

Xiao et al. [15], health and skill-related physical fitness strongly relate to the competitive ability of athletes. The study on the physical fitness of young tennis players identified speed, agility, power, strength, and flexibility as essential performance characteristics of the players.

According to Fernandez-Fernandez et al. [16], tennis players must possess agility, speed, aerobic capacity, endurance, and other physical fitness components that allow the execution of advanced shots to compete successfully against opponents. Physical fitness determines the player who wins and who loses, so ITF recommends all players

Table 2 Gender-based coping strategies

Mean ranks for coping strategies depending on gender				
	Gender	*N*	Mean rank	Sum of ranks
Coping with adversity	Males	19	19.82	376.50
	Females	17	17.03	289.50
	Total	36		
Peaking under pressure	Males	19	20.79	395.00
	Females	17	15.94	271.00
	Total	36		
Goal setting, mental preparation	Males	19	22.53	428.00
	Females	17	14.00	238.00
	Total	36		
Concentration	Males	19	16.55	314.50
	Females	17	20.68	351.50

undergo a physical fitness examination. The ITF testing evaluates the physical condition of the player to allow the setting of training programs and performance goals. The testing criteria are aerobic and anaerobic endurance tests, upper and lower body power tests, strength, speed, agility, and flexibility tests [17].

Exercise training has beneficial effects on physical fitness because tennis players require running skills to win. On average, athletes run 3 m every time they strike a ball, and they must run 8–12 m to earn a ball score. Agility enables the players to make a quick shift in the direction of their movement, either forward, backward, or sideways. They should also be flexible to enhance muscle strength and movement range [18]. Physical exercises help the body release endorphins, a chemical from the brain that helps relieve pain and stress. Exercise also minimises the production and release of cortisol and adrenaline hormones associated with stress. A daily exercise of thirty minutes has been proven to help people feel calmer, and therefore, it is encouraged [19].

Psychological Coping Strategies. According to Carrasco et al. [20], the environmental demands of professional sports affect athletes' psychological well-being. In their study involving 155 male professional tennis players aged 14.6 years on average from South American Tennis Federation tournaments, Carrasco et al. [20] found that the utilised preferred coping strategies determined the autonomy of the young athletes. The players utilised cognitive restructuring, emotional calmness, active planning, and social support as their coping strategies.

Social Support Coping Strategy. Social support is a significant factor in physical and psychological health maintenance. The social support received from family, friends, teammates, staff, and coaches positively affect athletes' behavioural, cognitive, and emotional aspects. The support can be emotional, tangible, or informational, coming from various people. Katagami and Tsuchiya [21] argue that receiving support is a predictor of the psychological well-being of athletes in and out of the sport. On the other hand, perceived support was found to correlate positively with self-schema. According to McCormack et al. [22], despite the high level of work engagement, sports cause burnout in the players. Social support has been identified as a critical resource that can preserve resources that are valued and enhance the performance and well-being of these players. Athletes are therefore encouraged to form peer support groups to prevent isolation. Social support has also been beneficial to athletes during injury rehabilitation and recovery as it gives them assurance and a sense of belonging [23].

Emotional Calmness Coping Strategy. Emotions dictate how a player performs during competition. Sports can bring mixed emotions such as fear, panic, anger, inspiration, frustrations, and exhilaration, so how players manage emotions affects their consistency and performance [20]. The biological instinct of fear and internal pressure can cause athletes to tense during training and competition. To maintain emotional calmness, players should try to remain focused, keep a healthy perspective, reflect, be compassionate to themselves, understand triggers, and remember.

Planning and Cognitive Restructuring Coping Strategy. Planning enables proper coordination amongst the players. It increases the efficiency of players and coaches, enhances sport performance, and help to control activities [20]. Cognitive restructuring helps players change how they think and embrace positive thinking with self-belief. It is a technique used to manage the stress of the sport. It helps athletes identify their stress triggers, understand how to respond to the stress emotionally, and empower them to apply constructive responses [20].

4.4 Studies that Applied a Mixed Method Approach

Athletes and players should ensure that they focus on what they are doing to achieve optimum performance. They should learn to avoid and ignore distractors that may come in the form of noise, weather, and movements because any distraction can have serious consequences. They should train themselves to focus by practising certain activities like listening to music and creating routines. Application of psychological coping strategies can be made through mental rehearsal, visualisation, and imagery which will help to heighten arousal and concentration levels. They can also practice relaxation through yoga, massage, meditation, and controlled breathing and employ goal setting for objective focussing.

Liew et al. [24] argue that mental toughness is one of the highest-ranked psychological characteristics that determine successful performance because athletes' mental ability contributes significantly to their success against their opponents. Mental toughness can be defined as an athlete's ability to cope with training and competitions and maintain concentration and resilience. Applied sports psychologists, coaches, and athletes consistently refer to mental toughness with success and positive outcomes in elite sports. Mental toughness enables athletes to attain high performance despite the underlying pressures, maintain emotional calmness, and bounce back in case of any mistake and obstacles. Mental toughness personality traits include; stress tolerance, level-headedness, resiliency, self-structure, thoroughness, and persistence. It can be used as a hook that attracts athletes to open dialogue on mental health importance and improve knowledge of critical issues such as stigma and symptoms [25].

The systemic review of fourteen studies by Liew et al. [24] revealed that there are common themes and shared strategies and experiences in developing mental toughness. Also that mentally tough athletes develop two specific skills. One is the ability to increase the flow of positive energy when faced with a crisis and the ability to think in ways that promote the right attitudes when solving problems, facing pressure, or during competition. These studies by Lewis used qualitative and quantitative methods, and Psychological Performance Inventory (PPI) was broadly used to measure mental toughness. This review noted that the PPI contains 42 items that measure mental toughness in seven dimensions including, attention and imagery control, motivation, self-confidence, visualisation, negative, and positive energy [24]. However, this approach is faced with criticism because psychometric evidence of PPI's hypothesised measurement model is not encouraging future use. The review focused on psychological variables such as hardiness, coping skills, resilience, and optimism. The result showed that a higher level of mental toughness determines the performance of both motor and cognitive skills. Players need to have a mental toughness because it provides unshakable self-belief and persistence, helping them cope effectively with pressure and retain concentration.

5 Studies on Coping Strategies and Career Variables

A study involving 83 university students in the Polish University Championship in tennis ($n = 32$) and alpine skiing ($n = 51$) consisting of 44 men and 39 women indicated that tennis players use coping actions more than alpine skiers in all actions. The study sought to identify coping strategies with stress caused by sport in both games. The participants were requested to fill questionnaire after finishing their match and describe the measures they had applied before and during the competition. Tennis players displayed their coping strategies as seeking support, logical analysis, distancing, mental distraction, venting unpleasant emotions, and resignations (Table 3). Participants who performed lowest tend to use mental distraction and resignation strategies. Tennis players have to continuously analyse situations and adjust tactics to adapt to their opponents on the tennis court. The logical analysis, therefore, becomes their domain-coping strategy. They also need to quickly forget unpleasant and unsuccessful actions and divert their attention to furthering successful actions. Their mental distraction against stressors helps to control their excitation. Tennis players also need to seek social support during break time from their coaches when they can relax and relax their muscles before continuing with the game. The study indicates that tennis players should avoid resigning, venting unpleasant emotions, and mental distraction instead of using task-focused strategies to achieve higher results [26].

Table 3 Stress coping strategies in tennis and skiing

Variable	Class of result			Discipline		ANOVA					
	Low	Medium	High	Tennis	Skiing	Discipline		Class of result		Interaction	
						F	*p*	*F*	*p*	*F*	*p*
Mental imagery	12.08 ± 2.72	11.51 ± 3.46	12.95 ± 3.59	13.00 ± 2.63	12.09 ± 3.65	1.022	0.314	1.412	0.248	2.420	0.093
Effort expenditure	9.88 ± 2.59	10.53 ± 2.66	11.10 ± 2.50	11.19 ± 2.57	10.52 ± 2.59	1.638	0.203	2.504[a]	0.086	2.298	0.105
Thought control	9.50 ± 2.74	10.84 ± 2.39	11.14 ± 2.68	10.63 ± 2.54	10.78 ± 2.69	0.206	0.651	4.531[b]	0.013	1.129	0.327
Seeking support	9.79 ± 4.18	10.65 ± 4.01	9.41 ± 3.98	11.53 ± 4.17	9.36 ± 3.86	6.711	0.011	1.313	0.273	0.136	0.873
Relaxation	12.13 ± 3.38	10.63 ± 3.51	12.13 ± 3.97	11.75 ± 3.83	11.59 ± 3.76	0.204	0.653	1.173	0.313	3.587	0.031
Logical analysis	8.04 ± 2.65	8.02 ± 2.69	7.71 ± 3.06	10.16 ± 1.94	7.13 ± 2.70	28.920	< 0.001	0.058	0.943	0.204	0.816
Distancing	7.67 ± 2.41	6.19 ± 2.79	5.90 ± 2.80	8.34 ± 2.91	5.66 ± 2.42	22.594	< 0.001	2.641	0.075	0.095	0.910
Mental distraction	10.17 ± 3.73	8.33 ± 3.50	7.37 ± 2.84	10.53 ± 3.91	7.44 ± 2.81	26.253	< 0.001	8.168	< 0.001	2.151	0.121
Venting of unpleasant emotions	10.71 ± 5.35	9.37 ± 4.51	8.11 + 4.49	14.31 ± 4.42	7.28 ± 3.36	91.960	< 0.001	3.385	0.037	2.463	0.089
Resignation	11.88 ± 4.28	9.98 ± 3.46	7.83 ± 2.84	11.16 ± 4.29	8.67 ± 3.24	17.004	< 0.001	16.675	< 0.001	5.745	0.004

[a] The results at the level of a tendency are italicised
[b] The statistically significant results are in bold

6 The Career Development of Tennis Players in Malaysia

ITF requires tennis Malaysia to send national teams to six events every year and field teams in multi-sport events such as SEA games. A squad of the best players is chosen to compete in Cup events, namely Federation and Junior Federation, Davis and Junior Davis, and boys and girls World Junior Tennis Championship. The association also heavily engages in junior development to develop young tennis players into technical and competitive players. Andersen et al. [27] argue that developing elite athletes is a multifaceted and multi-layered process. Malaysian players must be willing to learn how to play the game better and become more competitive. The junior development process facilitates the transition of the players from regional to national and international competitions hence becoming senior players. According to Rahim et al. [6], the junior players are provided with beginner's level coaching regularly and more competitions that help gauge each player's progress. Tournaments are used as the primary performance gauge of players in Malaysia. During tournaments, the players are provided with an opportunity to compete in each category nationally. The players gain more life experience and become more competitive, and their performance is gauged through a ranking system. The players test their skills against their opponents, identify strengths and weaknesses, and provide drills and exercises for improvement through the help of a coach. Tennis officials also benefit from more skills to lead the players better. The players are coached by certified and recognised coaches who are consistent and with a high degree of experience and knowledge. As of 2019, there were 232 certified coaches, 95 in level 1, 127 in level 2, 10 in level 3, and 55 certified tennis officials [6].

6.1 Study Characteristics

The studies included in this review covered a period of 38 years (1984–2022). The studies included participants who were juniors and adults, such as college and professional athletes. Most of the studies were from the English tennis setting. The sports distribution included tennis and skiing. Initial interest was in understanding the coping concept and its application in sports. The study took a deductive approach from coping strategies in sport to specific strategies in tennis.

7 Conclusion, Discussion and Recommendations for Future Research

Players apply many coping strategies in sport, depending on the specific situation. The players apply physical, psychological, emotional, cognitive, social, and planning coping strategies. Physical activities and exercises were a strong coping strategy against mental fatigue. It ensures sustained attention and cardiovascular and musculoskeletal system development and enhances learning and memory. It improves an individual's quality of life and academic achievement, enhances muscle strength, and enhances the release of stress-relieving hormones. DUTA academy players have adopted physical activity as a coping strategy where they perform aerobics and running during practices. PA has helped the DUTA team to possess agility, speed, aerobic capacity, endurance, and other physical fitness components, which are crucial for the game. They have also adopted coachability and training through the help of the Director of coaching, head coach, and two assistant coaches. The coaches guide many junior players to achieve their highest game potential and instil life values that enable them to respect and believe in themselves with positive thinking. The players have also adopted planning as a coping strategy through which they can book coaches, competitors, and competitions. The strategy enables the players to save time and get enough time to engage in work and other assignments and achieve their goals because they are strategically positioned. The emotional coping strategy enables the DUTA team not to worry and achieve mental toughness. The players learn to tolerate stress and become level-headed, resilient, thorough, and persistent. Their cognitive and motor skills are enhanced, which helps the players to cope with pressure and retain concentration effectively. DUTA players have also adopted logical analysis and distancing strategies that enable them to avoid conflict and logical reasoning on how to solve their problems. The players understand the need for social support as per the recommendation of ITF. DUTA academy encourages its players to have a social life besides the sport. Friends, teammates, coaches, and family support are highly encouraged at the academy. During classes, tennis players have time to socialise with each other and offer peer support. During competitions, the players enjoy the company of their fans which encourage them to attain higher performance. DUTA academy has embraced equality by providing both male and female tennis classes. The studies have indicated that male players have higher coping abilities than females, so training and coaching are done separately to ensure that each player's abilities are maximally utilised. Players at DUTA academy have adopted different coping strategies, positively impacting their consistency and performance. They train and learn to overcome fear, panic, anger, inspiration, frustrations, and exhilarations. DUTA players enhance performance through these coping strategies, and their well-being is maintained, thus making the academy one of the best tennis academies in Malaysia.

The reviewed literatures reveals that most of the coping studies done in the sports discipline have concentrated much on stress and the psychological dimension of the athletes. Few studies concentrate on other aspects, such as coping with injury and other health challenges. The evidence also reveals that PPI's hypothesised measurement model is not encouraging future use, and also, there are limited studies on tennis coping strategies, especially in Malaysia. The study, therefore, recommends further studies that will explore how players can cope with injury and health issues. It also recommends further research on new models of mental toughness and more studies on tennis in Malaysia and the coping strategies applied by the players.

References

1. Crocker PRE, Tamminen KA, Gaudreau P (2015) Coping in sport. In: Mellalieu SD, Hanton S (eds) Contemporary advances in sport psychology: a review. Routledge
2. Nakajima T, Hiratsuka M, Ito S, Konno A (2017) Aerodynamic characteristics and PIV analyses concerning tennis balls. IOP Conf Ser Mater Sci Eng 249(1)
3. Clark C (1937) Table tennis: out of the basement—into the schools. J Health Phys Educ 8(7):426–451
4. Lyndhurst G, Barry SL (2019) Tennis|rules, history, prominent players, & facts. In: Encyclopædia Britannica [online]. Available https://www.britannica.com/sports/tennis
5. Brouwers J, Sotiriadou P, De Bosscher V (2015) Sport-specific policies and factors that influence international success: the case of tennis. Sport Manag Rev 18(3):343–358
6. Rahim MRA, Shapie MNM, Abdullah NM, Parnabas V, Hasan H, Khir ZM (2021) Review on the transformation of tennis development in Malaysia. MJSSR 17(1):73–88
7. DUTA International Tennis Academy (2022) Tennismalaysia.com
8. Folkman S, Lazarus RS (1984) Stress, appraisal, and coping. Springer, New York
9. Frydenberg E (2014) Coping research: Historical background, links with emotion, and new research directions on adaptive processes. Aust J Psychol 66(2):82–92
10. Hajar MS, Rizal H, Kuan G (2019) Effects of physical activity on sustained attention: a systematic review. Sci Med 29(2):1–14
11. Sarter M, Givens B, Bruno JP (2001) The cognitive neuroscience of sustained attention: where top-down meets bottom-up. Brain Res Rev 35(2):146–160
12. Faber LG, Maurits NM, Lorist MM (2012) Mental fatigue affects visual selective attention. PLoS One 7(10):e48073
13. Roland AO (2017) Effects of acute physical activity on sustained attention-the possible influence of natural environments (Master's thesis, NTNU)
14. Cosma G, Chiracu A, Stepan R, Cosma A, Nanu C, Păunescu C (2020) Impact of coping strategies on sports performance. J Phys Educ Sport 20(3):1380–1385
15. Xiao W, Geok SK, Bai X, Bu T, Wazir, MRN, Talib O, Liu W, Zhan C (2022) Effect of exercise training on physical fitness among young tennis players: a systematic review. Front Publ Health 10
16. Fernandez-Fernandez J, Sanz-Rivas D, Sanchez-Muñoz C, Pluim BM, Tiemessen I, Mendez-Villanueva A (2009) A comparison of the activity profile and physiological demands between advanced and recreational veteran tennis players. J Strength Cond Res 23(2):604–610
17. FITNESS TESTING (2019) Itftennis.com
18. Munivrana G, Filipčić A, Filipčić T (2015) Relationship of speed, agility, neuromuscular power, and selected anthropometrical variables and performance results of male and female junior tennis players. Coll Antropol 39(Supplement 1):109–116
19. Brennan D (2021) Benefits of sports for mental health. WebMD
20. Carrasco AER, Campbell RZ, López AL, Poblete IL, García-Mas A (2013) Autonomy, coping strategies and psychological well-being in young professional tennis players. Span J Psychol 16:E75
21. Katagami E, Tsuchiya H (2016) Effects of social support on athletes' psychological well-being: the correlations among received support, perceived support, and personality. Psychol 7(13):1741
22. McCormack HM, MacIntyre TE, O'Shea D, Campbell MJ, Igou ER (2015) Practicing what we preach: investigating the role of social support in sport psychologists' well-being. Front Psychol 6:1854
23. Rehmer LN (2021) The role of social support during injury recovery, rehabilitation, and return to play. Honors Theses:3–21
24. Liew GC, Kuan G, Chin NS Hashim HA (2019) Mental toughness in sport. Ger J Exerc Sport Res 49(4):381–394

25. Gucciardi DF, Hanton S, Fleming S (2017) Are mental toughness and mental health contradictory concepts in elite sport? A narrative review of theory and evidence. J Sci Med Sport 20(3):307–311
26. Knittel M, Guszkowska M (2016) Strategies of coping with stress and the sport results of alpine skiers and tennis players. Biomed Human Kinetics 8(1):44
27. Andersen S, Houlihan B, Ronglan LT (2015) Managing elite sport systems. Taylor and Francis, Florence

Design of the Physical Fitness Detection System of Young Children Based on Cloud Computing

Haiyan Zhang, Yee Cheng Kueh, and Garry Kuan

Abstract As the flowers of the motherland, children's health status is directly related to their growth and improvement, and even affects the future improvement of the country. Early childhood education is an indispensable link in their growth process, and children's sports are a very important part of early childhood education, which requires the attention of the whole society. The purpose of this paper is to design a physical fitness detection system for young children based on cloud computing. In the study, the independent samples *t*-test was used to assign the children in the experimental group and the control group to compare the improvement level of physical fitness, so as to realize the physical fitness detection system of children. The experimental results show that the children in the experimental group that passed the skills and physical fitness were significantly better than the children in the control group using the physical fitness test, and their movements were more coherent and coordinated. Further, various physical qualities show significant improvement.

Keywords Cloud computing · Physical fitness of young children · Physical fitness detection · System design

1 Introduction

Physical fitness pays attention to the cultivation of the comprehensive health quality of the human body, and focuses on the excavation of human sports potential, which can better reflect the human body's ability to adapt to the environment [1]. The

H. Zhang (✉) · G. Kuan
Exercise and Sports Science Programme, School of Health Sciences, Universiti Sains Malaysia, 16150 Kubang Kerian, Malaysia
e-mail: zhanghaiyan@student.usm.my

Y. C. Kueh
Biostatistics and Research Methodology Unit, School of Medical Sciences, Universiti Sains Malaysia, 16150 Kubang Kerian, Malaysia

H. Zhang
Guangzhou Huali College, Guangzhou 511325, China

G. Kuan et al. (eds.), *Advancing Sports and Exercise via Innovation*, Lecture Notes in Bioengineering, https://doi.org/10.1007/978-981-19-8159-3_47

enthusiasm for participating in sports plays an important role [1]. Society and parents pay too much attention to the improvement of children's intelligence and professional quality, but there is no research on children's physical fitness. The choice of children's health improvement and exercise methods is worthy of attention.

Nowadays, when the national fitness movement is in full swing, many people ignore the special group of young children. The age of 3–6 years is the link between the improvements of preschool children, relatively speaking, their cognition and behavior already have a certain degree of subjectivity. Kryst's [2] study found that overweight individuals tend to have a stronger bone and more muscle mass than lean individuals. The aim of this study was to examine the immune system, bone and muscle mass, and body weight of obese and non-obese polish children and adolescents. The research participants consisted of girls and boys. Height, weight, triceps skin, humerus, femur, hip, mid-thigh, and upper arm circumferences were measured to estimate BMI, body frame index, bone, and muscle mass. Results were obtained for grip strength, standing long jump, sitting, and vehicle handling tests. The participants were classified as normal weight or overweight/obese according to Cole's cutoff value. Statistical differences were assessed using two-way ANOVA. Vargas AC conducted the study and reported that the people with intellectual disabilities have lower levels of physical fitness compared to their peers without intellectual disabilities because of higher levels of sedentary behaviors in this population. To understand the relationship between quality of life and physical health in adults with intellectual disabilities, ninety-six adults with intellectual disabilities were given quality of life questionnaires and physical fitness tests, including balance, muscle strength, flexibility, and aerobic condition. Adults with higher self-reported levels of quality of life had higher levels of physical fitness in terms of balance, muscle strength, and flexibility. In contrast, under aerobic conditions, there was no significant association with self-reported quality. These findings support the hypothesis that people with intellectual disabilities and low levels of physical fitness affect their quality of life. This knowledge is useful for improving treatment that promote the physical health of the population [3]. It also cultivate children's sense of movement and interest in movement, so as to improve children's participation in sports activities, which is conducive to children's overall physical and mental improvement.

The present study identifies the related research on physical fitness, the definition of core concepts of physical fitness, and the overview of cloud computing, including cloud computing improvement, cloud computing service types, and important characteristics of cloud computing. In the experiment, a physical fitness detection system for children aged 4–6 years was designed, and 60 children aged 4–6 years in a kindergarten in M city were selected as the experimental subjects. Logistic regression algorithm and independent sample *t*-test were used to analyze and compare the physical fitness improvement level of children in the experimental group and children in the control group.

2 Design and Research on the Physical Fitness Detection System of Young Children Based on Cloud Computing

2.1 *Physical Fitness Related Research*

In the early childhood, with the increase of age, the results tested according to the standard of physical fitness also improved to a certain extent, and it was proposed that the amount and intensity of physical activity have beneficial effects to the human body [4]. The more proficient the students' operational skills, the higher the cardiovascular endurance, upper-body muscle strength and endurance level, and the higher the level of physical fitness. Mastering proficient operational skills and maintaining a healthy level of physical fitness are the expected outcomes of teaching at the elementary level.

The cultivation of healthy physical fitness can be divided into five stages: first is a happy exercise experience; the second is to obtain the correct exercise method; the third is to try different sports to form your own habits; the fourth is to master your own state; the fifth is own training plan and can be adjusted in time. The complete model of school health and fitness education is divided into three parts: the bottom layer, activities of daily life; the middle layer with intensive aerobic, flexibility, muscle strength, and other exercises; and the top layer of static cultivation activities, reading, etc. Creating an environment for students to stimulate interest in sports requires the attention of parents and schools [5]. Physical activity, fitness, and athletic ability improve in concert during childhood and complement each other. Physical fitness is the mediating factor between the two, and the degree of correlation increases with age. Physical activity can promote the improvement of motor skills and physical fitness. Early learning and accumulation of various exploratory sports experiences can promote the formation and improvement of basic movement patterns in children.

The level of physical fitness is of great significance to the health of the human body. It is well reported that to physical fitness promote physical health and sports performance [6]. Certain concepts related to physical fitness should be injected into the training of young children and adolescents to ensure that training pays more attention to health and students' feelings [6].

2.2 *Definition of Core Concepts of Physical Fitness*

Physical fitness refers to a certain amount of energy to cope with emergencies after completing various tasks in daily life, combined with comprehensive investigation and judgement in terms of skill improvement and metabolic index [7]. Physical fitness is the top priority of physical exercise research, including cardiorespiratory fitness, physical fitness, flexibility, muscle strength, and muscle endurance. Focusing on the cultivation of lifestyle and study habits at early childhood will improve good physical fitness in early childhood, so that young children can grow up healthily and promote the health of their bodies. Physical fitness is now more in line with the view that the body is always full of energy and willing to do physical exercise after dealing with daily life. This means that good physical condition is the key to getting people to participate in social exercise and improve their physical condition and healthy lifestyle.

What is physical fitness: Promoting physical fitness through a range of physical activities ensures that the body can successfully perform a variety of tasks as it adapts to the environment. The physical fitness of young children has two parts: physical health and mental health. Through a scientific and reasonable physical exercise, it is helpful to improve children's physical quality and comprehensive ability and lay a good foundation for their future improvement. Curriculum mode refers to the structure and basic framework of the curriculum system, as well as educational activities established under the guidance of certain educational concepts or educational theories. Under normal circumstances, it can be called a model, that is, a model that foresees other educational activities [8]. A very general and instructive lesson plan with some level of advancement. Physical fitness is the embodiment of human's physical and mental adaptability in today's life. Physical fitness means that everyone should choose the most suitable exercise method and amount of exercise, which can achieve the effect of improving physical fitness in different situations. Physical fitness of young children is a subdivision proposed by kindergartens and children's health management organizations in various regions, aiming to improve children's physical strength, self-protection ability, environmental adaptability and psychological ability through scientific and reasonable physical training. It provides self-regulation and a good physical foundation for future study, life, and work. Only good physical health from childhood can ensure the healthy growth of young children and promote healthy physical improvement.

2.3 *Overview of Cloud Computing*

(1) Improvement of cloud computing

Cloud computing is an emerging Internet-based service computing model improved on the basis of traditional virtualization technology, parallel technology, and grid technology in recent years [9]. Through cloud computing, computing resources, and storage resources connected to the Internet can be shared, and cloud users can be given access rights flexibly and on demand. The resources in the cloud resource pool can be flexibly allocated and released, so cloud computing is an ideal resource service. As an on-demand computing, Internet-based service model that can provide shared computing resources and data, the cloud can be accessed and expanded on a pervasive and on-demand basis. The Internet era is an era of exponential expansion of users. With the comprehensive expansion of business, the current information era has entered an era of "cloud computing and big data". The explosive growth of data makes users not satisfied with the computing processing of a single device, and instead entrust data processing to the cloud. For cloud resource providers, one of the advantages of cloud computing is dynamic scalability. By dynamically adding and deleting nodes, the operation of the cloud center can still maintain a normal state on-demand service. Because cloud providers need to spend a certain amount of manpower and material resources to maintain the normal operation of their cloud centers, the current mainstream cloud providers have adopted a method of charging services. Taking the website business as an example, users and enterprises can purchase a certain number of virtual machines or virtual machine services to host their websites and only need to pay the running cost of the virtual machines when the website is running. Of course, if you need to configure a higher virtual machine, you can purchase the corresponding high-quality package. This rapid resource purchase and deployment can meet the needs of various users in a timely and effective manner and low cost. With the reduction of the cost of hardware production, the prices of various physical devices supporting the cloud center are also accepted by various cloud manufacturers. The high degree of perfection and integration of the underlying software of cloud services, on the one hand, makes the management of the cloud more and more user-friendly, and on the other hand, it also reduces the management cost of the cloud. Cloud computing has improved recently, and its main architecture has undergone many adjustments. Overall, cloud computing is a need for service-oriented architecture.

(2) Regarding the types of cloud computing services, they can generally be divided into the following three types

Basic settings as a service: By providing basic equipment on demand, it usually refers to the service mode of virtual machines with certain configurations. Those who provide this kind of service are called IaaS providers. Well-known IaaS providers are Amazon EC2, Flexiscale, OpenStack, CloudStack, etc.

Platform as a Service: A service model that provides platform-level resources, including operating systems and supported software improvement environments. The more well-known PaaS providers are Google App Engine, Microsoft Windows Azure, etc.

Application software as a service: a service model that provides applications on demand through the Internet. Well-known SaaS providers include SAP Business, Rackspace, etc. According to the service level, a service-oriented cloud computing architecture can be established.

By virtualizing computer resources, storage resources, network resources, and integrating resources, a resource pool is constructed. Then, in the upper layer, design a mode that is convenient for users to use, and encapsulate resources into Web services. In addition, cloud computing also needs to integrate existing resources and have certain necessary structures such as resource scheduling and security management. The module is responsible for security management to ensure the security of the cloud computing system from being attacked, free from intrusion threats, and protect multiple nodes. At the same time, cloud computing can meet the needs of users and enterprises to use applications and store some data in the cloud. In summary, cloud computing extends the concept of converging infrastructure with shared resources [10].

(3) Cloud computing has the following important characteristics

On a large scale: In the current cloud era, well-known cloud service providers such as Amazon, Google, Alibaba, Baidu, etc., are all enterprises with more than one million servers. By satisfying a certain order of server clusters, these vendors are free to provide the necessary resource services to their users.

Virtualization: On top of traditional virtualization technology, cloud virtualization is a perfect interpretation of virtualization technology. By virtualizing hardware resources, operating systems, and specific applications, cloud computing transforms actual physical devices such as CPUs, disks, and networks into logical existence that can be controlled. The virtualized device can pass through a series of given interfaces, which makes it easier for users to configure requirements from a huge resource pool.

Reliability: Before the advent of cloud computing, it was either virtualization technology, grid technology or other. In the related art, there is no relatively unified solution for reliability. After the advent of the cloud era, taking Google as an example, the multi-level data copy fault tolerance advocated by them is based on the premise of sacrificing hardware resources to preserve user service quality. In addition, the idea of isomorphic calling between nodes satisfies the basic guarantee of data recovery from damage. The establishment of cloud data centers, through strict access control and ever-changing security protection technologies, can enable cloud users to safely place data in the cloud without worrying about being hacked or destroyed [11].

Versatility: The emergence of the cloud is, to a certain extent, a good package for platforms and applications with strong heterogeneity. Through the cloud package, in the same resource pool, enterprises and individual users can create different platforms and host different applications on it. Since the implementation details are transparent, users only need to focus on the resource acquisition level. Tasks related to resource allocation can be handed over to the cloud for self-management.

3 Investigation and Design of the Physical Fitness Detection System of Young Children Based on Cloud Computing

3.1 Research Content and Objects

Research content: Physical fitness detection system for children aged 4–6 years old. The subjects selected for this study were 60 children aged 4–6 in a kindergarten in M city. Experimental grouping: The selected 60 children have not received systematic physical training before the experiment. The control group was a kindergarten with a total of 30 students; the experimental group was a sports hall with a total of 30 children. Since the gender differences of 4–6-year-olds have little effect on their athletic ability and will not have a significant impact on the arrangement of training content and test results, no gender group matching is performed in this paper.

The specific sports test contents are as follows: 10-m reentry running, standing long jump, tennis throwing, continuous jumping with both feet, sitting forward bending, and walking on balance beam.

3.2 *Independent Samples* T-*Test*

Independent samples *t*-test, also known as the group *t*-test, is suitable for the completely random design of the comparison of two-sample means. It is generally used for normally distributed two sets of data. Multiple sets of data use one-way ANOVA, and the normality test should be conducted before the *t*-test. The function formula is as follows:

① analysis comparing mean independent sample *t*-test
② adds test variables and group variables, respectively, and then defines groups: experimental and control groups
③ analytic result

The *t*-test statistic for the independent samples is:

$$t = \frac{\overline{X}_1 - \overline{X}_2}{\sqrt{\frac{(n_1-1)S_1^2+(n_2-1)S_2^2}{n_1+n_2-2}\left(\frac{1}{n_1}+\frac{1}{n_2}\right)}}$$

S_1^2 and S_2^2 are two-sample variance; n_1 and n_2 are two sample sizes.

4 Analysis and Research on the Physical Fitness Detection System of Young Children Based on Cloud Computing

4.1 *Comparison of Physical Fitness Improvement Levels of Children in Children Group*

In order to verify that the continuity and speed of the continuous jumping of the feet of the children in the experimental group were improved after the intervention, the upper limbs could coordinate with the lower limb's movements, and the body could remain stable during jumping movements; the children in the experimental group walked on the balance beam. Stand upright, overcome fear, and move forward independently, and move your feet alternately and smoothly. Therefore, the result shows that the stability training in physical fitness can improve the balance ability of young children. Table 1 and Fig. 1 show the specific results of the experimental precursor fitness test of the experimental group and the control group:

The data showed that the experimental group and the control group had comparable results in each test of pre-intervention pre-existing fitness. After independent sample *t*-test, it was concluded that there was no significant mean difference of physical fitness between the two groups of children before the intervention ($p > 0.05$).

Table 1 Test data of young children's physical fitness experiment

Test item	Experimental group	Control group	*t*-test	*p*-value
Ten-meter return run (s)	8.93	8.79	0.635	0.612
Standing long jump (cm)	80.36	80.01	0.985	0.418
Tennis throwing (m)	4.21	3.98	0.715	0.654
Feet continuous jump (s)	8.06	7.96	0.151	0.536
Sitting position forward flexion (cm)	9.04	8.84	0.102	0.684
Walk the balance beam (s)	10.31	10.02	0.320	0.987

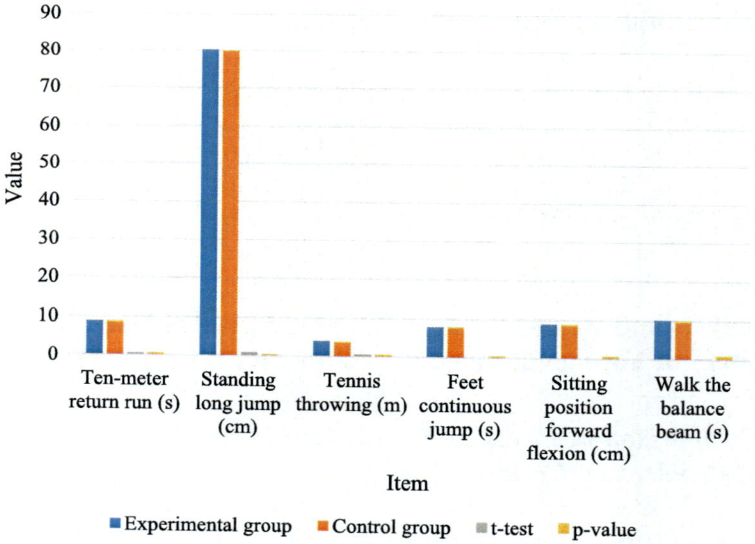

Fig.1 Experimental precursor fitness test results of young children in the experimental group and the control group

4.2 Comparative Analysis of Physical Fitness Levels in Children Group After the Experiment

In order to test the physical fitness level of the children after the experiment through the independent sample t-test, the physical fitness test results are shown in Table 2 and Fig. 2.

The data showed that there was a big gap in physical fitness performance between the experimental group and the control group after the intervention. The independent samples t-test shows that there was a significant mean difference in the standing long jump, tennis throwing, and double-legged continuous jumping performance of the experimental group after the intervention compared with the control group ($p < 0.01$). Also, compared with the control group, there was a significant mean difference in

Table 2 Physical fitness test results of experimental and control children

Test item	Experimental group	Control group	t-test	p-value
Ten-meter return run (s)	10.02	9.31	0.058	0.015
Standing long jump (cm)	109.61	89.47	5.641	0.005
Tennis throwing (m)	6.01	4.08	5.751	0.004
Feet continuous jump (s)	8.96	8.04	0.353	0.002
Sitting position forward flexion (cm)	11.3	9.31	0.003	0.984
Walk the balance beam (s)	9.98	11.12	0.080	0.051

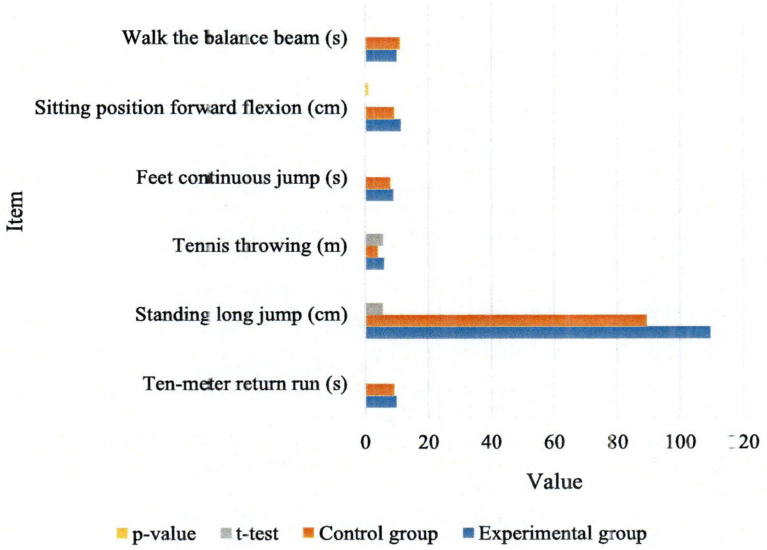

Fig.2 Physical fitness test results after the experiment between two class

the scores of children's 10-m reentry running and walking on the balance beam ($p < 0.05$); there was no significant mean difference between the children's sitting and forward bending scores between the experimental group and the control group ($p > 0.05$).

To sum up, the children in the experimental group that have undergone the skills and physical fitness test performed significantly better than the children in the control group in the physical fitness test, and their movements were more coherent and coordinated. Furthermore, various physical qualities were significantly improved in the experimental group than the control group.

5 Conclusions

The improvement of children's physical quality has certain difficulties. Children have certain athletic ability and form a certain social cognition, but they are less restrictive in behavior and lack understanding of the consequences of their own behavior. Therefore, they need more guidance to regulate their positive and negative behaviors. Physical fitness courses must improve the children's physical and mental quality, and at the same time, they must meet the interests of children. The improvement of physical fitness courses in kindergartens should be strengthened. On the one hand, it is necessary to improve the existing kindergarten curriculum design, improve the physical quality of children, and promote the overall children's physical and mental health. On the other hand, it also has far-reaching significance for the healthy growth of children in my country.

References

1. Guerrant DI, Moore SR, Lima AM (2019) Association of early childhood diarrhea and cryptosporidiosis with impaired physical fitness and cognitive function four-seven years later in a poor urban community in northeast Brazil. Am J Trop Med Hygiene 61(5):707–713. https://doi.org/10.4269/ajtmh.1999.61.707
2. Kryst U, Egleń M, Woronkowicz A (2021) Skeletal and muscular robustness and physical fitness of Polish children and adolescents (3–18 years) with normal weight and overweight/obesity. Anthropol Anz 79(3):277–287
3. Vargas AC, Cruzado DP, Moya AR et al (2022) Relationship between quality of life and physical fitness in adults with intellectual disabilities. Adv Ment Health Intellect Disabil 16(1):44–52
4. Ucok K, Yalcinkaya H, Acay A (2018) Do patients with newly diagnosed type 2 diabetes have impaired physical fitness, and energy expenditures? Neth J Med 73(6):276–283
5. Stella M, Lucy-Joy W, Allana L (2019) Temporal trends and correlates of physical activity, sedentary behaviour, and physical fitness among school-aged children in Sub-Saharan Africa: a systematic review. Int J Environ Res Publ Health 11(3):3327–3359. https://doi.org/10.3390/ijerph110303327
6. Muntaner-Mas A, Palou P, Vidal-Conti J (2018) A mediation analysis on the relationship of physical fitness components, obesity, and academic performance in children. J Pediatr 198(1):90–97. https://doi.org/10.1016/j.jpeds.2018.02.068

7. Ishihara T, Morita N, Nakajima T(2018) Direct and indirect relationships of physical fitness, weight status, and learning duration to academic performance in Japanese school children. Euro J Sport Sci 18(10):286–286. https://doi.org/10.1080/17461391.2017.1409273
8. Jo G, Rossow-Kimball B, Lee Y (2018) Effects of 12-week combined exercise program on self-efficacy, physical activity level, and health related physical fitness of adults with intellectual disability. J Exercise Rehabil 14(2):175–182. https://doi.org/10.12965/jer.1835194.597
9. Macmillan C, Olivier B, Benjamin-Damons N (2021) Sport science lab screening protocol: the association between physical fitness parameters and injury among elite rugby players. Phys Ther Sport 52(1):272–279
10. Greg K, Denny M, Hardman RJ (2018) Physical fitness and aortic stiffness explain the reduced cognitive performance associated with increasing age in older people. J Alzheimer's Dis 63(4):1307–1316. https://doi.org/10.3233/JAD-171107
11. Duncan MJ, Eyre E, Noon MR (2021) Actual and perceived motor competence mediate the relationship between physical fitness and technical skill performance in young soccer players. Eur J Sport Sci 2:1–20

The Relationship Between Physical Exercise and Mental Health Promotion Among University Students in China

Wang Xiawei and Garry Kuan

Abstract The purpose of this article is to investigate the mental health status of Chinese university students, the influence of physical exercise on the mental health factors of Chinese university students, and the relationship mechanism between physical exercise and the mental health of Chinese university students, in order to provide a reference for promoting the mental health of Chinese university students. Methods: A stratified random sampling method was used to randomly select 1000 university students from a university in Henan Province, China using a questionnaire survey, and the valid sample data were 921 cases. The SPSS 26.0 software was used for correlation analysis of the collected sample data. Result: The assessment of physical exercise level of Chinese university students was based on the results rated by the Chinese National Student Physical Fitness Test Standard, in which 78 (8.46%) were excellent, 307 (33.29%) were good, 469 (50.88%) were passing, and 66 (7.37%) were failing. For the comparative analysis of the mental health, the excellent rate, good rate, and passing rate of physical test scores of male students were better than that of female students, whereas the scores of all the factors for symptom self-assessment scale (SCL-90) of female students were significantly higher than those of the male students, ($P < 0.05$). The comparative analysis between the mental health level of the students and the four types of sports performance, revealed that the scores of depression, anxiety, and terror in the SCL-90 with excellent, good, and passing physical fitness test were significantly lower than those with failing physical fitness test ($P < 0.05$). The correlation between physical exercise and the mental health of the students was analyzed and the scores of physical exercise (physical fitness test scores), depression, anxiety, and terror in the symptom self-assessment scale (SCL-90) were significant and negatively correlated ($P < 0.05$).

Keywords Physical exercise · China · University students · Mental health

W. Xiawei (✉) · G. Kuan
Exercise and Sports Science Programme, School of Health Sciences, Universiti Sains Malaysia, Kubang Kerian, Kelantan, Malaysia
e-mail: wangxiawei@student.usm.my

W. Xiawei
Zhengzhou Electric Power Technology College, Zhengzhou, Henan, China

G. Kuan et al. (eds.), *Advancing Sports and Exercise via Innovation*, Lecture Notes in Bioengineering, https://doi.org/10.1007/978-981-19-8159-3_48

1 Introduction

With the global economic development and changes in human lifestyle, insufficient physical exercise and mental health disorders have become global social problems, which have become the main causal factors of psychological disorders such as overweight and obesity, anxiety, depression, and panic [1]. As an important group for the future development of the world, the physical condition and mental health of university students have become a topic of concern for scholars from various countries [2].

Physical and psychological health are fundamental conditions for the overall development of university students [1]. Physical exercise not only strengthens the immune system, but also promotes psychological health [3]. It is therefore, important to study the mental health of university students because it improve their physical fitness, relieve tension, and treat their psychological sub-health. This study investigates the relationship between physical exercise and psychological health, and the role of physical exercise in influencing the psychological health in order to provide insights for the university students to actively and effectively participate in physical exercise and, promote their psychological health and overall health.

2 Methods

2.1 Participants

The study examines takes the relationship between physical exercise and mental health among university students in China. The respondents were university students from Henan Province, China, and their ages ranged from were generally 18 to21 years old. The study adopted stratified random sampling, and a total of 1000 students were enrolled and given the study questionnaires. 979 questionnaires were collected, of which 921 were valid, with a return rate of 94%, of which 689 were male students and 232 were female students. They consisted of 329 first year students, 274 males and 55 females; 308s year students, 257 males and 51 females; and 284 third year students, 158 males and 126 females.

2.2 Fundamental Movement Skill

The study was a cross-sectional survey using a self-administered questionnaire to assess the participant's fitness test scores. The physical fitness test scores were determined using the four-level scoring system of China's National Student Physical Health Standards, which comprehensively assesses students' physical fitness level in terms of physical form, physical function, and physical quality. It is an evaluation tool that serve as a mean to promote the development of students' physical health and motivation for physical exercise participation [4]. The physical exercise score in this paper is classified as excellent: total score of 90 and above, good: total score of 80–89; pass: total score of 60–79; and fail: total score of 60 or below. The total number of samples collected in this survey was 1000 students in total, but those with complete responses were 921 students.

The SCL-90 was used to investigate the mental health level of the university students, with a total of 90 items within nine factors rated using a five-point scale ($1 < 2 < 3 < 4 < 5$), ranging from asymptomatic to severe. The questionnaire was based on the Symptom Self-Rating Scale (SCL-90) developed by L. R. Derogatis, which is currently the most widely used scale for detecting mental disorders and mental illnesses worldwide [5].

2.3 *Data Analysis*

All the data obtained were entered into an Excel sheet for preliminary collation and then entered into SPSS 26.0 software for analysis. Independent t-test analysis was performed to determine the mean difference of the mental health level between male and female students, and between the four types of sports performance. Person's correlation analysis was performed to determine the relationship between physical exercise and students' mental health. Descriptive results were expressed as mean ± standard deviation. A $p < 0.05$ was considered a statistically significant difference.

3 Results

3.1 *Mental Health Level of University Students*

The mental health level of university students was investigated by the Symptom Self-Rating Scale (SCL-90), and the number of valid test cases was 921, among which 689 were male and 232 were female. As long as the score of one factor of SCL-90 is greater than or equal to 3, it is considered that there may be moderate severity of psychological problems [6]. A total of 104 people have psychological problems, accounting for 11.35% of the total, and this kind of people belong to the key population of research attention.

3.2 Physical Exercise of University Students

The amount of physical exercise assessment was based on the students' physical fitness test results. The total number of students that report their physical fitness test results was 921, among which 689 were male and 232 were female. The result shows that 78 (8.46%) had excellent physical fitness test results; 307 (33.29%) had good physical fitness test results; 469 (50.88%) had passing physical fitness test results, and 66 (7.37%) had a failing physical fitness test.

3.3 Comparative Analysis of the Mental Health Levels of Male and Female University Students

For comparison of the sample genders, the excellent, good, and passing scores of physical fitness test for the male students were better than that of the female students, while the difference between the failing physical fitness tests scores of male and female students was not significant. Female university students scored higher on SCL-90 in somatization, obsession, interpersonal sensitivity, depression, anxiety, hostility, terror, delusion, and psychoticism than the male students ($p < 0.05$). All the results are displayed in Table 1.

Table 1 Comparative analysis of factor scores on the mental health level scale for male and female university students

Gender	Female	Male	*t-test*	*p-value*
No. (n)	232	689		
F1 somatization	1.55 ± 0.25	1.43 ± 0.20	5.549	0.000
F2 obsession	1.90 ± 0.35	1.75 ± 0.39	4.726	0.000
F3 interpersonal sensitivity	1.78 ± 0.44	1.61 ± 0.36	4.113	0.000
F4 depression	1.73 ± 0.41	1.66 ± 0.47	3.628	0.000
F5 anxiety	1.59 ± 0.31	1.52 ± 0.21	4.561	0.000
F6 hostility	1.58 ± 0.33	1.55 ± 0.28	0.923	0.216
F7 terror	1.40 ± 0.23	1.39 ± 0.18	1.392	0.049
F8 delusion	1.52 ± 0.29	1.48 ± 0.30	0.969	0.158
F9 psychoticism	1.62 ± 0.33	1.43 ± 0.31	5.903	0.000

Table 2 Comparative analysis of scores on the symptom self-assessment scale for college students in four categories of sports performance

Sports performance	Excellent	Good	Passing	Failing
No. (n)	78	307	469	66
F1 somatization	1.38 ± 0.36	1.39 ± 0.38	1.41 ± 0.42	1.45 ± 0.43
F2 obsession	1.79 ± 0.69	1.81 ± 0.65	1.83 ± 0.59	1.85 ± 0.54
F3interpersonal sensitivity	1.59 ± 0.54	1.61 ± 0.49	1.65 ± 0.47	1.67 ± 0.43
F4 depression	1.48 ± 0.85	1.50 ± 0.74	1.53 ± 0.62	1.61 ± 0.56*,**,***
F5 anxiety	1.49 ± 0.63	1.50 ± 0.48	1.53 ± 0.47	1.55 ± 0.45*,**,***
F6 hostility	1.39 ± 0.38	1.40 ± 0.36	1.42 ± 0.35	1.45 ± 0.28
F7 terror	1.32 ± 0.43	1.36 ± 0.32	1.40 ± 0.28	1.46 ± 0.33*,**,***
F8 delusion	1.52 ± 0.39	1.53 ± 0.36	1.55 ± 0.34	1.57 ± 0.42
F9 psychoticism	1.50 ± 0.42	1.51 ± 0.39	1.53 ± 0.32	1.57 ± 0.35

Note * indicates $P < 0.05$ for failing compared to passing, ** indicates $P < 0.05$ for failing compared to good, ***indicates $P < 0.05$ for failing compared to excellent

3.4 Comparative Analysis of the Mental Health Levels of University Students in the Four Sports Performance Categories

In the physical fitness test, the sport performance scores of somatization, obsession, interpersonal sensitivity, hostility, delusion, and psychoticism in the SCL-90 were no significantly different in the excellent, good and passing category, whereas the scores of depression, anxiety, and terror in the SCL-90 were significantly different in the excellent, good, and passing category (Table 2).

3.5 Correlation Between Physical Exercise and Students' Mental Health

Physical exercise (physical fitness test scores) was significant an negatively correlated with the scores of the three factors of depression, anxiety, and terror SCL-90 (Table 3).

Table 3 Correlation analysis of physical activity and mental health of university students

Sports performance	*r*	*t-test*	*p-value*
F4 depression	– 0.128	2.983	0.001
F5 anxiety	– 0.092	5.389	0.000
F7 terror	– 0.076	6.118	0.000

4 Discussion

4.1 Physical Exercise and Mental Health of Male and Female Students

This study shows that the physical exercise and mental health results of 921 university students—the excellent rate, good rate and passing rate of physical fitness test results of male students—are better than that of female students. This concludes that the level of physical exercise among university students is still low level especially among females. The high academic pressure on college students, heavy professional and practical training courses can make the college students have less time to participate in physical exercise [7]. Also, there are other factors such as inadequate environment of physical activity, which restrict students' physical exercise activities. The physical fitness test scores of the male students were significantly higher than the female students, which is in line with the basic characteristics of male and female sports, with male students' awareness of physical exercise and physical exercise behavior higher than that of the female students. Male students are generally more athletic than female students and are more willing to participate in high-intensity and high-density sports activities, such as running, basketball, and football, while female students prefer low-intensity sports with small intervals, such as yoga, table tennis, and aerobics.

The students' mental health showed that female students scored significantly higher on the Symptom Self-Rating Scale (SCL-90) for somatization, obsession, interpersonal sensitivity, depression, anxiety, hostility, terror, delusion, and psychoticism than the male students ($p < 0.05$). It can be concluded that on one hand that male students have a stronger interest in sports, are more athletically gifted than female students, and prefer sports activities with a high level of exercise, while female students are more fragile and sensitive and are more prone to psychological problems as they are unable to vent and release pressure when facing stress. On the other hand, the items that girls participate in physical exercise do not match the characteristics of the physical side items, which to a certain extent confirms the fact that girls lack physical exercise and produce anxiety, depression and sensitivity in the face of study, employment, interpersonal and relationship, and the psychological pressure cannot be released, all of which affect the psychological health level of female university students.

Therefore, when teaching physical education and organizing campus activities, universities need to focus on students who fail in physical fitness, and should pay more attention to the mental health of female students to actively guide them to participate in physical exercise, formulate reasonable and effective mental health guidance programs, and to provide effective methods and strategies for students with low mental health status.

4.2 *Comparative Analysis of the Mental Health Level of University Students with Four Types of Physical Test Scores*

In this study, it was found that among the four categories of the college students, namely, excellent, good, passing, and failing; the excellent rate was only 8.46% with only 78 students, and good rate was 33.29% with 307 students, highlighting the poor effect of students' physical exercise, which school leaders should pay great attention to. On the Self-Rating Symptom Scale for College Students (SCL-90), in the three factors of depression, anxiety, and terror, the students with excellent, good, and passing scores in physical education are significantly lower than those in failing scores of physical fitness test.

It can be concluded that physical exercise indirectly affects students' body and mind, and the lack of high intensity, high density, and high amount of physical activity during school studies triggers students' uneasy psychology. Depression and anxiety are frequent among male and female university students, and terror is easily focused on female students, because they have high mood fluctuations influenced by the outside world. A targeted improvement of physical exercise can alleviate and prevent psychological disorders [8]. In addition, somatization, compulsion, interpersonal sensitivity, hostility, delusions, and psychoticism scores were not statistically significantly different. The survey data showed that there was no direct effect of the six factors of somatization, obsession, interpersonal sensitivity, hostility, delusions, and psychoticism on the four types of physical test scores, but according to the actual interview, some students believed that these psychological aspects had some correlation with physical exercise.

4.3 *Analysis of the Correlation Between Physical Exercise and Mental Health*

According to this study, the correlation between physical exercise (physical fitness test scores) and the scores of depression, anxiety, and terror in the SCL-90 was significant and negative. Therefore, it shows that college students' participation in sports and exercise can significantly improve their mental health level, and can alleviate their depression, anxiety, terror, and other psychological factors. A moderate amount of physical activity can help reduce students' depression, anxiety and terror levels. Also, active physical exercise can improve their self-confidence, relax their bodies and minds, release their stress and vent their negative psychological emotions [9]. In addition, physical exercise by university students can promote the dynamic balancing mechanism of the nervous system and hormone levels, which in turn produces inhibitory hormones on negative psychological conditions such as psychological depression and emotional anxiety, ensuring a good endosomal secretion state, emotional stability, and physical and mental relaxation. In their daily study life, university students with poor physical fitness must pay attention to their psychological state, strengthen physical exercise, release bad emotions and effectively carry out the psychological treatment of physical exercise, so as to improve their mental health.

5 Conclusion

This study shows that physical exercise and psychological health level of college students are positively correlated. To a certain extent, physical exercise can promote the psychological health of college students and improve their bad psychological emotions. On the whole, the surveyed university students have low sports participation, low physical exercise status, and poor mental health. The psychological problems of university students today are the focus of research by scholars from various countries. Hence, it is important to study the relationship between physical exercise (sports performance) and psychological health of university students, and to take effective means to intervene according to the research results, so as to guide students to actively participate in physical exercise, improve their psychological health and overall well-being.

Acknowledgements This research was supported Key Scientific Research Project Plan of Colleges and Universities of Henan Province in China with Project Code: 22B890007.

References

1. Li D, Li F, Liu Y, Jia Y, Li W, Yao Y, Wan Z (2021) Interpretation of the 2020 World Health Organization guidelines on exercise and sedentary behaviour. Chin J Clin Thoracic Cardiovascular Surg (04):376–383
2. Zhang L (2019) Empirical research on the relationship between physical activity, sedentary behavior and physical health of adolescents. J Guangzhou Sport Univ (03):101–104. https://doi.org/10.13830/j.cnki.cn44-1129/g8.2019.03.025
3. Zhang M (2016) Influence of physical exercise on the mental health and interpersonal relationships of adolescents. J Nanjing Sport Inst (05):88–94. https://doi.org/10.15877/j.cnki.nsic.2016.05.016
4. Zheng Y, Fang A, Cai J, Deng S (2009) Comparative study of national students constitutional health criterion and students constitutional health criterion (on trial). China Sport Sci (07):92–96. https://doi.org/10.16469/j.css.2009.07.003
5. Liu Y, Wu S, Li Y, Shao F, Su J, Liu X (2018) A survey of mental symptoms of Chinese population based on SCL-90. Chin Mental Health J (05):437–441
6. Jin H, Wu W (1986) Primary analysis of SCL-90 assessment in Chinese normal subjects. Chin J Nervous Mental Dis (05):260–263
7. Basso JC, Suzuki WA (2017) The effects of acute exercise on mood, cognition, neurophysiology, and neurochemical pathways: a review. Brain Plast (2). https://doi.org/10.3233/BPL-160040

8. Rose Lauren T, Soundy A (2020) The positive impact and associated mechanisms of physical activity on mental health in underprivileged children and adolescents: an integrative review. Behavioral Sci (11). https://doi.org/10.3390/bs10110171
9. Fredrickson BL (2001) The role of positive emotions in positive psychology. Am Psychol (3). https://doi.org/10.1037/0003-066x.56.3.218
10. Yew JM, Kueh YC, Norsa'adah B, Leong FW, Tang HY, Kuan G (2022) A path model of the relationship between mood, exercise behavior, coping, and mental health among Malaysians during the COVID-19 pandemic. Int J Environ Res Public Health (19):5939. https://doi.org/10.3390/ijerph19105939